AF581677

The Physical Properties of the Groups of Galaxies

The Physical Properties of the Groups of Galaxies

Editors

Lorenzo Lovisari
Stefano Ettori

MDPI • Basel • Beijing • Wuhan • Barcelona • Belgrade • Manchester • Tokyo • Cluj • Tianjin

Editors
Lorenzo Lovisari
Istituto Nazionale di Astrofisica
Italy

Stefano Ettori
Istituto Nazionale di Astrofisica
Italy

Editorial Office
MDPI
St. Alban-Anlage 66
4052 Basel, Switzerland

This is a reprint of articles from the Special Issue published online in the open access journal *Universe* (ISSN 2218-1997) (available at: https://www.mdpi.com/journal/universe/special_issues/PPGG).

For citation purposes, cite each article independently as indicated on the article page online and as indicated below:

LastName, A.A.; LastName, B.B.; LastName, C.C. Article Title. *Journal Name* **Year**, *Volume Number*, Page Range.

ISBN 978-3-0365-1773-5 (Hbk)
ISBN 978-3-0365-1774-2 (PDF)

Contents

About the Editors

Lorenzo Lovisari (Dr.) is a junior researcher at the Astrophysics and Space Science Observatory (OAS) of Bologna in Italy. He is an X-ray observer interested in the physics of the hot plasma in groups and clusters of galaxies. He obtained his PhD in Physics in 2010 at the University of Innsbruck studying "Metal Distribution in Galaxy Clusters". Since then, he held several postdoctoral positions, at Argelander-Institut für Astronomie in Bonn (Germany) and at Center for Astrophysics — Harvard & Smithsonian in Cambridge (USA). His main research interests focused on the scaling properties of galaxy systems and on their chemical enrichment. He is also very involved in scientific activities through giving talks, publishing articles, organizing workshops, and being part of several international collaborations.

Stefano Ettori (Dr.) is Primo Ricercatore at the Astrophysics and Space Science Observatory (OAS) of Bologna in Italy. He works on the formation and evolution of galaxy clusters, focusing on the physical properties of the Intra-Cluster Medium and on the use of galaxy clusters as cosmological probes. He obtained his PhD in 1998 at IoA, University of Cambridge (UK). He was a research associate at IoA and then fellow at ESO (Garching, Germany) before becoming staff researcher at INAF OAS in Bologna in 2004. He is involved as key member in several international collaborations (chair of SWG1 for ESA L-mission Athena; co-lead of WP-7 "Astrophysics of Clusters of Galaxies" for ESA M-mission Euclid; co-PI of CHEX-MATE).

 MDPI

Editorial

The Physical Properties of the Groups of Galaxies

Lorenzo Lovisari [1,2,*] and Stefano Ettori [1,3]

1 INAF—Osservatorio di Astrofisica e Scienza dello Spazio di Bologna, via Piero Gobetti 93/3, 40129 Bologna, Italy; stefano.ettori@inaf.it
2 Center for Astrophysics | Harvard & Smithsonian, 60 Garden Street, Cambridge, MA 02138, USA
3 INFN, Sezione di Bologna, viale Berti Pichat 6/2, 40127 Bologna, Italy
* Correspondence: lorenzo.lovisari@inaf.it

Citation: Lovisari, L.; Ettori, S. The Physical Properties of the Groups of Galaxies. *Universe* **2021**, 7, 254. https://doi.org/10.3390/universe7080254

Received: 14 July 2021
Accepted: 19 July 2021
Published: 21 July 2021

Publisher's Note: MDPI stays neutral with regard to jurisdictional claims in published maps and institutional affiliations.

Galaxy groups consist of a few tens of galaxies bound in a common gravitational potential. They dominate the number count of the halo mass function and contain a significant fraction of the overall universal baryon budget. Being less massive than clusters, the energy that is supplied by supernovae and active galactic nuclei (AGNs) to the hot intragroup medium (IGrM) can easily exceed their gravitational binding energy. Thus, it is expected that these non-gravitational mechanisms have a strong effect on the distribution of of the baryons, making galaxy groups ideal targets to constrain the mechanisms governing the cooling–heating balance. The net effect of the various feedback processes in action in the gravitational potential wells is to change the radial distribution of the energy and mass in groups, affecting the correlations between their observed properties. Therefore, they are key to our understanding of how the bulk of matter in the Universe accretes and forms hierarchical structures and how different sources of feedback affect their gravitational collapse.

Despite their crucial role in cosmic structure formation and evolution, galaxy groups have received less attention compared to massive clusters. This is in part due to the challenges (e.g., faint X-ray emission, low number of galaxies in optical, low S/N in SZ) associated with their detection, observation, and characterization. With the advent of eROSITA (launched in 2019), many thousands of galaxy groups will be detected by X-ray, complementing on-going and future optical (DES, Euclid, Vera Rubin), SZ (SPT-3G, ACT), and radio (LOFAR, MeerKAT) surveys, paving the way towards the exploitation of the next generation of X-ray instruments (onboard XRISM—expected to fly by 2023—and the ESA L2 mission Athena—expected to fly in the 2030s).

To foster progress in the field of the physical properties of galaxy groups, facilitating effective cross-communication among observers, theorists, and simulators, we organized a Special Issue (https://www.mdpi.com/journal/universe/special_issues/PPGG (accessed on 21 July 2021)) dedicated to the physical properties of galaxy groups. We aimed to collect and organize the latest developments in our understanding of these systems and present future prospects from both observational and theoretical points of view. This Special Issue includes five manuscripts, which we summarize briefly in the following.

- *"Properties of Fossil Groups of Galaxies", by Aguerri and Zarattini [1]*

Fossil groups are an ostensibly special class of galaxy systems. They are objects dominated by a single, bright, elliptical galaxy and are thought to be the latest stage in the evolution of galaxy groups. Their properties differ from the one of other galaxy groups, and since it is likely that they did not experience recent major mergers, they should represent archetypal old undisturbed systems, and are therefore important systems to study. In the review, we show the main observational and theoretical works demonstrating that these systems fall very well in the current theory of structure formation in the Universe.

- *"Scaling Properties of Galaxy Groups", by Lovisari et al. [2]*

The scaling relations are the result of the different physical processes at work in the intracluster medium and provide an important tool to study its thermodynamic history.

In fact, the various processes that govern the formation of galaxy groups are suspected of systematically increasing the intrinsic scatter of the groups relations and changing their integrated properties. We overview the most recent studies on the X-ray scaling relations, obtained at the galaxy group scale, and their relations with optical properties and the supermassive Black Hole (SMBH) mass.

- *"Feedback from Active Galactic Nuclei in Galaxy Groups", by Eckert et al. [3]*

  The formation and evolution of the physical properties in groups are a direct consequence of the interplay between galaxy evolution, the development of the intragroup medium, and feedback. Many authors have argued that feedback from SMBHs plays a crucial role in regulating the star formation rates of massive galaxies and suppressing the onset of catastrophic cooling by carving cavities and driving shocks across the medium. We review the current observational evidence for AGN feedback in nearby galaxy groups with observations at X-ray, radio, and millimeter wavelengths and describe the theoretical advances made in recent years to interpret the heating–cooling cycle.

- *"Simulating Groups and the IntraGroup Medium: The Surprisingly Complex and Rich Middle Ground between Clusters and Galaxies", by Oppenheimer et al. [4]*

  The influence of the feedback processes is complex and difficult to model and to reproduce in simulations. However, cosmological simulations have enabled breakthroughs in our understanding of the gas and stellar contents of groups and of the impact of groups for cosmological parameter estimation. The review focuses on how groups process their baryons in a cosmological context, discussing the current limitations and the perspectives for improving the theoretical modeling in the near future.

- *"The Metal Content of the Hot Atmospheres of Galaxy Groups", by Gastaldello et al. [5]*

  Metals play a central role in the thermodynamic balance of galaxy systems by sustaining the cooling of their environment by means of spectral line emissions. DUe to the shallower gravitational potential of groups, feedback effects leave important marks on their gas and metal contents. Therefore, the shape of the abundance profiles can be used to investigate the impact of the feedback in the IGrM. We review the status of the metal abundance measurements in the IGrM and the progress made by simulations to reproduce and interpret those measurements.

Funding: L.L. and S.E. acknowledge financial contribution from the contracts ASI-INAF Athena 2015-046-R.0, ASI-INAF Athena 2019-27-HH.0, "Attività di Studio per la comunità scientifica di Astrofisica delle Alte Energie e Fisica Astroparticellare" (Accordo Attuativo ASI-INAF n. 2017-14-H.0), and from INAF "Call per interventi aggiuntivi a sostegno della ricerca di main stream di INAF".

Conflicts of Interest: The authors declare no conflict of interest.

References

1. Aguerri, J.A.L.; Zarattini, S. Properties of Fossil Groups of Galaxies. *Universe* **2021**, *7*, 132. [CrossRef]
2. Lovisari, L.; Ettori, S.; Gaspari, M.; Giles, P.A. Scaling Properties of Galaxy Groups. *Universe* **2021**, *7*, 139. [CrossRef]
3. Eckert, D.; Gaspari, M.; Gastaldello, F.; Le Brun, A.; O'Sullivan, E. Feedback from Active Galactic Nuclei in Galaxy Groups. *Universe* **2021**, *7*, 142. [CrossRef]
4. Oppenheimer, B.D.; Babul, A.; Bahé, Y.; Butsky, I.S.; McCarthy, I.G. Simulating Groups and the IntraGroup Medium: The Surprisingly Complex and Rich Middle Ground between Clusters and Galaxies. *Universe* **2021**, *7*, 209. [CrossRef]
5. Gastaldello, F.; Simionescu, A.; Mernier, F.; Biffi, V.; Gaspari, M.; Sato, K.; Matsushita, K. The Metal Content of the Hot Atmospheres of Galaxy Groups. *Universe* **2021**, *7*, 208. [CrossRef]

 universe

MDPI

Review

Scaling Properties of Galaxy Groups

Lorenzo Lovisari [1,2,*,†], Stefano Ettori [1,3,†], Massimo Gaspari [1,4,†] and Paul A. Giles [5,†]

1 INAF—Osservatorio di Astrofisica e Scienza dello Spazio di Bologna, via Piero Gobetti 93/3, 40129 Bologna, Italy; stefano.ettori@inaf.it (S.E.); massimo.gaspari@inaf.it (M.G.)
2 Center for Astrophysics | Harvard & Smithsonian, 60 Garden Street, Cambridge, MA 02138, USA
3 INFN, Sezione di Bologna, Viale Berti Pichat 6/2, 40127 Bologna, Italy
4 Department of Astrophysical Sciences, Princeton University, 4 Ivy Lane, Princeton, NJ 08544, USA
5 Department of Physics and Astronomy, University of Sussex, Falmer, Brighton BN1 9QH, UK; P.A.Giles@sussex.ac.uk
* Correspondence: lorenzo.lovisari@inaf.it
† All authors contributed equally to this work.

Abstract: Galaxy groups and poor clusters are more common than rich clusters, and host the largest fraction of matter content in the Universe. Hence, their studies are key to understand the gravitational and thermal evolution of the bulk of the cosmic matter. Moreover, because of their shallower gravitational potential, galaxy groups are systems where non-gravitational processes (e.g., cooling, AGN feedback, star formation) are expected to have a higher impact on the distribution of baryons, and on the general physical properties, than in more massive objects, inducing systematic departures from the expected scaling relations. Despite their paramount importance from the astrophysical and cosmological point of view, the challenges in their detection have limited the studies of galaxy groups. Upcoming large surveys will change this picture, reassigning to galaxy groups their central role in studying the structure formation and evolution in the Universe, and in measuring the cosmic baryonic content. Here, we review the recent literature on various scaling relations between X-ray and optical properties of these systems, focusing on the observational measurements, and the progress in our understanding of the deviations from the self-similar expectations on groups' scales. We discuss some of the sources of these deviations, and how feedback from supernovae and/or AGNs impacts the general properties and the reconstructed scaling laws. Finally, we discuss future prospects in the study of galaxy groups.

Keywords: galaxy groups; X-ray and optical observations; intragroup medium/plasma; active galactic nuclei; hydrodynamical simulations

 check for updates

Citation: Lovisari, L.; Ettori, S.; Gaspari, M.; Giles, P.A. Scaling Properties of Galaxy Groups. *Universe* **2021**, *7*, 139. https://doi.org/10.3390/universe7050139

Academic Editor: Francesco Shankar

Received: 31 March 2021
Accepted: 28 April 2021
Published: 10 May 2021

1. Introduction

Following the hierarchical scenario of structure formation, galaxy systems form through episodic mergers of small mass units. The less massive ones (often referred as groups) are the building blocks for the most massive ones (clusters), and trace the filamentary components of the large-scale structure (e.g., Eke et al. [1]). However, the distinction between groups and clusters is quite loose and no universal definition exists in the literature. Also, because the halo mass function is continuous, a naive starting point would be to not single out the low-mass end objects. Nonetheless, these poor systems have some notable differences (e.g., lack of dominance of the gas mass over the stellar/galactic component; Giodini et al. [2]) with respect to their more massive counterpart and they cannot be simply considered their scaled-down versions.

A conventional "rule of thumb" definition is to label systems of less than 50 galaxies as groups and above as clusters. More in general, galaxy groups have been broadly classified into three main classes based on their optical and physical characteristics: poor/loose groups, compact groups, and fossil groups (e.g., Eigenthaler and Zeilinger [3]). Poor/loose groups are aggregate of galaxies with a space density of $\sim 10^{-5}$ Mpc^{-3} (e.g., Nolthenius and White [4]). Compact groups are small and relatively isolated systems of typically 4–10 galaxies with a space density of $\sim 10^{-6}$ Mpc^{-3} (e.g., Hickson [5]). Fossil groups are objects dominated by a single bright elliptical galaxy (a formal definition is provided in Jones et al. [6]). Early studies (e.g., Helsdon and Ponman [7]) showed that subsamples of loose and compact groups share the same scaling relations. Thus, in this review, we do not make distinction between poor/loose and compact groups, and hereafter we simply refer to them as galaxy groups. The properties of fossil groups are instead discussed in the companion review by Aguerri et al. However, since the optical properties are not always available, a threshold of $M \sim 10^{14} M_{\odot}$, corresponding to a temperature of 2–3 keV, is also often used to classify these systems. We will show later that this threshold roughly corresponds to the temperature for which there is a significant change in the X-ray emissivity.

Despite the crucial role played by groups in cosmic structure formation and evolution, they have received less attention compared to massive clusters. One of the reasons is that typical groups contain only a few bright galaxies in their inner regions, making very difficult to detect them in optical with a relatively good confidence. A much easier method of detecting them is to study the X-ray emission from the hot intragroup medium (IGrM). The detection of hot plasma carries witness that galaxy groups (and clusters) are not simple conglomerate of galaxies put together by projection effects, but real physical systems which are undergoing some degree of virialization. Galaxy groups often show lower and flatter X-ray surface brightness than clusters (e.g., Ponman et al. [8], Sanderson et al. [9]). Therefore, the physical properties of the gas derived for galaxy groups are presumably less robust than the properties derived for galaxy clusters. Nonetheless, they represent a more common environment because the mass function of virialized systems, which describes the number density of clusters above a threshold mass M, is higher at lower masses (with a factor of $\sim$30/210/1500 more objects in the mass range $M_{500} = 10^{13}\ M_{\odot} - M_1$ than in $M_{500} > M_1$, and $M_1 = 1/2/5 \times 10^{14} M_{\odot}$ at $z = 0$; see, e.g., [10]). Hence, the detection and characterization of galaxy groups is especially important for astrophysical and cosmological studies.

1.1. Galaxy Groups and Astrophysics

Galaxy groups cover the intermediate mass range between large elliptical galaxies and galaxy clusters and contain the bulk of all galaxies and baryonic matter in the local Universe (e.g., Tully [11], Fukugita et al. [12], Eke et al. [1]). Because of that, they are crucial for understanding the effects of the local environment on galaxy formation and evolution processes. Moreover, the feedback from supernovae (SNe) and supermassive black holes (SMBHs) is expected to significantly alter the properties of these systems being the energy input associated with these sources comparable to the binding energies of groups (e.g., Brighenti and Mathews [13], McCarthy et al. [14], Gaspari et al. [15]). However, the relative contributions of the different feedback processes are still a matter of debate and it will constitute a major subject of research for the next decade. These factors make galaxy groups great laboratories to understand the complex baryonic physics involved, and to study the differences with their massive counterpart. For instance, we know that the fraction of strong cool-cores (CC; i.e., systems with a central cooling time $t_{\rm cool} < 1$ Gyr, as described in Hudson et al. [16]), weak cool-cores ($1 < t_{\rm cool} < 7.7$ Gyr), and non-cool-cores (NCC; $t_{\rm cool} > 7.7$ Gyr) objects at the group scale are similar to those in galaxy clusters (Bharadwaj et al. [17]). However, O'Sullivan et al. [18] found that the CC fraction increases dramatically when the samples are restricted to low-temperature systems (i.e., kT<1.5 keV) showing a correlation between system temperature and CC status. Bharadwaj et al. [17] also found that brightest group galaxies have a higher stellar mass than brightest cluster galaxies, suggesting that there is less gas available to feed the SMBHs. Recent results suggest that the IGrM and intracluster medium (ICM) are also providing a source of gas which feeds and grows the central SMBHs, in particular leading to novel scaling relations between the SMBH mass and the X-ray properties of their host gaseous halos (e.g., Bogdán et al. [19], Gaspari et al. [20], Lakhchaura et al. [21]).

These findings imply an interplay between the feedback mechanisms connected with the SMBHs and the macro-scale halos, which could explain some features of cosmological simulations driving a relative break of the L_X–T_X and L_X–M relations at low temperatures (e.g., McCarthy et al. [14], Sijacki et al. [22], Puchwein et al. [23], Fabjan et al. [24], Le Brun et al. [25]). This deviation is often attributed to active galactic nucleus (AGN) feedback (e.g., Planelles et al. [26], Gaspari et al. [27], Truong et al. [28]).

The properties described above have an important effect on the correlation between different physical quantities. For instance, it is well established that CC and NCC objects populate different regions of the X-ray luminosity space of any scaling relations (e.g., Markevitch [29], Pratt et al. [30], Mittal et al. [31], Bharadwaj et al. [32], Mantz et al. [33], Lovisari et al. [34]). Therefore, a change in the fraction of CC/NCC systems as a function of the temperature (mass) will have an impact to the slope, normalization, and scatter of the observed scaling relations. Hence, it is crucial to have a full coverage for the whole sample to minimize the systematic errors due to the incompleteness. In fact, if all the missing objects happen to belong to one of the subsamples (e.g., NCC), the normalization (and the scatter) of the studied scaling relations will be wrong. Moreover, the CC/NCC fraction of systems in a sample depends on the selection function and may not be representative of the underlying population. For instance, X-ray selected samples are known to be biased toward centrally peaked and relaxed systems, in particular in the low-mass regime (Eckert et al. [35]). In fact, recent results by O'Sullivan et al. [18], who analyzed a sample of optically selected groups, show that $\sim$20% of X-ray bright groups (probably the most disturbed ones, or with no concentrated CC) in the local Universe may have been missed. Thus, the scaling relations of galaxy groups (and clusters) are the result of the various processes that govern the formation and evolution of these systems making them ideal targets for studying the effect of the interplay between galaxy evolution, the development of the IGrM, and feedback.

1.2. Galaxy Groups and Cosmology

Clusters of galaxies have proven to be remarkably effective probes of cosmology (e.g., [36–46]). However, since galaxy groups represent a large fraction of the number density of virialized systems, their impact might be relevant (in particular, on the reconstruction of halo mass function). For instance, recent results of the Dark Energy Survey (DES) collaboration show that the σ_8–Ω_M posteriors have a 5.6σ tension with *Planck* CMB results, and a 2.4σ tension with galaxy clustering and cosmic shear results [47]. The cause of this tension is thought to reside at the low-mass (low richness) end of the cluster population, specifically, clusters with a richness of $\lambda < 30$ (corresponding to $\sim 10^{14}$ $M_\odot$). The removal of low richness systems from the analysis significantly reduces the tension with comparative cosmological probes. However, various tests undertaken in Abbott et al. [48] suggest that the discrepancy is probably due to the modeling of the weak-lensing signal rather than the group and cluster abundance. The mass calibration for the DESY1 analysis is based upon a stacked weak-lensing analysis, through application of the weak-lensing–richness relation [49]. This relation is derived over the full richness range, which would not account for any deviations at the low-mass end. Furthermore, since the mass analysis relies on stacked quantities, information on scatter in mass with richness is lost and must be informed from external relations. In the case of the DESY1 analysis, the mass scatter information is inferred from the temperature–richness relation using X-ray data [50]. This scatter is assumed constant with richness, which again, could evolve as a function of richness. The investigation of these effects will become of critical importance as the low-mass end of the mass scales are increasingly probed by future surveys (e.g., those constructed from the Legacy Survey of Space and Time undertaken by the *Vera C. Rubin Observatory*).

Excluding low-mass systems significantly reduces the cosmological parameter constraints. Thus, despite the important complications present at the group scale, it is becoming generally appreciated that galaxy groups should be included in the cosmological analysis. To use them to constrain the cosmological parameters we need a good knowledge of the selection function to properly correct for the incompleteness, otherwise studies employing the cluster mass function may find lower Ω_M and/or σ_8 values than the true values. This scenario is supported by the finding of Schellenberger and Reiprich [44], who showed how the increasing incompleteness of parent samples in the low-mass regime together with a steeper L_X–M relation observed for groups, can lead to biased cosmological parameters. It is worth noticing that if a large fraction of galaxy systems is missed, then the tension between cluster counts and primary CMB cosmological constraints may become less severe.

Most of the upcoming large surveys will push the measurements down to the low-mass regime. Thus, to fully exploit the future datasets to constrain the cosmological parameters, we need to properly characterize the properties of galaxy groups and the differences with galaxy clusters, accounting for the different selection effects, and estimating the amplitude of the various biases.

1.3. This Review

In this work, we present an overview of the most recent studies on scaling relations between several integrated observed quantities of galaxy groups, and complement/update the previous reviews in the field by, e.g., Mulchaey [51] and Sun [52]. The review is organized as follows. In Section 2, we derive the self-similar X-ray scaling relations and overview the observed deviations. In Section 3, we discuss the relations between X-ray and optical properties. In Section 4, we discuss the relation between the SMBH mass and the global group quantities. In Section 5, we shortly discuss the most relevant upcoming missions and their expected contribution to the field. In Section 6 we provide our final remarks.

2. X-ray Scaling Relations

2.1. Theoretical Expectations

The X-ray scaling relations for galaxy systems were derived by Kaiser [53], based on the simple assumption that the thermodynamic properties of the ICM are only determined by gravity (i.e., gas just follow the dark matter collapse). Since gravity is scale free, this model predicts that objects of different sizes are the scaled version of each other. For that reason, this model is often referred as self-similar, and the derivation of the predicted relations has been extensively covered in the literature (e.g., Kitayama and Suto [54], Bryan and Norman [55], Voit [56], Maughan et al. [57], Borgani et al. [58], Böhringer et al. [59], Ettori [60], Giodini et al. [61], Maughan [62], Ettori [63], Ettori et al. [64]). Here, we only provide a brief review of the standard derivation of the self-similar scaling relations for massive systems, and then extend them, when necessary, to the low-mass regime where gas physics is playing a significant role.

In the self-similar scenario, two galaxy systems which have formed at the same time have the same mean density. Hence,

$$\frac{M_{\Delta_z}}{R^3_{\Delta_z}} = \text{constant} \tag{1}$$

where M_{Δ_z} is the mass contained within the radius R_{Δ_z}, encompassing a mean density Δ_z times the critical density of the Universe $\rho_c(z)$, so that $M_{\Delta_z} \propto \rho_c(z)\Delta_z R^3_{\Delta_z}$. The critical density of the Universe scales with redshift as $\rho_c(z) = \rho_{c(z=0)}E^2(z)$, where $E(z) = H_z/H_0$ describes the evolution of the Hubble parameter with redshift z.

During the gravitational collapse, the gas density increases, and a shock propagates outward from the cluster center and heats the gas. After the passage of the shock, IGrM and ICM can be considered in hydrostatic equilibrium, so the temperature T_X provides an estimate of the gravitational potential well (i.e., $T_X \propto GM_{\Delta_z}/R_{\Delta_z} \propto R^2_{\Delta_z}$), and therefore of the total mass of the cluster:

$$M_{\Delta_z} \propto T_X{}^{3/2}. \tag{2}$$

In the self-similar scenario, where the gas fraction, f_g, of galaxy groups and clusters is universal, one expects for the total gas mass, M_g, a similar dependence on the gas temperature: $M_g \propto T_x{}^{3/2}$.

The hot gas in galaxy systems is typically described as an optically thin plasma in collisional ionisation equilibrium. Its X-ray emissivity (i.e., the energy emitted per time and volume) is equal to

$$\epsilon = n_e\, n_p\, \Lambda(T_x, Z_\odot), \tag{3}$$

where n_e and n_p are the number densities of electrons and protons, respectively, that are related to the gas mass density ρ_g through the relation $\rho_g = \mu m_p(n_e + n_p)$, μ is the mean molecular weight ($\sim$0.6 for a plasma with solar abundance), m_p is the proton mass, and $\Lambda(T_x, Z_\odot)$ is the cooling function which depends on the mechanism of the emission[1] and on the considered energy window. At high temperatures (i.e., kT > 3 keV) the main mechanism of emission is thermal bremsstrahlung and the cooling function in the full energy band mainly depends only on T_x (i.e., $\Lambda(T_x, Z_\odot) \propto T_x^{1/2}$). Thus, for sufficiently massive systems the bolometric X-ray luminosity (i.e., 0.01–100 keV band) is given by

$$L_{x,bol} \propto \int \epsilon\, dV \propto n_p^2\, T_x^{1/2} R^3 \propto f_g^2\, T_x^2 \propto T_x^2 \tag{4}$$

with the last scaling obtained assuming a constant gas fraction as predicted by the self-similar scenario. By combining Equations (2) and (4) one obtains the well-known relation between bolometric luminosity and total mass (i.e., $L_{x,bol} \propto M^{4/3}$).

In the literature the X-ray luminosities are also often provided in soft energy bands (e.g., 0.1–2.4 or 0.5–2 keV), more representative of the bandpass covered by current (and past) X-ray facilities used for the study of groups and clusters. In Figure 1 (left panel), we show that for massive systems with typical cluster abundance the X-ray emissivity in soft band is almost independent of the system temperature (e.g., for $Z = 0.3Z_\odot$ the change in ϵ between 3 and 10 keV plasmas in the 0.5–2 keV band is <10% for given emission measure), so that $L_{x,soft} \propto T_x^{3/2}$, and hence using Equation (2), $L_{x,soft} \propto M$.

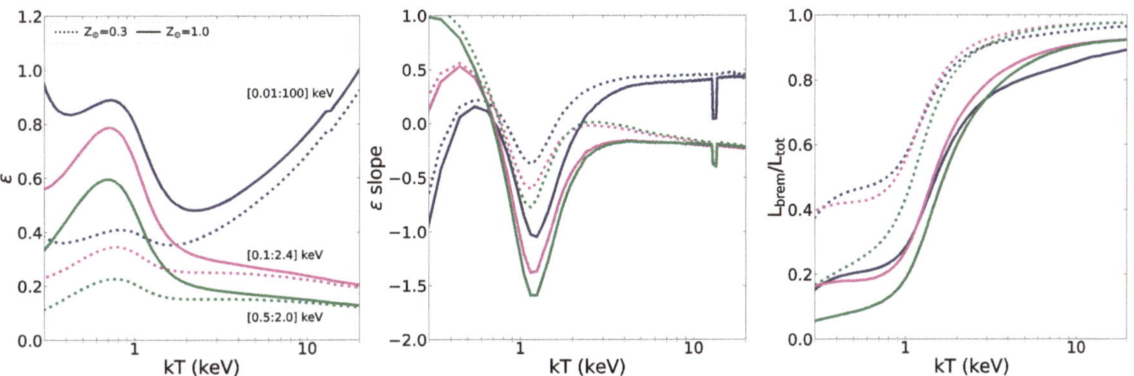

Figure 1. **(left panel)**: total X-ray emissivity as function of the plasma temperature in different energy bands (bolometric in blue, 0.1–2.4 keV in magenta, and 0.5–2 keV in green). The curves are calculated using an APEC (Smith et al. [69]) model (v3.0.9) in XSPEC (Arnaud [70]) for two different values of metallicity: 1.0 (solid lines) and 0.3 times the solar abundance as in Asplund et al. [71]. All curves are normalized to the bolometric emissivity at $k_B T_x$ = 20 keV with $Z = 1\, Z_\odot$. **(middle panel)**: the emissivity slope as a function of temperature showing the impact of the different $Z_\odot$ and T_x in the low-temperature regime. **(right panel)**: bremsstrahlung emission fraction (L_{brem}/L_{tot}) as a function of the temperature, illustrating the increasing contribution of line emission to the total luminosity for low-temperature plasmas.

[1] Three main processes contribute to the X-ray emission: thermal bremsstrahlung (due to the deflection of a free electron by the electric field of a ion), recombination (due to the capture of an electron by an ion), and two-photon decay (due to the changing of the quantum level of an electron in an ion). See details in the reviews from, e.g., Sarazin [65], Peterson and Fabian [66], Kaastra et al. [67], and Böhringer and Werner [68].

However, the gas fraction is not constant, with a difference of almost a factor of two between groups and clusters (e.g., Vikhlinin et al. [72], Gonzalez et al. [73], Gastaldello et al. [74], Pratt et al. [30], Dai et al. [75], Gonzalez et al. [76], Lovisari et al. [77], Eckert et al. [78]; see also the companion reviews by Eckert et al. and Oppenheimer et al.). Moreover, at low temperatures, line cooling becomes very important, and the emissivity (both in soft and bolometric bands) becomes strongly abundance ($Z_{\odot}$) and temperature dependent. In Figure 1 (left and middle panels) we show the dependence of the emissivity on the temperature and metallicity for widely used energy bands for scaling relations, clearly showing that a simple scaling cannot be derived. In Table 1, we provide the dependence for a set of interesting cases.

Table 1. Emissivity dependence on T_x and $Z_{\odot}$ for different temperature regimes and energy bands.

E Band	T Range	ϵ Slope ($Z = 0.3Z_{\odot}$)	ϵ Slope ($Z = 0.5Z_{\odot}$)	ϵ Slope ($Z = 1.0Z_{\odot}$)
bol	0.4−0.7	+0.20	+0.16	+0.11
	0.4–2.0	−0.00	−0.14	−0.34
	0.4–3.0	+0.06	−0.07	−0.26
	0.4–10.0	+0.20	+0.11	−0.03
	0.7–2.0	−0.11	−0.30	−0.58
	0.7–3.0	+0.01	−0.15	−0.40
	0.7–10.0	+0.20	+0.10	−0.06
	2.0–10.0	+0.40	+0.36	+0.28
	3.0–10.0	+0.43	+0.41	+0.35
0.1–2.4	0.4−0.7	+0.44	+0.42	+0.39
	0.4–2.0	−0.04	−0.19	−0.42
	0.4–3.0	−0.04	−0.18	−0.39
	0.4–10.0	−0.06	−0.16	−0.31
	0.7–2.0	−0.29	−0.52	−0.84
	0.7–3.0	−0.22	−0.40	−0.68
	0.7–10.0	−0.16	−0.27	−0.45
	2.0–10.0	−0.08	−0.12	−0.20
	3.0–10.0	−0.10	−0.12	−0.17
0.5–2	0.4−0.7	+0.63	+0.56	+0.50
	0.4–2.0	−0.03	−0.23	−0.48
	0.4–3.0	−0.02	−0.21	−0.45
	0.4–10.0	−0.04	−0.17	−0.35
	0.7–2.0	−0.38	−0.65	-1.00
	0.7–3.0	−0.27	−0.50	−0.81
	0.7–10.0	−0.18	−0.32	−0.52
	2.0–10.0	−0.06	−0.11	−0.22
	3.0–10.0	−0.07	−0.11	−0.18

The complexity of the emissivity function in the low-temperature regime may lead to a wrong interpretation of the results of scaling relation studies. In fact, it is conventional to compare the slopes of the scaling relations obtained with sample of groups to the self-similar predictions derived for massive clusters. However, if there is no feedback (i.e., the relations follow the self-similar predictions), then the L_X–T_X and L_X–M relations should flatten at low temperatures and masses. Thus, without accounting for the increasing contribution of the line emission in the low-temperature regime, one could interpret the agreement between group and cluster relations such that feedback processes play a negligible role in shaping the IGrM. Thus, the impact of the feedback could be underestimated. To visualize the contribution of line emission as function of the temperature, we follow the simple approach of Zou et al. [79] in which we measure the luminosity (L_{tot}) in different energy bands (i.e., bolometric, 0.1–2.4, and 0.5–2) of APEC spectra with a metal abundance of $Z_{\odot}$ = 1.0 (not rare at the center of galaxy groups, see companion review by Gastaldello et al.), and then setting $Z_{\odot}$ = 0 without changing any other parameters to approximate the luminosity of the pure bremsstrahlung component (L_{brem}). We repeated the exercise for a more standard

$Z_{\odot} = 0.3$. The results are shown in Figure 1 (right panel) where it is clear the significant contribution of line emission to the total luminosity in the low-temperature regime. Thus, the luminosity–temperature and luminosity–mass relations can be approximated as $L_X \propto T_X^{1.5+\gamma}$ and $L_X \propto M^{1+\gamma}$, where γ is the slope of the X-ray emissivity in the considered energy band (e.g., soft or bolometric) and temperature range covered by the systems in the studied dataset (see Table 1). It follows that the self-similar L_X–T_X and L_X–M relations for galaxy groups are expected to be significantly flatter than the ones for galaxy clusters. It is also worth noticing that even for massive systems with $Z_{\odot} = 0.3$ there is a ~5% contribution from line emission. Thus, the bolometric emissivity slope is smaller than 0.5 (i.e., the value one gets from pure bremsstrahlung emission) with the net effect being that the correct self-similar expectation becomes $L_{X,bol} \propto T_X^{\sim 1.9}$.

The abundance and temperature dependence of the X-ray emissivity at low temperatures need to be taken into account when determining the luminosities of galaxy groups. Normally, the luminosities are estimated applying a conversion factor to the observed count rates to obtain the X-ray fluxes. From Figure 1 it is clear that this conversion factor in the low-temperature regime depends strongly on the metallicity of the system. Given the observed temperature and abundance gradients in groups (e.g., Rasmussen and Ponman [80], Sun et al. [81], Mernier et al. [82], Lovisari and Reiprich [83]; see also the companion review by Gastaldello et al.), a possible strategy is to use the observed profiles of temperature, abundance, and surface brightness to estimate the luminosity in each radial bin obtained during the spectral analysis (e.g., Sun [52], Lovisari et al. [77]). Sun [52] pointed-out that although the average luminosities (soft band or bolometric) only change by ~5% when the overall values of temperature and abundance are used in the conversion instead of the profiles, the scatter increases by 10–15%. This is an important point to keep in mind when using survey data (e.g., ROSAT, eROSITA) for which simple assumptions such as isothermality and single overall abundance are chosen to obtain an estimate of the luminosity.

The dependence of the cooling function on the metallicity also implies that the use of different abundance tables can lead to different estimates of the rest-frame X-ray luminosities. Typically, one recovers the source count rate within a given radius from the surface-brightness profile, and then obtain the X-ray flux by setting the normalization of a thermal model (with proper temperature and metallicity) to match the observed count rate. However, the shape of the thermal model (which depends only on the abundance for a given temperature and column density) can diverge at lower and higher energies than the ones used to derive the surface brightness. To visualize the impact, we ran a set of simulations in which the normalization of the thermal model for systems at $z = 0.02$ (i.e., median redshift of the current local group samples, see Table 2) was set in order to match a count rate of 1 count/sec in the 0.5–2 keV energy band (i.e., the bandpass where many X-ray facilities have most of their effective area, and often used to derive the surface-brightness profiles) for each abundance table. Then, we estimated the luminosity in different energy bands. In Figure 2 (top panels) we show the impact on the estimated luminosity as function of the system temperature and common abundance tables. There is a very good agreement in the 0.5–2 keV band luminosity, regardless of the abundance table used for the analysis. Instead, small differences (i.e., in the order of a few percent) in the 0.1–2.4 keV band and bolometric luminosities arise for low-temperature systems (i.e., $kT \lesssim 1$ keV) when the abundances of (Grevesse and Sauval [84], GRSA), (Asplund et al. [71], ASPL), or (Lodders et al. [85], LODD) are used. The disagreement is much more significant (i.e., up to ~10%) when the luminosities are estimated with the abundance table by (Anders and Grevesse [86], ANGR). Most of the differences are due to the much higher Fe abundance in ANGR with respect to the other tables investigated here. When the Fe abundance of ANGR is set to the value of ASPL (leaving unchanged all the other ANGR abundances) the estimated luminosities are in much better agreement (see the dashdot lines in the middle panels of Figure 2). The reason for the differences highlighted in Figure 2 is that by switching the abundance table we change the emissivity and the relative contribution of the line emission with respect to the bremsstrahlung emission (see Figure 1). The difference between the rest-frame luminosity estimated with one or another table tends to increase at higher redshifts (see bottom panels of Figure 2). However, unless very high redshifts are considered, the effect is usually smaller than a few percent. In general, the soft-bands (in particular the 0.5–2 keV band) are the ones showing a smaller impact on the estimated luminosity by switching abundance table and should be preferred for galaxy groups studies. Although in most cases the effect is relatively small, it can lead to systematic effects and should be kept in mind when comparing independent literature results.

Another very useful quantity to describe the IGrM and ICM is the entropy which is generated during the hierarchical assembly process. In X-ray studies of galaxy groups and clusters, the entropy is usually defined as

$$K = k_B T_X \, n_e^{-2/3} \tag{5}$$

where k_B is the Boltzmann constant. Entropy is conserved during adiabatic processes and it is only modified by processes changing the physical characteristics of the gas. Entropy increases when heat energy is introduced and decreases when radiative cooling carries heat energy away (e.g., [56]), keeping a record of the energy injection and dissipation in the intracluster gas. Thus, entropy measurements provide a useful tool for our understanding of the thermodynamic history of galaxy groups and clusters. Gas entropy in galaxy groups shows a significant excess to that achievable by pure gravitational collapse (e.g., Ponman et al. [8], Lloyd-Davies et al. [87], Ponman et al. [88], Finoguenov et al. [89], Sun et al. [81], Johnson et al. [90], Panagoulia et al. [91]), indicating a substantial IGrM heating often ascribed to non-gravitational processes. In fact, due to the shallower potential well of the small systems, it is expected that the energy released by past star formation and AGN activities leaves a clear imprint on the thermodynamic properties of IGrM and ICM (see companion reviews by Eckert et al. and Oppenheimer et al.). An effect that can be seen in both the integrated properties (i.e., in the scaling relations) and in the shape of the entropy profiles which are expected to follow $K \propto R^{1.1}$.

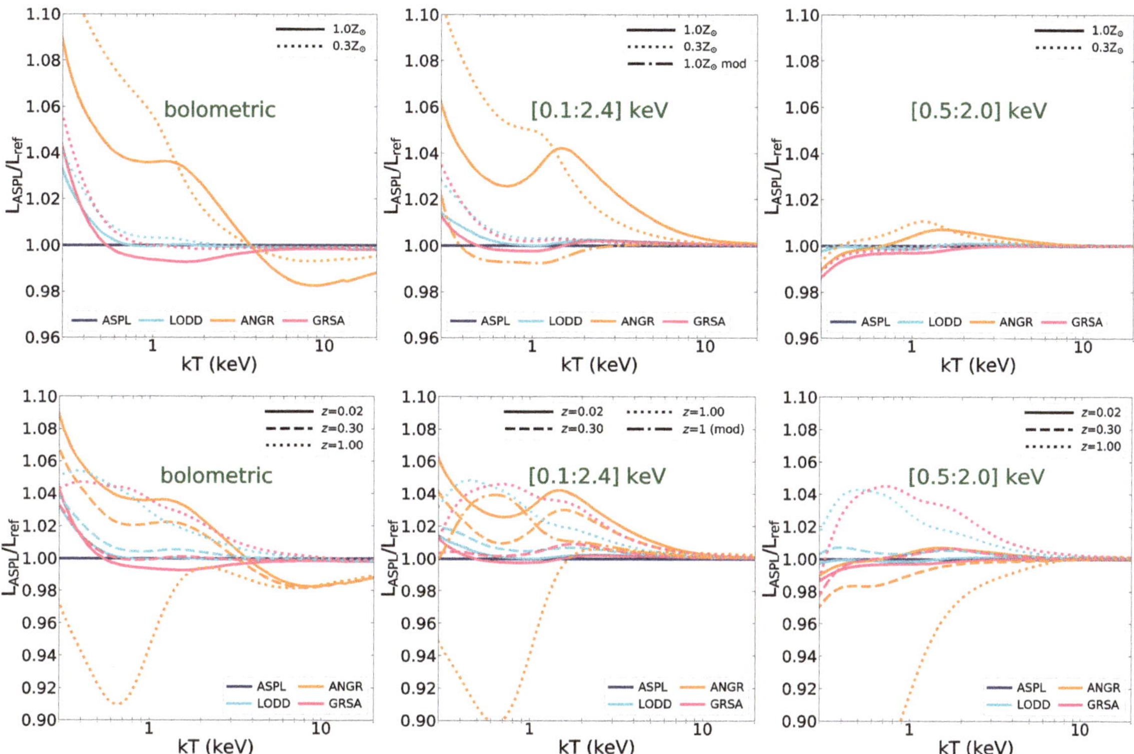

Figure 2. Ratio between the rest-frame L_X (**left panels**: bolometric, **middle panels**: 0.1–2.4 keV band, **right panels**: 0.5–2 keV band) derived using the abundances as in Asplund et al. [71] and L_X obtained with the abundances of (Grevesse and Sauval [84], pink), (Lodders et al. [85], turquoise), and (Anders and Grevesse [86], orange), respectively. In the **top panels** we show the results for systems at $z = 0.02$ with a metallicity of 1.0 (solid lines) or 0.3 times the solar abundance. The dashdot line in the **middle panels** refers to the simulations with a modified table of ANGR in which the Fe abundance was set equal to the value of ASPL, showing that indeed most of the differences arise from the significant discrepancy in Fe between ANGR and the other tables. The 0.5–2 keV band (i.e., the band used to rescale the APEC normalization) provide the best agreement between the different tables. In the **bottom panels**, we show the impact on L_X of changing abundance table for systems at different redshifts: $z = 0.02$ (solid line), $z = 0.3$ (dashed), and $z = 1$ (dotted, dashdotted). The simulations were performed with $Z = 1Z_{\odot}$. The plot shows how the differences increases with z, although the effect is generally smaller than $\sim$5% unless very high z are considered. The residual difference between ASPL and the modified ANGR table for high z objects is due to the differences in elements other than Fe (e.g., C, N, O, Ne, Mg, Si, S).

2.2. Observed Scaling Relations

The L_X–T_X relation involves two of the easiest quantities that can be derived using X-ray data. It was one of the first X-ray correlations to be studied and is still one of the most disputable scaling law between integrated observed properties. In fact, there have been conflicting reports in the literature about whether the relation for groups behaves as the one derived for massive clusters (i.e., whether groups are simply scaled-down versions of clusters or not). It has been clear for many years that the $L_{X,bol}$–T_X relation for massive systems does not scale self-similarly (see, e.g., Giodini et al. [61] for a review about the relation for galaxy clusters), with slopes significantly higher than 2. Although pioneering studies of the relation for galaxy groups suggested considerably steeper slopes (i.e., slopes larger than 4; Helsdon and Ponman [92], Helsdon and Ponman [7], Xue and Wu [93]), later investigations found relations only slightly steeper than the ones for clusters (e.g., Osmond and Ponman [94], Shang and Scharf [95], Eckmiller et al. [96], Sun [52], Lovisari et al. [77]). In Figure 3, we show a compilation of data for the L_X–T_X relation taken from recent studies of galaxy groups observed with *XMM–Newton* and *Chandra*, and in Table 2 we list the best-fit slopes from these and other studies. The results show that indeed the slope obtained for poor systems (i.e., kT $<$ 3 keV) is consistent to the one derived for the more massive clusters (with hints of a slightly different normalization that cause a flattening when all the systems are fitted together). However, since the L_X–T_X relation is expected to flatten in the low-mass regime (see Section 2.1) these results clearly indicate a more significant contribution of the non-gravitational processes at the group scale. In fact, feedback processes (e.g., AGN heating) are expected to increase the entropy of the gas reducing its density (and hence the X-ray luminosity, steepening the relation). For massive systems, the binding energy is so large that only the very central regions are affected, and the integrated properties of galaxy clusters remain essentially unchanged. Conversely, at the group scale the gas can be easily removed towards or beyond the virial radius modifying their global properties. The agreement between the L_X–T_X relation of groups and clusters seems to stand also when the Malmquist bias (i.e., the preferential detection of intrinsically bright objects) is accounted for in galaxy groups studies as previously done for massive clusters. The Malmquist bias is expected to flatten the observed X-ray relations because only objects above a certain flux value are considered (either because one enforces an observational threshold or because faint systems are not detected). The correction needed to recover the underlying relation depends on the real intrinsic scatter with larger values requiring a larger correction (i.e., if the scatter increases in the low-mass regime then the magnitude of the flattening is larger than for galaxy clusters). Although an attempt to correct this bias in cluster scaling relations has been provided in many different X-ray studies (e.g., Ikebe et al. [97], Stanek et al. [98], Pacaud et al. [99], Vikhlinin et al. [38], Pratt et al. [30], Mittal et al. [31], Schellenberger and Reiprich [100]), there are only a few papers providing the correction in the galaxy group regime. For instance, Lovisari et al. [77] analyzing an X-ray flux-selected sample of local groups showed an increase of the $L_{0.1-2.4}$–T_X relation slope after correcting for the Malmquist bias. A similar result was obtained by Bharadwaj et al. [32], who estimated a correction for an archival sample of groups observed with *Chandra*. In contrast to these results, are the finding by Kettula et al. [101] and Zou et al. [79] who did not find any significant steepening after the bias correction. However, all the bias corrected relations obtained in the different studies show a great agreement (once they are converted into the same energy band). This agreement may suggest that the observational discrepancies arise from differences in the sample selection (which might cause the sample to be more or less biased). Once the biases are accounted for, then the results are not sensitive to the initial choices. Zou et al. [79] showed that even once selection biases are taken into account the $L_{X,bol}$–T_X relation at the group scale is consistent with the one for clusters. This finding confirms the stronger impact of the non-gravitational processes in the low-mass regime (otherwise a flattening should be observed). Of course, these corrections work under the assumption that the X-ray selected samples are representative of the underlying population which might not be the case as suggested by, e.g., Rasmussen et al. [102], Anderson et al. [103], Andreon et al. [104], and O'Sullivan et al. [18] who argued that the X-ray surveys miss a large fraction of galaxy systems. One possible reason for this incompleteness is related to the source detection algorithms mostly based on sliding cell detection methods. These algorithms work efficiently at finding point-like sources but has difficulties in detecting extended features, especially for nearby objects and for sources close to the detection limit (e.g., Valtchanov et al. [105]; see also Šuhada et al. [106] for a performance comparison between sliding cell and wavelet detection algorithms). Xu et al. [107], using a method optimized for the extended source detection, found a large number of new group candidates which are not included in any existing X-ray or Sunyaev-Zel'dovich (SZ) cluster catalogs. If studies are restricted to groups that are a priori known to be X-ray bright and which properties may be quite different from those of optically selected groups, as argued by Miniati et al. [108], then our view could be significantly biased.

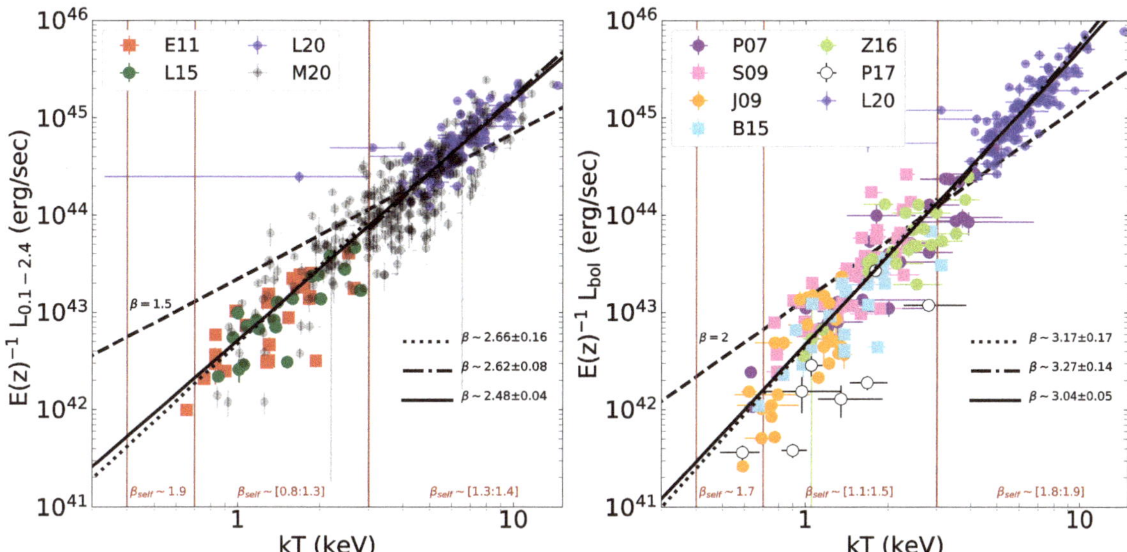

Figure 3. *XMM–Newton* (circles) and *Chandra* (squares) measurements of the $L_{0.1-2.4}$–T_x (**left panel**) and $L_{x,bol}$–T_x (**right panel**) relations for different samples of groups: (Eckmiller et al. [96], E11), (Lovisari et al. [77], L15), (Sun et al. [81], S09), (Johnson et al. [90], J09), (Bharadwaj et al. [32], B15), (Zou et al. [79], Z16), (Pearson et al. [109], P17). The groups measurements are compared with the ones from X-ray-selected (Migkas et al. [110], M20) and SZ-selected (Lovisari et al. [34], L20) cluster samples. We note that different studies used different atomic models (including APEC v1.3.1 which provide a significantly different modeling of the Fe-L line with respect to newer versions). The luminosities are all within R_{500} while temperatures are obtained in different regions (see Table 2). The *Chandra* measurements are converted to *XMM–Newton*-like temperatures using the relations given in Schellenberger et al. [111]. Empty symbols are from optically selected samples. The lines represent the fitted relation for $T_x < 3$ keV systems (dotted), $T_x > 3$ keV systems (dashed-dotted), all systems (solid), and are compared with the case predicted by the self-similar scenario (dashed). The fits have been performed with LIRA (Sereno [112]) assuming self-similar time evolution and, conservatively, without the scatter on the X variable, and are meant for visualization purposes only. In brown we provide the expected values for the slope using the dependence of the emissivity tabulated in Table 1 for different ranges of temperatures (showed as brown vertical lines).

Table 2. Overview of the most recent published scaling relations for galaxy groups based on *XMM–Newton* and *Chandra* data.

Relation	N	kT (keV)	z Range	slope$_{self}$	slope$_{obs}$	slope$_{LIRA}$	Reference	Note
L–T$_{exc}$	26	0.6–3.0	0.012–0.049	[0.9:1.1]	2.25 ± 0.21	3.27 ± 0.26	E11	$a \dagger \Box \uparrow\uparrow$
L–T$_{exc}$ BC	20	0.9–2.8	0.012–0.034	[0.7:1.2]	2.86 ± 0.29	-	L15	$a \dagger \Box \downarrow\downarrow$
L–T$_{exc}$	20	0.9–2.8	0.012–0.034	[0.7:1.2]	2.05 ± 0.32	2.90 ± 0.36	L15	$a \dagger \Box \downarrow\downarrow$
L$_{exc}$–T$_{exc}$ BC	12	1.7–8.2	0.1–0.47	[1.3:1.5]	2.52 ± 0.17	-	K15	$a \ddagger \triangle \downarrow\downarrow$
L$_{exc}$–T$_{exc}$	12	1.7–8.2	0.1–0.47	[1.3:1.5]	2.65 ± 0.17	2.47 ± 1.23	K15	$a \ddagger \triangle \downarrow\downarrow$
L–T$_{exc}$ BC	26	0.6–3.6	0.012–0.049	[1.2:1.6]	3.20 ± 0.26	-	B15	$c \dagger \Box \uparrow\uparrow$
L–T$_{exc}$	26	0.6–3.6	0.012–0.049	[1.2:1.6]	2.17 ± 0.26	3.11 ± 0.54	B15	$c \dagger \Box \uparrow\uparrow$
L–T BC	23	1.0–3.9	0.03–0.147	[0.8:1.3]	2.79 ± 0.33	-	Z16	$b \ddagger \Diamond \uparrow\uparrow$
L–T BC	23	1.0–3.9	0.03–0.147	[1.2:1.6]	3.29 ± 0.33	-	Z16	$c \ddagger \Diamond \uparrow\uparrow$
L–T	23	1.0–3.9	0.03–0.147	[1.2:1.6]	3.28 ± 0.33	2.92 ± 0.25	Z16	$c \ddagger \uparrow\uparrow$
L$_{exc}$–T$_{exc}$	23	1.0–3.9	0.03–0.147	[1.2:1.6]	3.81 ± 0.46	3.46 ± 0.45	Z16	$c \ddagger \Diamond \uparrow\uparrow$
L–M$_{HE}$	26	0.6–3.0	0.012–0.049	[0.4:0.8]	1.34 ± 0.18	1.47 ± 0.43	E11	$a \dagger \uparrow\uparrow$
L–M$_{HE}$ BC	20	0.9–2.8	0.012–0.034	[0.2:0.7]	1.66 ± 0.22	-	L15	$a \dagger \downarrow\downarrow$
L–M$_{HE}$	20	0.9–2.8	0.012–0.034	[0.2:0.7]	1.32 ± 0.24	1.68 ± 0.32	L15	$a \dagger \downarrow\downarrow$
L$_{exc}$–M$_{WL}$	12	1.7–8.2	0.1–0.47	[0.8:0.9]	1.43 ± 0.16	1.52 ± 0.73	K15	$a \ddagger \downarrow\downarrow$
L–M$_{WL}$ BC	105	0.6–6.0	0.054–1.033	[0.4:0.8]	1.07 ± 0.37	-	S20	$b \ddagger \downarrow\downarrow$
M$_{HE}$–T$_{exc}$	43	0.7–2.7	0.012–0.122	1.5	1.67 ± 0.15	1.75 ± 0.14	S09	$\Diamond \uparrow\uparrow$
M$_{HE}$–T$_{exc}$	26	0.6–3.0	0.012–0.049	1.5	1.68 ± 0.20	1.87 ± 0.37	E11	$\Box \uparrow\uparrow$
M$_{WL}$–T$_{exc}$	10	1.2–4.6	0.124–0.834	1.5	1.71 ± 0.49	1.46 ± 0.58	K13	$\odot \downarrow\downarrow$
M$_{HE}$–T$_{exc}$	20	0.9–2.8	0.012–0.034	1.5	1.65 ± 0.07	1.61 ± 0.10	L15	$\Box \downarrow\downarrow$
M$_{WL}$–T$_{exc}$ BC	12	1.7–8.2	0.1–0.47	1.5	1.52 ± 0.17	-	K15	$\triangle \downarrow\downarrow$
M$_{WL}$–T$_{exc}$	12	1.7–8.2	0.1–0.47	1.5	1.68 ± 0.17	1.22 ± 0.82	K15	$\triangle \downarrow\downarrow$
M$_{WL}$–T$_{300}$	76	0.6–6.0	0.044–1.002	1.5	1.33 ± 0.75	1.14 ± 0.32	U20	$\boxtimes \downarrow\downarrow$
M$_{HE}$–Y$_X$	43	0.7–2.7	0.012–0.122	0.6	0.56 ± 0.03	0.71 ± 0.24	S09	$\uparrow\uparrow$
M$_{HE}$–Y$_X$	26	0.6–3.0	0.012–0.049	0.6	0.53 ± 0.06	0.58 ± 0.19	E11	$\uparrow\uparrow$
M$_{HE}$–Y$_X$	20	0.9–2.8	0.012–0.034	0.6	0.60 ± 0.03	0.58 ± 0.04	L15	$\downarrow\downarrow$
M$_g$–M$_{HE}$ *	43	0.7–2.7	0.012–0.122	1	1.14 ± 0.03	0.97 ± 0.21	S09	$\uparrow\uparrow$
M$_g$–M$_{HE}$	26	0.6–3.0	0.012–0.049	1	1.38 ± 0.18	1.22 ± 0.44	E11	$\uparrow\uparrow$
M$_g$–M$_{HE}$	20	0.9–2.8	0.012–0.034	1	1.09 ± 0.08	1.11 ± 0.10	L15	$\downarrow\downarrow$
M$_g$–M$_{WL}$	118	0.6–6.0	0.054–1.033	1	1.35 ± 0.30	-	S20	$\downarrow\downarrow$
K–T$_{exc}$	43	0.7–2.7	0.012–0.122	1	0.83 ± 0.20	-	S09	$\Diamond \uparrow\uparrow$

The subscripts *exc, 300, HE*, and *WL* indicate properties derived excluding the core, within R < 300 kpc, under the assumption of hydrostatic equilibrium, and with weak-lensing analysis, respectively. BC indicates the relations corrected for selection effects. The slope of the L_x–T_x relation predicted by the self-similar scenario have been obtained as $L \propto T^{1.5+\gamma}$ where γ is the slope of the X-ray emissivity in the considered energy band (e.g., soft or bolometric) and temperature range covered by the systems analyzed in each work (see Table 1). Since the X-ray emissivity strongly depends on the metallicity we provide the extreme values obtained with $Z_\odot$ = 0.3 and $Z_\odot$ = 1.0. The slope of the L_x–M relation is obtained similarly as $L_x \propto M^{1+\gamma}$. The values for slope$_{LIRA}$ have been obtained by fitting each dataset with LIRA (Sereno [112]) assuming self-similar time evolution with scatter on both variables and with the following pivot values: 3 keV, 10^{44} erg/sec, 2×10^{14} M$_\odot$, 10^{13} M$_\odot$, 10^{14} M$_\odot$ for T_x, L_x, M, M_g, and Y_X respectively. Before fitting the L_x–T_x relation, *Chandra* temperatures have been converted into *XMM–Newton*-like temperatures. Note: *a*, *b*, and *c* refer to L_x obtained in the 0.1–2.4 keV, 0.5–2 keV, and bolometric band; † and ‡ indicate L_x obtained with *ROSAT* or *XMM* data, while $\uparrow\uparrow$ and $\downarrow\downarrow$ indicate if *Chandra* or *XMM* data have been used for the analysis. $\Box$, $\triangle$, and $\Diamond$, indicate that the core-excised region was not a fixed fraction of R_{500}, or fixed to $0.1R_{500}$ or $0.15R_{500}$, while $\odot$ and $\boxtimes$ indicate the region 0.1–0.5R_{500} and R < 300 kpc, respectively. * Relation derived fitting together the groups with a sample of clusters. References: (Sun et al. [81], S09), (Eckmiller et al. [96], E11), (Kettula et al. [113], K13), (Lovisari et al. [77], L15), (Bharadwaj et al. [32], B15), (Kettula et al. [101], K15), (Zou et al. [79], Z16), (Umetsu et al. [114], U20), (Sereno et al. [115], S20).

Beside the selection biases there are other issues complicating the comparison between different studies and between systems with different temperatures (masses). The first is the cross-calibration uncertainty between different instruments. For instance, (Schellenberger et al. [111], see also Nevalainen et al. [116]) showed that the cluster temperatures derived with *XMM–Newton* are systematically lower than those obtained with *Chandra*. To complicate this issue is the temperature dependence of this difference. Fortunately, in the low-temperature regime the differences are relatively small, a result that seems to hold also when including *Suzaku* data (e.g., Kettula et al. [113]). Although some caution is still needed, one can expect that calibration differences do not significantly affect the derived relations at the group scale. However, the impact of the calibrations needs to be taken into account when comparing the results obtained for sample of groups and sample of clusters. Another issue, pointed out by Osmond and Ponman [94], is related to the flattening of the fitted relation because the scatter in $\log{(T_X)}$ will be asymmetric (assuming that the scatter in temperature is symmetric) with larger scatter towards low $\log{(T_X)}$. Moreover, if the quality of the data is homogeneous across the sample, the statistical errors are expected to be larger in systems with low luminosities, which also tend to flatten the fitted relation. Finally, each study employs a different fitting algorithm (each with pros and cons) and treatment of the scatter and selection biases which impact the final results (see, e.g., Lovisari et al. [34]). To remove this last uncertainty and provide comparable results we fit the published data with the same fitting method (i.e., using LIRA; Sereno [112]) and assumptions (e.g., self-similar redshift evolution). The results are given in Table 2.

The correlation between X-ray luminosity and gas temperature reflects the fact that a deeper potential well (leading to a higher T_X) generally contains more hot gas (leading to a higher L_X). However, it has been shown that the gas fraction varies as function of the total mass with galaxy groups showing almost a factor of two lower gas fraction than galaxy clusters. Since the X-ray luminosity is proportional to the amount of gas in the IGrM and ICM, a change in the gas content in low-mass systems translates into a lower luminosity with the effect of steepening the L_X–T_X relation. Anyway, the mass dependence of the gas fraction seems to vanish in the outer regions (e.g., Sun et al. [81]) implying that the low gas fraction observed in groups is mainly due to the low gas fraction of groups within $\sim R_{2500}$. This weak ability of the groups to retain the gas in the inner regions is probably a consequence of their shallow gravitational potential and thus of the increasing contribution of different non-gravitational processes. These mechanisms are expected to provide an extra heating to the gas preventing the gas from falling toward the center, and by that, reducing gas density and X-ray emissivity in the cores. The effect is expected to play a significant role in poor systems leading to the steepening of the L_X–T_X relation, as observed, and of the L_X–σ_V relation which, however, is not currently supported by observations (see Section 3.2). Supporting this scenario is the fact that when the core regions of galaxy clusters are ignored (i.e., by removing the regions where non-gravitational processes are expected to affect more the gas properties) the slope of the L_X–T_X relation is more in agreement with the self-similar prediction. This is also in agreement with the suggestion by Mittal et al. [31] that, for low-temperature systems (i.e., $kT < 2.5$ keV), AGN heating becomes more important than ICM cooling (which is the dominant mechanism in massive clusters). Colafrancesco and Giordano [117] suggested that intracluster magnetic fields can also affect more strongly the gas properties in the low-mass regime, resulting in an effective steepening of the scaling relations. In fact, as shown in Colafrancesco and Giordano [117], the magnetic pressure tends to counterbalance part of the gravitational pull of the cluster preventing the gas from a further infalling. Thus, the presence of a magnetic field determines the final distribution of the gas density resulting in a less concentrated core (i.e., leading to a lower luminosity). The effect is mild for massive systems (due to their large gravitational potential) but is relevant in the group regime. The presence of the magnetic field is also expected to decrease the temperature because of the additional magnetic field energy term that needs to be included in the virial theorem. However, since galaxy groups and clusters are not isolated systems, the presence of an external pressure induced by the infalling gas from filaments, would tend to compensate the decrease of T_X caused by the magnetic field.

The non-gravitational heating implies the existence of an entropy floor (i.e., an excess of entropy with respect to the level referable to the gravity only) calling for some energetic mechanisms that can be summarized in three classes: preheating, local heating, and cooling (see the companion review by Eckert et al. for detailed description of these mechanisms). Indeed, entropy in excess with respect to that achievable by pure gravitational collapse is observed in the inner regions of groups and poor clusters (e.g., Mahdavi et al. [118],

Finoguenov et al. [89], Sun et al. [81], Johnson et al. [90], Panagoulia et al. [91]). The excess is found to be radial and mass-dependent, being smaller for massive systems and extending to larger radii in low-mass objects. Moreover, Johnson et al. [90] found that the excess is higher for groups with higher feedback (roughly estimated assuming that both the integrated feedback from SNe and AGNs scale with the stellar mass). Since entropy is expected to remain unchanged when neglecting non-gravitational processes, in the self-similar scenario it simply scales with the gas temperature. The finding by Sun et al. [81] shows that the slope of the relation depends on the scaled radius at which the measurement is taken, and groups behave more regularly in the outer regions (e.g., beyond R_{2500}) than in the core. This was already pointed-out by Ponman et al. [88]. Thus, the slope of the K-T_X relation approaches the self-similar value at R_{500}, where there is no significant entropy excess above the entropy baseline (see the companion review by Eckert et al.). This agrees with the finding by Pratt et al. [119] for a sample of galaxy clusters.

Less studied than the L_X–T_X relation, but of paramount importance for cosmological studies, is the relationship between X-ray luminosity and total mass (i.e., L_X–M). This is particularly true for shallow X-ray surveys because it can be used to directly convert the easiest to derive observable (luminosity requires only source detection and redshift information) to the total mass. The calibration of this relation down to the low-mass regime will allow the breaking of the degeneracy between Ω_m and σ_8 (e.g., Reiprich and Böhringer [36]). A large number of observations of galaxy clusters (e.g., see Mantz et al. [120], Schellenberger and Reiprich [44], Mantz et al. [121], Bulbul et al. [122], and Lovisari et al. [34] for recent studies; we refer to Böhringer et al. [59] and Giodini et al. [61] for older investigations) found that most of the values for the relation slope range from 1.4 to 1.9, steeper than the self-similar prediction of 4/3 suggesting that the luminosity is affected by non-gravitational processes. Unfortunately, the literature in the low-mass regime is still quite limited (in the left panel of Figure 4 we show a compilation of recent galaxy groups studies). Eckmiller et al. [96] found a slope of 1.34 ± 0.18 and suggested that the single power-law modeling of the relation holds also for low-mass objects (see also the illustrative fit in Figure 4). However, given the considerations provided in the previous section, for the sample analyzed by Eckmiller et al. [96] the luminosity should scale with the total mass to the power of [0.4:0.8] (see Table 2) significantly lower than the finding by Eckmiller et al. [96]. A similar slope was also obtained by Lovisari et al. [77] but, after correcting for the selection biases, the corrected slope is steeper (i.e., 1.66 ± 0.22) than the observed one (i.e., 1.32 ± 0.24). making the deviations from the self-similar prediction even larger than what observed for galaxy clusters. This behavior can be explained by a gradual steepening of the true underlying L_X–M relation towards the low-mass regime. This implies that as expected, also the luminosity of groups is heavily affected by non-gravitational processes. However, one should keep in mind that a mass-dependent bias in the total mass can also affect the shape of the L_X–M relation. Because of the difficulties to distinguish the low-temperature emitting gas of these systems from the galactic foreground, the properties of galaxy groups can usually be observed out to a smaller radial extent than what is done for galaxy clusters. Thus, an estimate of the group masses at R_{500} requires an extrapolation for most of the systems making the groups more prone to biases. Moreover, mass biases can also arise from the assumptions of hydrostatic equilibrium and spherical symmetry which are not valid for most of the systems. One way to overcome the problem is to use the weak-lensing masses which are expected to provide a less biased view of the true masses. Because of the difficulties to obtain shear maps for low-mass systems the first attempts have been performed via stacking analysis. Leauthaud et al. [123] stacked the weak-lensing measurements of a sample of X-ray-selected galaxy groups and found an L_X–M_{200} relation in agreement with the finding of galaxy clusters. This result suggests that the L_X–M_{200} relation is well described by a single power-law down to the low-mass regime. However, the lensing analysis of galaxy groups by [101] led to a shallower slope although the agreement in the low-mass regime of the two relations is fairly good, with significant tension appearing only at high masses (i.e., above a few 10^{14} $M_\odot$). With the advent of multi-wavelength surveys, which uniformly scan large areas of sky, there has been significant progress in the weak-lensing analysis of large samples of galaxy groups. For instance, in the XXL framework (see Pierre et al. [124]), the L_X–M_{WL} relation was investigated by Sereno et al. [115] who found a quite good agreement with previous X-ray studies. The results suggest that the measured hydrostatic bias is consistent with a small role of non-thermal pressure. However, due to the large uncertainties associated with the derived weak-lensing masses a large deviation from hydrostatic equilibrium cannot be completely excluded and further investigations with larger samples and higher quality data are required to make progress in the field.

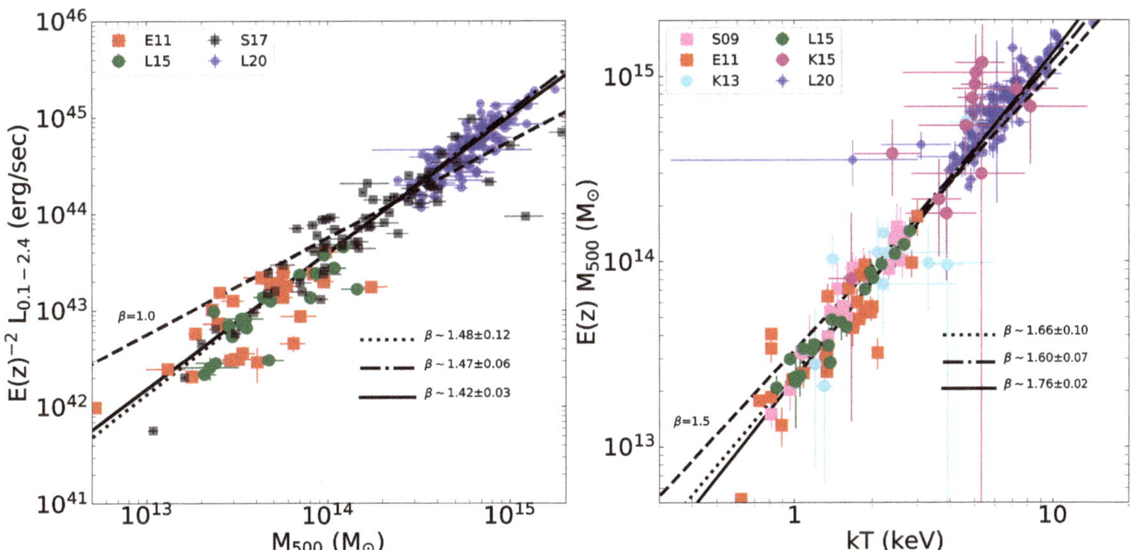

Figure 4. *XMM–Newton* (circles) and *Chandra* (squares) measurements of the $L_{0.1-2.4}$–M (**left panel**) and M–T_X (**right panel**) relations for different samples of groups: (Eckmiller et al. [96], E11), (Lovisari et al. [77], L15), (Sun et al. [81], S09), (Kettula et al. [113], K13). The groups measurements are compared with the ones from an X-ray-selected (Schellenberger and Reiprich [100], S17) and an SZ-selected (Lovisari et al. [34], L20) cluster sample. Luminosities and masses are all within R_{500} while temperatures are obtained in different regions (see Table 2). Please note that different studies used different methods to estimate the total masses. The lines represent the fitted relation for $M < 10^{14} M_\odot$ systems (dotted), $M > 10^{14} M_\odot$ systems (dashed-dotted), all systems (solid), and are compared with the case predicted by the self-similar scenario (dashed). Because of the strong covariance between M and T_X we did not convert the *Chandra* temperatures to *XMM–Newton*-like temperatures.

In contrast to the L_X–T_X and L_X–M relations, the M–T_X relation is expected to follow the same behavior for galaxy groups and clusters, under the assumption that the gas temperature reflects the depth of the underlying potential well. However, while the assumption is probably reasonable for many clusters (at least for the most relaxed ones) it may not be strictly true for groups where the gas has probably been significantly heated by non-gravitational processes. For this reason, the expectation is that the scatter should increase in the low-mass regime where the global temperature is not insensitive to the details of the heating/cooling processes as in the high-mass regime. However, these processes are definitely more important in the inner regions with their effect fading at large distances from the center. Nonetheless, some simulations suggested that the gas removed by AGN activity in groups can affect the gas properties out to several Mpc (e.g., Schaye et al. [125]), potentially affecting also cosmic shear measurements (e.g., Semboloni et al. [126]). Thus, it is important to define the region within which the characteristic cluster temperature is determined. This is not trivial because, for instance, we have evidence that the central drop (i.e., the region in the center of relaxed clusters showing a significant temperature decline, probably caused by radiative cooling), typically present in relaxed clusters, does not scale uniformly with the mass (Hudson et al. [16]). However, a common practice is to exclude the regions within $0.15R_{500}$.

There had been a lot of studies investigating the M–T_x relation before the *Chandra* and *XMM–Newton* era. Many of them (e.g., Finoguenov et al. [127], Sanderson et al. [9], and references therein) suggested that the low- and high-mass end of the relation is characterized by different slopes with the cross-over temperature between the two regimes at ∼3 keV. However, most of these studies could not constrain the gas properties (i.e., gas density and temperature gradients) at large radii making the estimated hydrostatic masses more prone to biases. Thanks to *Chandra* and *XMM–Newton*, the measurements could be extended to larger fraction of R_{500} reducing the impact of the extrapolation. Sun et al. [81] found a relation only slightly steeper than the prediction of the self-similar scenario. Both Eckmiller et al. [96] and Lovisari et al. [77] found that the slope for galaxy groups is consistent with the one of galaxy clusters but with a normalization 10–30% lower. The net effect of this finding is a steepening of the relation when groups and clusters are fitted together (see, e.g., right panel of Figure 4). Indeed, due to the limited field-of-view of *Chandra* and the high and variable *XMM–Newton* background level, X-ray measurements are still not tracing well the outer regions (despite the improvement with respect to previous missions, measurements extend out to R_{500} only for a few systems). If the density profiles of groups are steepening at large radii then the masses could be underestimated explaining the lower normalization of the M–T_x relation. Another possible issue is that the samples analyzed in the above-mentioned papers are biased toward relaxed systems. Thus, if relaxed and disturbed systems do not share the same relation (e.g., because hydrostatic masses are more biased for disturbed systems) the relative fraction of relaxed/disturbed groups can impact the normalization (and possibly the slope) of the observed M–T_x relations. Lovisari et al. [34] showed that this is probably not the case for massive systems, but a dedicated study in the group regime is still missing. Focusing on the slopes, the results of the most recent papers on galaxy groups agree with the results for galaxy clusters for which most of the slope values range from 1.5 to 1.7 (see Table 2). This agreement suggests a small impact of the non-gravitation processes to this relation. Again, before overinterpreting these results, one of the key questions is to assess if the level of mass bias in these systems is similar to the one of galaxy clusters. For instance, (Kettula et al. [128], see also Kettula et al. [101]) argued that the hydrostatic mass bias at 1 keV reaches a level of 30%-50%, higher than what usually observed for galaxy clusters (e.g., by calibrating the hydrostatic masses with other mass proxies, such as weak-lensing or velocity dispersion). However, the sample consists of only 10 galaxy groups and the dynamical state of the systems is not discussed. A much larger sample was investigated by Umetsu et al. [114] who found the relation to be consistent within statistical uncertainties with the self-similar expectations. However, the uncertainties of the individual weak-lensing mass measurements in the group regime are still quite large and tighter constraints are needed in the future to exclude deviations from self-similarity. An increasing mass bias in the low-mass regime would not be fully unexpected. In fact, in the unmagnetized case, the viscosity scales as $T_X^{5/2}$ (Spitzer [129]) favoring the development of strong turbulences. Acting as additional pressure support against gravity, turbulent motions may increase the mass bias. Anyway, the IGrM is magnetized and the real magnitude of the turbulence is still unknown.

Beside the shape of the scaling relations another important information is given by the scatter (i.e., the dispersion around the best-fit). Minimizing the scatter of the scaling relations is of paramount importance to obtain accurate constraints on cosmological parameters which are dominated by uncertainties in the mass–observable relations. Moreover, understanding the scatter in the relations is the key to pinpoint the physical processes at play in the group regime. However, the measurement errors for most of the groups are large, thus the intrinsic scatter is not well constrained, yet. Because of that, refs. [77,81] were not able to constrain the intrinsic scatter in their samples. [96] instead suggested that the scatter in galaxy groups (kT $<$ 3 keV) is much larger than the one derived for the HIFLUGCS sample (Reiprich and Böhringer [36]).

A mass proxy which has been shown by simulations to bear a low scatter is the Y_X parameter (i.e., the product of the gas temperature and the gas mass; see Kravtsov et al. [130]). Sun et al. [81] were the first to investigate the M–Y_X relation in the low-mass regime finding that a single power-law model can fit very well both galaxy groups and clusters. This result was also confirmed by Eckmiller et al. [96] and Lovisari et al. [77]. Unfortunately, given the sample size and the relatively large measurements errors, the intrinsic scatter is not properly constrained. However, both Sun et al. [81] and Eckmiller et al. [96] suggested that the scatter of the M–Y_X relation is almost half of the M–T_x relation. The findings by Eckmiller et al. [96] also suggest that the scatter for galaxy groups is significantly higher than for galaxy clusters.

The self-similar model also predicts the X-ray scaling relations to be redshift-dependent (e.g., Giodini et al. [61] and references therein), reflecting the decrease with time of the mean density of the Universe. Non-gravitational processes are expected to affect the evolution of the X-ray scaling relations because of the increasing importance of such processes to the energy budget of galaxy systems as a function of redshift. Unfortunately, although groups are more common than clusters, because of their fainter and cooler nature it is more difficult to detect them over the background, especially at higher redshifts. Thus, due to the big challenges to detect large and representative samples of galaxy groups beyond the local Universe the literature on this subject is very limited. The few studies (e.g., Jeltema et al. [131], Pacaud et al. [99], Alshino et al. [132], Umetsu et al. [114], Sereno et al. [115]) which have tried to address the evolution of the X-ray properties of galaxy groups did not find convincing evidence for such evolution. A characterization of the evolution of the scaling relations also on galaxy group scales is one of the goal of the next-generation instruments (such as Athena; see Section 5).

3. Optical Scaling Relations

Due to the low X-ray flux at the group scale, there is high probability that X-ray selected samples are biased toward groups with rich IGrM. Moreover, since the luminosity strongly depends on the metallicity (see Section 2.1), variations in the metal abundance between groups (possibly related to their feedback history) can significantly impact the selection function. Thus, it is advantageous to explore scaling relations between an X-ray property that can be measured relatively well in the low count regime (e.g., X-ray luminosity or gas temperature) and an optical property that can be used as a proxy for the group mass (e.g., velocity dispersion, optical band luminosity).

3.1. Velocity Dispersion

The velocity dispersion (σ_v) of galaxy groups (and indeed clusters) can be used to estimate dynamical masses via the application of the virial theorem. Furthermore, the velocity of member galaxies complements X-ray information about the cluster morphology projected onto the sky. For example, studying the luminosity–velocity dispersion (L_x–σ_v) relation provides an understanding of the dynamical properties of galaxy clusters and their impact on the scaling relations.

One of the most commonly used estimators of the velocity dispersion at the group regime is via the use of the *gapper* estimator from [133]. Of critical importance at the group scale, the *gapper* estimator is unbiased when using low numbers of member galaxies (down to $\sim$10 members, e.g., [134]), and is robust against outliers. The *gapper* velocity estimator (σ_v) is given by

$$\sigma_v = \frac{\sqrt{\pi}}{N(N-1)} \sum_{i=1}^{N-1} w_i g_i, \tag{6}$$

where, for ordered velocity measurements, the gaps between each velocity pair are defined as $g_i = v_{i+1} - v_i$ (for $i = 1, 2, 3..., N - 1$), as well as Gaussian weights defined as $w_i = i(N - i)$.

As stated above, one can study the L_X–σ_V relation to understand dynamical properties and the impact on scaling relations. In Equation (4), it is shown that in the self-similar scenario the bolometric luminosity is expected to scale with the gas temperature as $L_{X,bol} \propto T_X^2$. Under the consideration that both the cluster/group hot gas and galaxies feel the same potential, assuming that they both have the same kinetic energy, the temperature can be converted to velocity dispersion using

$$\beta = \frac{\sigma_V^2 \mu m_p}{k_B T_X} \approx 1, \tag{7}$$

where the parameter β is the ratio of the specific energy in galaxies to the specific energy in the hot gas. Using Equation (7) and the self-similar scaling of $L_{X,bol}$ and T_X above, the self-similar scaling of velocity dispersion and X-ray properties can be given by

$$L_{X,bol} \propto \sigma_V^4, \tag{8}$$

$$T_X \propto \sigma_V^2. \tag{9}$$

However, because of the behavior of the X-ray emissivity in the low-temperature regime, the dependence of the luminosity on the temperature is more complicated (see discussion in Section 2.1) and can be approximated as $L_X \propto T_X^{1.5+\gamma}$, where γ is the slope of the X-ray emissivity in the considered energy band (e.g., soft or bolometric) and temperature range (see Table 1). Using this γ dependent relation, it follows that

$$L_X \propto \sigma_V^{3+2\gamma}. \tag{10}$$

For temperatures lower than 3 keV, the value of γ is negative (unless very cool systems are considered), implying that the expected L_X–σ_V relation for galaxy groups is shallower than what is predicted for galaxy clusters (e.g., Equation (8)).

3.2. The Luminosity-Velocity Dispersion Relation

At the cluster scale, generally, it has been found that the observed luminosity–velocity dispersion (L_X–σ_V) relation follows, or is slightly steeper than, the expectation of Equation (8) (e.g., [135–140]). Furthermore, at the cluster scale, studies of the L_X–σ_V relation now use samples of clusters numbering in the high hundreds (e.g., [141], using 755 clusters to investigate the L_X–σ_V relation). Studies of the L_X–σ_V relation at the group scale attempt to compare the form of the relation at the high-mass regime to investigate differences at these two mass scales (e.g., to probe the effect of AGN feedback processes at high and low masses). Early studies comparing the slope of the relation between the two mass regimes provided a mixed picture, with studies finding groups have a flatter (e.g., [7,93,94]) or consistent (e.g., Ponman et al. [142], Mulchaey and Zabludoff [136], Mahdavi and Geller [137]) relation than their high-mass counterparts. Figure 5 (left-panel) provides a (non-comprehensive) compilation of the slope of the L_X–σ_V relation from various studies in the literature. The solid horizontal line represents the dividing line between studies using clusters (top half) and groups (bottom half). The $L_{X,bol} \propto \sigma_V^4$ expectation is given by the vertical dashed line. Although there appears to be a clear division between the slopes for groups and clusters, many of the group scale studies compare to the usual $L_{X,bol} \propto \sigma_V^4$ expectation at the cluster scale. As shown in Equation (10), the scaling can be given by $L_X \propto \sigma_V^{3+2\gamma}$, with γ dependent on the X-ray emissivity and energy band used. If we assume a group temperature range of 0.7–3.0 keV, then given the range of emissivities in Table 1, the bolometric scaling in the group regime becomes $L_{X,bol} \propto \sigma_V^{[2.2:3.0]}$. This range is highlighted by the blue shaded region in Figure 5 (left-panel) at the group scale. For comparison, using these same arguments, the bolometric scaling for clusters (assuming 3.0–10.0 keV) becomes $L_{x,bol} \propto \sigma_V^{[3.7:3.9]}$ (again highlighted by the blue shaded region in Figure 5, appropriate to the cluster scale). Considering the above, studies investigating the group scale relation can indeed be considered consistent with the self-similar expectation (e.g., [94]). Although this is the case, many authors note caveats when studying groups, which are discussed below.

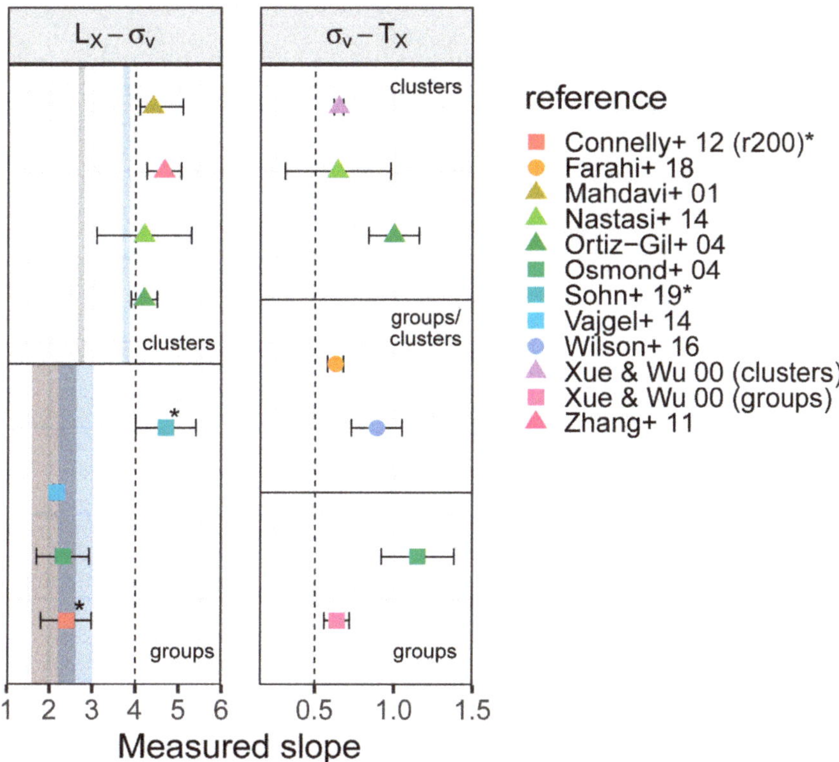

Figure 5. Compilation of the measured slopes of the luminosity-velocity dispersion (L_X–σ_V, **left panel**) and velocity dispersion-temperature (σ_V–T_X, **right panel**) relation in the literature. In each case, the dashed vertical line highlights the usual self-similar expectation on the slope for each relation (Equation (8), **left panel**, and Equation (9), **right panel**). The horizontal lines represent a dividing line between the mass scales used for the comparison in each relation. As shown in Section 3.1 when the L_X–σ_V scaling can be given as $L_X \propto \sigma_V^{3+2\gamma}$ (with γ dependent on the energy band and emissivity). The blue shaded region represents the self-similar expectation when considering bolometric luminosities for clusters (assuming T_X = 3.0–10.0 keV, $L_X \propto \sigma_V^{[3.7:3.9]}$) and groups (assuming T_X = 0.7–3.0 keV, $L_X \propto \sigma_V^{[2.2:3.0]}$). The grey shaded region represents the self-similar expectation when considering 0.1–2.4 keV luminosities for clusters (assuming T_X = 3.0–10.0 keV, $L_X \propto \sigma_V^{[2.7:2.8]}$) and groups (assuming T_X = 0.7–3.0 keV, $L_X \propto \sigma_V^{[1.6:2.6]}$). Unless otherwise stated, references consider (bolometric) luminosities and temperatures derived within an estimate of R_{500}. Please note that the [143] relation is based upon an analysis using 7 bins of using the full sample of 74 systems. * refers to references using 0.1–2.4 keV band luminosities.

Many early studies were based upon ensemble collections of groups that can lead to biases in the derived scaling relations. In recent years, studies of the L_X–σ_V relation have used groups selected over contiguous survey regions. One such study was performed by [144].

Groups were selected from regions of the Canadian Network for Observational Cosmology Field Galaxy Redshift Survey 2 (CNOC2, [145]) that were covered by *XMM–Newton* and *Chandra* observations, totaling 0.2 and 0.3 deg^2 contiguous areas of two fields of the CNOC2 survey. Using X-ray selected groups with high quality redshift information, they find a slope of the L_X–σ_V of $2.40^{+0.58}_{-0.60}$ (including groups with lower quality redshift information yields a slope of $1.35^{+0.42}_{-0.47}$). Although initial inspection of the value of the slopes would imply the slope is shallower than the self-similar expectation (as noted in [144]), we note that the luminosities are reconstructed in the 0.1–2.4 keV band (from the flux in the 0.5–2 keV band, and by correcting for extension and K-correction as described in Finoguenov et al. [146]). Assuming the scaling follows $L_X \propto \sigma_V^{3+2\gamma}$, then for luminosities in the 0.1–2.4 keV band, the scaling can be given by $L_X \propto \sigma_V^{[2.0:2.7]}$ (depending on the metallicity of the groups). This expectation is shown in Figure 5 (left-panel), highlighted by the grey shaded region at the group regime. The slope determined by [144] is coincident with this scaling. Therefore, if the energy band is considered, the [144] relation is consistent with the self-similar expectation. Hence, it can be assumed that the groups studied in [144] are consistent with the cluster scale (assuming clusters follow the self-similar expectation). Another study using contiguous fields is presented in [143], using groups selected from the 2 deg^2 Cosmic Evolution Survey (COSMOS, [147]). This study constructs a catalog of galaxy groups based upon those identified in [148], using *XMM–Newton* and *Chandra* observations of the COSMOS field, reaching an X-ray flux limit of $\sim 10^{-15}$ erg s^{-1} cm^{-2}. Ref. [143] use galaxy redshift information from a wide variety of surveys in the literature and associate them with the X-ray detected groups, compiling a final sample of 146 groups with at least three spectroscopic members. Based upon a cleaned sample of 74 groups, ref. [143] showed that the relation for individual groups appears to follow a shallower relation than clusters (consistent with that found by previous studies, e.g., [137]). However, they note that this trend may be affected by a small number of groups that appear to have anomalously low velocity dispersions (at $\sigma_V \lesssim 125$ km s^{-1}) for their measured X-ray luminosity (discussed further in Section 3.4). To overcome this, Ref. [143] estimated the median velocity dispersion for 7 bins of groups created from the 74 groups in their sample, finding a slope of $L_X \propto \sigma_V^{4.7\pm0.7}$. This study considers luminosities in the 0.1–2.4 keV band, therefore, as discussed above, the measured slope is in fact steeper than the self-similar expectation.

Although contiguous regions have been used to study the L_X–σ_V relation, an extremely small number of studies have attempted to correct for X-ray selection biases. One study that attempts to do so is presented in [149], using 14 groups with at least 5 galaxy members selected from the 9 deg^2 X-Boötes survey [150]. They find that the group scale relation is consistent with the self-similar expectation. To test the effects of Malmquist bias on the observed relations, ref. [149] determined the limiting X-ray luminosity in two survey volumes (z = 0.20 and z = 0.35). The resulting relations are consistent with the sample relation, with the authors concluding the sample may not be dominated by Malmquist bias effects. However, due to the associated large error on each relation, making this conclusion is challenging and requires the construction of larger samples. Another use of contiguous surveys, particularly those covered by multiple wavelengths, is the possibility to compare the relations derived using groups selected via multiple selection methods (e.g., X-ray, optical). Furthermore, the use of optically selected groups allows one to estimate the form of scaling relations independent of the usual X-ray selection biases (e.g., [104]). The study by [144], as detailed above, also constructed a sample of 38 optically (spectroscopically) selected groups. Using this optically selected sample, they derive a slope of the L_X–σ_V relation of $1.78^{+0.60}_{-0.54}$. Given the large uncertainties on the measured slopes, the comparison of the X-ray and optically selected samples is somewhat limited (note that the comparison is not affected by the energy band used, as discussed above, since they are consistent between the X-ray and optically selected samples).

3.3. The Velocity Dispersion-Temperature Relation

Velocity dispersion and gas temperature are two independent probes of the depth of the cluster potential well, estimated by using baryons as tracers. Therefore, this relation can provide useful information about the effect of non-gravitational processes, which are responsible for the deviation from thermal equilibrium of the IGrM and ICM. Hence, it is useful to compare group and cluster relations to investigate the differences between these mass scales. Figure 5 (right-panel) shows a compilation (again, a non-comprehensive picture) of the slope of the σ_V-T_X relation from studies in the literature. The horizontal lines represent the division between studies using groups (bottom section), clusters (top section) and those using systems which straddle the group/cluster regime (middle section). Based upon Equation (7), it is expected that the velocity dispersion of the galaxies should scale with the square-root of the temperature of the gas, $\sigma_V \propto T_X^{1/2}$. In the context of clusters, various studies have found that the σ_V–T_X relation has a slope steeper than the self-similar expectation (e.g., [138,151]), with others finding a steeper slope but with errors too large to confirm a deviation (e.g., [140]). Unfortunately, the study of the σ_V–T_X relation for groups is somewhat limited in the literature. An early investigation presented in [94] showed evidence for steepening of the relation at the group scale (where they find a slope of 1.15 ± 0.23). However, they caution that there is both large uncertainties on the measured X-ray temperatures and a large amount of scatter observed in the relation, which could be the cause of tension with previous studies attempting to investigate any steepening of the relation for groups. Although [94] found evidence for a steepening of the relation, they remarked that a comparison cluster-based relation passes through the center of the group relation data, and represents adequately the cluster-based relation. However, recent studies of the relation, especially at the group scale, become scarce. One recent study of the σ_V–T_X is presented in [151], making use of groups/clusters detected serendipitously in the *XMM* Cluster Survey [152]. Using 19 groups/clusters with redshifts $z < 0.5$, spanning the temperature range $1.0 \lesssim T_X \lesssim 5.5$ keV (with 50% of clusters with a temperature < 3 keV), the σ_V–T_X relation is found to have a slope of 0.89 ± 0.16. Although again steeper than the self-similar expectation, the result is somewhat shallower (although not significant) than that presented in [94]. Finally, the last relation considered is that given in [153], which investigated the σ_V–T_X relation for a sample of X-ray selected clusters detected in the XXL survey [124]. Clusters were selected from the 25 deg^2 XXL-N region, with spectroscopic data compiled from a range of surveys (see [154], for full details of the spectroscopic coverage). Using a sample of 132 clusters (the majority of which have $T_x < 3$ keV), ref. [153] found a relation of the form $\sigma_V \propto T_X^{0.63 \pm 0.05}$. Please note that the relation is fitted using an ensemble maximum likelihood method, with fitted slope in tension with the self-similar expectation. Since both velocity dispersion and X-ray temperature scales with total mass, one can combine the information to determine a useful mass calibration (e.g., [153]). However, consideration must be given to velocity anisotropies during mass modeling using velocity information, which can vary for loose, compact and virialized groups [155]. However, corrections based upon halo concentration have been developed (e.g., [156]). The study of the σ_V-T_X relation can also be probed down to the galaxy scale. Ref. [157] used galaxies from the volume-limited MASSIVE survey [158], to study the relation between galaxy kinematics (σ_e) and X-ray temperature. Ref. [157] found a relation of the form $T_X \propto \sigma_e^{1.3-1.8}$ (note the inverse of the relation as discussed above), noted as being marginally flatter than the self-similar expectation.

As discussed in Section 3.4, AGN feedback and its effects on the ICM could result in deviations from self-similarity, in particular, the steepening observed above in the T_X–σ_V. Although an observational consensus on the magnitude of the deviation from self-similarity at the group scale compared to the cluster scale has yet to be reached, simulations have indeed shown a mass dependence (e.g., Le Brun et al. [159], Farahi et al. [160], Truong et al. [28]). The deviations discussed for the L_X–σ_V and σ_V–T_X are thought to arise due to the effects of AGN feedback on the ICM (as shown in simulations, e.g., [161]), which has little effect on the galaxy velocity dispersions. Furthermore, Ref. [157] measured a median value of $\beta = 0.6$ for their galaxy sample, suggesting the galaxies have undergone, or still in the process of, additional heating due to, e.g., AGN feedback, as discussed above.

3.4. Low Velocity Dispersion Groups

One observation made by various authors studying the L_X–σ_V relation, is the presence of low velocity dispersion groups (appearing at $\sigma_V \lesssim 200$ km s^{-1}) that have a high X-ray luminosity in comparison to their σ_V (conversely, it can be stated that these groups have a low σ_V for their L_X). These low velocity outliers have been noted in various studies in the literature (e.g., [137,162]), attributed as the cause of the flattening of the L_X–σ_V relation at the group scale (e.g., Vajgel et al. [149], Sohn et al. [143]). Although it has been shown in Section 3.2 that the group scale relations may be consistent with self-similar predictions when accounting for the differing emissivity, the presence of these outliers are extreme cases. A physical interpretation of the low σ_V outliers is therefore currently lacking. Furthermore, the presence of these outliers remains somewhat of a mystery if one considers the effects of AGN feedback on the intragroup medium. During an AGN outburst, gas will be removed from the group potential, hence lowering the group overall X-ray luminosity. With the group velocity dispersion unaffected by this process, the expectation would be that the group should have a lower L_X for a given σ_V, contrary to this outlier population. It could therefore be argued that it is in fact the velocity dispersion that have been underestimated for these groups. Potential explanations for the presence of these low σ_V outliers were given in [162]. It is postulated the cause could be: (i) through dynamical friction, energy is transferred from a large orbiting body to the sea of dark matter particles through which it moves; (ii) due to tidal interactions, the orbital energy may be converted into internal energy of the galaxies; and (iii) the orbital motion happens in the plane of the sky, therefore contributing little to the line-of-sight velocity dispersion. Although a current physical interpretation is lacking, the presence of low velocity dispersion outliers could be due to X-ray selections effects (e.g., Eddington and Malmquist biases). Extreme outliers (e.g., [143]) may not be attributed to selection; however various relations involving L_x when using X-ray selected samples characteristically show a flattening when not account for selection (e.g., [163,164]). In fact, the preferential selection of higher luminosity groups for a given velocity dispersion (i.e., Malmquist bias), leads to the presence of "moderate" outliers. Ref. [104] argues that an unbiased sample of clusters can be obtained when selecting clusters from optical properties and therefore able to probe the full range of scatter and the true form of the relation. As stated, [144] have used an optically selected sample of clusters to investigate the form of the L_X–σ_V relation; however, they find constancy in both the form and scatter of the X-ray and optically selected group samples (due to the large errors on the scaling parameters). This sample only covered an area of 0.5 deg^2, therefore the comparison of optically and X-ray selected samples over overlapping contiguous fields requires further attention to truly probe the differences in selection.

3.5. Stellar Gas Content of Galaxy Groups

The gas mass fraction of clusters can be used as a probe of cosmology (e.g., Allen et al. [37], Ettori et al. [165], Mantz et al. [166], Schellenberger and Reiprich [44]). However, as mentioned in the previous sections, it has been shown that the fraction decreases as a function of total mass. Interestingly, the opposite is true for the stellar mass fraction ($f_{stars} = M_{stars}/M_{tot}$), with an increasing stellar mass fraction as a function of decreasing total mass (e.g., Lin et al. [167], Gonzalez et al. [73], Giodini et al. [2], Behroozi et al. [168], Zhang et al. [169], Leauthaud et al. [170], Laganá et al. [171], Chiu et al. [172], Decker et al. [173]). To investigate this trend, much effort has been afforded to the study of the gas mass and stellar mass content in groups and clusters (e.g., to determine star formation rates). One such observation is that the stellar mass has a correlation with the halo mass with a slope <1 (see discussion below). This has the implication that at the group scale, star formation is more efficient. One early study that specifically used groups to constrain the form of the stellar mass–halo mass relation (M_{stars}–M) and the group f_{stars}

is that of [2]. An X-ray selected sample of groups was constructed from the COSMOS survey, in which X-ray extended sources were detected based upon a wavelet detection routine [174]. Mean photometric redshifts were assigned to each candidate and checked against available spectroscopic redshifts from zCOSMOS [175]. After quality checks, a final sample of 91 groups were used to constrain the form of the M_{stars}–M relation. Masses were estimated based upon a stacked weak-lensing analysis [123] and the construction of a L_X–M_{200} relation, from which the catalog masses were estimated (note that M_{500} masses were used in the final analysis, estimated from the M_{200} assuming an NFW profile -Navarro et al. [176], Navarro et al. [177], and constant concentration, $c = 5$). Within R_{500} the M_{stars}–M relation was found to follow a form of $M_{stars} \propto M^{0.81\pm0.11}$ and a stellar mass fraction of the form $f_{stars} \propto M^{-0.26\pm0.09}$ (extending this to higher masses with the inclusion of clusters, the form follows a relation of $f_{stars} \propto M^{-0.37\pm0.04}$). More recent studies have used increased area X-ray surveys. The *XMM Blanco* Cosmology Survey (XMM-BCS, [106]) covers 12 deg^2 of the sky with *XMM–Newton*, and was used by [178] to study the form and evolution of the M_{stars}–M relation using 46 groups/clusters within a mass and redshift range of $(2 \lesssim M \lesssim 25) \times 10^{13}$ $M_\odot$ and $0.1 \lesssim z \lesssim 1.02$, respectively. The M_{stars}–M relation is fitted including an evolutionary redshift term, with parameters estimated by evaluating a likelihood based upon observing a cluster with observed properties (L_X and M_{stars}) given a mass, redshift, L_X–M relation (mass calibration) and the M_{stars}–M relation. The likelihood is weighted by the mass function, with full details given in [179]. The fitted relation has the form $M_{stars} \propto M^{0.69\pm0.15}(1+z)^{-0.04\pm0.47}$, again consistent with previous results showing a shallower than unity slope of the relation. Furthermore, these results indicate little evolution in the stellar content with stellar mass fraction staying constant out to $z \simeq 1$. In Figure 6, we plot the M_{stars}–M relation for various results obtained in the literature (namely [2,178,180–182]). We note that no attempt has been made to correct for the differences in mass calibration used in the various studies. For reference, two constant stellar mass fractions are given by the black dashed lines. As discussed above, the relations for [2,178] are derived at the group scale, whereas [182] straddles the high-mass groups/low-mass cluster regime (see below) and [180] used primarily high-mass clusters (note that the relation plotted here includes clusters from Gonzalez et al. [76], as detailed in Kravtsov et al. [180]). All the relations have a slope less than unity, and show the trend of decreasing stellar mass fraction from the group to cluster regime. The observed relations are also consistent with that found in simulations. Results obtained from the IllustrisTNG simulations show the same trend in stellar mass (shown by the red dot-dashed line in Figure 6, taken from [181]).

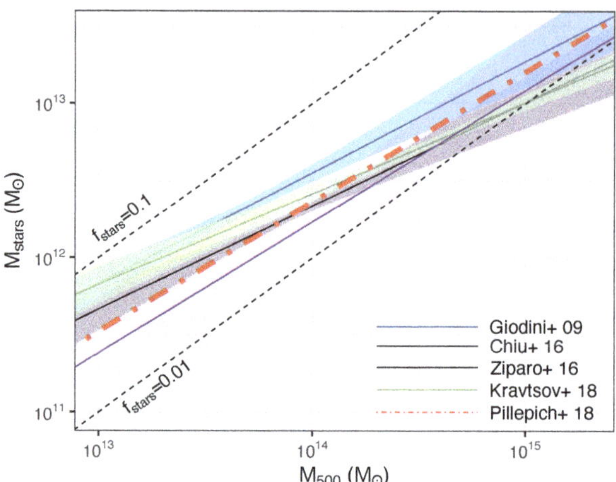

Figure 6. The M_{stars}–M relation of various studies in the literature. Two lines of stellar mass fraction are highlighted by the dashed lines. Please note that the [182] relation is derived from the conversion of L_K to M_{stars} assuming a constant mass-to-light ratio of 0.73 (as used in [182]).

Due to the difficulty of measuring the stellar masses of groups directly, requiring deep observations, it is beneficial to use a proxy for the stellar mass. One such proxy as a tracer of the stellar mass is the K-band luminosity (L_K), as shown in various studies (e.g., [167,183]). The use of L_K as a stellar mass proxy was investigated in [182] using a sample of 20 groups/clusters selected from the *XXL* survey. The clusters were selected from the overlap of the *XXL* and CFHTLS, using clusters with an individual weak-lensing mass estimate. Ref. [182] found a relation of the form $L_K \propto M_{WL}^{0.85^{+0.35}_{-0.27}}$, which while shallower than unity, a slope of 1 cannot be ruled out. Furthermore, when combined with a sample of high-mass clusters from LoCuSS, ref. [182] measured a slope of $1.05^{+0.16}_{-0.14}$. The relation derived for the *XXL* sample is shown in Figure 6 (purple line, with the shaded light purple region highlighting the 1σ uncertainty), which is derived from the L_K–M_{WL} relation assuming a constant mass-to-light ratio of 0.73 (as adopted in [182]).

4. The Role of SMBHs: Observed Scaling Relations and Predictions via HD Simulations

As introduced in Section 1, the evolution of the IGrM filling galaxy groups cannot be merely understood in isolation as giant self-similar gaseous spheres. Particularly in the last decade, a wide range of evidence has accumulated showing that the SMBHs at the center of each galaxy group are tightly co-evolving with the hot X-ray halo. Such co-evolution works in both directions: the hot-halo acts as an active atmosphere and reservoir of gas which recurrently feeds the central SMBH (Gaspari et al. [184], Prasad et al. [185], Voit et al. [186], Temi et al. [187], Tremblay et al. [188], Gaspari et al. [189], Rose et al. [190], Storchi-Bergmann and Schnorr-Müller [191]). In turn, the SMBH re-ejects back large amount of mass and energy (in particular via jets and outflows; e.g., Tombesi et al. [192], Sądowski and Gaspari [193], Fiore et al. [194]), thus re-heating and re-shaping the IGrM via bubbles, shocks, and turbulence up to the group outskirts (McNamara and Nulsen [195], Fabian [196], Gitti et al. [197], Brighenti et al. [198], Gaspari [199], Liu et al. [200], Yang et al. [201], Wittor and Gaspari [202], Voit et al. [203]). Although the small-scale AGN self-regulation thermodynamics/kinematics is respectively tackled in the companion Eckert/Gastaldello et al. reviews, here we focus on the macro-scale integrated (X-ray) IGrM properties and group scaling relations, which complete and complement Sections 2 and 3. Furthermore, we compare with high-resolution and hydrodynamical (HD) simulations, in particular to discuss what the X-ray scaling relations can constrain and tell us in terms of the baryonic physics shaping the IGrM.

Figure 7 shows several key macro X-ray halo scaling relations, which are usually employed in cosmological studies (see Section 2), but now plotted against the SMBH mass $M_\bullet$, which is also an integrated property. These SMBH masses are retrieved only via robust direct measurements, i.e., resolving the stellar or gas kinematics within the SMBH influence region (e.g., via HST). The current largest sample correlated with the available X-ray hot gas properties is presented by Gaspari et al. [20], which includes central galaxies and satellites, with morphological types such as ellipticals (blue circles), lenticulars (green), and a few spirals (cyan). The 85 systems span a range of $M_{500} \sim 3 \times 10^{12} - 3 \times 10^{14}$ $M_\odot$, with most systems in the group regime ($T_X \sim 1$ keV) and a few in the poor or cluster tails. The companion Eckert et al. review shows that the $M_\bullet$ correlation with T_X is significantly tighter than the classical optical scaling, such as the Magorrian relation (e.g., Kormendy and Ho [204], Saglia et al. [205]), with intrinsic scatter down to 0.2 dex, in particular within the circumgalactic and core region. Here, in Figure 7 we show the other key X-ray properties integrated up to R_{500}, namely the plasma X-ray luminosity (in the 0.3–7.0 keV band), gas mass, total mass (gas plus stars plus dark matter), gas density, Compton parameter, and gas fraction. All the fits parameter–including the intercept, slope, scatter, and correlation coefficient–are shown in the top-left inset. The related Bayesian analysis (Gaspari et al. [20]) shows the 1-σ intrinsic scatter as light red bands with the dotted lines enveloping the rare 3-σ loci. As indicated by all correlation coefficients, even the macro-scale IGrM (several 100 kpc to Mpc scale) is tightly linked to the central $M_\bullet$. The tighter correlations are those involving the gas mass/luminosity, X-ray Compton parameter Y_X, and total mass, while the loosest one is that with the gas density. It is interesting to note that using the core radius (or smaller) as extraction radius (not shown) leads to similar results, except that the total mass scatter increases by 0.1 dex, with the gas properties emerging as dominant drivers (in particular M_{gas} and f_{gas}). In other words, we suggest using the R_{500} scaling to probe the total mass, while smaller extraction radii to probe gas mass (and related properties).

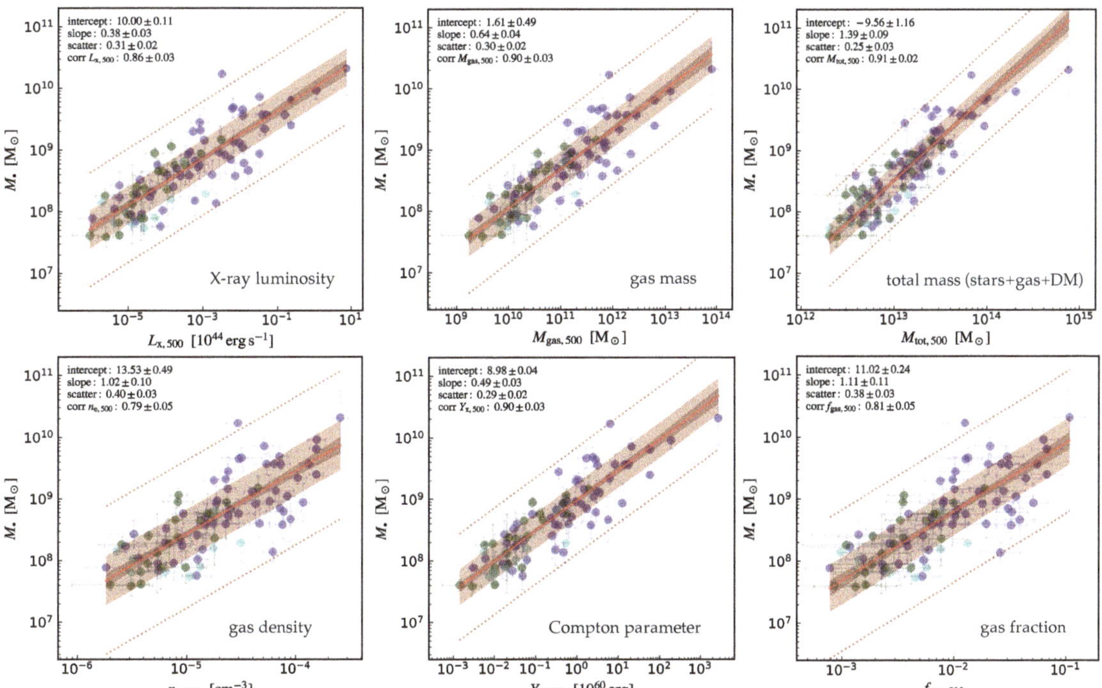

Figure 7. Scaling relations between the central (dynamical/direct) SMBH mass and key macro X-ray halo properties—adapted from Gaspari et al. [20]. **Top-left** to **bottom-right panels**: gas X-ray luminosity (in the 0.3–7.0 keV band), gas

The X-ray correlations shown in Figure 7 are important to probe models of galaxy group evolution. A key debated topic in current extragalactic astrophysics is which mode of accretion feeds internally the IGrM and eventually the central SMBH. In hot accretion modes (usually Bondi or ADAF; e.g., Bondi [206], Narayan and Fabian [207]), the larger the thermal entropy of the gas, the stronger the feeding is stifled, as the inflowing gas must overcome the hot-halo thermal pressure, increasing toward the center. This would induce negative correlations with the IGrM properties, which are ruled out by the slopes shown in Figure 7. Conversely, cold-mode accretion—typically in chaotic form due to the turbulent IGrM condensation generating randomly colliding clouds (e.g., Gaspari et al. [184], Prasad et al. [185], Voit [208], Olivares et al. [209])—would produce major positive and tight correlations with the gas mass and X-ray luminosity (e.g., the cooling rate is $\propto L_x$). Therefore, X-ray correlations favor chaotic cold accretion (CCA) over hot mode accretion. Hierarchical mergers (of both SMBHs and galaxies) are another channel to potentially grow such correlations. However, cosmological simulations (Bassini et al. [210], Truong et al. [211]) show this to be effective only at the high-mass end. Moreover, Figure 7 shows that all the mass scaling is either sub- or super-linear, far off from any simple self-similar expectation. In other words, a positive baseline due to hierarchical assembly is present, but gas feeding (dominated by CCA, in terms of mass) substantially shapes the slope and scatter of such $M_\bullet$ correlations over the long-term evolution. Overall, observed scaling relations of macro X-ray halo properties (shown in Sections 2 and 3) cannot be thought as disjointed from scaling relations of micro properties (e.g., $M_\bullet$), since both systems are tightly co-evolving and intertwined through the several billion years evolution and over 10 orders of magnitude in spatial scale (cf. the diagram in Gaspari et al. [212] linking the micro, meso, and macro scales). As striking as it appears, such scaling relations allow us to convert back and forth between vastly different scales, depending on the availability of either the micro (Section 4) or macro (Section 2) properties for each detected galaxy group.

The X-ray scaling relations presented in Section 2 can be also leveraged to test feedback models in large-scale simulations or to calibrate semi-analytic models of group evolution, thus giving us hints on the dominant baryonic processes in the IGrM (e.g., Puchwein et al. [23], McCarthy et al. [14], Kravtsov and Borgani [213], Tremmel et al. [214]). Figure 8 shows the key impact of archetypal feedback models on the evolution of the diffuse hot atmospheres (Gaspari et al. [27]). The filled black points indicate the Gyr evolution of the hot halos IGrM and ICM as it suffers recurrent injections of either anisotropic mechanical energy via jets (left column) or a strong impulsive thermal quasar-like blast (right column). Evidently, the latter model has a dramatic impact on the main L_x–T_x relation (even when the core is excised), producing a catastrophic evacuation of gas that lowers luminosities by 3 orders of magnitude, especially toward lower-mass group regime ($T_x < 1$ keV). Such quasar-like models are inconsistent with the observed X-ray scaling relations, in particular those probing the very low-mass regime via stacking analysis (e.g., Anderson et al. [103] shown via empty circles and solid line fit). Conversely, a tight self-regulation (e.g., achieved via CCA feeding) and a flickering injection via gentle AGN jets can preserve the hot-halo throughout the several 100 outburst cycles. The bottom panels show indeed that the initial cool-core (magenta contour) can be preserved even in less-bound halos, such as poor galaxy groups, without becoming overheated above half of the Hubble time. Such overheating is instead catastrophic for an impulsive AGN blast injection, transforming all hot halos into perennial non-cool-core systems, which is ruled out by observations finding groups to have almost universally a low central $t_{\rm cool}$ (Sun et al. [81], Babyk et al. [215]). Such self-regulated, gentle SMBH feedback has thus become a staple for subgrid models of cosmological simulations which can reproduce other tight scaling relations without any major break at the group scale, such as the M–Y_x or M–T_x computed over R_{500} (e.g., Planelles et al. [216], Truong et al. [28], Weinberger et al. [217]). For comparisons with further cosmological simulations, we refer the interested reader to the companion reviews by Oppenheimer et al. and Eckert et al. (Section 5).

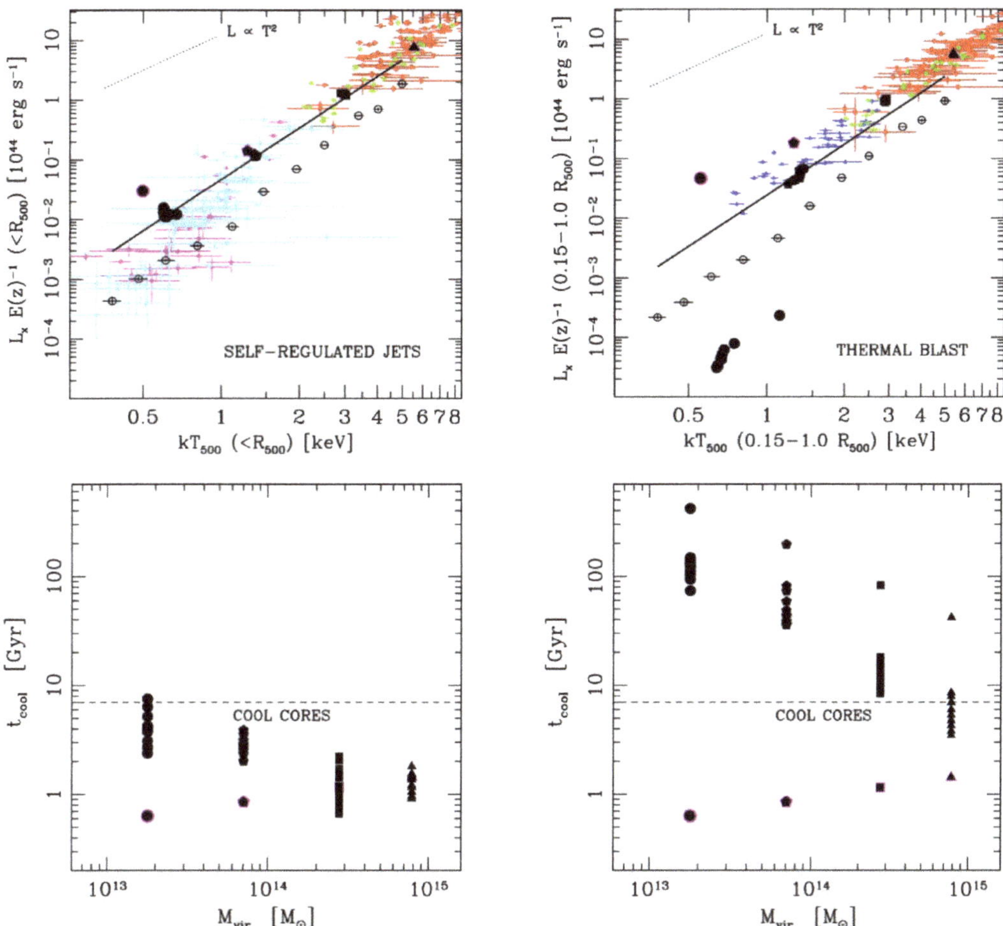

Figure 8. Effects of different baryonic models in shaping the evolution of the hot halos, in particular the X-ray luminosity–temperature relation (**top panels**) and cool-coreness via the central cooling time (**bottom panels**); adapted from Anderson et al. [103] and Gaspari et al. [27]. The colored individual objects in the top panels are from a wide range of observational works (Helsdon and Ponman [92], Osmond and Ponman [94], Mulchaey et al. [218], Pratt et al. [30], Sun et al. [81], Maughan et al. [219]; luminosities are extracted mostly in the 0.5–2 keV band). The empty circles and solid line show the raw Anderson et al. [103] stacking analysis and the unbiased fit, respectively. The filled black points show evolutionary tracks in large-scale HD simulations (Gaspari et al. [27]) implementing self-regulated AGN jets (**left**) or strong thermal blast feedback (**right**), preserving or evacuating the surrounding diffuse gaseous halo, respectively. The initial state is marked with magenta contour. Evacuation and overheating becomes particularly dramatic in low-mass, less-bound groups.

5. Galaxy Groups with the Next-Generation Instruments

Over the next decade, dedicated survey instruments will increase the number of known groups and clusters out to high redshift, constraining the scenario for their formation and evolution. Examples include *eROSITA* in X-rays, *Vera Rubin Observatory* and *Euclid* in the Optical/Infrared, and several "Stage 3" ground-based mm-wave observatories. The SZ-effect surveys, in particular, will break new ground by providing robustly selected, large catalogs of clusters at $z > 1.5$, as well as the first informative absolute mass calibration from CMB-cluster lensing. All future observatories list the baryonic mass and energy distribution on groups' scales resolved up to redshift ∼2 and beyond, when they first appeared as collapsed X-ray bright structures, as one of their main scientific goals.

Currently, a big step forward in the collection and characterization of low-mass systems is expected from the ongoing observations of the all X-ray sky with the *extended ROentgen Survey with an Imaging Telescope Array* (*eROSITA*[2], Predehl et al. [220]). eROSITA is operating in the X-ray energy band (0.2–10 keV) at L2 orbit on-board the 'Spectrum-Roentgen-Gamma' (SRG) satellite. *eROSITA* has a spatial resolution comparable to the *XMM–Newton* one, a similar effective area at low energies, but a wider field of view, while it will be 20–30 times more sensitive than the *ROSAT* sky survey in the soft band and will provide the first all-sky imaging survey in the hard band. Optimizing galaxy group and cluster detection has been one of the most important tasks during the mission preparation (e.g., Clerc et al. [221], and Käfer et al. [222] in particular for a detection and characterization through ICM outskirts that reduces possible biases due the peaked X-ray emission associated with cool cores). During its 4-yr-long all-sky survey, with an average exposure of 2.5 ks (whereas the average exposure in the ecliptic plane region is ∼1.6 ks), eROSITA is planned to deliver a sample of about 3 million active galactic nuclei (AGNs) and about 125,000 galaxy systems (mostly groups) detected with more than 50 photons and $M_{500c} > 10^{13} M_{\odot}/h$ up to redshift ∼1 (median: $z \sim 0.3$) [223–227]. Almost all groups (and clusters) detected with *eROSITA* will lack sufficient X-ray photons to accurately constrain temperature and mass profiles (Borm et al. [225]). Thus, cosmological studies using group and cluster of galaxies to be detected with *eROSITA*, will rely heavily on a detailed understanding of the scaling relations where systematic effects would have to be factored in to ensure that the cosmological applications of these relations are not hampered. Hence, a thorough investigation of these systems, to understand the interplay between the development of the hot IGrM and feedback processes, becomes highly important, not only for cosmology but also to understand complex baryonic physics. Moreover, to reach the planned goals of 1σ errors of 1%, 1%, 7%, and 25% on σ_8, Ω_m, w_0, and w_a, respectively, the critical passages will be: (i) a better knowledge (by a factor of ∼4) of the parameters describing the L_x–M relation to improve the constraints on σ_8 and Ω_m, and (ii) a lower mass threshold to enlarge the analyzed sample to reduce the statistical uncertainties in DE sector.

The physics of IGrM and ICM will be the main scientific driver for the exposures with the *Advanced Telescope for High-ENergy Astrophysics* (*Athena*[3]), the X-ray observatory mission selected by ESA as the second L(large)-class mission (due for launch in early 2030s) within its Cosmic Vision programme to address the Hot and Energetic Universe scientific theme. Among the main scientific goals, *Athena* will have the capabilities to find evolved groups of galaxies with $M_{500c} > 5 \times 10^{13} M_{\odot}$ and hot gaseous atmospheres at $z > 2$. For about ten of those, a global gas temperature estimate is expected to be measurable [228]. *Athena* will determine the magnitude of the injection of non-gravitational energy into the IGrM and ICM as a function of cosmic epoch by measuring the structural properties (e.g., the entropy profiles) out to R_{500}, and their evolution up to $z \sim 2$, for a sample of galaxy groups and clusters, improving significantly the constraints, presently unknown, on the evolution of the scaling relations between bulk properties of the hot gas [228,229]. In local systems, *Athena* will be also in condition to determine the occurrence and impact of AGN feedback phenomena by searching for ripples in surface brightness in the cores of a statistical sample of objects. Using temperature-sensitive line ratios, *Athena*'s observations will trace how much gas is at each temperature in the cores of these systems, providing a complete description of the gas heating-cooling balance [230] and transport processes such as turbulence and diffusion (Cucchetti et al. [231], Roncarelli et al. [232], Mernier et al. [233]).

2 http://www.mpe.mpg.de/erosita/, accessed on 3 May 2021.
3 https://www.the-athena-x-ray-observatory.eu/, accessed on 3 May 2021.

Presently, concepts funded for study by NASA for consideration in the 2020 Astrophysics Decadal Survey, *Lynx*[4] (as high-energy flagship mission) and *AXIS*[5] (as probe-class mission) are proposing to investigate with sub-arcsecond resolution over a FoV of 400–500 arcmin^2 the X-ray sky, improving this capability of a factor $\sim$100 with respect to *Chandra* ACIS-I. Their predicted low background level and capability to resolve embedded and background AGN will allow the tracking of group and cluster emissions at very low surface-brightness values. For example, *AXIS* is expected to reach a flux limit of $\sim 1 \times 10^{-16}$ erg/s/cm^2 (0.5–2 keV) over the 50 deg^2 of the proposed Wide Survey (e.g., [234]), providing the detection of thousands of groups and clusters, and evidence of merging and effects of feedback resolved even at high-z. With a larger collection of instruments, *Lynx* will be also able to resolve the thermodynamic and kinematic structure of systems at $z \approx 2$, as well as determine the role of feedback from AGN and stars.

Complementary data will be provided from the ongoing (and planned) SZ surveys. *SPT-3G*[6] will extend the work of *SPT-SZ* by covering a nearly identical area of 2500 deg^2 but with noise levels about 12, 7, and 20 times lower at 95, 150, and 220 GHz, respectively. This will enhance the sensitivity, allowing to a reduction of the mass limit and extending the redshift coverage with respect to *SPT-SZ*. About 5000 clusters with $M_{500c} \gtrsim 10^{14} M_\odot$ at a signal-to-noise > 4.5 (corresponding to a 97% purity threshold) are expected by the completion of the survey (2023; [235]). The next-generation ground-based cosmic microwave background experiment *CMB-S4*[7], with a planned beginning of science operations in 2029, will build catalogs more than an order of magnitude larger than current ones, lowering the mass limit M_{500c} to 6–8 $\times 10^{13}$ $M_\odot$ at $z > 0.3$ and being especially adept at finding the most distant groups and clusters. Large catalogs of low-mass systems together with the progress on the measurement of the thermal SZ power spectrum will open a new window into groups.

In the optical and near-infrared bands, space missions (*Euclid*[8]-from 2022- and *Nancy Grace Roman Space Telescope*[9]-formerly *WFIRST*; launch date: 2025) and ground-based missions (*Vera Rubin Observatory*[10] and the 4-metre Multi-Object Spectroscopic Telescope, *4MOST*[11]) will map the large-scale structures over more than 15,000 deg^2, extending the current catalogs of systems with $M_{500c} > 5 \times 10^{13} M_\odot$ (see, e.g., results from DES[12] in [48]) by orders of magnitude, in particular at high ($z > 1$) redshifts (e.g., [236]). Of particular interest for the measurement of the velocity dispersions, is the *WFIRST* and *4MOST* observatories. The *4MOST* observatory has been designed as a survey instrument at the forefront, with plans underway to combine the power of *4MOST* with *eROSITA* [237] to provide dynamical mass estimates for $\sim$10000 clusters at redshift $z < 0.6$ and masses $> 10^{14}$ $M_\odot$. $4MOST$ will also provide spectroscopic confirmation of $eROSITA$ detected groups at redshifts <0.2 down to a mass limit of 10^{13} $M_\odot$. Additionally, the wide area vista extragalactic survey (WAVES, [238]) being planned using $4MOST$, is aiming to perform the WAVES-Wide and WAVES-Deep surveys, allowing for the construction of optically selected groups catalogs. The WAVES-WIDE(-DEEP) surveys aiming to cover an area of $\sim$70 (1200) deg^2, identifying $\sim$50,000 (20,000) dark matter halos down to a mass of 10^{14} (10^{11}) $M_\odot$ and out to a redshift of $z_{phot} \lesssim 0.2$ (0.8). As with X-ray selected objects, optically selected groups are physically heterogenous systems (e.g., see the dynamical analysis by Zheng and Shen [239] for a sample of compact groups). However, it is possible that the physical processes at work in the IGrM of optically- and X-ray-selected systems are different. Thus,

4 https://www.lynxobservatory.com/, accessed on 3 May 2021.
5 http://axis.astro.umd.edu, accessed on 3 May 2021.
6 https://pole.uchicago.edu/, accessed on 3 May 2021.
7 https://cmb-s4.org/, accessed on 3 May 2021.
8 https://www.euclid-ec.org/, accessed on 3 May 2021.
9 https://roman.gsfc.nasa.gov/, accessed on 3 May 2021.
10 https://www.lsst.org, accessed on 3 May 2021.
11 https://www.4most.eu/cms/, accessed on 3 May 2021.
12 https://www.darkenergysurvey.org/, accessed on 3 May 2021.

the comparison of group samples selected via distinct methods can shed light on these physical phenomena.

6. Final Remarks

Bridging the gap in mass between field galaxies and massive clusters, galaxy groups are key systems to make progress in our understanding of structure formation and evolution. Thanks to the current generation of X-ray satellites, together with dedicated hydrodynamical simulations, there have been significant improvement in our comprehension of the interplay between the hot ambient gas, radiative cooling and feedback due to, e.g., AGN activity and SNe winds, in particular in the central regions. Indeed, the thermodynamic structure of galaxy groups is more complex than in massive galaxy clusters, with the physics associated with non-gravitational processes playing a significant role in shaping their general properties. The scaling relations capture the result of the various (thermal and non-thermal) processes and show that galaxy groups are not simply the scaled-down versions of rich clusters. Thanks to the enlarged catalogs of low-mass systems that the current (and upcoming) wide surveys at X-ray, millimeter, and optical wavelengths will provide, such scaling relations can be measured with very high precision. The comparison between results obtained from differently selected samples will shed light on the intrinsic properties of the groups' population.

Due to the complexity of the X-ray emitting processes in the low-temperature regime, and of how AGN heating impacts the general properties of the core of poor systems, the interpretation will depend on the specific choices of the individual analyses. For instance, many X-ray studies on groups (and clusters) provide integrated measurements within a certain aperture. However, the definition of such aperture is often quite different, and the comparison between the various works is not always straightforward. In the future, it is desirable to provide the global properties using a unified definition of the regions that are efficiently accessible from observations. This will ease the comparison between different observational and theoretical studies, improving our understanding of the physical processes at work in the complex group regime. Of course, in each study there are good reasons to use a specific energy band or definition of region of interest. However, regardless of the choices made in each paper to reach specific goals, we suggest to also provide, whenever possible, both global and core-excised properties within R_{500}. Although there is evidence that the cool-core radius does not scale uniformly with the virial radius, we think that the common choice of excising $r < 0.15R_{500}$ is a good starting point. For the rest-frame luminosities, we have shown that the 0.5–2 keV band is less sensitive to the choice of the abundance table and is easily accessible for all the current and future facilities (differently from the 0.1–2.4 keV band which extend to a regime where, for instance, *Chandra* and *XMM–Newton* are not well calibrated and also the choice of the abundance table start to play a role as discussed in Section 2.1). However, to ease the comparison with the literature it is useful to also provide the rest-frame bolometric and 0.1–2.4 keV band luminosities. Finally, until R_{500} will be routinely achieved for most of the systems in the low-mass regime, we also suggest providing the properties at R_{2500} (i.e., $\sim 0.5R_{500}$). Of course, there are further complications (e.g., the impact on the temperature of using a certain abundance table or spectral code, Lovisari et al. [77]; the choice of the column density, Lovisari and Reiprich [83]; the fitting technique, Balestra et al. [240]) which play a relevant role in the low-mass (but not only) systems. Nonetheless, starting to set standard definitions will definitely help the analysis in this critical regime.

Author Contributions: Conceptualization, L.L., S.E., M.G. and P.A.G.; resources, L.L., S.E., M.G. and P.A.G.; data curation, L.L., S.E., M.G. and P.A.G.; writing—original draft preparation, L.L., S.E., M.G. and P.A.G.; writing—review and editing, L.L., S.E., M.G. and P.A.G.; visualization, L.L., S.E., M.G. and P.A.G.; supervision, L.L.; project administration, L.L. All authors have read and agreed to the published version of the manuscript.

Funding: L.L. and S.E. acknowledge financial contribution from the contracts ASI-INAF Athena 2015-046-R.0, ASI-INAF Athena 2019-27-HH.0, "Attività di Studio per la comunità scientifica di Astrofisica delle Alte Energie e Fisica Astroparticellare" (Accordo Attuativo ASI-INAF n. 2017-14-H.0), and from INAF "Call per interventi aggiuntivi a sostegno della ricerca di main stream di INAF". M.G. acknowledges partial support by NASA Chandra GO8-19104X/GO9-20114X and HST GO-15890.020-A grants. P.A.G. acknowledges support from the UK Science and Technology Facilities Council via grants ST/P000525/1 and ST/T000473/1.

Institutional Review Board Statement: Not applicable.

Informed Consent Statement: Not applicable.

Data Availability Statement: Not applicable.

Acknowledgments: The authors thank the anonymous referees for useful comments and suggestions that helped to improve and clarify the presentation of this work. We thank M. Sun for providing the luminosity values for the group sample analysed in his work. We also thank MNRAS and the AAS, together with the authors of the corresponding publications, for granting permission to use images published in their journals.

Conflicts of Interest: The authors declare no conflict of interest.

References

1. Eke, V.R.; Baugh, C.M.; Cole, S.; Frenk, C.S.; Norberg, P.; Peacock, J.A.; Baldry, I.K.; Bland-Hawthorn, J.; Bridges, T.; Cannon, R.; et al. Galaxy groups in the 2dFGRS: The group-finding algorithm and the 2PIGG catalogue. *MNRAS* **2004**, *348*, 866–878. [CrossRef]
2. Giodini, S.; Pierini, D.; Finoguenov, A.; Pratt, G.W.; Boehringer, H.; Leauthaud, A.; Guzzo, L.; Aussel, H.; Bolzonella, M.; Capak, P.; et al. Stellar and Total Baryon Mass Fractions in Groups and Clusters Since Redshift 1. *ApJ* **2009**, *703*, 982–993. [CrossRef]
3. Eigenthaler, P.; Zeilinger, W.W. The Search for Fossil Groups of Galaxies. *Astron. Nachr.* **2007**, *328*, 699.
4. Nolthenius, R.; White, S.D.M. Groups of galaxies in the CfA survey and in cold dark matter universes. *MNRAS* **1987**, *225*, 505–530. [CrossRef]
5. Hickson, P. Systematic properties of compact groups of galaxies. *ApJ* **1982**, *255*, 382–391. [CrossRef]
6. Jones, L.R.; Ponman, T.J.; Horton, A.; Babul, A.; Ebeling, H.; Burke, D.J. The nature and space density of fossil groups of galaxies. *MNRAS* **2003**, *343*, 627–638. [CrossRef]
7. Helsdon, S.F.; Ponman, T.J. Are X-ray properties of loose groups different from those of compact groups? *MNRAS* **2000**, *319*, 933–938. [CrossRef]
8. Ponman, T.J.; Cannon, D.B.; Navarro, J.F. The thermal imprint of galaxy formation on X-ray clusters. *Nature* **1999**, *397*, 135–137. [CrossRef]
9. Sanderson, A.J.R.; Ponman, T.J.; Finoguenov, A.; Lloyd-Davies, E.J.; Markevitch, M. The Birmingham-CfA cluster scaling project—I. Gas fraction and the M-T_X relation. *MNRAS* **2003**, *340*, 989–1010. [CrossRef]
10. Despali, G.; Giocoli, C.; Angulo, R.E.; Tormen, G.; Sheth, R.K.; Baso, G.; Moscardini, L. The universality of the virial halo mass function and models for non-universality of other halo definitions. *MNRAS* **2016**, *456*, 2486–2504. [CrossRef]
11. Tully, R.B. Nearby Groups of Galaxies. II. an All-Sky Survey within 3000 Kilometers per Second. *ApJ* **1987**, *321*, 280. [CrossRef]
12. Fukugita, M.; Hogan, C.J.; Peebles, P.J.E. The Cosmic Baryon Budget. *ApJ* **1998**, *503*, 518–530. [CrossRef]
13. Brighenti, F.; Mathews, W.G. Heated Cooling Flows. *ApJ* **2002**, *573*, 542–561. [CrossRef]
14. McCarthy, I.G.; Schaye, J.; Ponman, T.J.; Bower, R.G.; Booth, C.M.; Dalla Vecchia, C.; Crain, R.A.; Springel, V.; Theuns, T.; Wiersma, R.P.C. The case for AGN feedback in galaxy groups. *MNRAS* **2010**, *406*, 822–839. [CrossRef]
15. Gaspari, M.; Brighenti, F.; Temi, P. Mechanical AGN feedback: Controlling the thermodynamical evolution of elliptical galaxies. *MNRAS* **2012**, *424*, 190–209. [CrossRef]
16. Hudson, D.S.; Mittal, R.; Reiprich, T.H.; Nulsen, P.E.J.; Andernach, H.; Sarazin, C.L. What is a cool-core cluster? a detailed analysis of the cores of the X-ray flux-limited HIFLUGCS cluster sample. *Astron. Astrophys.* **2010**, *513*, A37, [CrossRef]
17. Bharadwaj, V.; Reiprich, T.H.; Schellenberger, G.; Eckmiller, H.J.; Mittal, R.; Israel, H. Intracluster medium cooling, AGN feedback, and brightest cluster galaxy properties of galaxy groups. Five properties where groups differ from clusters. *Astron. Astrophys.* **2014**, *572*, A46, [CrossRef]
18. O'Sullivan, E.; Ponman, T.J.; Kolokythas, K.; Raychaudhury, S.; Babul, A.; Vrtilek, J.M.; David, L.P.; Giacintucci, S.; Gitti, M.; Haines, C.P. The Complete Local Volume Groups Sample—I. Sample selection and X-ray properties of the high-richness subsample. *MNRAS* **2017**, *472*, 1482–1505. [CrossRef]
19. Bogdán, Á.; Lovisari, L.; Volonteri, M.; Dubois, Y. Correlation between the Total Gravitating Mass of Groups and Clusters and the Supermassive Black Hole Mass of Brightest Galaxies. *ApJ* **2018**, *852*, 131, [CrossRef]
20. Gaspari, M.; Eckert, D.; Ettori, S.; Tozzi, P.; Bassini, L.; Rasia, E.; Brighenti, F.; Sun, M.; Borgani, S.; Johnson, S.D.; et al. The X-Ray Halo Scaling Relations of Supermassive Black Holes. *ApJ* **2019**, *884*, 169, [CrossRef]

21. Lakhchaura, K.; Truong, N.; Werner, N. Correlations between supermassive black holes, hot atmospheres, and the total masses of early-type galaxies. *MNRAS* **2019**, *488*, L134–L142, [CrossRef]
22. Sijacki, D.; Springel, V.; Di Matteo, T.; Hernquist, L. A unified model for AGN feedback in cosmological simulations of structure formation. *MNRAS* **2007**, *380*, 877–900. [CrossRef]
23. Puchwein, E.; Sijacki, D.; Springel, V. Simulations of AGN Feedback in Galaxy Clusters and Groups: Impact on Gas Fractions and the L_X-T Scaling Relation. *ApJ* **2008**, *687*, L53–L56, [CrossRef]
24. Fabjan, D.; Borgani, S.; Tornatore, L.; Saro, A.; Murante, G.; Dolag, K. Simulating the effect of active galactic nuclei feedback on the metal enrichment of galaxy clusters. *MNRAS* **2010**, *401*, 1670–1690. [CrossRef]
25. Le Brun, A.M.C.; McCarthy, I.G.; Schaye, J.; Ponman, T.J. Towards a realistic population of simulated galaxy groups and clusters. *MNRAS* **2014**, *441*, 1270–1290. [CrossRef]
26. Planelles, S.; Borgani, S.; Fabjan, D.; Killedar, M.; Murante, G.; Granato, G.L.; Ragone-Figueroa, C.; Dolag, K. On the role of AGN feedback on the thermal and chemodynamical properties of the hot intracluster medium. *MNRAS* **2014**, *438*, 195–216. [CrossRef]
27. Gaspari, M.; Brighenti, F.; Temi, P.; Ettori, S. Can AGN Feedback Break the Self-similarity of Galaxies, Groups, and Clusters? *ApJ* **2014**, *783*, L10, [CrossRef]
28. Truong, N.; Rasia, E.; Mazzotta, P.; Planelles, S.; Biffi, V.; Fabjan, D.; Beck, A.M.; Borgani, S.; Dolag, K.; Gaspari, M.; et al. Cosmological hydrodynamical simulations of galaxy clusters: X-ray scaling relations and their evolution. *MNRAS* **2018**, *474*, 4089–4111. [CrossRef]
29. Markevitch, M. The L_X-T Relation and Temperature Function for Nearby Clusters Revisited. *ApJ* **1998**, *504*, 27–34. [CrossRef]
30. Pratt, G.W.; Croston, J.H.; Arnaud, M.; Böhringer, H. Galaxy cluster X-ray luminosity scaling relations from a representative local sample (REXCESS). *Astron. Astrophys.* **2009**, *498*, 361–378. [CrossRef]
31. Mittal, R.; Hicks, A.; Reiprich, T.H.; Jaritz, V. The L_X - T_{vir} relation in galaxy clusters: Effects of radiative cooling and AGN heating. *Astron. Astrophys.* **2011**, *532*, A133, [CrossRef]
32. Bharadwaj, V.; Reiprich, T.H.; Lovisari, L.; Eckmiller, H.J. Extending the L_X - T relation from clusters to groups. Impact of cool core nature, AGN feedback, and selection effects. *Astron. Astrophys.* **2015**, *573*, A75, [CrossRef]
33. Mantz, A.B.; Allen, S.W.; Morris, R.G.; von der Linden, A.; Applegate, D.E.; Kelly, P.L.; Burke, D.L.; Donovan, D.; Ebeling, H. Weighing the giants- V. Galaxy cluster scaling relations. *MNRAS* **2016**, *463*, 3582–3603. [CrossRef]
34. Lovisari, L.; Schellenberger, G.; Sereno, M.; Ettori, S.; Pratt, G.W.; Forman, W.R.; Jones, C.; Andrade-Santos, F.; Randall, S.; Kraft, R. X-Ray Scaling Relations for a Representative Sample of Planck-selected Clusters Observed with XMM-Newton. *ApJ* **2020**, *892*, 102, [CrossRef]
35. Eckert, D.; Molendi, S.; Paltani, S. The cool-core bias in X-ray galaxy cluster samples. I. Method and application to HIFLUGCS. *Astron. Astrophys.* **2011**, *526*, A79. [CrossRef]
36. Reiprich, T.H.; Böhringer, H. The Mass Function of an X-Ray Flux-limited Sample of Galaxy Clusters. *ApJ* **2002**, *567*, 716–740. [CrossRef]
37. Allen, S.W.; Rapetti, D.A.; Schmidt, R.W.; Ebeling, H.; Morris, R.G.; Fabian, A.C. Improved constraints on dark energy from Chandra X-ray observations of the largest relaxed galaxy clusters. *MRAS* **2008**, *383*, 879–896. [CrossRef]
38. Vikhlinin, A.; Kravtsov, A.V.; Burenin, R.A.; Ebeling, H.; Forman, W.R.; Hornstrup, A.; Jones, C.; Murray, S.S.; Nagai, D.; Quintana, H.; et al. Chandra Cluster Cosmology Project III: Cosmological Parameter Constraints. *ApJ* **2009**, *692*, 1060–1074. [CrossRef]
39. Rozo, E.; Wechsler, R.H.; Rykoff, E.S.; Annis, J.T.; Becker, M.R.; Evrard, A.E.; Frieman, J.A.; Hansen, S.M.; Hao, J.; Johnston, D.E.; et al. Cosmological Constraints from the Sloan Digital Sky Survey maxBCG Cluster Catalog. *ApJ* **2010**, *708*, 645–660. [CrossRef]
40. Mantz, A.B.; von der Linden, A.; Allen, S.W.; Applegate, D.E.; Kelly, P.L.; Morris, R.G.; Rapetti, D.A.; Schmidt, R.W.; Adhikari, S.; Allen, M.T.; et al. Weighing the giants—IV. Cosmology and neutrino mass. *MNRAS* **2015**, *446*, 2205–2225. [CrossRef]
41. de Haan, T.; Benson, B.A.; Bleem, L.E.; Allen, S.W.; Applegate, D.E.; Ashby, M.L.N.; Bautz, M.; Bayliss, M.; Bocquet, S.; Brodwin, M.; et al. Cosmological Constraints from Galaxy Clusters in the 2500 Square-degree SPT-SZ Survey. *ApJ* **2016**, *832*, 95. [CrossRef]
42. Planck Collaboration Planck 2015 results. XXIV. Cosmology from Sunyaev-Zeldovich cluster counts. *Astron. Astrophys.* **2016**, *594*, A24. [CrossRef]
43. Böhringer, H.; Chon, G.; Fukugita, M. The extended ROSAT-ESO Flux-Limited X-ray Galaxy Cluster Survey (REFLEX II). VII. The mass function of galaxy clusters. *Astron. Astrophys.* **2017**, *608*, A65. [CrossRef]
44. Schellenberger, G.; Reiprich, T.H. HICOSMO: Cosmology with a complete sample of galaxy clusters—II. Cosmological results. *MNRAS* **2017**, *471*, 1370–1389. [CrossRef]
45. Pacaud, F.; Pierre, M.; Melin, J.B.; Adami, C.; Evrard, A.E.; Galli, S.; Gastaldello, F.; Maughan, B.J.; Sereno, M.; Alis, S.; et al. The XXL Survey. XXV. Cosmological analysis of the C1 cluster number counts. *Astron. Astrophys.* **2018**, *620*, A10. [CrossRef]
46. Bocquet, S.; Dietrich, J.P.; Schrabback, T.; Bleem, L.E.; Klein, M.; Allen, S.W.; Applegate, D.E.; Ashby, M.L.N.; Bautz, M.; Bayliss, M.; et al. Cluster Cosmology Constraints from the 2500 deg^2 SPT-SZ Survey: Inclusion of Weak Gravitational Lensing Data from Magellan and the Hubble Space Telescope. *ApJ* **2019**, *878*, 55. [CrossRef]
47. Abbott, T.M.C.; Abdalla, F.B.; Alarcon, A.; Aleksić, J.; Allam, S.; Allen, S.; Amara, A.; Annis, J.; Asorey, J.; Avila, S.; et al. Dark Energy Survey year 1 results: Cosmological constraints from galaxy clustering and weak lensing. *Phys. Rev. D* **2018**, *98*, 043526, [CrossRef]

48. Abbott, T.M.C.; Aguena, M.; Alarcon, A.; Allam, S.; Allen, S.; Annis, J.; Avila, S.; Bacon, D.; Bechtol, K.; Bermeo, A.; et al. Dark Energy Survey Year 1 Results: Cosmological constraints from cluster abundances and weak lensing. *Phys. Rev. D* **2020**, *102*, 023509. [CrossRef]
49. McClintock, T.; Varga, T.N.; Gruen, D.; Rozo, E.; Rykoff, E.S.; Shin, T.; Melchior, P.; DeRose, J.; Seitz, S.; Dietrich, J.P.; et al. Dark Energy Survey Year 1 results: Weak lensing mass calibration of redMaPPer galaxy clusters. *MNRAS* **2019**, *482*, 1352–1378. [CrossRef]
50. Farahi, A.; Chen, X.; Evrard, A.E.; Hollowood, D.L.; Wilkinson, R.; Bhargava, S.; Giles, P.; Romer, A.K.; Jeltema, T.; Hilton, M.; et al. Mass variance from archival X-ray properties of Dark Energy Survey Year-1 galaxy clusters. *MNRAS* **2019**, *490*, 3341–3354. [CrossRef]
51. Mulchaey, J.S. X-ray Properties of Groups of Galaxies. *ARA&A* **2000**, *38*, 289–335. [CrossRef]
52. Sun, M. Hot gas in galaxy groups: Recent observations. *New J. Phys.* **2012**, *14*, 045004, [CrossRef]
53. Kaiser, N. Evolution and clustering of rich clusters. *MNRAS* **1986**, *222*, 323–345. [CrossRef]
54. Kitayama, T.; Suto, Y. Semianalytic Predictions for Statistical Properties of X-Ray Clusters of Galaxies in Cold Dark Matter Universes. *ApJ* **1996**, *469*, 480, [CrossRef]
55. Bryan, G.L.; Norman, M.L. Statistical Properties of X-Ray Clusters: Analytic and Numerical Comparisons. *ApJ* **1998**, *495*, 80–99. [CrossRef]
56. Voit, G.M. Tracing cosmic evolution with clusters of galaxies. *Rev. Mod. Phys.* **2005**, *77*, 207–258. [CrossRef]
57. Maughan, B.J.; Jones, L.R.; Ebeling, H.; Scharf, C. The evolution of the cluster X-ray scaling relations in the Wide Angle ROSAT Pointed Survey sample at $0.6 < z < 1.0$. *MNRAS* **2006**, *365*, 509–529. [CrossRef]
58. Borgani, S.; Diaferio, A.; Dolag, K.; Schindler, S. Thermodynamical Properties of the ICM from Hydrodynamical Simulations. *Space Sci. Rev.* **2008**, *134*, 269–293. [CrossRef]
59. Böhringer, H.; Dolag, K.; Chon, G. Modelling self-similar appearance of galaxy clusters in X-rays. *Astron. Astrophys.* **2012**, *539*, A120, [CrossRef]
60. Ettori, S. The generalized scaling relations for X-ray galaxy clusters: The most powerful mass proxy. *MNRAS* **2013**, *435*, 1265–1277. [CrossRef]
61. Giodini, S.; Lovisari, L.; Pointecouteau, E.; Ettori, S.; Reiprich, T.H.; Hoekstra, H. Scaling Relations for Galaxy Clusters: Properties and Evolution. *Space Sci. Rev.* **2013**, *177*, 247–282. [CrossRef]
62. Maughan, B.J. PICACS: Self-consistent modelling of galaxy cluster scaling relations. *MNRAS* **2014**, *437*, 1171–1186. [CrossRef]
63. Ettori, S. The physics inside the scaling relations for X-ray galaxy clusters: Gas clumpiness, gas mass fraction and slope of the pressure profile. *MNRAS* **2015**, *446*, 2629–2639. [CrossRef]
64. Ettori, S.; Lovisari, L.; Sereno, M. From universal profiles to universal scaling laws in X-ray galaxy clusters. *Astron. Astrophys.* **2020**, *644*, A111, [CrossRef]
65. Sarazin, C.L. X-ray emission from clusters of galaxies. *Rev. Mod. Phys.* **1986**, *58*, 1–115. [CrossRef]
66. Peterson, J.R.; Fabian, A.C. X-ray spectroscopy of cooling clusters. *Phys. Rep.* **2006**, *427*, 1–39. [CrossRef]
67. Kaastra, J.S.; Paerels, F.B.S.; Durret, F.; Schindler, S.; Richter, P. Thermal Radiation Processes. *Space Sci. Rev.* **2008**, *134*, 155–190. [CrossRef]
68. Böhringer, H.; Werner, N. X-ray spectroscopy of galaxy clusters: Studying astrophysical processes in the largest celestial laboratories. *Astron. Astrophys. Rev.* **2010**, *18*, 127–196. [CrossRef]
69. Smith, R.K.; Brickhouse, N.S.; Liedahl, D.A.; Raymond, J.C. Collisional Plasma Models with APEC/APED: Emission-Line Diagnostics of Hydrogen-like and Helium-like Ions. *ApJ* **2001**, *556*, L91–L95, [CrossRef]
70. Arnaud, K.A. XSPEC: The First Ten Years. In *Astronomical Data Analysis Software and Systems V*; Astronomical Society of the Pacific Conference Series; Jacoby, G.H., Barnes, J., Eds.; NASA/GSFC: Greenbelt, MD, USA, 1996; Volume 101, p. 17.
71. Asplund, M.; Grevesse, N.; Sauval, A.J.; Scott, P. The Chemical Composition of the Sun. *Annu. Rev. Astron. Astrophys.* **2009**, *47*, 481–522. [CrossRef]
72. Vikhlinin, A.; Kravtsov, A.; Forman, W.; Jones, C.; Markevitch, M.; Murray, S.S.; Van Speybroeck, L. Chandra Sample of Nearby Relaxed Galaxy Clusters: Mass, Gas Fraction, and Mass-Temperature Relation. *ApJ* **2006**, *640*, 691–709. [CrossRef]
73. Gonzalez, A.H.; Zaritsky, D.; Zabludoff, A.I. A Census of Baryons in Galaxy Clusters and Groups. *ApJ* **2007**, *666*, 147–155. [CrossRef]
74. Gastaldello, F.; Buote, D.A.; Humphrey, P.J.; Zappacosta, L.; Bullock, J.S.; Brighenti, F.; Mathews, W.G. Probing the Dark Matter and Gas Fraction in Relaxed Galaxy Groups with X-Ray Observations from Chandra and XMM-Newton. *ApJ* **2007**, *669*, 158–183. [CrossRef]
75. Dai, X.; Bregman, J.N.; Kochanek, C.S.; Rasia, E. On the Baryon Fractions in Clusters and Groups of Galaxies. *ApJ* **2010**, *719*, 119–125. [CrossRef]
76. Gonzalez, A.H.; Sivanandam, S.; Zabludoff, A.I.; Zaritsky, D. Galaxy Cluster Baryon Fractions Revisited. *ApJ* **2013**, *778*, 14, [CrossRef]
77. Lovisari, L.; Reiprich, T.H.; Schellenberger, G. Scaling properties of a complete X-ray selected galaxy group sample. *Astron. Astrophys.* **2015**, *573*, A118, [CrossRef]
78. Eckert, D.; Ettori, S.; Coupon, J.; Gastaldello, F.; Pierre, M.; Melin, J.B.; Le Brun, A.M.C.; McCarthy, I.G.; Adami, C.; Chiappetti, L.; et al. The XXL Survey. XIII. Baryon content of the bright cluster sample. *Astron. Astrophys.* **2016**, *592*, A12, [CrossRef]

79. Zou, S.; Maughan, B.J.; Giles, P.A.; Vikhlinin, A.; Pacaud, F.; Burenin, R.; Hornstrup, A. The X-ray luminosity-temperature relation of a complete sample of low-mass galaxy clusters. *MNRAS* **2016**, *463*, 820–831. [CrossRef]
80. Rasmussen, J.; Ponman, T.J. Temperature and abundance profiles of hot gas in galaxy groups—I. Results and statistical analysis. *MNRAS* **2007**, *380*, 1554–1572. [CrossRef]
81. Sun, M.; Voit, G.M.; Donahue, M.; Jones, C.; Forman, W.; Vikhlinin, A. Chandra Studies of the X-Ray Gas Properties of Galaxy Groups. *ApJ* **2009**, *693*, 1142–1172. [CrossRef]
82. Mernier, F.; de Plaa, J.; Kaastra, J.S.; Zhang, Y.Y.; Akamatsu, H.; Gu, L.; Kosec, P.; Mao, J.; Pinto, C.; Reiprich, T.H.; et al. Radial metal abundance profiles in the intra-cluster medium of cool-core galaxy clusters, groups, and ellipticals. *Astron. Astrophys.* **2017**, *603*, A80, [CrossRef]
83. Lovisari, L.; Reiprich, T.H. The non-uniformity of galaxy cluster metallicity profiles. *MNRAS* **2019**, *483*, 540–557. [CrossRef]
84. Grevesse, N.; Sauval, A.J. Standard Solar Composition. *Space Sci. Rev.* **1998**, *85*, 161–174. [CrossRef]
85. Lodders, K.; Palme, H.; Gail, H.P. Abundances of the Elements in the Solar System. *arXiv* **2009**, arXiv:0901.1149.
86. Anders, E.; Grevesse, N. Abundances of the elements: Meteoritic and solar. *Geochim. Cosmochim. Acta* **1989**, *53*, 197–214. [CrossRef]
87. Lloyd-Davies, E.J.; Ponman, T.J.; Cannon, D.B. The entropy and energy of intergalactic gas in galaxy clusters. *MNRAS* **2000**, *315*, 689–702. [CrossRef]
88. Ponman, T.J.; Sanderson, A.J.R.; Finoguenov, A. The Birmingham-CfA cluster scaling project—III. Entropy and similarity in galaxy systems. *MNRAS* **2003**, *343*, 331–342. [CrossRef]
89. Finoguenov, A.; Ponman, T.J.; Osmond, J.P.F.; Zimer, M. XMM-Newton study of $0.012 < z < 0.024$ groups—I. Overview of the IGM thermodynamics. *MNRAS* **2007**, *374*, 737–760. [CrossRef]
90. Johnson, R.; Ponman, T.J.; Finoguenov, A. A statistical analysis of the Two-Dimensional XMM-Newton Group Survey: The impact of feedback on group properties. *MNRAS* **2009**, *395*, 1287–1308. [CrossRef]
91. Panagoulia, E.K.; Fabian, A.C.; Sanders, J.S. A volume-limited sample of X-ray galaxy groups and clusters—I. Radial entropy and cooling time profiles. *MNRAS* **2014**, *438*, 2341–2354. [CrossRef]
92. Helsdon, S.F.; Ponman, T.J. The intragroup medium in loose groups of galaxies. *MNRAS* **2000**, *315*, 356–370. [CrossRef]
93. Xue, Y.J.; Wu, X.P. The L_X-T, L_X-σ, and σ-T Relations for Groups and Clusters of Galaxies. *ApJ* **2000**, *538*, 65–71. [CrossRef]
94. Osmond, J.P.F.; Ponman, T.J. The GEMS project: X-ray analysis and statistical properties of the group sample. *MNRAS* **2004**, *350*, 1511–1535. [CrossRef]
95. Shang, C.; Scharf, C. A Low-Redshift Galaxy Cluster X-Ray Temperature Function Incorporating Suzaku Data. *ApJ* **2009**, *690*, 879–890. [CrossRef]
96. Eckmiller, H.J.; Hudson, D.S.; Reiprich, T.H. Testing the low-mass end of X-ray scaling relations with a sample of Chandra galaxy groups. *Astron. Astrophys.* **2011**, *535*, A105, [CrossRef]
97. Ikebe, Y.; Reiprich, T.H.; Böhringer, H.; Tanaka, Y.; Kitayama, T. A new measurement of the X-ray temperature function of clusters of galaxies. *Astron. Astrophys.* **2002**, *383*, 773–790. [CrossRef]
98. Stanek, R.; Evrard, A.E.; Böhringer, H.; Schuecker, P.; Nord, B. The X-Ray Luminosity-Mass Relation for Local Clusters of Galaxies. *ApJ* **2006**, *648*, 956–968. [CrossRef]
99. Pacaud, F.; Pierre, M.; Adami, C.; Altieri, B.; Andreon, S.; Chiappetti, L.; Detal, A.; Duc, P.A.; Galaz, G.; Gueguen, A.; et al. The XMM-LSS survey: The Class 1 cluster sample over the initial 5 deg^2 and its cosmological modelling. *MNRAS* **2007**, *382*, 1289–1308. [CrossRef]
100. Schellenberger, G.; Reiprich, T.H. HICOSMO—Cosmology with a complete sample of galaxy clusters—I. Data analysis, sample selection and luminosity-mass scaling relation. *MNRAS* **2017**, *469*, 3738–3761. [CrossRef]
101. Kettula, K.; Giodini, S.; van Uitert, E.; Hoekstra, H.; Finoguenov, A.; Lerchster, M.; Erben, T.; Heymans, C.; Hildebrandt, H.; Kitching, T.D.; et al. CFHTLenS: Weak lensing calibrated scaling relations for low-mass clusters of galaxies. *MNRAS* **2015**, *451*, 1460–1481. [CrossRef]
102. Rasmussen, J.; Ponman, T.J.; Mulchaey, J.S.; Miles, T.A.; Raychaudhury, S. First results of the XI Groups Project: Studying an unbiased sample of galaxy groups. *MNRAS* **2006**, *373*, 653–665. [CrossRef]
103. Anderson, M.E.; Gaspari, M.; White, S.D.M.; Wang, W.; Dai, X. Unifying X-ray scaling relations from galaxies to clusters. *MNRAS* **2015**, *449*, 3806–3826. [CrossRef]
104. Andreon, S.; Serra, A.L.; Moretti, A.; Trinchieri, G. The amazing diversity in the hot gas content of an X-ray unbiased massive galaxy clusters sample. *Astron. Astrophys.* **2016**, *585*, A147, [CrossRef]
105. Valtchanov, I.; Pierre, M.; Gastaud, R. Comparison of source detection procedures for XMM-Newton images. *Astron. Astrophys.* **2001**, *370*, 689–706. [CrossRef]
106. Šuhada, R.; Song, J.; Böhringer, H.; Mohr, J.J.; Chon, G.; Finoguenov, A.; Fassbender, R.; Desai, S.; Armstrong, R.; Zenteno, A.; et al. The XMM-BCS galaxy cluster survey. I. The X-ray selected cluster catalog from the initial 6 deg^2. *Astron. Astrophys.* **2012**, *537*, A39, [CrossRef]
107. Xu, W.; Ramos-Ceja, M.E.; Pacaud, F.; Reiprich, T.H.; Erben, T. A new X-ray-selected sample of very extended galaxy groups from the ROSAT All-Sky Survey. *Astron. Astrophys.* **2018**, *619*, A162, [CrossRef]
108. Miniati, F.; Finoguenov, A.; Silverman, J.D.; Carollo, M.; Cibinel, A.; Lilly, S.J.; Schawinski, K. The X-Ray Zurich Environmental Study (X-ZENS). II. X-Ray Observations of the Diffuse Intragroup Medium in Galaxy Groups. *ApJ* **2016**, *819*, 26, [CrossRef]

109. Pearson, R.J.; Ponman, T.J.; Norberg, P.; Robotham, A.S.G.; Babul, A.; Bower, R.G.; McCarthy, I.G.; Brough, S.; Driver, S.P.; Pimbblet, K. Galaxy And Mass Assembly: Search for a population of high-entropy galaxy groups. *MNRAS* **2017**, *469*, 3489–3504. [CrossRef]
110. Migkas, K.; Schellenberger, G.; Reiprich, T.H.; Pacaud, F.; Ramos-Ceja, M.E.; Lovisari, L. Probing cosmic isotropy with a new X-ray galaxy cluster sample through the L_X-T scaling relation. *Astron. Astrophys.* **2020**, *636*, A15, [CrossRef]
111. Schellenberger, G.; Reiprich, T.H.; Lovisari, L.; Nevalainen, J.; David, L. XMM-Newton and Chandra cross-calibration using HIFLUGCS galaxy clusters . Systematic temperature differences and cosmological impact. *Astron. Astrophys.* **2015**, *575*, A30, [CrossRef]
112. Sereno, M. A Bayesian approach to linear regression in astronomy. *MNRAS* **2016**, *455*, 2149–2162. [CrossRef]
113. Kettula, K.; Nevalainen, J.; Miller, E.D. Cross-calibration of Suzaku/XIS and XMM-Newton/EPIC using galaxy clusters. *Astron. Astrophys.* **2013**, *552*, A47, [CrossRef]
114. Umetsu, K.; Sereno, M.; Lieu, M.; Miyatake, H.; Medezinski, E.; Nishizawa, A.J.; Giles, P.; Gastaldello, F.; McCarthy, I.G.; Kilbinger, M.; et al. Weak-lensing Analysis of X-Ray-selected XXL Galaxy Groups and Clusters with Subaru HSC Data. *ApJ* **2020**, *890*, 148, [CrossRef]
115. Sereno, M.; Umetsu, K.; Ettori, S.; Eckert, D.; Gastaldello, F.; Giles, P.; Lieu, M.; Maughan, B.; Okabe, N.; Birkinshaw, M.; et al. XXL Survey groups and clusters in the Hyper Suprime-Cam Survey. Scaling relations between X-ray properties and weak lensing mass. *MNRAS* **2020**, *492*, 4528–4545. [CrossRef]
116. Nevalainen, J.; David, L.; Guainazzi, M. Cross-calibrating X-ray detectors with clusters of galaxies: An IACHEC study. *Astron. Astrophys.* **2010**, *523*, A22, [CrossRef]
117. Colafrancesco, S.; Giordano, F. Structure and evolution of magnetized clusters: Entropy profiles, S - T and L_X - T relations. *Astron. Astrophys.* **2007**, *466*, 421–435. [CrossRef]
118. Mahdavi, A.; Finoguenov, A.; Böhringer, H.; Geller, M.J.; Henry, J.P. XMM-Newton and Gemini Observations of Eight RASSCALS Galaxy Groups. *ApJ* **2005**, *622*, 187–204. [CrossRef]
119. Pratt, G.W.; Arnaud, M.; Piffaretti, R.; Böhringer, H.; Ponman, T.J.; Croston, J.H.; Voit, G.M.; Borgani, S.; Bower, R.G. Gas entropy in a representative sample of nearby X-ray galaxy clusters (REXCESS): Relationship to gas mass fraction. *Astron. Astrophys.* **2010**, *511*, A85, [CrossRef]
120. Mantz, A.B.; Allen, S.W.; Morris, R.G.; Schmidt, R.W. Cosmology and astrophysics from relaxed galaxy clusters—III. Thermodynamic profiles and scaling relations. *MNRAS* **2016**, *456*, 4020–4039. [CrossRef]
121. Mantz, A.B.; Allen, S.W.; Morris, R.G.; von der Linden, A. Centre-excised X-ray luminosity as an efficient mass proxy for future galaxy cluster surveys. *MNRAS* **2018**, *473*, 3072–3079. [CrossRef]
122. Bulbul, E.; Chiu, I.N.; Mohr, J.J.; McDonald, M.; Benson, B.; Bautz, M.W.; Bayliss, M.; Bleem, L.; Brodwin, M.; Bocquet, S.; et al. X-Ray Properties of SPT-selected Galaxy Clusters at $0.2 < z < 1.5$ Observed with XMM-Newton. *ApJ* **2019**, *871*, 50, [CrossRef]
123. Leauthaud, A.; Finoguenov, A.; Kneib, J.P.; Taylor, J.E.; Massey, R.; Rhodes, J.; Ilbert, O.; Bundy, K.; Tinker, J.; George, M.R.; et al. A Weak Lensing Study of X-ray Groups in the Cosmos Survey: Form and Evolution of the Mass-Luminosity Relation. *ApJ* **2010**, *709*, 97–114. [CrossRef]
124. Pierre, M.; Pacaud, F.; Adami, C.; Alis, S.; Altieri, B.; Baran, N.; Benoist, C.; Birkinshaw, M.; Bongiorno, A.; Bremer, M.N.; et al. The XXL Survey. I. Scientific motivations—XMM-Newton observing plan—Follow-up observations and simulation programme. *Astron. Astrophys.* **2016**, *592*, A1, [CrossRef]
125. Schaye, J.; Dalla Vecchia, C.; Booth, C.M.; Wiersma, R.P.C.; Theuns, T.; Haas, M.R.; Bertone, S.; Duffy, A.R.; McCarthy, I.G.; van de Voort, F. The physics driving the cosmic star formation history. *MNRAS* **2010**, *402*, 1536–1560. [CrossRef]
126. Semboloni, E.; Hoekstra, H.; Schaye, J.; van Daalen, M.P.; McCarthy, I.G. Quantifying the effect of baryon physics on weak lensing tomography. *MNRAS* **2011**, *417*, 2020–2035. [CrossRef]
127. Finoguenov, A.; Reiprich, T.H.; Böhringer, H. Details of the mass-temperature relation for clusters of galaxies. *Astron. Astrophys.* **2001**, *368*, 749–759. [CrossRef]
128. Kettula, K.; Finoguenov, A.; Massey, R.; Rhodes, J.; Hoekstra, H.; Taylor, J.E.; Spinelli, P.F.; Tanaka, M.; Ilbert, O.; Capak, P.; et al. Weak Lensing Calibrated M-T Scaling Relation of Galaxy Groups in the COSMOS Field^{sstarf}. *ApJ* **2013**, *778*, 74, [CrossRef]
129. Spitzer, L. *Physics of Fully Ionized Gases*; Courier Corporation: North Chelmsford, MA, USA, 1962.
130. Kravtsov, A.V.; Vikhlinin, A.; Nagai, D. A New Robust Low-Scatter X-Ray Mass Indicator for Clusters of Galaxies. *ApJ* **2006**, *650*, 128–136. [CrossRef]
131. Jeltema, T.E.; Mulchaey, J.S.; Lubin, L.M.; Rosati, P.; Böhringer, H. X-Ray Properties of Intermediate-Redshift Groups of Galaxies. *ApJ* **2006**, *649*, 649–660. [CrossRef]
132. Alshino, A.; Ponman, T.; Pacaud, F.; Pierre, M. Evolution of the X-ray profiles of poor clusters from the XMM-LSS survey. *MNRAS* **2010**, *407*, 2543–2556. [CrossRef]
133. Beers, T.C.; Flynn, K.; Gebhardt, K. Measures of Location and Scale for Velocities in Clusters of Galaxies—A Robust Approach. *AJ* **1990**, *100*, 32. [CrossRef]
134. Ruel, J.; Bazin, G.; Bayliss, M.; Brodwin, M.; Foley, R.J.; Stalder, B.; Aird, K.A.; Armstrong, R.; Ashby, M.L.N.; Bautz, M.; et al. Optical Spectroscopy and Velocity Dispersions of Galaxy Clusters from the SPT-SZ Survey. *ApJ* **2014**, *792*, 45, [CrossRef]
135. Quintana, H.; Melnick, J. The correlation between X-ray luminosity and velicity dispersion in clusters of galaxies. *AJ* **1982**, *87*, 972–979. [CrossRef]

136. Mulchaey, J.S.; Zabludoff, A.I. The Properties of Poor Groups of Galaxies. II. X-Ray and Optical Comparisons. *ApJ* **1998**, *496*, 73–92. [CrossRef]
137. Mahdavi, A.; Geller, M.J. The L_X-σ Relation for Galaxies and Clusters of Galaxies. *ApJ* **2001**, *554*, L129–L132, [CrossRef]
138. Ortiz-Gil, A.; Guzzo, L.; Schuecker, P.; Böhringer, H.; Collins, C.A. The X-ray luminosity-velocity dispersion relation in the REFLEX cluster survey. *MNRAS* **2004**, *348*, 325–332. [CrossRef]
139. Zhang, Y.Y.; Andernach, H.; Caretta, C.A.; Reiprich, T.H.; Böhringer, H.; Puchwein, E.; Sijacki, D.; Girardi, M. HIFLUGCS: Galaxy cluster scaling relations between X-ray luminosity, gas mass, cluster radius, and velocity dispersion. *Astron. Astrophys.* **2011**, *526*, A105. [CrossRef]
140. Nastasi, A.; Böhringer, H.; Fassbender, R.; de Hoon, A.; Lamer, G.; Mohr, J.J.; Padilla, N.; Pratt, G.W.; Quintana, H.; Rosati, P.; et al. Kinematic analysis of a sample of X-ray luminous distant galaxy clusters. The $L_X - \sigma_v$ relation in the z > 0.6 universe. *Astron. Astrophys.* **2014**, *564*, A17, [CrossRef]
141. Kirkpatrick, C.C.; Clerc, N.; Finoguenov, A.; Damsted, S.; Ider Chitham, J.; Kukkola, A.E.; Gueguen, A.; Furnell, K.; Rykoff, E.; Comparat, J.; et al. SPIDERS: An overview of the largest catalogue of spectroscopically confirmed x-ray galaxy clusters. *MNRAS* **2021**, *503*, 5763–5777. [CrossRef]
142. Ponman, T.J.; Bourner, P.D.J.; Ebeling, H.; Böhringer, H. A ROSAT survey of Hickson's compact galaxy groups. *MNRAS* **1996**, *283*, 690–708. [CrossRef]
143. Sohn, J.; Geller, M.J.; Zahid, H.J. A Spectroscopic Census of X-Ray Systems in the COSMOS Field. *ApJ* **2019**, *880*, 142, [CrossRef]
144. Connelly, J.L.; Wilman, D.J.; Finoguenov, A.; Hou, A.; Mulchaey, J.S.; McGee, S.L.; Balogh, M.L.; Parker, L.C.; Saglia, R.; Henderson, R.D.E.; et al. Exploring the Diversity of Groups at 0.1 < z < 0.8 with X-Ray and Optically Selected Samples. *ApJ* **2012**, *756*, 139, [CrossRef]
145. Carlberg, R.G.; Yee, H.K.C.; Morris, S.L.; Lin, H.; Sawicki, M.; Wirth, G.; Patton, D.; Shepherd, C.W.; Ellingson, E.; Schade, D. The CNOC2 field galaxy redshift survey. *Philos. Trans. R. Soc. Lond. Ser. A* **1999**, *357*, 167, [CrossRef]
146. Finoguenov, A.; Guzzo, L.; Hasinger, G.; Scoville, N.Z.; Aussel, H.; Böhringer, H.; Brusa, M.; Capak, P.; Cappelluti, N.; Comastri, A.; et al. The XMM-Newton Wide-Field Survey in the COSMOS Field: Statistical Properties of Clusters of Galaxies. *ApJS* **2007**, *172*, 182–195. [CrossRef]
147. Scoville, N.; Aussel, H.; Brusa, M.; Capak, P.; Carollo, C.M.; Elvis, M.; Giavalisco, M.; Guzzo, L.; Hasinger, G.; Impey, C.; et al. The Cosmic Evolution Survey (COSMOS): Overview. *ApJS* **2007**, *172*, 1–8. [CrossRef]
148. George, M.R.; Leauthaud, A.; Bundy, K.; Finoguenov, A.; Tinker, J.; Lin, Y.T.; Mei, S.; Kneib, J.P.; Aussel, H.; Behroozi, P.S.; et al. Galaxies in X-Ray Groups. I. Robust Membership Assignment and the Impact of Group Environments on Quenching. *ApJ* **2011**, *742*, 125, [CrossRef]
149. Vajgel, B.; Jones, C.; Lopes, P.A.A.; Forman, W.R.; Murray, S.S.; Goulding, A.; Andrade-Santos, F. X-Ray-selected Galaxy Groups in Boötes. *ApJ* **2014**, *794*, 88. [CrossRef]
150. Murray, S.S.; Kenter, A.; Forman, W.R.; Jones, C.; Green, P.J.; Kochanek, C.S.; Vikhlinin, A.; Fabricant, D.; Fazio, G.; Brand, K.; et al. XBootes: An X-Ray Survey of the NDWFS Bootes Field. I. Overview and Initial Results. *ApJS* **2005**, *161*, 1–8. [CrossRef]
151. Wilson, S.; Hilton, M.; Rooney, P.J.; Caldwell, C.; Kay, S.T.; Collins, C.A.; McCarthy, I.G.; Romer, A.K.; Bermeo, A.; Bernstein, R.; et al. The XMM Cluster Survey: Evolution of the velocity dispersion-temperature relation over half a Hubble time. *MNRAS* **2016**, *463*, 413–428. [CrossRef]
152. Romer, A.K.; Viana, P.T.P.; Liddle, A.R.; Mann, R.G. A Serendipitous Galaxy Cluster Survey with XMM: Expected Catalogue Properties and Scientific Applications. *Astrophys. J.* **1999**.
153. Farahi, A.; Guglielmo, V.; Evrard, A.E.; Poggianti, B.M.; Adami, C.; Ettori, S.; Gastaldello, F.; Giles, P.A.; Maughan, B.J.; Rapetti, D.; et al. The XXL Survey: XXIII. The Mass Scale of XXL Clusters from Ensemble Spectroscopy. *Astron. Astrophys.* **2018**, *620*, A8, [CrossRef]
154. Adami, C.; Giles, P.; Koulouridis, E.; Pacaud, F.; Caretta, C.A.; Pierre, M.; Eckert, D.; Ramos-Ceja, M.E.; Gastaldello, F.; Fotopoulou, S.; et al. The XXL Survey. XX. The 365 cluster catalogue. *Astron. Astrophys.* **2018**, *620*, A5, [CrossRef]
155. Mamon, G. Dynamical Theory of groups and Clusters of Galaxies. *arXiv* **1993**, arXiv:astro-ph/9308032.
156. Mamon, G.A.; Biviano, A.; Boué, G. MAMPOSSt: Modelling Anisotropy and Mass Profiles of Observed Spherical Systems—I. Gaussian 3D velocities. *MNRAS* **2013**, *429*, 3079–3098. [CrossRef]
157. Goulding, A.D.; Greene, J.E.; Ma, C.P.; Veale, M.; Bogdan, A.; Nyland, K.; Blakeslee, J.P.; McConnell, N.J.; Thomas, J. The MASSIVE Survey. IV. The X-ray Halos of the Most Massive Early-type Galaxies in the Nearby Universe. *ApJ* **2016**, *826*, 167, [CrossRef]
158. Ma, C.P.; Greene, J.E.; McConnell, N.; Janish, R.; Blakeslee, J.P.; Thomas, J.; Murphy, J.D. The MASSIVE Survey. I. A Volume-limited Integral-field Spectroscopic Study of the Most Massive Early-type Galaxies within 108 Mpc. *ApJ* **2014**, *795*, 158, [CrossRef]
159. Le Brun, A.M.C.; McCarthy, I.G.; Schaye, J.; Ponman, T.J. The scatter and evolution of the global hot gas properties of simulated galaxy cluster populations. *MNRAS* **2017**, *466*, 4442–4469. [CrossRef]
160. Farahi, A.; Evrard, A.E.; McCarthy, I.; Barnes, D.J.; Kay, S.T. Localized massive halo properties in BAHAMAS and MACSIS simulations: scalings, lognormality, and covariance. *MNRAS* **2018**, *478*, 2618–2632. [CrossRef]
161. Kar Chowdhury, R.; Chatterjee, S.; Lonappan, A.I.; Khandai, N.; Di Matteo, T. Cosmological Simulation of Galaxy Groups and Clusters. I. Global Effect of Feedback from Active Galactic Nuclei. *ApJ* **2020**, *889*, 60. [CrossRef]

162. Helsdon, S.F.; Ponman, T.J.; Mulchaey, J.S. Chandra Observations of Low Velocity Dispersion Groups. *ApJ* **2005**, *618*, 679–691. [CrossRef]
163. Mantz, A.; Allen, S.W.; Ebeling, H.; Rapetti, D.; Drlica-Wagner, A. The observed growth of massive galaxy clusters—II. X-ray scaling relations. *MNRAS* **2010**, *406*, 1773–1795. [CrossRef]
164. Giles, P.A.; Maughan, B.J.; Dahle, H.; Bonamente, M.; Landry, D.; Jones, C.; Joy, M.; Murray, S.S.; van der Pyl, N. Chandra measurements of a complete sample of X-ray luminous galaxy clusters: the luminosity-mass relation. *MNRAS* **2017**, *465*, 858–884. [CrossRef]
165. Ettori, S.; Morandi, A.; Tozzi, P.; Balestra, I.; Borgani, S.; Rosati, P.; Lovisari, L.; Terenziani, F. The cluster gas mass fraction as a cosmological probe: A revised study. *Astron. Astrophys.* **2009**, *501*, 61–73. [CrossRef]
166. Mantz, A.B.; Allen, S.W.; Morris, R.G.; Rapetti, D.A.; Applegate, D.E.; Kelly, P.L.; von der Linden, A.; Schmidt, R.W. Cosmology and astrophysics from relaxed galaxy clusters—II. Cosmological constraints. *MNRAS* **2014**, *440*, 2077–2098. [CrossRef]
167. Lin, Y.T.; Mohr, J.J.; Stanford, S.A. Near-Infrared Properties of Galaxy Clusters: Luminosity as a Binding Mass Predictor and the State of Cluster Baryons. *ApJ* **2003**, *591*, 749–763. [CrossRef]
168. Behroozi, P.S.; Conroy, C.; Wechsler, R.H. A Comprehensive Analysis of Uncertainties Affecting the Stellar Mass-Halo Mass Relation for $0 < z < 4$. *ApJ* **2010**, *717*, 379–403. [CrossRef]
169. Zhang, Y.Y.; Laganá, T.F.; Pierini, D.; Puchwein, E.; Schneider, P.; Reiprich, T.H. Star-formation efficiency and metal enrichment of the intracluster medium in local massive clusters of galaxies. *Astron. Astrophys.* **2011**, *535*, A78, [CrossRef]
170. Leauthaud, A.; George, M.R.; Behroozi, P.S.; Bundy, K.; Tinker, J.; Wechsler, R.H.; Conroy, C.; Finoguenov, A.; Tanaka, M. The Integrated Stellar Content of Dark Matter Halos. *ApJ* **2012**, *746*, 95, [CrossRef]
171. Laganá, T.F.; Martinet, N.; Durret, F.; Lima Neto, G.B.; Maughan, B.; Zhang, Y.Y. A comprehensive picture of baryons in groups and clusters of galaxies. *Astron. Astrophys.* **2013**, *555*, A66, [CrossRef]
172. Chiu, I.; Mohr, J.J.; McDonald, M.; Bocquet, S.; Desai, S.; Klein, M.; Israel, H.; Ashby, M.L.N.; Stanford, A.; Benson, B.A.; et al. Baryon content in a sample of 91 galaxy clusters selected by the South Pole Telescope at $0.2 < z < 1.25$. *MNRAS* **2018**, *478*, 3072–3099. [CrossRef]
173. Decker, B.; Brodwin, M.; Abdulla, Z.; Gonzalez, A.H.; Marrone, D.P.; O'Donnell, C.; Stanford, S.A.; Wylezalek, D.; Carlstrom, J.E.; Eisenhardt, P.R.M.; et al. The Massive and Distant Clusters of WISE Survey. VI. Stellar Mass Fractions of a Sample of High-redshift Infrared-selected Clusters. *ApJ* **2019**, *878*, 72, [CrossRef]
174. Vikhlinin, A.; McNamara, B.R.; Forman, W.; Jones, C.; Quintana, H.; Hornstrup, A. A Catalog of 200 Galaxy Clusters Serendipitously Detected in the ROSAT PSPC Pointed Observations. *ApJ* **1998**, *502*, 558–581. [CrossRef]
175. Lilly, S.J.; Le Fèvre, O.; Renzini, A.; Zamorani, G.; Scodeggio, M.; Contini, T.; Carollo, C.M.; Hasinger, G.; Kneib, J.P.; Iovino, A.; et al. zCOSMOS: A Large VLT/VIMOS Redshift Survey Covering $0 < z < 3$ in the COSMOS Field. *ApJS* **2007**, *172*, 70–85. [CrossRef]
176. Navarro, J.F.; Frenk, C.S.; White, S.D.M. The Structure of Cold Dark Matter Halos. *ApJ* **1996**, *462*, 563, [CrossRef]
177. Navarro, J.F.; Frenk, C.S.; White, S.D.M. A Universal Density Profile from Hierarchical Clustering. *ApJ* **1997**, *490*, 493–508. [CrossRef]
178. Chiu, I.; Saro, A.; Mohr, J.; Desai, S.; Bocquet, S.; Capasso, R.; Gangkofner, C.; Gupta, N.; Liu, J. Stellar mass to halo mass scaling relation for X-ray-selected low-mass galaxy clusters and groups out to redshift $z \approx 1$. *MNRAS* **2016**, *458*, 379–393. [CrossRef]
179. Liu, J.; Mohr, J.; Saro, A.; Aird, K.A.; Ashby, M.L.N.; Bautz, M.; Bayliss, M.; Benson, B.A.; Bleem, L.E.; Bocquet, S.; et al. Analysis of Sunyaev-Zel'dovich effect mass-observable relations using South Pole Telescope observations of an X-ray selected sample of low-mass galaxy clusters and groups. *MNRAS* **2015**, *448*, 2085–2099. [CrossRef]
180. Kravtsov, A.V.; Vikhlinin, A.A.; Meshcheryakov, A.V. Stellar Mass—Halo Mass Relation and Star Formation Efficiency in High-Mass Halos. [CrossRef]
181. Pillepich, A.; Nelson, D.; Hernquist, L.; Springel, V.; Pakmor, R.; Torrey, P.; Weinberger, R.; Genel, S.; Naiman, J.P.; Marinacci, F.; et al. First results from the IllustrisTNG simulations: The stellar mass content of groups and clusters of galaxies. *MNRAS* **2018**, *475*, 648–675. [CrossRef]
182. Ziparo, F.; Smith, G.P.; Mulroy, S.L.; Lieu, M.; Willis, J.P.; Hudelot, P.; McGee, S.L.; Fotopoulou, S.; Lidman, C.; Lavoie, S.; et al. The XXL Survey. X. K-band luminosity—Weak-lensing mass relation for groups and clusters of galaxies. *Astron. Astrophys.* **2016**, *592*, A9, [CrossRef]
183. Muzzin, A.; Yee, H.K.C.; Hall, P.B.; Lin, H. Near-Infrared Properties of Moderate-Redshift Galaxy Clusters. II. Halo Occupation Number, Mass-to-Light Ratios, and Ω_m. *ApJ* **2007**, *663*, 150–163. [CrossRef]
184. Gaspari, M.; Ruszkowski, M.; Oh, S.P. Chaotic cold accretion on to black holes. *MNRAS* **2013**, *432*, 3401–3422. [CrossRef]
185. Prasad, D.; Sharma, P.; Babul, A. Cool Core Cycles: Cold Gas and AGN Jet Feedback in Cluster Cores. *ApJ* **2015**, *811*, 108. [CrossRef]
186. Voit, G.M.; Meece, G.; Li, Y.; O'Shea, B.W.; Bryan, G.L.; Donahue, M. A Global Model for Circumgalactic and Cluster-core Precipitation. *ApJ* **2017**, *845*, 80. [CrossRef]
187. Temi, P.; Amblard, A.; Gitti, M.; Brighenti, F.; Gaspari, M.; Mathews, W.G.; David, L. ALMA Observations of Molecular Clouds in Three Group-centered Elliptical Galaxies: NGC 5846, NGC 4636, and NGC 5044. *ApJ* **2018**, *858*, 17. [CrossRef]
188. Tremblay, G.R.; Combes, F.; Oonk, J.B.R.; Russell, H.R.; McDonald, M.A.; Gaspari, M.; Husemann, B.; Nulsen, P.E.J.; McNamara, B.R.; Hamer, S.L.; et al. A Galaxy-scale Fountain of Cold Molecular Gas Pumped by a Black Hole. *ApJ* **2018**, *865*, 13. [CrossRef]

189. Gaspari, M.; McDonald, M.; Hamer, S.L.; Brighenti, F.; Temi, P.; Gendron-Marsolais, M.; Hlavacek-Larrondo, J.; Edge, A.C.; Werner, N.; Tozzi, P.; et al. Shaken Snow Globes: Kinematic Tracers of the Multiphase Condensation Cascade in Massive Galaxies, Groups, and Clusters. *ApJ* **2018**, *854*, 167. [CrossRef]
190. Rose, T.; Edge, A.C.; Combes, F.; Gaspari, M.; Hamer, S.; Nesvadba, N.; Peck, A.B.; Sarazin, C.; Tremblay, G.R.; Baum, S.A.; et al. Constraining cold accretion on to supermassive black holes: Molecular gas in the cores of eight brightest cluster galaxies revealed by joint CO and CN absorption. *MNRAS* **2019**, *489*, 349–365. [CrossRef]
191. Storchi-Bergmann, T.; Schnorr-Müller, A. Observational constraints on the feeding of supermassive black holes. *Nat. Astron.* **2019**, *3*, 48–61. [CrossRef]
192. Tombesi, F.; Cappi, M.; Reeves, J.N.; Nemmen, R.S.; Braito, V.; Gaspari, M.; Reynolds, C.S. Unification of X-ray winds in Seyfert galaxies: From ultra-fast outflows to warm absorbers. *MNRAS* **2013**, *430*, 1102–1117. [CrossRef]
193. Sądowski, A.; Gaspari, M. Kinetic and radiative power from optically thin accretion flows. *MNRAS* **2017**, *468*, 1398–1404. [CrossRef]
194. Fiore, F.; Feruglio, C.; Shankar, F.; Bischetti, M.; Bongiorno, A.; Brusa, M.; Carniani, S.; Cicone, C.; Duras, F.; Lamastra, A.; et al. AGN wind scaling relations and the co-evolution of black holes and galaxies. *Astron. Astrophys.* **2017**, *601*, A143. [CrossRef]
195. McNamara, B.R.; Nulsen, P.E.J. Mechanical feedback from active galactic nuclei in galaxies, groups and clusters. *New J. Phys.* **2012**, *14*, 055023, [CrossRef]
196. Fabian, A.C. Observational Evidence of Active Galactic Nuclei Feedback. *Annu. Rev. Astron. Astrophys.* **2012**, *50*, 455–489. [CrossRef]
197. Gitti, M.; Brighenti, F.; McNamara, B.R. Evidence for AGN Feedback in Galaxy Clusters and Groups. *Adv. Astronomy* **2012**, *2012*, 1–24. [CrossRef]
198. Brighenti, F.; Mathews, W.G.; Temi, P. Hot Gaseous Atmospheres in Galaxy Groups and Clusters Are Both Heated and Cooled by X-Ray Cavities. *ApJ* **2015**, *802*, 118. [CrossRef]
199. Gaspari, M. Shaping the X-ray spectrum of galaxy clusters with AGN feedback and turbulence. *MNRAS* **2015**, *451*, L60–L64, [CrossRef]
200. Liu, W.; Sun, M.; Nulsen, P.; Clarke, T.; Sarazin, C.; Forman, W.; Gaspari, M.; Giacintucci, S.; Lal, D.V.; Edge, T. AGN feedback in galaxy group 3C 88: Cavities, shock, and jet reorientation. *MNRAS* **2019**, *484*, 3376–3392. [CrossRef]
201. Yang, H.Y.K.; Gaspari, M.; Marlow, C. The Impact of Radio AGN Bubble Composition on the Dynamics and Thermal Balance of the Intracluster Medium. *ApJ* **2019**, *871*, 6, [CrossRef]
202. Wittor, D.; Gaspari, M. Dissecting the turbulent weather driven by mechanical AGN feedback. *MNRAS* **2020**, *498*, 4983–5002. [CrossRef]
203. Voit, G.M.; Bryan, G.L.; Prasad, D.; Frisbie, R.; Li, Y.; Donahue, M.; O'Shea, B.W.; Sun, M.; Werner, N. A Black Hole Feedback Valve in Massive Galaxies. *ApJ* **2020**, *899*, 70, [CrossRef]
204. Kormendy, J.; Ho, L.C. Coevolution (Or Not) of Supermassive Black Holes and Host Galaxies. *ARA&A* **2013**, *51*, 511–653. [CrossRef]
205. Saglia, R.P.; Opitsch, M.; Erwin, P.; Thomas, J.; Beifiori, A.; Fabricius, M.; Mazzalay, X.; Nowak, N.; Rusli, S.P.; Bender, R. The SINFONI Black Hole Survey: The Black Hole Fundamental Plane Revisited and the Paths of (Co)evolution of Supermassive Black Holes and Bulges. *ApJ* **2016**, *818*, 47. [CrossRef]
206. Bondi, H. On spherically symmetrical accretion. *MNRAS* **1952**, *112*, 195. [CrossRef]
207. Narayan, R.; Fabian, A.C. Bondi flow from a slowly rotating hot atmosphere. *MNRAS* **2011**, *415*, 3721–3730. [CrossRef]
208. Voit, G.M. A Role for Turbulence in Circumgalactic Precipitation. *ApJ* **2018**, *868*, 102. [CrossRef]
209. Olivares, V.; Salome, P.; Combes, F.; Hamer, S.; Guillard, P.; Lehnert, M.D.; Polles, F.L.; Beckmann, R.S.; Dubois, Y.; Donahue, M.; et al. Ubiquitous cold and massive filaments in cool core clusters. *Astron. Astrophys.* **2019**, *631*, A22, [CrossRef]
210. Bassini, L.; Rasia, E.; Borgani, S.; Ragone-Figueroa, C.; Biffi, V.; Dolag, K.; Gaspari, M.; Granato, G.L.; Murante, G.; Taffoni, G.; et al. Black hole mass of central galaxies and cluster mass correlation in cosmological hydro-dynamical simulations. *Astron. Astrophys.* **2019**, *630*, A144, [CrossRef]
211. Truong, N.; Pillepich, A.; Werner, N. Correlations between supermassive black holes and hot gas atmospheres in IllustrisTNG and X-ray observations. *MNRAS* **2021**, *501*, 2210–2230. [CrossRef]
212. Gaspari, M.; Tombesi, F.; Cappi, M. Linking macro-, meso- and microscales in multiphase AGN feeding and feedback. *Nature Astronomy* **2020**, *4*, 10–13. [CrossRef]
213. Kravtsov, A.V.; Borgani, S. Formation of Galaxy Clusters. *ARA&A* **2012**, *50*, 353–409. [CrossRef]
214. Tremmel, M.; Karcher, M.; Governato, F.; Volonteri, M.; Quinn, T.R.; Pontzen, A.; Anderson, L.; Bellovary, J. The Romulus cosmological simulations: A physical approach to the formation, dynamics and accretion models of SMBHs. *MNRAS* **2017**, *470*, 1121–1139. [CrossRef]
215. Babyk, I.V.; McNamara, B.R.; Nulsen, P.E.J.; Russell, H.R.; Vantyghem, A.N.; Hogan, M.T.; Pulido, F.A. A Universal Entropy Profile for the Hot Atmospheres of Galaxies and Clusters within R_{2500}. *ApJ* **2018**, *862*, 39, [CrossRef]
216. Planelles, S.; Fabjan, D.; Borgani, S.; Murante, G.; Rasia, E.; Biffi, V.; Truong, N.; Ragone-Figueroa, C.; Granato, G.L.; Dolag, K.; et al. Pressure of the hot gas in simulations of galaxy clusters. *MNRAS* **2017**, *467*, 3827–3847. [CrossRef]

217. Weinberger, R.; Springel, V.; Pakmor, R.; Nelson, D.; Genel, S.; Pillepich, A.; Vogelsberger, M.; Marinacci, F.; Naiman, J.; Torrey, P.; et al. Supermassive black holes and their feedback effects in the IllustrisTNG simulation. *MNRAS* **2018**, *479*, 4056–4072. [CrossRef]
218. Mulchaey, J.S.; Davis, D.S.; Mushotzky, R.F.; Burstein, D. An X-Ray Atlas of Groups of Galaxies. *ApJS* **2003**, *145*, 39–64. [CrossRef]
219. Maughan, B.J.; Giles, P.A.; Randall, S.W.; Jones, C.; Forman, W.R. Self-similar scaling and evolution in the galaxy cluster X-ray luminosity-temperature relation. *MNRAS* **2012**, *421*, 1583–1602. [CrossRef]
220. Predehl, P.; Andritschke, R.; Arefiev, V.; Babyshkin, V.; Batanov, O.; Becker, W.; Böhringer, H.; Bogomolov, A.; Boller, T.; Borm, K.; et al. The eROSITA X-ray telescope on SRG. *arXiv* **2020**, arXiv:2010.03477.
221. Clerc, N.; Ramos-Ceja, M.E.; Ridl, J.; Lamer, G.; Brunner, H.; Hofmann, F.; Comparat, J.; Pacaud, F.; Käfer, F.; Reiprich, T.H.; et al. Synthetic simulations of the extragalactic sky seen by eROSITA. I. Pre-launch selection functions from Monte-Carlo simulations. *Astron. Astrophys.* **2018**, *617*, A92, [CrossRef]
222. Käfer, F.; Finoguenov, A.; Eckert, D.; Clerc, N.; Ramos-Ceja, M.E.; Sanders, J.S.; Ghirardini, V. Toward the low-scatter selection of X-ray clusters. Galaxy cluster detection with eROSITA through cluster outskirts. *Astron. Astrophys.* **2020**, *634*, A8, [CrossRef]
223. Merloni, A.; Predehl, P.; Becker, W.; Böhringer, H.; Boller, T.; Brunner, H.; Brusa, M.; Dennerl, K.; Freyberg, M.; Friedrich, P.; et al. eROSITA Science Book: Mapping the Structure of the Energetic Universe. *arXiv* **2012**, arXiv:1209.3114.
224. Pillepich, A.; Porciani, C.; Reiprich, T.H. The X-ray cluster survey with eRosita: Forecasts for cosmology, cluster physics and primordial non-Gaussianity. *MNRAS* **2012**, *422*, 44–69. [CrossRef]
225. Borm, K.; Reiprich, T.H.; Mohammed, I.; Lovisari, L. Constraining galaxy cluster temperatures and redshifts with eROSITA survey data. *Astron. Astrophys.* **2014**, *567*, A65, [CrossRef]
226. Zandanel, F.; Fornasa, M.; Prada, F.; Reiprich, T.H.; Pacaud, F.; Klypin, A. MultiDark clusters: Galaxy cluster mock light-cones, eROSITA, and the cluster power spectrum. *MNRAS* **2018**, *480*, 987–1005. [CrossRef]
227. Pillepich, A.; Reiprich, T.H.; Porciani, C.; Borm, K.; Merloni, A. Forecasts on dark energy from the X-ray cluster survey with eROSITA: constraints from counts and clustering. *MNRAS* **2018**, *481*, 613–626. [CrossRef]
228. Pointecouteau, E.; Reiprich, T.H.; Adami, C.; Arnaud, M.; Biffi, V.; Borgani, S.; Borm, K.; Bourdin, H.; Brueggen, M.; Bulbul, E.; et al. The Hot and Energetic Universe: The evolution of galaxy groups and clusters. *arXiv* **2013**, arXiv:1306.2319.
229. Ettori, S.; Pratt, G.W.; de Plaa, J.; Eckert, D.; Nevalainen, J.; Battistelli, E.S.; Borgani, S.; Croston, J.H.; Finoguenov, A.; Kaastra, J.; et al. The Hot and Energetic Universe: The astrophysics of galaxy groups and clusters. *arXiv* **2013**, arXiv:1306.2322.
230. Croston, J.H.; Sanders, J.S.; Heinz, S.; Hardcastle, M.J.; Zhuravleva, I.; Bîrzan, L.; Bower, R.G.; Brüggen, M.; Churazov, E.; Edge, A.C.; et al. The Hot and Energetic Universe: AGN feedback in galaxy clusters and groups. *arXiv* **2013**, arXiv:1306.2323.
231. Cucchetti, E.; Pointecouteau, E.; Peille, P.; Clerc, N.; Rasia, E.; Biffi, V.; Borgani, S.; Tornatore, L.; Dolag, K.; Roncarelli, M.; et al. Athena X-IFU synthetic observations of galaxy clusters to probe the chemical enrichment of the Universe. *Astron. Astrophys.* **2018**, *620*, A173, [CrossRef]
232. Roncarelli, M.; Gaspari, M.; Ettori, S.; Biffi, V.; Brighenti, F.; Bulbul, E.; Clerc, N.; Cucchetti, E.; Pointecouteau, E.; Rasia, E. Measuring turbulence and gas motions in galaxy clusters via synthetic Athena X-IFU observations. *Astron. Astrophys.* **2018**, *618*, A39, [CrossRef]
233. Mernier, F.; Cucchetti, E.; Tornatore, L.; Biffi, V.; Pointecouteau, E.; Clerc, N.; Peille, P.; Rasia, E.; Barret, D.; Borgani, S.; et al. Constraining the origin and models of chemical enrichment in galaxy clusters using the Athena X-IFU. *Astron. Astrophys.* **2020**, *642*, A90, [CrossRef]
234. Marchesi, S.; Gilli, R.; Lanzuisi, G.; Dauser, T.; Ettori, S.; Vito, F.; Cappelluti, N.; Comastri, A.; Mushotzky, R.; Ptak, A.; et al. Mock catalogs for the extragalactic X-ray sky: Simulating AGN surveys with ATHENA and with the AXIS probe. *Astron. Astrophys.* **2020**, *642*, A184, [CrossRef]
235. Benson, B.A.; Ade, P.A.R.; Ahmed, Z.; Allen, S.W.; Arnold, K.; Austermann, J.E.; Bender, A.N.; Bleem, L.E.; Carlstrom, J.E.; Chang, C.L.; et al. SPT-3G: A next-generation cosmic microwave background polarization experiment on the South Pole telescope. In *Millimeter, Submillimeter, and Far-Infrared Detectors and Instrumentation for Astronomy VII*; Society of Photo-Optical Instrumentation Engineers (SPIE) Conference Series; Holland, W.S., Zmuidzinas, J., Eds.; International Society for Optics and Photonics: London, UK, 2014; Volume 9153, p. 91531P, [CrossRef]
236. Sartoris, B.; Biviano, A.; Fedeli, C.; Bartlett, J.G.; Borgani, S.; Costanzi, M.; Giocoli, C.; Moscardini, L.; Weller, J.; Ascaso, B.; et al. Next generation cosmology: Constraints from the Euclid galaxy cluster survey. *MNRAS* **2016**, *459*, 1764–1780. [CrossRef]
237. Finoguenov, A.; Merloni, A.; Comparat, J.; Nandra, K.; Salvato, M.; Tempel, E.; Raichoor, A.; Richard, J.; Kneib, J.P.; Pillepich, A.; et al. 4MOST Consortium Survey 5: eROSITA Galaxy Cluster Redshift Survey. *Messenger* **2019**, *175*, 39–41. [CrossRef]
238. Driver, S.P.; Liske, J.; Davies, L.J.M.; Robotham, A.S.G.; Baldry, I.K.; Brown, M.J.I.; Cluver, M.; Kuijken, K.; Loveday, J.; McMahon, R.; et al. 4MOST Consortium Survey 7: Wide-Area VISTA Extragalactic Survey (WAVES). *Messenger* **2019**, *175*, 46–49. [CrossRef]
239. Zheng, Y.L.; Shen, S.Y. Compact Groups of Galaxies in Sloan Digital Sky Survey and LAMOST Spectral Survey. II. Dynamical Properties of Isolated and Embedded Groups. *ApJ* **2021**, *911*, 105, [CrossRef]
240. Balestra, I.; Tozzi, P.; Ettori, S.; Rosati, P.; Borgani, S.; Mainieri, V.; Norman, C.; Viola, M. Tracing the evolution in the iron content of the intra-cluster medium. *Astron. Astrophys.* **2007**, *462*, 429–442. [CrossRef]

MDPI

Review

Feedback from Active Galactic Nuclei in Galaxy Groups

Dominique Eckert [1,*,†], Massimo Gaspari [2,3,†], Fabio Gastaldello [4,†], Amandine M. C. Le Brun [5,†] and Ewan O'Sullivan [6,†]

1 Department of Astronomy, University of Geneva, Ch. d'Ecogia 16, CH-1290 Versoix, Switzerland
2 INAF—Osservatorio di Astrofisica e Scienza dello Spazio di Bologna, Via Piero Gobetti 93/3, I-40129 Bologna, Italy; massimo.gaspari@inaf.it
3 Department of Astrophysical Sciences, Princeton University, 4 Ivy Lane, Princeton, NJ 08544, USA
4 IASF—Milano, INAF, Via A. Corti 12, I-20133 Milano, Italy; fabio.gastaldello@inaf.it
5 LUTh, UMR 8102 CNRS, Observatoire de Paris, PSL Research University, Université Paris Diderot, 5 Place Jules Janssen, 92190 Meudon, France; amandine.le-brun@obspm.fr
6 Center for Astrophysics, Harvard & Smithsonian, 60 Garden Street, Cambridge, MA 02138, USA; eosullivan@cfa.harvard.edu
* Correspondence: Dominique.Eckert@unige.ch
† All authors contributed equally to this work.

Abstract: The co-evolution between supermassive black holes and their environment is most directly traced by the hot atmospheres of dark matter halos. The cooling of the hot atmosphere supplies the central regions with fresh gas, igniting active galactic nuclei (AGN) with long duty cycles. Outflows from the central engine tightly couple with the surrounding gaseous medium and provide the dominant heating source preventing runaway cooling by carving cavities and driving shocks across the medium. The AGN feedback loop is a key feature of all modern galaxy evolution models. Here, we review our knowledge of the AGN feedback process in the specific context of galaxy groups. Galaxy groups are uniquely suited to constrain the mechanisms governing the cooling–heating balance. Unlike in more massive halos, the energy that is supplied by the central AGN to the hot intragroup medium can exceed the gravitational binding energy of halo gas particles. We report on the state-of-the-art in observations of the feedback phenomenon and in theoretical models of the heating-cooling balance in galaxy groups. We also describe how our knowledge of the AGN feedback process impacts galaxy evolution models and large-scale baryon distributions. Finally, we discuss how new instrumentation will answer key open questions on the topic.

Keywords: black holes; galaxy groups; elliptical galaxies; intragroup medium/plasma; active nuclei; X-ray observations; hydrodynamical and cosmological simulations

Citation: Eckert, D.; Gaspari, M.; Gastaldello, F.; Le Brun, A.M.C.; O'Sullivan, E. Feedback from Active Galactic Nuclei in Galaxy Groups. *Universe* **2021**, *7*, 142. https://doi.org/10.3390/universe7050142

Academic Editor: Francesco Shankar

Received: 31 March 2021
Accepted: 29 April 2021
Published: 11 May 2021

1. Introduction

Structure formation in the Universe operates as a bottom-up process in which small halos formed at high redshift progressively merge and accrete the surrounding material to form the massive halos we see today [1]. Given the evolution of the halo mass function, the peak of the mass density in the current Universe occurs in halos of $\sim 10^{13} M_\odot$—the *galaxy group* regime. At the current epoch, galaxy groups are the building blocks of the structure formation process and, thus, they occupy a key regime in the evolution of galaxies. Typical $L_\star$ galaxies exist within groups rather than within isolated halos [2]. The galaxy stellar mass function exhibits a cut-off at $M_\star \sim 10^{11} M_\odot$ [3], corresponding to the central galaxies of galaxy groups, brightest group galaxies (BGGs). Abundance matching studies show that the star formation efficiency reaches a maximum at $M_h \sim 10^{12} M_\odot$ and decreases at both higher and lower masses [4–6]. At the high-mass end, non-gravitational energy input is needed to quench star formation and reproduce the shape of the stellar mass function [7].

Feedback from active supermassive black holes (SMBH) is currently the favored mechanism for regulating the star formation activity in massive galaxies, explaining the observed shape of the stellar mass function, and quenching catastrophic cooling flows. Outflows and jets from the central active galactic nuclei (AGN) interact with the surrounding hot intragroup medium (IGrM) and release a large amount of energy, which prevents the gas reservoir from cooling efficiently and fueling star formation (for previous reviews, see McNamara and Nulsen [8]; Fabian [9]; Gitti et al. [10]; Gaspari et al. [11]). All of the modern galaxy evolution models include a prescription for AGN feedback to reproduce the shape of the galaxy luminosity function and the halo baryon fraction (see the companion review by Oppenheimer et al.). Earlier attempts at reproducing these observables with supernova feedback resulted in catastrophic cooling and largely overestimated the stellar content of groups (e.g., [12]). State-of-the-art cosmological simulations all implement sub-grid prescriptions for AGN feedback, either in the form of thermal, isotropic feedback, or in the form of mechanical, directional feedback. The adopted feedback scheme strongly affects the gas properties of galaxy groups. Indeed, strong feedback raises the entropy level of the surrounding gas particles, which can lead to a global depletion of baryons in group-scale halos. Cosmological simulations are now facing the important challenge of reproducing at the same time realistic galaxy populations and gas properties.

The imprint of AGN feedback is most easily observed through high-resolution X-ray observations of nearby galaxy groups and clusters. Bubbles of expanding energetic material carve cavities in the gas distribution, spatially coinciding with energetic AGN outflows that are traced by their radio emission. Supersonic outflows also drive shock fronts permeating the surrounding IGrM and distributing heat across the environment. In parallel, extended Hα nebulae demonstrate the existence of efficient gas cooling from the hot phase, thereby feeding the central SMBH. Recent ALMA observations at millimetric wavelengths also provide evidence for large amounts of cool gas at the vicinity of the SMBH. Thus, multi-wavelength observations of the cores of nearby massive structures allow us to investigate, in detail, the balance between gas cooling and AGN heating.

While a great deal of attention has been devoted to studying these phenomena in the most massive nearby clusters, such as Perseus (see [8] for a review), observations of similar quality only exist for a handful of groups, such as NGC 5044 [13] and NGC 5813 [14]. Detailed observations of galaxy groups are crucial for our understanding of feedback processes, as the physical conditions differ from those of galaxy clusters in several important ways. First and foremost, the ratio of feedback energy to gravitational energy is different from that of clusters. The AGN energy input is sometimes parameterized as $\dot{E}_{\rm feed} \approx \epsilon_f \epsilon_r \dot{M}_{\rm BH} c^2$ with ϵ_r ~10% the energy output of the BH and ϵ_f the coupling efficiency between the BH outflows and the surrounding medium. On the other hand, the gravitational binding energy is a strong function of halo mass, $E_{\rm bind} \propto M_h^2$. If the coupling efficiency only depends on the physical properties of the gas and the feedback loop has a long duty cycle, then the total integrated feedback energy $E_{\rm feed}$ becomes comparable to $E_{\rm bind}$ or even exceeds it in group-scale halos, whereas it remains substantially lower in the most massive systems. Similarly, the radius inside, which non-gravitational energy dominates over gravitational energy, is comparably larger in group-scale halos. The impact of AGN in groups is spread over much larger volumes and it can even lead to a depletion of baryons within the virial radius. On top of that, the radiative cooling function experiences a transition of regime between the temperature range of clusters and that of groups. For temperatures that are greater than ~3 keV, the plasma is almost completely ionized and the Bremsstrahlung process dominates. For temperatures of ~1 keV, line cooling dominates, which makes the radiative losses comparably more important. Therefore, the radiative cooling time can become much shorter than the Hubble time, even at relatively low gas densities, and the supply of gas to the SMBH can be sustained more easily. For all of these reasons, studying the feedback loop across a wide range of halo masses is necessary for informing our theoretical models.

Here, we review the current state of the art in our knowledge of the AGN feedback process in the specific case of galaxy groups. For the purpose of this review, we define galaxy groups as galaxy concentrations with halo masses in the range 10^{13}–$10^{14} M_\odot$ and with an X-ray bright intragroup medium (IGrM). Such masses correspond to virial temperatures of $\sim$0.5–2 keV. Most of the processes that are discussed in this review are also relevant in the case of X-ray bright isolated elliptical galaxies and massive spirals with $kT \sim 0.3$–0.5 keV. Whenever it is appropriate, we will discuss halos of lower masses as well. The paper is organized, as follows. In Section 2, we describe why AGN feedback is presently thought to be a key ingredient in the evolution of galaxy groups and how invoking AGN feedback can solve a number of overarching issues in galaxy evolution. In Section 3, we review the current observational evidence for AGN feedback in nearby galaxy groups, with our main focus on observations at X-ray and radio wavelengths and additional information coming from millimeter and Hα observations. Section 4 summarizes the theoretical framework that was put into place to interpret the heating/cooling cycle and the main physical processes involved, with a specific focus on the galaxy group regime. Section 5 discusses how AGN feedback is implemented in modern cosmological simulations and its impact on the evolution and large-scale distribution of the baryonic component of the Universe. We conclude our review in Section 6 with a short presentation of the most relevant upcoming experiments and their expected contribution to the field.

2. The Need for AGN Feedback in Galaxy Evolution

Even though the first indications that AGN feedback could be one of the missing elements in theories of galaxy formation and evolution are already nearly half a century old [15–17], theoretical and observational studies of the process are still in their infancy. In the last fifteen years, AGN feedback has emerged as the most promising solution to a number of overarching problems in galaxy formation and evolution in both semi-analytic models and cosmological hydrodynamical simulations (e.g., [18–22]). In short, the main issues are: cosmic 'downsizing' [23], i.e., the observation that the majority of both star formation and AGN activity took place before redshift $z \sim 1$ (e.g., [24,25]), the shape of the galaxy stellar mass function at the high-mass end, the gas content of massive galaxies, groups and clusters, the absence of a cooling flow, the deviations from self-similarity of gas scaling relations, and the entropy floor (the latter three are discussed in detail in Section 3.1). AGN feedback progressively appeared to be a promising solution to solve each of these problems separately, eventually leading to the realization that these issues could in fact be seen as different facets of the same problem (e.g., [19,26,27]). In this section, we highlight the main reasons why AGN feedback plays a central role in galaxy evolution models, specifically at the scale of galaxy groups.

2.1. The Shape of the Galaxy Stellar Mass Function

White and Rees [17] presented one of the first models of galaxy formation in a cosmological context (see also [16,28]). The authors proposed a two-stage process in which galaxies form via radiative cooling of the baryons within halos that had already formed via gravitational collapse of the collisionless dark matter. The authors also argued that an additional non-gravitational process, called feedback, was needed to avoid the overproduction of faint galaxies as compared to observations. Indeed, in that model, the galaxy stellar mass function would simply be a scaled-down version of the dark matter halo mass function (see the discussion in, e.g., [26] illustrated by their Figure 1 and Figure 1 of this review). Later studies pointed out that the scaled halo mass function model overpredicts the abundance of the most massive galaxies when compared to observations (e.g., [26,29–35]). In short, galaxy formation is fundamentally an inefficient process as only a small fraction of the Universe's baryons are in the form of stars (about 10 per cent; e.g., [36–38]) and the star formation efficiency strongly depends on halo mass. Thus, the mass of the dark matter halo plays a fundamental role in shaping the galaxies that it contains. Observational evidence for this halo-mass dependency of the efficiency of galaxy formation was first obtained using galaxy group catalogues [39,40].

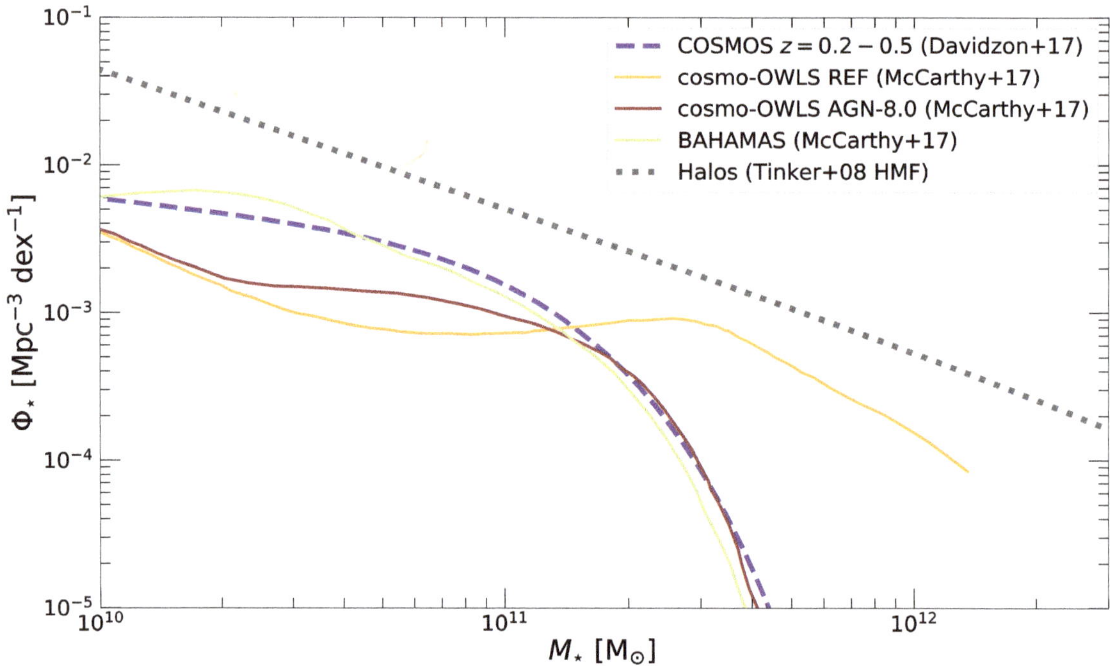

Figure 1. Galaxy stellar mass function (GSMF) $\Phi_\star(M_\star) = \frac{dN}{dM_\star dV}$ in observations and simulations. The dashed purple curve shows the double-Schechter fit to the local GSMF measured in the COSMOS survey [3]. The solid curves show the predictions of the cosmo-OWLS/BAHAMAS model [12,41] in the case with only stellar feedback (REF, yellow), cosmo-OWLS with AGN feedback (AGN-8.0, maroon), and the latest BAHAMAS model (green), in which the feedback model was tuned to reproduce jointly the GSMF and the gas fraction. For comparison, the dotted gray curve shows the Tinker et al. [42] halo mass function in *Planck* cosmology scaled by $f_b = \Omega_b/\Omega_m$, which highlights what one expects to see in the case in which each halo is populated by a galaxy with $M_\star = f_b M_h$.

To illustrate the first point, in Figure 1 we show the local galaxy stellar mass function (GSMF) $\Phi_\star(M_\star)$ that is measured within the COSMOS survey [3] modeled using a double Schechter function. The dashed line shows the Tinker et al. [42] halo mass function with the masses scaled by the universal baryon fraction $f_b = \Omega_b/\Omega_m$, which produces the GSMF one would expect to see if every halo was populated by a single galaxy and all the available baryon content had been converted into stars. The observed GSMF vastly differs from the scaled halo mass function both in shape and normalization. The lower normalization implies that the star formation efficiency is much less than 100%, whereas the steep decline at high masses shows that the growth of galaxies does not follow the structure formation process. Attempts at reproducing the shape of the GSMF with feedback from supernovae and star formation were unsuccessful, as the injected energy was insufficient to offset cooling and regulate the star formation efficiency [26]. In Figure 1, we compare the observed GSMF with the predictions of hydrodynamical cosmological simulations from the cosmo-OWLS and BAHAMAS suites [12,41] implementing several prescriptions for baryonic physics. The cosmo-OWLS run, including cooling, star formation, and stellar feedback (labelled REF) suffers from overcooling, and, thus, it overpredicts the observed abundance of massive galaxies ($M_\star \gtrsim 10^{12} M_\odot$) by several orders of magnitude. Conversely, including AGN feedback allows the model to closely reproduce the abundance of massive galaxies. AGN feedback is implemented by releasing a fraction of the rest-mass energy of cooling gas particles within the surrounding environment, which reheats the gas and regulates the star formation efficiency (see Section 5.1 for details). The effect of AGN feedback kicks in around $M_\star \sim 2$–$3 \times 10^{11} M_\odot$ corresponding to the stellar masses of BGGs, which highlights the crucial role that is played by the galaxy group regime.

Following the early works highlighting the discrepancy between the observed GSMF and the structure formation theory, it took over 30 years to pinpoint the most likely source of feedback. Early models of non-gravitational heating invoked a 'pre-heating' of the baryonic content before the epoch of formation of massive haloes, but did not specify the source of the energy injection (e.g., [31,43,44]). Because these models cannot predict the impact on the galaxy population, but only on the intra-group medium, they are not discussed any further here (but, see Section 3.1 for a detailed discussion). Models implementing gas cooling without feedback were able to reproduce the breakdown of self-similarity that was observed in X-ray selected samples, but this came at the price of greatly overpredicting the abundance of galaxies (e.g., [45–50] for a review). In the late 1990s and early 2000s, teams working with both semi-analytic models and hydrodynamical simulations, who had started to model the effects of cooling, star formation, and stellar feedback, came to the realization that supernova heating, while being a suitable explanation for the inefficiency of galaxy formation at the low-mass end, could neither solve the remaining problem with the high-mass end of the galaxy stellar mass function (if anything it made it worse; see e.g., Menci and Cavaliere [51]; Bower et al. [52]; Benson et al. [26]) nor reproduce the properties of the gas in massive galaxies, groups, and clusters (e.g., [53–55]). Indeed, the amount of energy that is injected by supernovae is insufficient to eject gas from the potential wells of groups and clusters, even if one assumes that the feedback is one hundred per cent efficient (e.g., [56–58]), and such efficiencies were in conflict with contemporary observations of galactic outflows (e.g., [59]). Heating from thermal conduction was also investigated (e.g., [26,60–62]) and eventually ruled out, as it required that the conduction coefficient should exceed the Spitzer rate expected for a fully ionized plasma. By contrast, the energy that is injected by the supermassive black holes at the center of galaxies is, in principle, sufficient to eject gas from the potential wells of groups and clusters (e.g., [18,26,63]). Additionally, the existence of a feedback loop between supermassive black holes and galaxy formation provides an attractive solution to explain the observed correlation between galaxy and black hole properties (see Sections 2.2 and 5.3 for details). Galaxy groups are the best astrophysical laboratories for studying the impact of various feedback mechanisms, since they have managed to retain enough hot gas to allow for a study of the impact of feedback on the IGrM, while, at the same time, representing a transitional regime for the stellar properties of galaxies. We will discuss this point in more detail in the remainder of this review.

2.2. Co-Evolution between Black Hole Mass and Galaxy Properties

Since the late 1990s, it has become clear that central SMBH co-evolves with the properties of their host galaxies. Thanks to the increasingly precise measurements of SMBH masses obtained through spatially resolved dynamical measurements of stars and gas at the BH's vicinity, it is now well established that the vast majority of galaxies host a central SMBH with a mass that correlates with the properties of its host galaxy (see [64] for an extensive review). The SMBH mass was found to correlate with a galaxy's near-infrared luminosity L_K, i.e., with the galaxy's integrated stellar content, and with the velocity dispersion σ_v of the stars in the bulge, which is often used as a proxy for the halo mass [64–70]. SMBH masses scale with galaxy properties as $M_{\rm BH} \propto \sigma_{\rm e}^{4.5}$ and $M_{\rm BH} \propto L_K^{1.1}$. While it is still unclear whether the relations between SMBH mass and galaxy properties are fundamental or derive from correlations between hidden variables, the existence of these scaling relations implies that the processes leading to the growth of BH are tightly linked with the evolution of the host halo. Widespread AGN feedback provides a natural explanation for the existence of SMBH scaling relations [7,18,71]. The radiative and mechanical energy output of AGN outbursts can, in principle, be sufficient to expel cold gas away from galaxy bulges, thereby, at the same time, regulating the AGN accretion rate and the star formation activity of the host galaxy. Feedback from AGN has the potential to directly link the properties of supermassive black holes and their host galaxies. The coupling of the energy released by the formation of the supermassive black hole to the surrounding forming galaxy should lead to a relationship that is close to the observed $M_{\rm BH} - \sigma_{\rm e}$ relation [63,72]. The generality of these arguments imply that this result may be reasonably independent of the specific details of the feedback model.

While the $M_{\rm BH} - L_K$ and $M_{\rm BH} - \sigma_{\rm e}$ relations are now well established, the typical intrinsic scatter of optical/stellar relations remains substantial (even in early-type galaxies), $\epsilon \sim 0.4$–0.5 dex, especially for stellar luminosities and masses (e.g., Saglia et al. [73]). Moreover, optical extraction radii are necessarily limited to galactic half-light radii $R_{\rm e}$ ($\sim$2–5 kpc), neglecting the key role of the host halo of the group. It has been suggested that SMBH properties are fundamentally linked with the mass of the host halo [74,75], and that the $M_{\rm BH} - L_K$ and $M_{\rm BH} - \sigma_v$ relations arise as a byproduct from the scaling relations between halo mass and optical galaxy properties. Several recent studies seem to confirm that the mass of the central SMBH is more tightly related to the temperature of the host gaseous halo, i.e., the global gravitational potential and hot-halo processes [76–78]. We discuss this point, in detail, in Section 5.3.

3. Observational Evidence

3.1. X-ray Observations

3.1.1. Feedback-Induced Hydrodynamical Features

The X-ray observatories in orbit for the past 20 years, *Chandra* and XMM-*Newton*, have revolutionized our understanding of the cores of relaxed galaxies, groups, and clusters, which show a highly peaked X-ray emission from a hot interstellar medium whose radiative cooling time is often less than 1 Gyr. Soon after the launch of XMM-*Newton* and *Chandra* it was realized that the gas in the central regions of nearby groups and clusters does not efficiently cool from the X-ray phase, condense, and flow toward the center, as expected from the original 'cooling flow' model [79]. Spectroscopic observations with *Chandra* and XMM-*Newton* have established that there is little evidence for emission from gas cooling below $\sim T_{vir}/3$ [80–82]. Precisely where the gas should be cooling most rapidly, it appears not to be cooling at all. This effect is known as the 'cooling flow problem' (e.g., [83]).

Therefore, a compensating heat source must resupply the radiative losses, and many possibilities have been proposed, including thermal conduction (e.g., [84]), energy released by mergers (e.g., [85,86]), or by supernovae (e.g., [87]); see Section 2.1. However, feedback from the central AGN was rapidly established as the most appealing solution to the problem. There is, in fact, clear observational evidence for AGN heating as the majority of brightest cluster galaxies of cool-core clusters and groups host a radio loud AGN (e.g., [88,89]) and, following the launch of *Chandra*, disturbances, such as shocks, ripples, and cavities, have been found in the central atmospheres of many clusters, groups, and elliptical galaxies (e.g., [90–98]). The cavities, which appear as X-ray surface brightness depressions, have been interpreted as bubbles of low-density relativistic plasma inflated by radio jets, displacing the thermal gas and causing PdV heating (e.g., [99]). Weak shocks that are associated with outbursts, long expected in models of jet-fed radio lobes [100], were also finally detected in deep *Chandra* observations, for example, in M87 [93], Hydra A [101], and MS 0735+7241 [102]. The energies available from the AGN were found to be not only comparable to those that are needed to stop gas from cooling, but the mean power of the outbursts was well correlated with the radiative losses from the IGrM [103].

Given the lower surface brightness of groups, in the early days of the *Chandra* era, the study of AGN feedback in these systems did not progress at the same pace as for more massive clusters [8]. However, the situation has improved in more recent times, with a number of studies addressing sizable samples of groups and characterizing the cavities in their IGrM [104–107]. Table A1 in Appendix A presents a list of groups with known cavities being detected using high-resolution X-ray observations. Deep *Chandra* X-ray data are now available for a number of galaxy groups (HCG 62, NGC 5044, NGC 5813). In particular, Randall et al. [14] presented the results of a very deep (650 ks) observation of NGC 5813, the longest such observation available to date. In Figure 2, we show the *Chandra* image and temperature map of NGC 5813, revealing an impressive number of feedback-induced features. Multiple pairs of X-ray cavities can be observed on both sides of the nucleus, indicating that the system has undergone several consecutive AGN outbursts inflating powerful expanding bubbles. Two pairs of concentric shock fronts were discovered perpendicular to the jet axis. The passage of the shock fronts reheats the IGrM, as evidenced by the higher temperatures that were measured in the post-shock regions (see the right-hand panel of Figure 2).

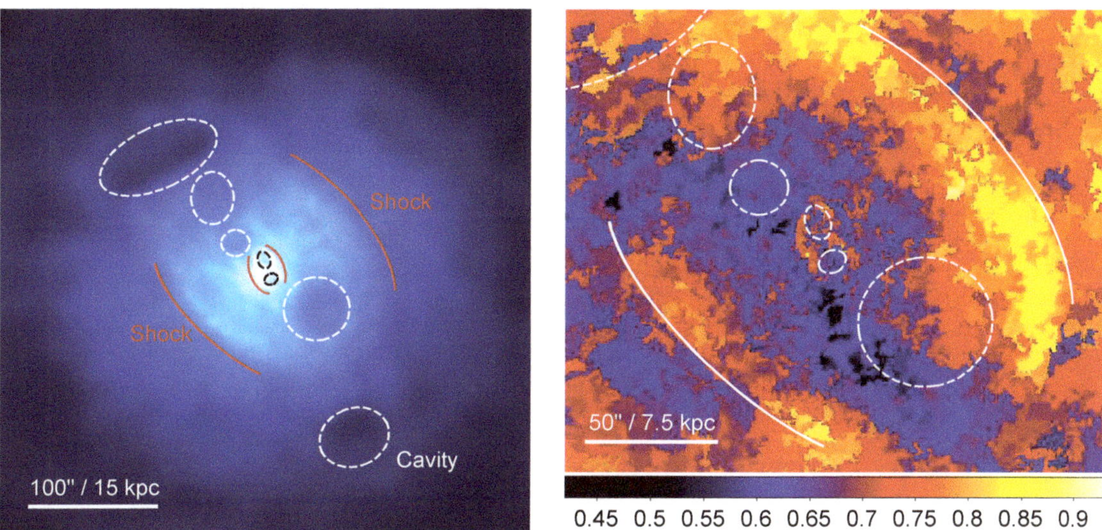

Figure 2. Cavities and shocks in the NGC 5813 galaxy group. The **left panel** shows an adaptively smoothed *Chandra* 0.5–2 keV image of the group, with cavities marked by dashed ellipses and two pairs of shock fronts by solid curved lines. The **right panel** shows a temperature map (in units of keV) with the cavities and outer shock fronts marked. Note the shock-heated gas (red and yellow) behind the outer shock fronts and in the shocked rims of the innermost set of cavities. Images drawn from the *Chandra* Early-type Galaxy Atlas [108].

Because cavities are the most commonly detected feedback-related structures, AGN energy input is usually gauged from their properties, as briefly summarized below. The energy that is required to inflate radio bubbles creating cavities in the X-ray emitting gas is usually expressed as the enthalpy, i.e., the sum of the work done to carve out the cavity and the internal energy of the radio lobes:

$$H = E_{\rm int} + pV = \frac{\gamma}{\gamma - 1} pV \tag{1}$$

where p is the pressure of the surrounding IGrM, V is the volume of the cavity, and γ is the ratio of the specific heats of the plasma filling the cavities. If the plasma is relativistic, $\gamma = 4/3$ and $H = 4\ pV$; if it is non-relativistic, $\gamma = 5/3$ and $H = 2.5\ pV$. The exact composition of the cavities is still unknown, even though X-ray, radio ([109] and references therein), and SZ observations [110] suggest that it is likely a mixture of the two species. The age of the cavity is the other key physical quantity that can be estimated from observations. As summarized by Bîrzan et al. [92], several age estimators have been suggested:

(a) the *sonic time*, i.e., the time that is required by the cavity to reach its projected distance R at the speed of sound,

$$t_s = R/c_s \tag{2}$$

with $c_s = (\gamma kT/\mu m_H)^{1/2}$, μ the mean atomic weight of the plasma, and m_H the proton mass;

(b) the *refill time* that is required by the gas to refill the displaced volume as the cavity rises upward,

$$t_{\rm ref} \approx \sqrt{r/g} \tag{3}$$

where r is the radius of the cavity and $g = GM(< R)/R^2$ is the gravitational acceleration at the cavity position;

(c) the *buoyancy time*, i.e., the time that iis required for the cavity to rise buoyantly at its terminal velocity,

$$t_{\rm buoy} = R/v_t \approx R\sqrt{SC/2gV} \tag{4}$$

where V and S are the volume and the cross-section of the cavity, respectively, and $C = 0.75$ is the drag coefficient [111]

Cavity ages that are estimated using these three methods usually agree within a factor of two, with the buoyancy times typically in between the shorter sonic time and the longer refill time ([10] and references therein). Dividing the enthalpy H by the characteristic timescale provides an observational estimate of the cavity power P_{cav}. The cavity power is a lower limit to the mechanical power of the AGN, given the paucity of detected shocks and other possible sources of energy feedback, such as sound waves.

Operationally, estimating P_{cav} requires a measurement of the geometry and size of the cavity and the pressure of the surrounding ICM. The total cavity power can be compared with the gas luminosity inside the cooling radius, $L_{\rm cool}$, which needs to be balanced by the AGN mechanical feedback. $L_{\rm cool}$ is usually defined as the total luminosity inside the regions where the cooling time is less than 7.7×10^9 yrs [112], although different thresholds exist in the literature (e.g., 3 Gyr, [113]). $L_{\rm cool}$ is estimated by deprojecting the X-ray temperature and emissivity profiles and computing the corresponding bolometric luminosity [96]. A number of possible biases and systematic errors can affect this apparently straightforward observational approach, as the detectability of cavities depends on the depth of the observation, the position of the cavity with respect to the plane of the sky, and uncertainties in the assumed geometry ([8,10,109] and the references therein). The exact impact of these observational uncertainties on the statistics of cavities in clusters and groups is yet to be quantified.

The $P_{cav} - L_{cool}$ relation has been investigated through the years in an increasing number of objects, ranging from ellipticals to groups and clusters ([10,109] and references therein). The general consensus is that the cavity power is enough to offset cooling given an average 4 pV injected energy per cavity and that the jet mechanical power correlates well with the cooling luminosity. In Figure 3, we show the relation between cooling luminosity and cavity power from a compilation of literature measurements [104,106,112]; see Table A1. Here, the total cavity power for each system was computed by summing up the power of each individual cavity. While at the high-mass end, the data are broadly consistent with an enthalpy $H = 4pV$, being typical of heating by a relativistic plasma, in the group regime, the cavity power is substantially higher. To quantify this effect, we fitted the $L_{\rm cool} - P_{\rm cav}$ relation with a power law using `PyMC3` [114]. The blue curve and shaded area show the best-fit relation, which reads

$$\log\left(\frac{P_{\rm cav}}{10^{43}\ {\rm erg/s}}\right) = (0.41 \pm 0.09) + (0.70 \pm 0.05)\log\left(\frac{L_{\rm cool}}{10^{43}\ {\rm erg/s}}\right) \tag{5}$$

with an intrinsic scatter of 0.51 ± 0.07 dex. The slope of the fitted relation is significantly shallower than unity, which would be the expected slope if the feedback efficiency is independent of the halo mass. At face value, this result implies that the feedback efficiency is higher in groups than in clusters, with the cavities injecting enough energy to overheat the cores and deplete the central regions from their gas content. We discuss this point, in detail, in Section 3.1.4.

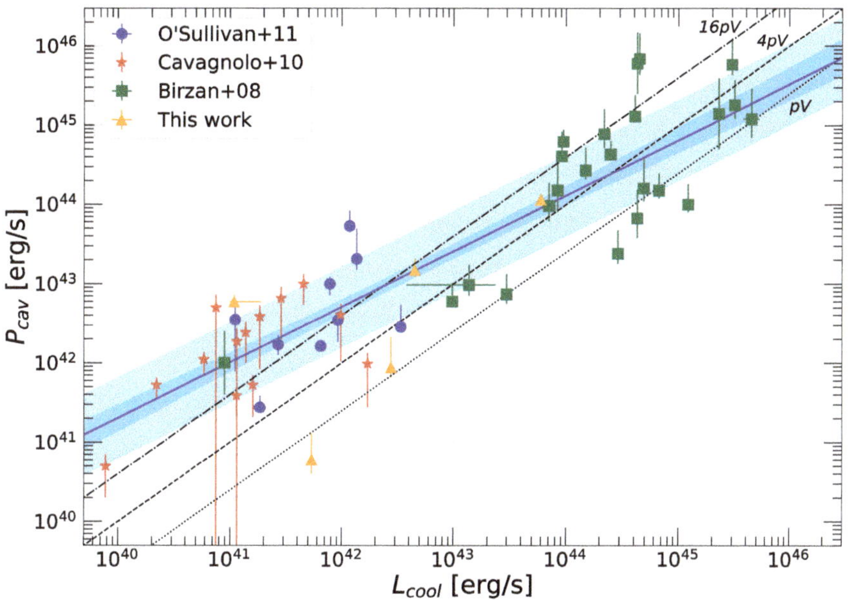

Figure 3. The relation between the luminosity within the cooling radius (L_{cool}) and the power injected by the cavities assuming $H = 4\ pV$ (P_{cav}) for several literature samples. The orange points were either recomputed for this work or collected from papers on individual objects; we refer to Table A1 for detailed references. The blue line shows a fit to the data using a power law with intrinsic scatter. The uncertainty on the fitted relation is indicated by the blue shaded area, whereas the cyan range indicates the intrinsic scatter around the relation.

The presence of shock fronts is another clear observational hydrodynamical feature that is caused by AGN feedback. The passage of a shock front compresses and heats the gas, raising its entropy and providing an effective heat input

$$\Delta Q \approx T\Delta S = T\Delta \ln K \tag{6}$$

with K the entropy index usually quoted as entropy by X-ray astronomers (see Section 3.1.2). Because of their transient nature, single weak shocks fail to compensate for the radiative losses in a cool core, but the cumulative effect of multiple shocks can be relatively important (see, e.g., the discussion in [109]). This is again highlighted by the exemplar case of NGC 5813 (see Figure 2), where each set of three cavities has been associated with an elliptical shock measured at 1 kpc, 10 kpc and 30 kpc, respectively, with Mach numbers $\mathcal{M}$ in the range 1.17–1.78. Generally speaking, the detected shock fronts are weak, i.e., their Mach number falls in the range $\mathcal{M}$ ∼1–2 [115]. This range can be understood given the typical evolution of the Mach number as a function of the fundamental parameters, total energy, and duration, of the AGN outbursts (see, for example, the discussion in [116] and Section 4.2). The cumulative heating effect of these successive shock fronts is sufficient to offset cooling within the inner 30 kpc. The shock energy can be estimated as

$$E_s = p_1 V_s (p_2/p_1 - 1) \tag{7}$$

where p_1 and p_2 are the pre- and post-shock pressures, respectively, and V_s is the volume that is enclosed by the shock (e.g., [14]). Liu et al. [115] made an exhaustive search for groups and clusters with detected shocks and studied the dependence of the shock energies and related Mach numbers on the cavity enthalpies (see Figure 4). The shock energies span almost seven orders of magnitude from 4×10^{61} erg s^{-1} in the cluster MS 0735 + 7421 [102] to 10^{55} erg s^{-1} in NGC 4552 [117], with group-scale objects in the range 10^{56}–10^{59} erg s^{-1} and Mach numbers all in the range 1–2 with the exception of Centaurus A. The shock energy is similar to the cavity energy (see the right-hand panel of Figure 4), suggesting that shocks and cavities may play a comparable role in supplying mechanical energy that is provided by the AGN (with the balance between the two mainly driven by the duration of the outburst, e.g., [116,118]). Other heating mechanisms that are discussed in Section 4.2, such as turbulent heating, have yet to be explored observationally at the group scale.

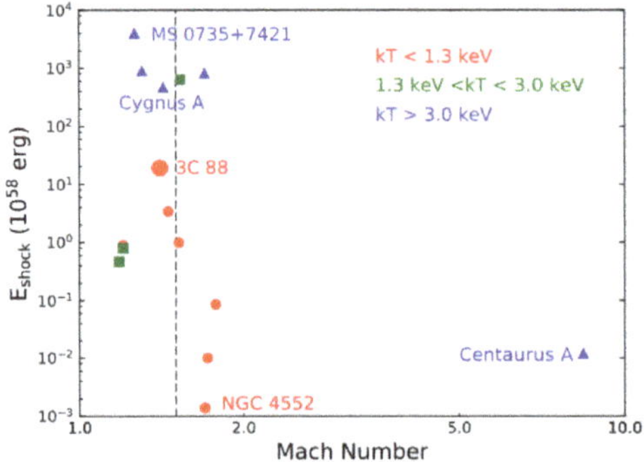

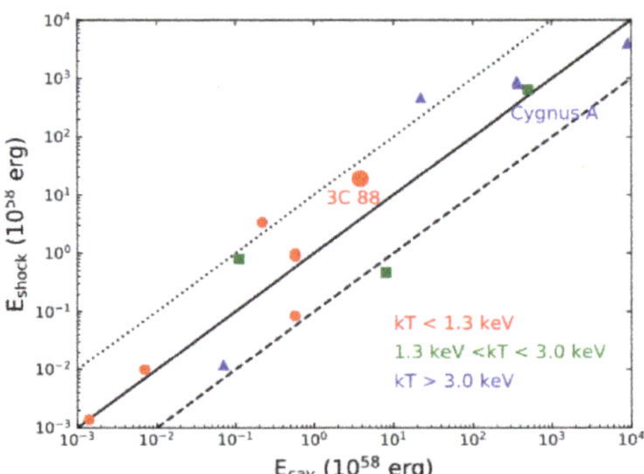

Figure 4. Shock energy versus Mach number (**left panel**) and shock energy versus cavity enthalpy (**right panel**) for groups and clusters with available shock energy in the literature. Figure reproduced from Liu et al. [115]. Objects with a red circle have $kT < 1.3$ keV, green square: $1.3 < kT < 3.0$ keV and blue triangles $kT > 3.0$ keV. The dashed line in the left panel marks $\mathcal{M} = 1.5$ and the dotted, solid, dashed lines in the right panel represent E_{shock}/E_{cav}=10, 1, 0.1 respectively. The objects considered in the plots are: Hydra A [101,119], MS 0735+7421 [102], Centaurus A [120], Cygnus A [92,121], 3C 444 [122], M87 [118], Abell 2052 [123], 3C 310 [124], NGC 4552 [117], NGC 4636 [125], HCG 62 [126], and NGC 5813 [14].

3.1.2. Non-Gravitational Feedback Energy and Entropy Profiles

One of the earliest pieces of evidence for non-gravitational feedback energy came from observed deviations from the self-similar scaling relations that are driven only by gravity [127], in particular the deviation of the observed luminosity-temperature relation from the predicted $L \propto T^2$ [128–130]; see the companion review by Lovisari et al. on the scaling relations of galaxy groups. Advocating a minimum entropy in the pre-collapse intergalactic medium to break self-similarity was one of the first attempts at a solution [43,44], with the result of bending the relation from self-similar at the scale of massive clusters to a steeper slope at the scale of groups. It was recognized that a given entropy level could be reached through different thermodynamic histories, and that the key insight would have been given by the sequence of adiabats through which baryons evolve: the excess entropy could have been achieved prior accretion to the collapsed halo (the external scenario) or in the higher density medium after accretion (internal scenario (e.g., [131,132]). A third option was initially considered while realizing that cooling alone could remove low entropy gas from the centers of halos producing a similar effect to non-gravitational heating (e.g., [46,133]).

Ponman et al. [134] provided the first observational evidence by measuring the entropy ($S = T/n_e^{2/3}$) at a fixed scaled radius (0.1 r_{virial}), and showing that it does not follow the expected linear trend with temperature (see the left-hand panel of Figure 5). *ROSAT* surface brightness profiles were combined with *Ginga* mean temperatures under the assumption of isothermality to derive entropy profiles for 25 objects, six objects with $T < 2$ keV

(NGC4261, NGC2300, NGC533, HCG97, HCG94, and HCG62), and objects, like MKW4, MKW3, AWM4, and AWM7, all in the range T = 2–4 keV. The authors concluded that the observational data were consistent with a scenario involving an external preheating mechanism through supernova winds, thereby raising the central entropy and enriching the medium in heavy elements. They also established that preheating would have a broader impact on the general picture of structure formation, as, for example, the level and timing of the heating required could not violate the constraints from the Lyman-α forest (see also [135]).

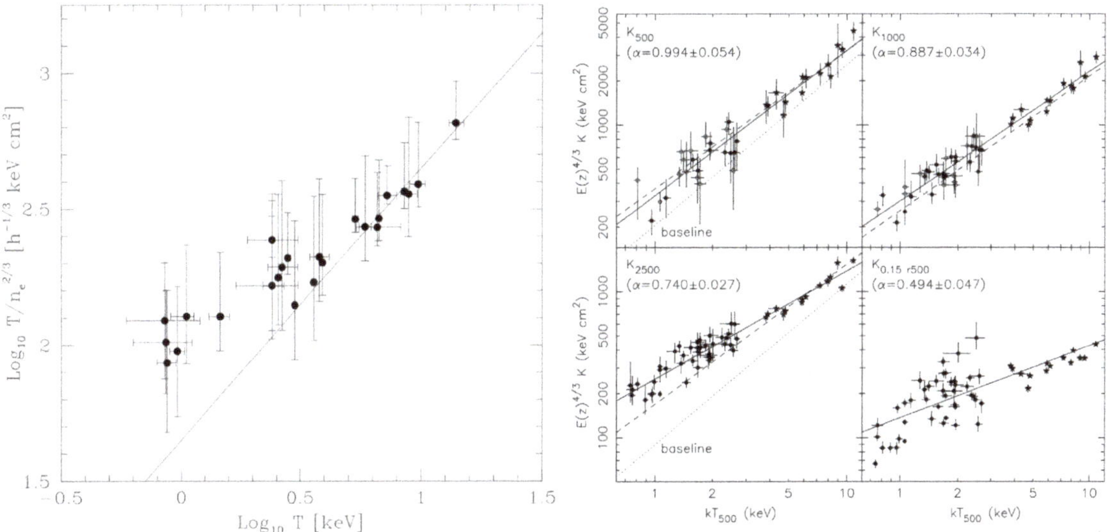

Figure 5. Left: The gas entropy at the fiducial radius of $0.1R_{virial}$ as a function of the temperature for the 25 systems in the sample of Ponman et al. [134]. The solid line shows the relation obtained from numerical non-radiative simulations [136]. Figure reproduced from Ponman et al. [134], arXiv author's version. **Right**: The same relation at different radii for the groups analyzed in the sample of Sun et al. [137], together with the sample of clusters in Vikhlinin et al. [138], both being analyzed with *Chandra* data. Figure reproduced from Sun et al. [137].

A key improvement with respect to this early result was the ability to go beyond the assumption of isothermality by constraining the temperature profiles exploiting the combination of *ROSAT* and *ASCA* data [139,140]. These studies confirmed that low-mass systems exhibit higher scaled entropy profiles. However, they did not show the large isentropic cores that were predicted by simple preheating models (e.g., [141]) and the high entropy excess in galaxy groups was found to extend to large radii (as also shown by the *ASCA* analysis of [142]).

The *Chandra* and XMM-*Newton* results have made the observational picture clearer and more solid. In the comprehensive work that was done by [137], an archival sample of 43 groups observed with *Chandra* with a temperature range of $kT_{500} = 0.7$–2.7 keV was analyzed, deriving detailed entropy profiles thanks to the superb spatial resolution of the satellite. The derived entropy-temperature scaling relations at six characteristic radii (30 kpc, 0.15 R_{500}, R_{2500}, R_{1500}, R_{1000}, and R_{500}) show a large intrinsic scatter at small radii, but already at R_{2500}, the scatter reaches a value of 10% and remains the same beyond this point. When combined with similar observations in the galaxy cluster regime (the sample of [138]), the slope of the relation is found to gradually approach the self similar value, steepening from 0.740 ± 0.027 at R_{2500} to 0.994 ± 0.054 at R_{500} (see the right-hand panel of Figure 5). The entropy ratios that were calculated with respect to the baseline entropy profile expected from purely gravitational processes [143] confirm an excess entropy, which is a function of mass and radius, with groups having higher ratios at small radii. The weighted mean ratio for groups decreases from 2.2 at R_{2500} to 1.6 at R_{500}. In general, the entropy profiles of groups have slopes (0.7–0.8) that are flatter than the self-similar expectation from pure gravitational processes of 1.1. Deep observations of nearby poor clusters with *Suzaku* (RX J1159, Humphrey et al. [144]; Virgo, Simionescu et al. [145]; UGC 03957, Thölken et al. [146]) have shown that the entropy excess can extend all the way to the system's virial radius. Similar observations on a larger sample of groups are needed to determine whether the high entropy of galaxy groups is a general feature that is linked to the AGN feedback phenomenon.

The XMM-*Newton* study that wsa performed by [147] on a sample of 29 groups based on the two-dimensional XMM-*Newton* Group Survey, 2DXGS [148,149] supplemented by groups from the sample of Mahdavi et al. [150] divided the objects into cool core (CC) and non-cool core (NCC) objects on the basis of the presence of a temperature gradient in the core, the first time this had been done for galaxy groups. The slope of the scaling relations of the entropy with temperature, incorporating the cluster sample of Sanderson et al. [151], at 0.1 R_{500} has a slope of 0.79 ± 0.06 consistent with the results of Sun et al. [137]. The entropy profiles of NCC groups show greater scatter than the CC sub-sample, and they have higher central entropies, in qualitative agreement with the results at the cluster scale (e.g., [152]). The excess entropy with respect to the baseline expected from gravitational processes cannot be reproduced by simple theoretical models of entropy modification, such as pure pre-heating or pure cooling, and the required mechanism should provide increasingly large entropy shifts for higher entropy gas.

Another significant advance in the study of entropy profiles of group-scale objects has been provided by the work of Panagoulia et al. [153], which analyzed the entropy profiles of 66 nearby groups and clusters drawn from a volume-limited sample of 101 objects that were assembled from the NORAS and REFLEX catalogues. The study pointed out that the flattening of the entropy profiles at small radii found in previous studies [152] was mainly a matter of resolution and it could be affected by the presence of multi-temperature gas. In particular, for nearby groups, a broken power law model provides the best description of the entropy profiles, with an inner slope of 0.64 within the central 20 kpc. This finding was confirmed in the recent studies of Hogan et al. [154] and Babyk et al. [155], who found that the behavior of the entropy profiles in the inner regions of relaxed clusters and groups can be well described by a broken power law with $K(r) \propto R^{2/3}$, a break around 100 kpc ($\sim 0.1 R_{2500}$) and an outer slope of ~ 1.1 matching the predictions of gravitational collapse models [143]. Because the cooling time is $\propto K^{3/2}$, the non-existence of an entropy floor affects the interpretation of the SMBH feeding and feedback processes, as we will discuss in Section 3.1.3 and Section 4.

3.1.3. Thermal Instability Timescale Profiles

In the past decade, a substantial amount of work has been dedicated to understanding the triggering of AGN feedback in galaxy groups and clusters. The energy injected by the central AGN and the cooling of the IGrM appear to be closely balanced over the long-term, as we will review in Section 4. This reflects a tight relation between cooling/feeding (Section 4.1) and heating/feedback (Section 4.2) processes. Initial works suggested that the onset of thermal instability in hot halos and the triggering of runaway cooling can be expressed in terms of the ratio of the cooling time to the free-fall time [156–160]. The cooling time is usually expressed as the ratio of the thermal energy to radiative cooling rate,

$$t_{\rm cool} = \frac{(3/2)nk_{\rm B}T}{n_{\rm e}n_{\rm i}\Lambda} \approx \frac{3k_{\rm B}T}{n_{\rm e}\Lambda} \quad (8)$$

with Λ the cooling function (see Figure 10) and $n_{\rm i} \approx n_{\rm e}$ the IGrM ion number density. The free-fall time describes the timescale that is necessary for a gas particle to directly fall to the bottom of the potential well,

$$t_{\rm ff} = [2R/g(R)]^{1/2} \tag{9}$$

with $g(R)$ the local gravitational acceleration. If $t_{\rm cool} \gg t_{\rm ff}$, the gas particle is losing internal energy too slowly and radial oscillations are eventually damped. Conversely, as $t_{\rm cool}$ becomes comparable to $t_{\rm ff}$, the gas rapidly loses pressure support, developing runaway thermal instability, and eventually sinking radially onto the central galaxy and SMBH. This 'classical' thermal instability (and related TI-ratio) has been also referred to as 'raining' (Gaspari et al. [156]) or 'precipitation' [159], being analogous to early physics studies that are based on analytical approximations (e.g., Field [161]).

The observational constraints on thermal instability can be investigated by studying the radial profiles of such IGrM timescales. In the absence of cooling and non-gravitational heating ($t_{\rm cool} \gg 1$ Gyr), the cooling time follows a baseline universal profile set by the structure formation process [143,162]. Deviations from the baseline profile occur in the central regions, where $t_{\rm cool} \ll 1/H_0$ and radiative cooling losses can no longer be neglected. The left-hand panel of Figure 6 shows the cooling time profiles for a large sample of galaxy clusters and groups from the ACCEPT database [159,163]. All of the systems exhibiting evidence of multi-phase gas (e.g., Hα, CO) show a floor set by the threshold $t_{\rm cool}/t_{\rm ff} \sim 10$–30, such that, on average, the IGrM does not experience runaway cooling. Figure 6 (right panel) includes 40 galaxy groups and massive early-type galaxies with deep *Chandra* observations [155], showing that the ratio of cooling time to free-fall time reaches a floor at $t_{\rm cool}/t_{\rm ff} \sim 10$. In the inner regions where the entropy rises approximately as $K \propto R^{2/3}$, $t_{\rm cool}/t_{\rm ff}$ is approximately constant with an average value of $t_{\rm cool}/t_{\rm ff} \sim 30$ (blue line).

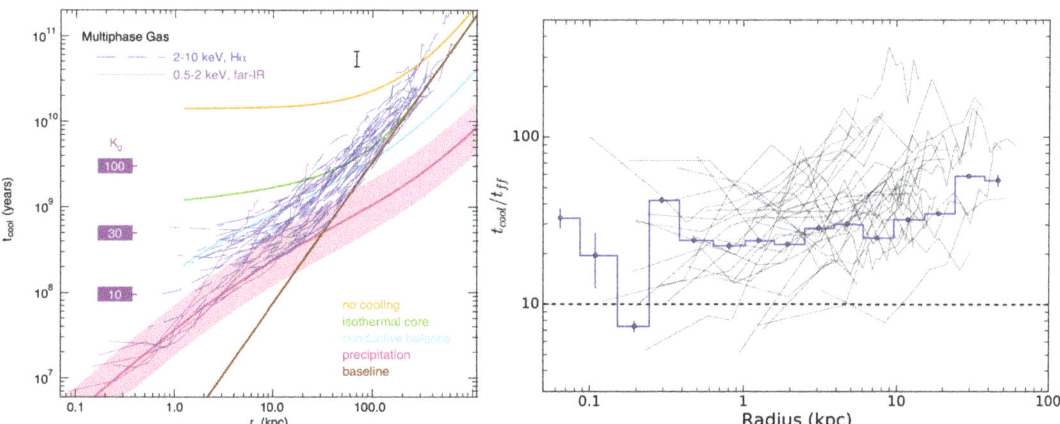

Figure 6. Observed timescales profiles in the IGrM (mainly related to TI). **Left**: Cooling time profiles for a sample of cool-core clusters (blue) and groups (magenta) hosting multiphase gas (Hα, CO; figure reproduced from Voit et al. [159], arXiv author's version). The observed profiles are compared with the 'baseline' structure formation profile (maroon) and the typical simulation threshold $t_{\rm cool}/t_{\rm ff} \sim 10$ (pink). **Right**: The ratio of the cooling time to free-fall time in a sample of groups and ellipticals (the figure reproduced from Babyk et al. [155]; the average is shown with a blue line).

On the one hand, the above results show that the TI-ratio is correlated with the presence of multiphase gas. On the other hand, it is clear that the threshold is puzzlingly not unity (as expected in classical thermal instability), and that there is a substantial intrinsic scatter, even in multiphase systems. However, classical thermal instability starts from idealized linear fluctuations and it does not account for key astrophysical processes, such as AGN feedback or mergers, both recurrently injecting substantial gas motions (e.g., turbulence) at small and large radii, respectively (e.g., Lau et al. [164]; see Section 4.2). In this more realistic IGrM case of 'turbulent' nonlinear TI, the key physical timescale is not the free-fall time, but rather the turbulence eddy turn-over time (Gaspari et al. [165]),

$$t_{\rm eddy} = \frac{2\pi R^{2/3} L^{1/3}}{\sigma_{v,L}}, \tag{10}$$

where σ_v is the turbulence velocity dispersion and L is the related injection scale. The velocity dispersion of the *ensemble* warm Hα-emitting gas should linearly correlate with σ_v of the hot IGrM, allowing for us to convert between the two, in particular by leveraging the higher spectral resolution of optical/IR telescopes. Rough estimates of the injection scale can be also obtained via the size of the ensemble warm gas filaments/nebulae, or via the AGN cavity diameter. In the presence of a turbulent halo, thermal instability develops chaotically and non-linearly in a very rapid way whenever $t_{\rm cool}/t_{\rm eddy} \sim 1$ [166–168]. Future X-ray microcalorimeter missions, like XRISM and Athena (see Section 6), will allow us to measure the eddy time directly and test, in more depth, the above scenarios. In Section 4, we discuss the related processes from a theoretical perspective.

3.1.4. Baryon Content

A key quantity for AGN feedback models in cosmological simulations is the total integrated baryon budget and its dependence on halo mass. While galaxy cluster halos are massive enough to retain all of their baryons (e.g., [169]), energy injection by AGN feedback can lead to an overall *depletion* of baryons all the way out to the virial radius. Observational studies have found that the gas fraction within R_{500} increases with halo mass [170–177]. Because an estimate of the gas density can be obtained from imaging data only, a lot of attention has been devoted to the study of gas density profiles [175,178,179]. Using a compilation of measurements from the literature, Sun [178] showed that, while the gas density of galaxy group cores is systematically lower than that of more massive systems, at R_{500}, the gas density is nearly independent of mass. Eckert et al. [175] studied the gas density profiles of the 100 brightest galaxy clusters and groups in the XMM-XXL survey. While, at galaxy clusters scales, the measured profiles show a well-defined core and a relatively steep decline in the outskirts, the density profiles of the selected groups exhibit a power-law behavior with a very flat index $n_e(r) \propto r^{-1.2}$, indicating that the gas fraction in spherical shells increases steeply with radius. It is believed that most of the gas has been evacuated from the inner regions under the influence of AGN feedback and displaced to larger radii, thereby explaining the observed shallow slopes [12,156].

In Figure 7, we present a compilation of published measurements of the hot gas fraction at R_{500} as a function of the corresponding halo mass. The gas fractions were derived from X-ray data under the assumption of hydrostatic equilibrium, with the exception of XMM-XXL and SPT-SZ. In the case of XMM-XXL [175], weak lensing measurements for a sub-sample of 35 clusters were used to calibrate the mass-temperature relation. SPT-SZ masses [180] were derived as an ensemble from a joint fit to the cosmological parameters and the relation between SZ observable and halo mass. All of the studies find that the hot gas fraction contained within R_{500} increases with halo mass roughly as $f_{\rm gas} \propto M_{500}^{0.2}$. Here, we provide a conservative estimate of the $f_{\rm gas} - M_{500}$ relation, which aims at encompassing all state-of-the-art observational studies and their uncertainties. To this aim, we collected the compilation of observational studies from Figure 7 and estimated in each mass bin the median and 90% confidence range of the data points. The resulting relation is shown as the gray band depicted in Figure 7. The gray band can be approximated as a power law, which reads

$$f_{\rm gas,500} = 0.079^{+0.026}_{-0.025} \times \left(\frac{M_{500}}{10^{14} M_\odot}\right)^{0.22^{+0.06}_{-0.04}}. \tag{11}$$

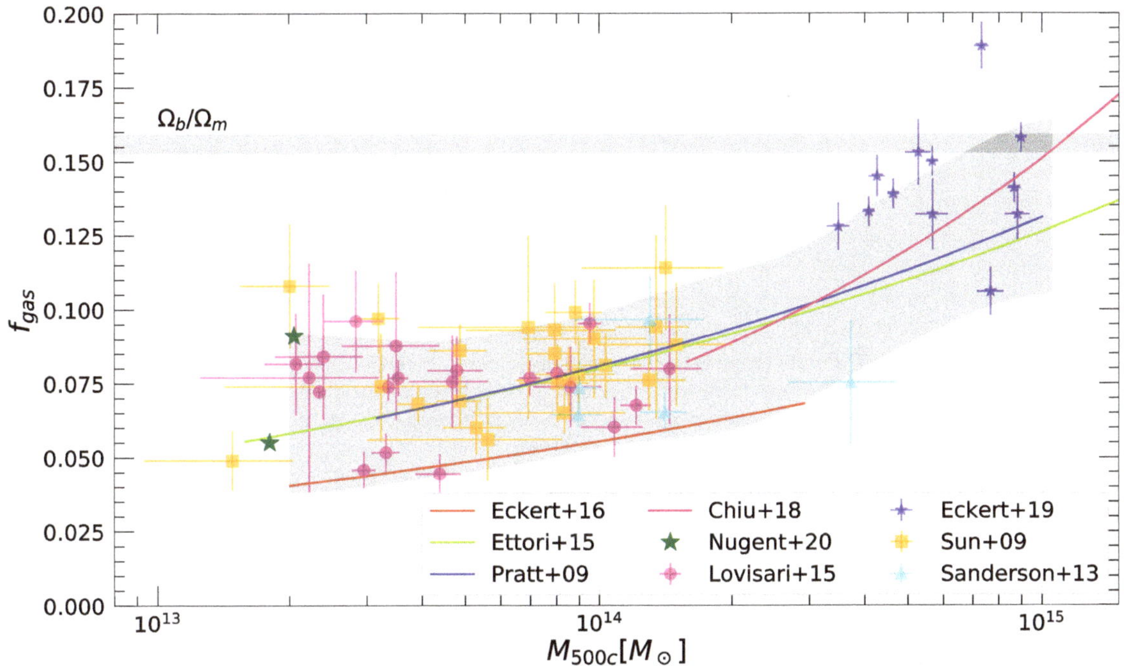

Figure 7. Compilation of existing measurements of the hot gas fraction at R_{500} in galaxy groups and clusters as a function of halo mass M_{500}. The data points show the galaxy group samples of Sun et al. [137] (orange), Lovisari et al. [174] (magenta), Sanderson et al. [181] (cyan), and Nugent et al. [177] (green). The data from the X-COP sample [169] at the high-mass end are shown as the blue points for comparison. The solid lines show the $f_{gas} - M$ relations that are derived from REXCESS (blue, [173]), XMM-XXL (red, [175]), SPT-SZ (magenta, [180]), and the literature sample of Ettori [176]. The gray shaded area shows the 90% confidence range encompassing the existing observational data and their corresponding uncertainties (see text).

While at the high-mass end, the gas fractions approach the cosmic baryon fraction, on galaxy group scales, the IGrM only contains about half of the baryons that are expected from the self-similar structure formation scenario. On the other hand, the stellar fraction $f_\star$ is a weak function of halo mass and decreases only slightly from 2–3% at $10^{13} M_\odot$ to 1–1.5% at $10^{15} M_\odot$ [5,6,172,175,180,182,183]. The weak dependence of the stellar fraction on halo mass is insufficient to compensate for the steeper dependence of the gas fraction, which results in a deficit of baryons in galaxy groups with respect to the cosmic baryon fraction. Here, we note that this result is independent of the hydrostatic equilibrium assumption adopted by most authors. Indeed, an additional non-thermal pressure term would lead to a slight underestimation of the mass in these studies (e.g., [184]), which, in turn, would result in the gas fraction being *overestimated* [169]. Thus, a high level of non-thermal pressure would render the lack of baryons in group-scale halos even more severe.

Here, we caution that the measurement of the gas fraction of group-scale halos is a difficult one and it is hampered by numerous systematic uncertainties. While halo mass estimates definitely represent the leading source of systematics, several other sources introduce potential systematic errors. In the temperature range of galaxy groups, line cooling renders the X-ray emissivity highly dependent on gas metallicity, which is difficult to measure away from group cores (see the review conducted by Gastaldello et al. within this issue). This can introduce uncertainties as large as 20% in the recovered gas mass [174]. Sample selection, usually based on *ROSAT* all-sky survey data, may bias the selected samples towards gas-rich systems, especially if the scatter at fixed mass is substantial [185,186].

Finally, most of the studies do not detect the X-ray emission all the way out to R_{500} (see e.g., Figure 8 of [137]) and must rely on extrapolation. For all of these reasons, the question of what is the exact baryon fraction of galaxy groups within R_{500} is still very much an open one, let alone within the virial radius.

3.2. Radio Observations

3.2.1. Interaction between Radio Sources and the IGrM

Radio surveys have made clear that the centers of galaxy groups and clusters are special locations for AGN (e.g., [89,187,188]), with group-central galaxies twice as likely to host radio-mode activity than non-central galaxies of equal mass out to $z > 1$. Deeper observations show that almost all the central galaxies of X-ray luminous groups host some radio emission [189,190], though in the local universe some of these may be contaminated by emission from low-level star formation [191]. The observations of nearby groups show a wide range of radio morphologies (e.g., [192]), with jet-mode feedback dominated by FR-I radio galaxies, as in clusters. Roughly one-third of X-ray luminous groups appear to host currently or recently active jet sources in their central galaxies [106] with typical jet powers in the range 10^{41}–10^{44} erg s^{-1} [190].

While cavities and shocks are the most accurate indicators of the impact of AGN feedback on the IGrM (see Section 3.1), current X-ray instruments have a limited ability to detect these features outside the high surface brightness cores of nearby groups. Radio studies offer an observationally cheaper way to measure feedback, particularly at higher redshifts. Radio galaxies are only periodically active and, once their AGN ceases to power them, their emission fades fastest at high frequencies. Therefore, low-frequency observations can be particularly effective at identifying older, dying radio sources, and measuring their full extent and luminosity (e.g., [193,194]). The radio spectrum can also give an indication of the properties of the source, most notably its age and the lobe pressure, which, for older sources, is usually in equilibrium with the surrounding IGrM. Combining radio and X-ray observations, we can observe multiple cycles of outbursts in individual groups, e.g., NGC 5813 and NGC 5044 (Figures 2 and 8, [14,195]). In particular, in Figure 8, we show the existing high-quality radio, X-ray, and Hα observations of NGC 5044 [195]. GMRT 235 MHz radio observations trace the oldest outburst via detached lobes and a bent, one-sided radio jet, while *Chandra* detects cavities on $\sim$5 kpc and $\sim$150 pc scales. Interestingly, the current radio jets, which are traced by high-resolution VLBA observations (bottom-left panel), are not aligned with the X-ray cavities, possibly indicating the precession of the jet axis with time.

The properties of group-central radio galaxies are closely linked to the IGrM. Both groups and clusters show a correlation between X-ray luminosity and the radio luminosity of the central source [189,196,197]. In clusters, central radio source luminosity is observed to be higher in systems with cooling times $<10^9$ yr [198] and, in groups, it appears that radio jets are more common in the central galaxies of groups with short central cooling times, low t_{cool}/t_{ff} ratios, and declining central temperature profiles [106]. However, perhaps the most important correlation is that between jet power, as determined from the enthalpy of AGN-inflated cavities, and radio luminosity. This P_{cav}-L_{radio} relation was first established for galaxy clusters by Bîrzan et al. [92,112] and later extended to early-type galaxies [104] and galaxy groups [199]. Although there is significant scatter in the relation, it offers a mechanism for determining the energy that is available from AGN feedback in the many systems where direct determination in the X-ray is impossible.

Applying the P_{cav}-L_{radio} relation to a large sample of SDSS groups and clusters with radio sources identified from the NVSS and FIRST surveys, Best et al. [89] showed that central radio galaxies dominate the heating of the IGrM within the cooling radius. They also found that the efficiency of AGN heating cannot be constant across the full mass range of groups and clusters; feedback must be less efficient in groups if they are not to be over-heated. Support for this result came from a study of groupsobserved in the COSMOS survey [200], which found that, factoring in the likely duty cycle of the AGN population, group-central radio galaxies can inject energies that are comparable to the binding energy of the IGrM. These results suggest that group-central AGN have the potential to drive gas out of the group core, and perhaps out of the group altogether, unless some mechanism reduces their effectiveness in heating the gas.

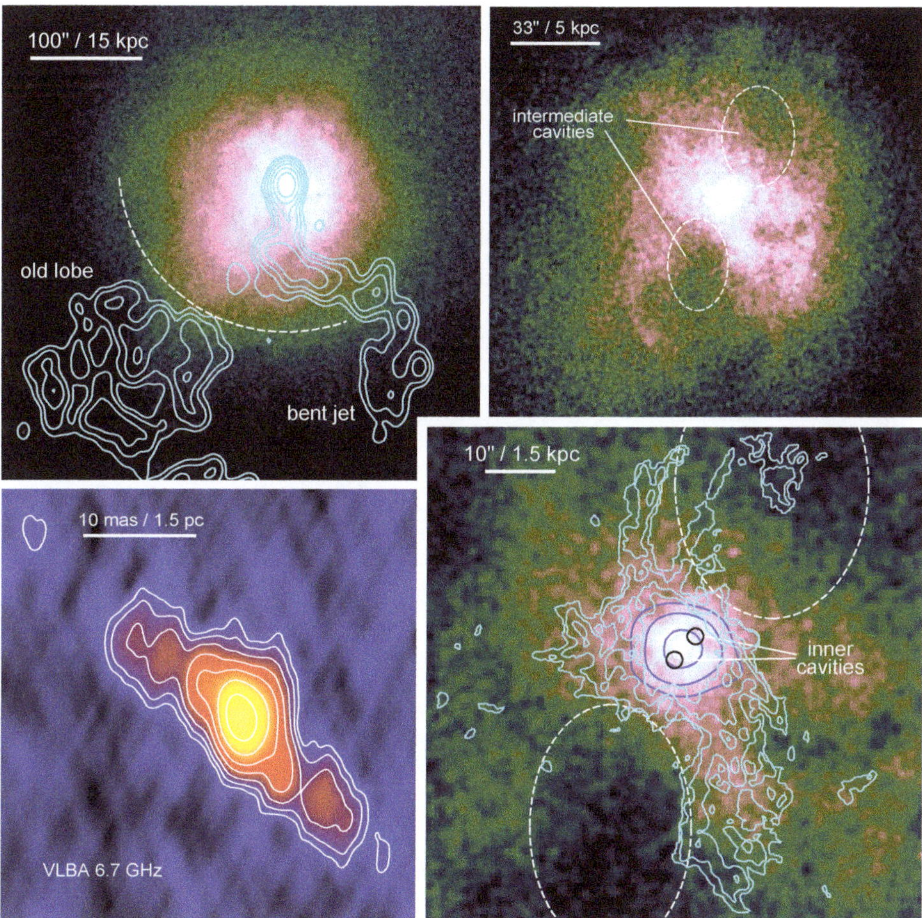

Figure 8. Multiple cycles of AGN feedback in the NGC 5044 galaxy group. The *upper left* panel shows the 0.5–2 keV *Chandra* image with GMRT 235 MHz contours overlaid. These reveal an old, bent radio jet and detached lobe from a prior AGN outburst, whose structure has been affected by the sloshing front marked with a dashed line. The *upper right* panel zooms in to show more detail of the complex of cavities and cool filaments in the group core. The *lower right* panel zooms in further, with contours showing the MUSE Hα (cyan) and ACA diffuse CO (blue) in the densest parts of the X-ray filaments and core. The *lower left* panel shows parsec-scale VLBA 6.7 GHz radio emission in the nucleus of the galaxy, evidence of a new cycle of AGN jet activity (adapted from [195]).

3.2.2. Giant Radio Galaxies

Because feedback studies at the group scale are largely limited to the X-ray bright cores of nearby groups where cavities are most easily identified, they have tended to focus on relatively small radio sources, with jet sizes of less than a few tens of kiloparsecs. However, the population of group-central radio galaxies includes much larger objects, some of which extend to very large radii, well beyond the cool core, and even into the outskirts of their groups. Pasini et al. [197] show that radio galaxies larger than 200 kpc are more common in groups than clusters, and the largest radio galaxies are located in groups, probably because the IGrM is less able to confine their growth than the ICM. There is also evidence from new radio surveys, with greater sensitivity to extended diffuse emission, that giant radio galaxies may be more common, in general, than previously believed [201].

Large radio sources pose a problem for AGN feedback models, in that they may inject a large fraction of their energy into the IGrM at large radii, rather than in the core, where it is needed to balance cooling. Even some medium sized sources appear to have jets that tunnel out of the cool core and inflate cavities outside it (e.g., NGC 4261 [202]). IC 4296 is an extreme example in which at least one cavity is confirmed, which hosts an FR-I radio galaxy whose 160 kpc diameter lobes extend out to a projected radius of $\sim$230 kpc [203]. This is far beyond the cool core (20–30 kpc radius) and about half of R_{500} for this $\sim$1 keV group. In such a system, while some of the energy that is involved in lobe inflation will likely have heated the core, the energy that is bound up in the relativistic particles and magnetic field of the radio lobes will likely be released at large radii, heating gas that is unlikely to contribute to fuelling the AGN. Other nearby examples of group-central giant FR-Is include NGC 315 and NGC 383 [192] and NGC 6251 [204]. As in clusters, group-central FR-II galaxies are uncommon, but not unknown (see, e.g., [196,205]). Their faster, more collimated jets likely provide feedback heating via shocks during expansion (as in, e.g., 3C 88, [115]), and lobe inflation will drive turbulence, but, as with the giant FR-Is, it is less clear how they affect the cooling region once they grow beyond it.

Therefore, giant radio galaxies pose a number of important questions for feedback models of groups. Do they provide feedback that can balance the rapid cooling in group cores, and if so how? The large sizes of these systems, particularly the FR-Is, implies that their jets have been active for very long periods. How do these sources stay active for so long?

3.3. *Multiwavelength Observations*

In the cool cores of galaxy clusters, many observations have shown evidence of material cooling from the hot atmosphere, in the form of highly multi-phase filamentary nebulae surrounding the central galaxy and containing gas and dust with temperatures ranging from $\sim 10^6$ K to a few $\times 10$ K. Some cool core galaxy groups show similar structures, although they are generally less luminous and are thus far less thoroughly explored. As of yet, few studies have specifically focused on BGGs, but samples of giant ellipticals provide a window on the group regime.

Hα emission from ionized gas with temperatures $\sim 10^4$ K may be the most accessible tracer of cooled material. Lakhchaura et al. [206] find that, in giant ellipticals, as in galaxy clusters (c.f. [207]), the presence of Hα emission is associated with high IGrM densities, short cooling times, low values of the thermal instability criterion, t_{cool}/t_{ff} (see Section 3.1.3), and disturbed X-ray morphologies, with the overlap between galaxies with and without detected Hα, suggesting that the transition between the two states can happen fairly easily. They also report a weak correlation between the mass of Hα-emitting gas and P_{cav}, as expected if the Hα traces cooling material, some of which will eventually fuel the AGN. Some of the best known X-ray bright groups contain examples of Hα filaments that are similar to those seen in clusters (e.g., [14,208–210]). As in clusters, the filaments are closely correlated with feedback-related structures, showing signs of having been drawn out behind, or wrapping around, radio lobes and cavities. In some cases they are located in cool X-ray filaments that show signs of being thermally unstable [209]. Figure 8 shows an example of this in the NGC 5044 group, where the Hα nebula is correlated with the brightest cool X-ray emission and it appears to wrap around the base of the intermediate-scale cavities. Spatially resolved spectroscopy shows that, while the inner parts of these Hα nebulae are generally cospatial with the stellar bodies of the BGGs, they do not rotate with the stars, supporting formation from the IGrM rather than stellar mass loss (e.g., [211–213]). It should be noted that, while BGGs do host some star formation (SF), their Hα nebulae are not tracing SF. McDonald et al. [214] studied the relation between the star formation rate inferred from infrared data and the X-ray cooling luminosity (Section 3.1) and found that the inferred star formation rates in BGGs are typically quenched by a factor 10–100 as compared to the pure cooling scenario.

Molecular gas in groups has been observed via multiple tracers. *Herschel* observations revealed [CII] emission from $\sim$100 K gas with a similar distribution to the Hα, and [CII]/Hα flux ratios indicating that both phases are powered by the same source [208]. *Spitzer* IRS spectra show rotational H_2 lines in the BGGs of some X-ray bright groups [215], tracing gas at a few $\times$100 K, and CN has been detected in absorption in a handful of cases via the millimeter-wave band [216]. The forthcoming *James Webb Space Telescope* will open an important observation window on H_2, which is likely the dominant mass component of the molecular phase. However, at present, emission from CO is our best tracer for this phase, allowing for us to examine the coolest, densest gas in the cooling regions of groups.

Babyk et al. [217] examine CO in a large sample of local ellipticals (many of which are BGGs) and find that the molecular gas mass M_{mol} is correlated with the density of the IGrM and its mass in the central 10 kpc, and that systems with $t_{cool} < 1$ Gyr at 10 kpc are more likely to contain molecular gas. They also find that M_{mol} is proportional to P_{cav}, confirming that the molecular gas is the fuel source for the central AGN. However, cooling from a surrounding hot halo is not the only source of gas for ellipticals. Davis et al. [218] use a combination of the ATLAS3D and MASSIVE samples to show that gas-rich mergers are an important source of molecular gas in these galaxies. The observations of smaller samples of BGGs find some of the same trends, and show that BGGs of X-ray bright, cool core groups are not the CO-richest systems [219,220]. BGGs of X-ray fainter groups can contain more CO (and HI), and it is more often located in disks, rather than filaments. This, again, emphasizes the importance of gas-rich mergers in groups, although IGrM cooling is likely still the more important process in the cool core groups in which AGN feedback is most often observed.

The BGGs of X-ray bright cool core groups generally seem to contain only a few $\times 10^6$ or $\times 10^7$ $M_\odot$ of molecular gas [220], which makes them challenging targets, even for ALMA. However, being nearer than typical cool core clusters, groups offer an opportunity to study individual molecular cloud associations within the cool core, rather than the overall filamentary structures. Three well-known systems have been studied in detail by ALMA, NGC 4636, NGC 5846 [221], and NGC 5044 [209,222]. The velocity dispersions of the molecular clumps that were observed in these systems suggest that they are not gravitationally bound, and they are likely collections of smaller, denser clouds, with more diffuse gas between them. Atacama Compact Array (ACA) observations of NGC 5044 show that a significant fraction of the molecular gas in the BGG is more diffuse than the clumps observed by ALMA [213], and a similar argument can be made for the other two groups by comparing the CO masses that are derived from ALMA and IRAM 30m observations. The denser CO clumps are generally located within filamentary structures visible at other wavelengths [221], and the extent and velocity distribution of the diffuse CO in NGC 5044 is similar to that of the Hα and [CII] emission, supporting the idea that all of the observed phases are material cooling from the IGrM. Figure 8 shows the diffuse CO emission in the group core, collocated with the peak of the Hα and X-ray emission. As with Hα, the CO in these ALMA-observed systems is cospatial with the stellar component, but shows little sign of rotation or velocity gradients, consistent with formation from the IGrM.

Intriguingly, ALMA studies of giant radio galaxies, some of them group-central systems, show a different CO morphology, with the molecular gas being located in compact disks [223,224]. The difference in cold gas morphology may indicate a difference in the fuelling of the AGN. There are examples of group-dominant giant radio galaxies that appear to be fed by cold gas (e.g., NGC 1167, [225]) and, given the importance of galaxy interactions in groups, the potential for fuelling by gas rich mergers cannot be ignored. With only a handful of group-dominant galaxies mapped thus far in molecular gas, there is a significant opportunity for the exploration of the mechanics of AGN fuelling in these important systems.

4. Theoretical Framework

Hot halos are a fascinating and crucial element of virialized systems in the Universe, which have been unveiled to be a fundamental engine for the growth and triggering of SMBH, despite the large difference in spatial and temporal scales, which span over nine orders of magnitude (commonly sub-divided in three major scales; see Figure 9: micro—meso—macro). Here, we review the AGN feedback process in terms of fundamental physics and why it is expected in the more theoretical framework of accretion out of the hot halo and onto SMBH, with a keen eye on galaxy groups. It is important to appreciate that AGN feedback is only half of the self-regulated cycle, which is bootstrapped via the AGN feeding, the key complementary mechanism on which we will also focus below, as shown in the summary diagram of Figure 9.

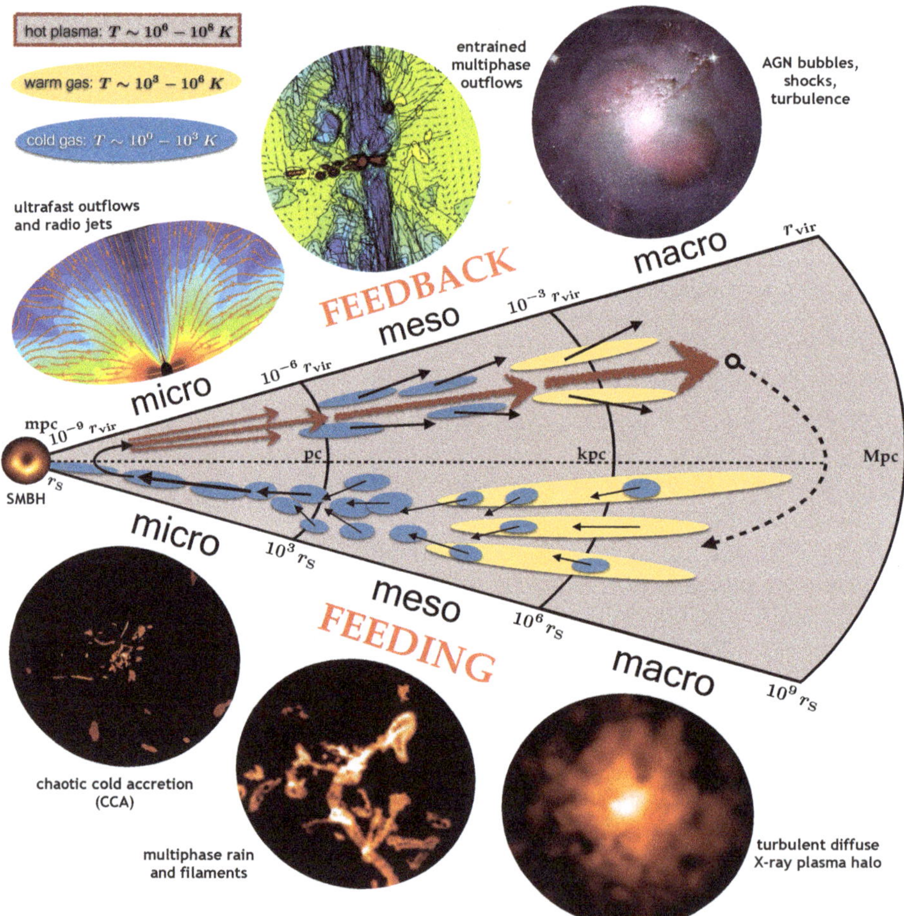

Figure 9. A schematic representation of the self-regulation loop necessary to fully link feeding and feedback processes over nine orders of magnitude in space (and time) and over the multiphase/multiband cascade, including X-rays (hot halo/outflows), the IR/optical (warm filaments), and radio/mm (molecular clouds) bands—reproduced from Gaspari et al. [11] (arXiv authors' version). In particular, the IGrM experiences relatively stronger top-down condensation rain compared with clusters, due to the lower central cooling times. Tightly related to such an enhanced feeding is the higher frequency of the AGN outflow/jet feedback events, which gently self-regulate each galaxy group for several billion years during the cosmic evolution.

While complex, non-linear thermo-hydrodynamical (THD) mechanisms are at play over the macro (kpc-Mpc), meso (pc-kpc), and micro scales (mpc-pc)—thus requiring expensive numerical 3D Eulerian simulations—it is useful here to understand the whole SMBH-halo system as a unified, co-evolving engine. In essence, such a global THD system can be described via the simple conservation of energy, or analogously via the (Lagrangian) entropy equation (Peterson and Fabian [83], Gaspari [226]):

$$U\,\frac{d}{dt}\ln K = \mathcal{H} - \mathcal{L}, \tag{12}$$

where $K = k_{\rm b}T/n^{\gamma-1}$ is the astrophysical entropy (with $n = n_{\rm e} + n_{\rm i} \approx 2\,n_{\rm e}$ the sum of the electron and ion number densities), $U = P/(\gamma-1)$ is the internal/thermal gas energy per unit volume ($\gamma = 5/3$ is the IGrM adiabatic index), $\mathcal{H}$ and $\mathcal{L}$ are the gas heating and cooling rates per unit volume (erg $\rm s^{-1}\,cm^{-3}$), respectively. A few immediate insights from Equation (12): the internal energy acts as an effective normalization knob (the larger the X-ray temperature times density, the stronger the required heating/AGN feedback, in absolute erg $\rm s^{-1}$ values); secondly, the macro entropy evolution is the sole result of the competition of heating and cooling processes, which translates in the competition between AGN feedback and feeding. Let us first discuss the (astro)physics and consequences of the cooling/feeding component of the cycle, i.e., $\mathcal{L}$.

4.1. AGN Feeding & Cooling Processes

The cooling process is very well understood from basic quantum physics and laboratory plasma/ionized gas experiments, with a radiative cooling loss [227] $\mathcal{L} = n_{\rm e}n_{\rm i}\,\Lambda(T,Z)$, where Λ is a cooling function varying with gas temperature and metallicity Z (∼0.6–1 $Z_\odot$ for group cores; Mernier et al. [228], cf. the companion Gastaldello et al. review). The hot IGrM experiences a significantly enhanced Λ due to the influence of line cooling (mostly recombination) taking over from the Bremsstrahlung/free-free emission ($\Lambda \propto T^{1/2}$), which instead shapes the more massive galaxy clusters ($T > 2$ keV), as shown in Figure 10.

Figure 10 depicts the three main (quasi)stable phases that arise during the top-down condensation cascade [166], especially during the feeding dominated stage of the AGN cycle. Assuming relatively slow motions over the group macro gravitational potential (quasi pressure equilibrium), Equation (12) can be approximated as (e.g., Pope [229])

$$H - L \approx \frac{c_{\rm s}^2}{\gamma-1}\,\dot{M}_{net}, \tag{13}$$

where $c_{\rm s}^2 = \gamma k_{\rm b}T/\mu m_{\rm p}$ is the (squared) IGrM sound speed (with $\mu \approx 0.62$ the mean atomic weight of the IGrM) and H/L is the gas heating/cooling power (or luminosity in erg $\rm s^{-1}$). Notably, any net cooling or heating will induce a net mass inflow ($\dot{M}_{\rm cool}$) or outflow rate ($\dot{M}_{\rm OUT}$) in the macro-scale halo, being denoted as $\dot{M}_{net}$; here, uppercase subscripts denote macro properties, while lowercase subscripts denote micro properties. Therefore, the thermal evolution of the IGrM is deeply intertwined with feeding/feedback processes, as intuitively anticipated above.

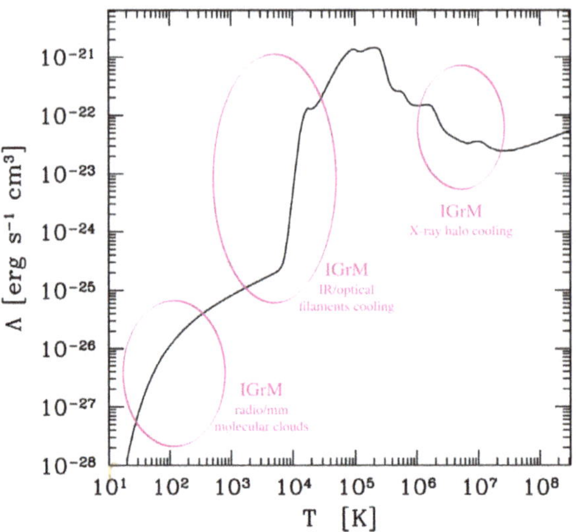

Figure 10. Multiwavelength cooling function for the IGrM (adapted from Gaspari et al. [166], unifying the atomic/plasma physics studies by Sutherland and Dopita [227], Dalgarno and McCray [230], Inoue and Inutsuka [231]; $Z = 1\,Z_\odot$), which was specifically used for a typical galaxy group akin to NGC 5044. Above the neutral hydrogen recombination ($T \sim 10^4$ K), the gas is fully ionized and in collisional ionization equilibrium; below this threshold, the gas becomes progressively less ionized ($\lesssim$1%), leading to the formation of neutral filaments and, subsequently, dense molecular clouds. The three magenta ellipses highlight the three key (semi)stable phases of the condensing IGrM, in particular during the feeding dominated part of the AGN cycle that is shown in Figure 9 (bottom insets).

In the absence of any heating, the entire hot atmosphere would rapidly condense and collapse, initiating from the inner denser radial regions (see Section 3.1). As noted in the previous sections, such massive cooling flows are not observed in our Universe, especially in galaxy groups. On the other extreme of (idealized) feeding models, the IGrM halo might experience pure cooling while having significant angular momentum. In this regime, the gas would condense through helical paths onto the equatorial plane, and there form a thin rotating multiphase disk [166]. While extended disks have been found in some BGGs (Hamer et al. [232], Juráňová et al. [233], Ruffa et al. [223]), such a scenario would induce both large (unobserved) cooling rates in X-ray spectra, as well as drastically reduced accretion rates onto the SMBH due to the preservation of high angular momentum and related centrifugal barrier.

Realistic IGrM atmospheres, instead, often reside in an intermediate THD regime, neither strongly rotating nor in a spherical cooling flow (David et al. [209], Lakhchaura et al. [206], O'Sullivan et al. [106], Temi et al. [221]—Section 3.3). Indeed, hot halos experience significant amount of turbulence, with an irreducible level of 3D turbulent velocity dispersion $\sigma_v \approx 100$–$300\,\mathrm{km\,s^{-1}}$, due to both the previous AGN feedback outbursts and the secular cosmological flows (e.g., Vazza et al. [234], Valentini and Brighenti [235]), as shown by high-resolution HD/cosmological simulations (Lau et al. [164], Gaspari et al. [236], Hillel and Soker [237], Weinberger et al. [238], Wittor and Gaspari [239]) and X-ray spectroscopy (Sanders and Fabian [240], Ogorzalek et al. [241], Hitomi Collaboration et al. [242]). While we review the kinematical features in a companion review (Gastaldello et al.), here we focus on its thermodynamical impact, namely the formation of *chaotic cold accretion* (CCA) and related multiphase rain, a key process driving the bulk of AGN feeding and, hence, the recurrent AGN feedback triggering. In a turbulent hot halo, chaotic multiscale eddies drive local perturbations in relative gas density proportionally to the turbulence sonic Mach number

($\delta n/n \propto \mathcal{M}_t$; Gaspari and Churazov [243], Zhuravleva et al. [244]). The relative increase in IGrM density produces in-situ enhanced radiative cooling ($\mathcal{L} \propto n^2$), thus leading to turbulent non-linear thermal instability (TI; Gaspari et al. [245], Voit [246]). It is important to note that such a chaotic instability is different from classical TI (Field [161], McCourt et al. [157], Pizzolato and Soker [247]), in the sense that direct non-linear fluctuations are seeded by chaotic motions, rather than growing from tiny linear amplitudes. This triggers a quick top-down condensation of localized (soft X-ray) patches to the first quasi-stable phase at $T \sim 10^4$ K, which is best traced via ionized line-emitting (e.g., Hα+[NII]; Gastaldello et al. [13], McDonald et al. [248], Werner et al. [208]) filaments or nebulae observed in optical/UV (e.g., see the synthetic image in the bottom middle inset of Figure 9). Sustained turbulent perturbations lead to the further condensation cascade onto the last stable and compact gas phase, molecular gas clouds (bottom left inset of Figure 9; see Section 3.3). Such cold clouds will then strongly and frequently collide inelastically within the meso/micro scale, cancelling angular momentum and, thus, feeding the central SMBH (hence, the 'CCA' nomenclature; Gaspari et al. [166]), with the consequent trigger of the next stage of AGN feedback (Section 4.2). CCA feeding recurrently boosts the accretion rates over 100 fold over the feeble and quiescent hot-mode (Bondi [249], Narayan and Fabian [250]) accretion, thereby overcoming the inefficiency of classical hot mode accretion.

On the macro scale, the hot halo can be assessed to reside or predicted to soon enter the CCA raining phase, whenever the ratio of the plasma cooling time and the turbulence eddy gyration/turnover time reaches unity. This reference dimensionless number is called C-ratio (from condensation or CCA; Gaspari et al. [165], Olivares et al. [167]),

$$C \equiv \frac{t_{\rm cool}}{t_{\rm eddy}} \sim 1, \tag{14}$$

where the cooling and turbulence timescales have been defined in Equations (8) and (10). A correlated ratio and thermal-instability threshold is the TI-ratio $\equiv t_{\rm cool}/t_{\rm ff} \lesssim 10$–$30$ (e.g., Gaspari et al. [156], Sharma et al. [158], Voit et al. [251]), where the free-fall time is defined in Equation (9). As introduced in Section 3.1.3, all three IGrM timescales can be constrained from X-ray or optical/IR datasets. While both the C-ratios and TI-ratios are valuable complementary tools, the simulations show that the C-ratio is the more direct physical criterion to apply to probe the onset and extent of *nonlinear* thermal instability (e.g., Gaspari et al. [165]; see also Figure 17). Indeed, unlike in classical linear TI, turbulence acts as an irreducible background of fluctuations over the whole IGrM (e.g., Lau et al. [164]). In particular, AGN bubbles are a key recurrent mechanism for inducing such fluctuations in the IGrM (e.g., McNamara et al. [252], Voit [253]). In this regard, it is not surprising that $t_{\rm cool}/t_{\rm ff}$ profiles show a large non-trivial deviation above unity, as well as a large intrinsic scatter (e.g., Singh et al. [254]; also see Figure 6). Figure 11 shows the average cooling and turbulence eddy time profiles (with scatter) for the galaxy group regime (Gaspari et al. [165]). Evidently, the crossing of $C \sim 1$ matches the dotted circle denoting the typical size of the condensed extended multiphase nebulae well (e.g., McDonald et al. [248]).

Together with mild C-ratios, a CCA-driven atmosphere—often found in the IGrM cores—is described by a low turbulent Taylor number (Juráňová et al. [168], Gaspari et al. [255]):

$${\rm Ta_t} \equiv \frac{v_{\rm rot}}{\sigma_v} \lesssim 1. \tag{15}$$

Given that the dominant galaxies of hot gas rich galaxy groups tend to be early-type (as do those galaxies that host their own extended hot halos), they have a fairly weak coherent gas rotational velocity $v_{\rm rot}$ (e.g., Caon et al. [256], Diehl and Statler [257]), unlike lower-mass/spiral galaxies. Therefore, a median IGrM long-term evolution is to oscillate between stages of strong CCA rain (${\rm Ta_t} \sim 0.3$–1, $C \sim 0.5$–1) and mild rain superposed on a clumpy disk (${\rm Ta_t} \sim 1$–3, $C \sim 1$–2). Evidently, extremes of strong rotation (${\rm Ta_t} \gg 1$) or overheated quiescence ($C \gg 1$) can lead to periods of disk- or Bondi-driven accretion, both experiencing highly suppressed inflow rates and feedback (albeit such periods must be short-lived to avert the cooling flow catastrophe). High-resolution HD simulations have shown that the chaotic behaviour of a CCA-driven halo is imprinted not only in the thermodynamical maps and kinematical properties, but also in the time-series spectra. Unlike quiescent and continuous hot modes (Bondi/ADAF), CCA drives a characteristic flicker/'pink' noise power spectrum (logarithmic slope of -1) in the Fourier space of frequencies f (Gaspari et al. [166]), thus generating strong self-similar variability on all temporal scales (the integral over df of this spectrum yields constant variance), from several Myr down to years and minutes, as ubiquitously observed in multiwavelength AGN lightcurves (Ulrich et al. [258], Peterson [259]). This triggers the second key part of the self-regulated loop, AGN feedback, the focus of the next section.

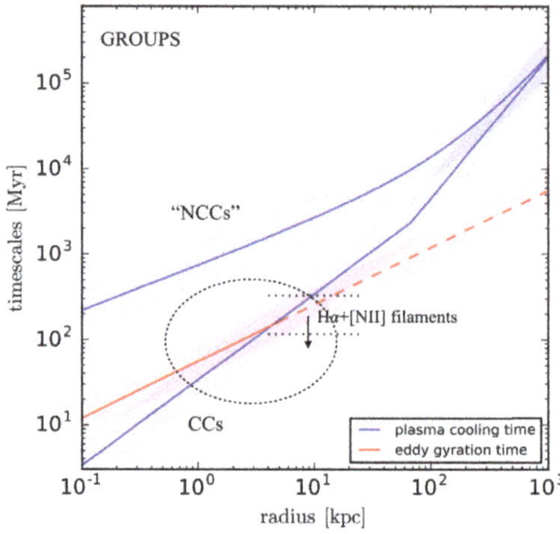

Figure 11. The average cooling time and turbulence eddy time profiles in the IGrM (with 90% confidence scatter bands), the latter constrained mainly via optical/IR telescopes (figure reproduced from Gaspari et al. [165]; group sub-sample). The dotted circle marks the size of the condensed warm nebular emission, which matches the $C \equiv t_{\rm cool}/t_{\rm eddy} \sim 1$ turbulent TI threshold.

4.2. AGN Feedback & Heating Processes

While pure cooling flows and catastrophic condensation are not detected in IGrM observations, strong overheating is equally ruled out, as virtually all galaxy groups exhibit central cooling times well below 1 Gyr due to the high efficiency of radiative cooling in their characteristic temperature range (see Figure 10). We note that massive galaxy clusters, instead, show a dominant population of non-cool-core systems with central cooling times above the Hubble time (e.g., [260,261]). The system would be in perfect thermal equilibrium in the absence of any heating and cooling terms, as inferred from Equation (12). However, an analogous configuration can be achieved if a heating process (macro AGN feedback) balances the cooling rate (macro condensation). This configuration is the more realistic state of observed galaxy groups, with the characteristic feature that the self-regulation process is intrinsically chaotic (from the macro down to micro scales; Section 4.1), hence only leading to a *statistical* thermal balance $\mathcal{H} \sim \langle \mathcal{L} \rangle$. Moreover, while Equation (12) formally allows for the entropy to decrease (pure cooling), fundamental THD physics dictates that entropy shall always increase in real systems (even over the ensemble Universe). With such intuition, we can already expect that the heating rate is as essential—if not eventually more vigorous—than the cooling component (Figure 9).

Before tackling the AGN feedback physical sub-processes, here we first discuss the key difference between groups and clusters. In Figure 12, we show an analysis of the potential impact of the stored SMBH energy versus the gravitational binding energy of the hot halo cores (from small groups to clusters), by leveraging the large sample of Gaspari et al. [77]

with both direct BH masses and extended hot halos detected. The potentially available SMBH mechanical energy is $E_{\rm BH} = \varepsilon_{\rm M} M_{\rm BH} c^2$, where $\varepsilon_{\rm M}$ is the macro mechanical efficiency (Gaspari and Sądowski [262]). We test for the now commonly used fixed $\varepsilon_{\rm M} \sim 10^{-3}$ (we will explore variations to this basic modeling further below). The gravitational binding energy is tightly related to the thermal energy via the virial theorem, $E_{\rm bind} \approx 2\,E_{\rm th} \propto M_{\rm gas} T_{\rm x}$. Here, we consider the integration over a large scale, $R < 0.15\,R_{500}$. Evidently, the linear regression fit (including the intrinsic scatter band) is significantly shallower than the dashed line of the one-to-one balance. In particular, the mechanical feedback energy that a SMBH can release could potentially overcome the core binding energy if released in a very short period of time—for instance, assuming a quasar-like/Sedov blast scenario. This would drastically overheat and evacuate the gaseous core atmosphere, becoming more serious toward the galaxy group regime and lower mass halos ($E_{\rm bind,c} < 10^{59}$ erg), where the feedback energy might even evacuate the entire gas virial region (Puchwein et al. [22], Gaspari et al. [263], as in Section 5.2). Because observations almost ubiquitously detect hot atmospheres, this indicates that the AGN feedback in groups shall be well self-regulated and relatively gentler than in massive galaxy clusters, which can sustain much stronger and impulsive AGN feedback deviations over the cosmic evolution.

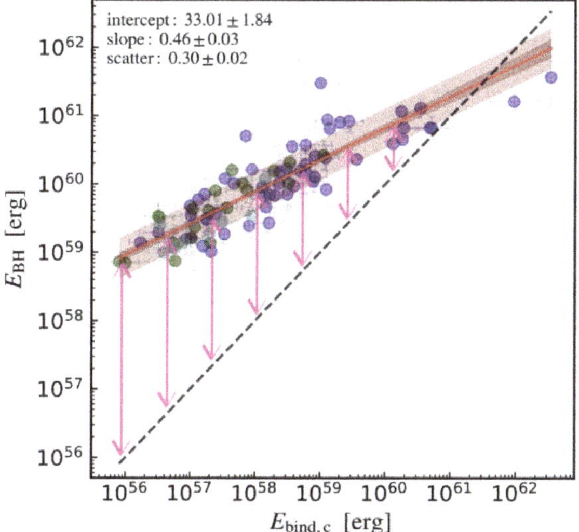

Figure 12. Available mechanical feedback energy of the central SMBH versus gravitational binding energy of the hot gas within the core of the host halo ($R \lesssim 0.15\,R_{500}$). The SMBH energy is $E_{\rm BH} = 10^{-3}\,M_{\rm BH} c^2$, while the binding energy is related to the thermal energy via the virial theorem $E_{\rm bind} \approx 2\,E_{\rm th} \propto M_{\rm gas} T_{\rm x}$. The 85 points are taken from Gaspari et al. [77], which include the observed direct/dynamical SMBH mass with the X-ray halo detected in the host group or cluster. The solid red curve shows a fit to the relation with a power law, with the 16–84 percentile interval being indicated by the red shaded area. The 1-σ intrinsic scatter is plotted as a light red band on top of the mean fit. The circle colors reflect the morphological type of the central galaxy: elliptical (blue), lenticular (green), and spiral (cyan). The black dashed line demarks the one-to-one energy equivalence, whereas the magenta arrows highlight the excess BH energy when compared to the binding energy.

Reaching a gentle self-regulation, while avoiding strong overheating, implies two major features of AGN feedback in galaxy groups and related improvements when compared with the above modeling. First, the conversion efficiency of accreted rest mass energy into feedback energy is expected to decline with lower halo mass: equating the macro AGN power to the gas X-ray radiative cooling rate requires to modify the above with

$\varepsilon_{\rm M} \sim 10^{-3}(T_{\rm x}/2\,{\rm keV})$ (Gaspari and Sądowski [262]; see also Equation (16)). This can be explained by the weaker macro-scale coupling of the AGN jets/outflows with the hot halo, as the IGrM atmospheres are more diffuse than the dense ICM counterparts. Second, in order to avoid the above evacuation outburst, such self-regulated AGN feedback has to be not only gentler, but significantly more frequent, i.e. with larger duty cycle (ratio of on/off activity). This is also naturally explained by the relatively lower cooling times (tens of Myr) in the inner IGrM regions (as compared with the ICM counterparts), due to the lower $T_{\rm x}$ and the substantially enhanced Λ via line emission (Figure 10). Both such key features have been extensively tested and retrieved by high-resolution HD simulations (Gaspari et al. [156], Sharma et al. [158], Gaspari et al. [264], Prasad et al. [265]), and found in observations (e.g., Best et al. [89], O'Sullivan et al. [106]).

While we have discussed the key characteristics and requirements of AGN heating, the next major question is: how is the AGN feedback energy propagated and dissipated within the IGrM? The problem is challenging, as it entails a wide range of scales and phases, from the milliparsec up to at least the 100 kpc region, as shown in Figure 9 (top insets). General-relativistic, radiative-magnetohydrodynamical simulations (GR-rMHD; Sądowski and Gaspari [266]) resolving radial distance of $\sim$500 $r_{\rm S}$ (Schwarzschild radii) show that the AGN triggered via CCA is able to transform the inner gravitational energy into wide ultrafast outflows (UFOs) with velocities $\sim$0.1c (top-left inset in Figure 9; Fukumura et al. [267], Tombesi et al. [268]). Under strong magnetic field tower and spin conditions (Tchekhovskoy et al. [269]), the AGN is also able to generate a very collimated relativistic (radio-emitting) jet, perpendicularly to the thick accretion torus. The above GR-rMHD simulations show that kinetic feedback appears to be present over both low and high Eddington ratios ($\dot{M}_{\rm BH}/\dot{M}_{\rm Edd} \equiv \dot{M}_{\rm BH}/[23\, M_\odot\, {\rm yr}^{-1}({\rm M}_{\rm BH}/10^9{\rm M}_\odot)]$), with a retrieved *micro* mechanical efficiency $\varepsilon_{\rm m} \simeq 0.03 \pm 0.01$. At variance, the radiative efficiency declines dramatically below $\varepsilon_{\rm r} \ll 0.01$ at $\dot{M}_{\rm BH}/\dot{M}_{\rm Edd} < 1\%$, which is the typical regime of local AGN in massive galaxies (Russell et al. [270]). Further, in order to achieve an efficacious macro self-regulation, the AGN feedback has to satisfy energy conservation (Costa et al. [271]) and related micro- to macro-scale power transfer (Gaspari and Sądowski [262]), such as

$$(P_{\rm out} \equiv \varepsilon_{\rm m} \dot{M}_{\rm BH} c^2) \;=\; (P_{\rm OUT} \equiv \varepsilon_{\rm M} \dot{M}_{\rm cool} c^2) \;\sim\; L_{cool}, \tag{16}$$

with L_{cool} the cooling luminosity (see Section 3.1). The discrepancy between the above *macro* and the larger *micro* efficiency is crucial: it implies that most of the accreted matter ($\dot{M}_{\rm out}/\dot{M}_{\rm cool} = (1-\varepsilon_{\rm M}/\varepsilon_{\rm m}) > 90\%$) is re-ejected back by the SMBH, as discussed above, driven mostly in the kinetic form of UFOs and relativistic jets. Such AGN outflows/jets propagate and percolate deeper into the meso-scale atmosphere and start to entrain progressively more IGrM, loading part of the surrounding gas mass and decreasing their velocity down to several 1000 km s^{-1} (e.g., Giovannini [272], Fiore et al. [273]).

The last missing tile of the self-regulated cycle is the macro AGN feedback deposition—a strongly debated topic since the launch of *Chandra* and XMM-*Newton* telescopes, as their angular resolution is mostly limited to the macro scale (Figure 9, top-right inset). While numerous physical mechanisms have been proposed to compensate macro cooling flows (e.g., McNamara and Nulsen [109]), heren we focus on the physics of the three major mechanisms that have been firmly established to be present in the majority of hot halos, particularly the IGrM (see the observational evidences in Section 3), namely: buoyant bubbles, shocks, and turbulence. While previous reviews tried to assess what is the dominant or sole driver of the AGN feedback, we show that the macro AGN feedback deposition is a strong nonlinear composition of at least three key processes. We can dissect such non-linearity and sub-processes via the local enstrophy analysis, which we define as the squared magnitude of the flow vorticity $\epsilon = \frac{1}{2}|\boldsymbol{\omega}|^2 \equiv \frac{1}{2}|\nabla \times \mathbf{v}|^2$. Neglecting the small dissipation term, the Lagrangian (tracer particle) framework leads to the following enstrophy evolution decomposition (Wittor and Gaspari [239]):

$$\frac{d\epsilon}{dt} = \underbrace{-2\epsilon(\nabla \cdot \mathbf{v})}_{F_{\rm com}} + \underbrace{2\epsilon\left(\frac{\omega}{|\omega|} \cdot \nabla\right)\mathbf{v} \cdot \frac{\omega}{|\omega|}}_{F_{\rm str}} + \underbrace{\frac{\omega}{\rho^2} \cdot (\nabla\rho \times \nabla P)}_{F_{\rm bar}}, \tag{17}$$

where the three right-hand-side (positive/negative) terms are compressions/rarefactions, stretching/squeezing motions, and baroclinicity, respectively.

The more that we want to zoom into the detailed processes of the AGN heating/weather, the more we need to rely on nonlinear HD simulations. Figure 13 shows a typical AGN-heating dominated period taken from a self-regulated AGN feedback simulation of meso AGN outflows/jets that consistently balance the macro cooling flow down to 1–10% $\dot{M}_{\rm cool,pure}$ (Gaspari et al. [156]). The 30 million tracer particles injected on top of the Eulerian grid can dissect the major components of the macro AGN feedback. First, the progressively slower and entrained outflow/jet inflates a pair of underdense cavities/bubbles, as its ram pressure is balanced by the surrounding IGrM thermal pressure (e.g., Brighenti et al. [274]). Such bubbles are often—albeit not universally—traced by radio synchrotron emission spilling from the micro/meso jets (see Section 3.2). Within their ellipsoidal volumes $V_{\rm b}$ (Shin et al. [107]), they contain a substantial amount of enthalpy $E_{\rm cav} \simeq 4P_{\rm b}V_{\rm b}$ in the purely relativistic case (see Section 3.1); dividing by the buoyancy time (Churazov et al. [111]), the related cavity power/heating rate is $H_{\rm cav} = E_{\rm cav}/t_{\rm buoy}$. Second, the bubbles are often encased within a cocoon shock, which is the result of the strong compressional motions of the expanding outflow and bubbles (see the thick blue contours in the second panel of Figure 13). At this stage, the shock Mach number has become already weakly transonic, $\mathcal{M} \sim 1$–2 (see Figure 4); as the AGN outflow recurrently ignites, they generate a series of weak shock ripples in the IGrM (Randall et al. [14], Liu et al. [115]), which heat the gas non-adiabatically via cumulative entropy jumps, with heating rate $H_{\rm shock} = (e_{\rm th}\,\Delta \ln K)/t_{\rm age}$ (where $e_{\rm th}$ is the specific thermal energy and $t_{\rm age}^{-1}$ is the frequency of shocks). Figure 13 shows that both processes are indeed present, although the relative heating ratio varies as a function of time, with shock heating being initially more vigorous toward the inner regions, while cavity deposition is more effective at 10–100 kpc radii.

Without the third component—subsonic turbulence (see also Section 4.1)—the final macro AGN feedback deposition would be either highly anisotropic (bubble pairs) or localized (thin shock jumps). Figure 13 (third panel) shows that the AGN feedback induces major turbulence/vorticity in a quasi isotropic manner. While the jet direction is a continuous source of enhanced turbulence, the whole IGrM core experiences a quasi irreducible level of turbulent motions ($\sigma_v \sim 100$–$300\,{\rm km\,s^{-1}}$; $\mathcal{M}_t \lesssim 0.5$). At the same time, the simulation shows that the (negative) rarefactions avoid the runaway accumulation of large vorticity by balancing the (positive) stretching term in a volume-filling way. The final panel finally shows that baroclinicity is negligible during the macro AGN feedback deposition, as subsonic turbulence is able to preserve the alignment of density and pressure gradients. It is important to note that, while turbulence provides a key source of isotropic *mixing* (with characteristic scale $t_{\rm eddy}$; Section 4.1), its subsonic nature implies that the heating rate ($H_{\rm turb} = \frac{1}{2}M_{\rm gas}\sigma_v^2/t_{\rm turb} \propto \sigma_v^3$, where $t_{\rm turb} = t_{\rm eddy}/\mathcal{M}_t^2$; Gaspari et al. [275]) is not only a fraction of the global cooling rate, but it also has a substantially delayed deposition time $t_{\rm turb} \gg t_{\rm cool}$ [237,276]. Alternatively, reorienting jets similar to the case of NGC 5044 (see Figure 8) may provide an alternative way of heating the gas in a quasi-isotropic way. Cielo et al. [277] presented simulations of AGN/IGrM interaction in the case of precessing jets and claimed that the distributed energy is sufficient for offsetting cooling and reproducing the features seen in real cool-core clusters.

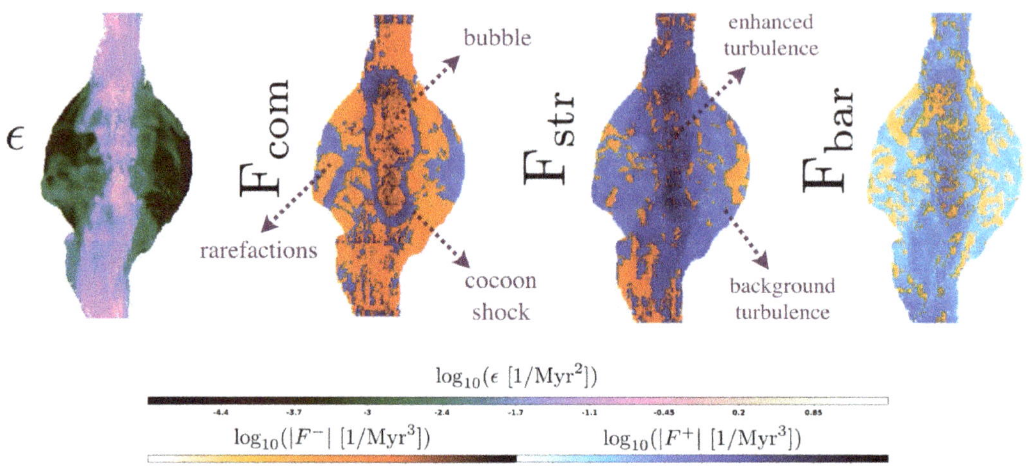

Figure 13. The macro AGN feedback transfer and deposition highlighted via the enstrophy ϵ decomposition (Equation (17)) into its main positive/negative components: compressions/rarefactions (second panel), stretching/squeezing motions (third), baroclinicity (fourth)—adapted from Wittor and Gaspari [239]. This is achieved via Lagrangian tracer particles on top of an adaptive-mesh-refinement HD simulation of self-regulated AGN outflows/jets in a central massive galaxy (Gaspari et al. [156]). The cycle of CCA rain, AGN outflow injection, bubble inflation, cocoon shock expansion, and turbulence cascade repeats self-similarly over several billion years, recurrently quenching the macro cooling flow.

In closing this theoretical section, we remark a few remaining important differences between galaxy groups and the more massive clusters. While we have discussed above that the IGrM shall be strongly self-regulated to avoid overheating/overcooling, this does not imply that groups are less variable than clusters. Indeed, the tails of the chaotic feeding/feedback loop can generate relatively more disruptive imprints in the less bound IGrM (e.g., Voit et al. [278]). This is reflected in the increased morphological diversity of groups (Sun et al. [137]) and larger intrinsic scatter of the scaling relations toward the low-mass regime, as found in the fundamental $L_x - T_x$ (Goulding et al. [279]; see the companion Lovisari et al. review) and $M_{BH} - T_x$ (see Section 5.3) relations. While, in absolute values, the AGN deposition radius is significantly larger in galaxy clusters (up to several 100 kpc), normalized to function of R_{500} the AGN feedback outliers can pierce through relatively larger regions of the less bound IGrM (e.g., Grossová et al. [203]). Interestingly, many elliptical galaxies (including non-centrals) show the presence of a mini-cool core with a size of ~ 1 kpc, which could represent the irreducible inner CCA condensation region, enabling the more frequent self-regulated AGN feedback discussed above for galaxy groups.

5. Impact of AGN Feedback on Large Scales

5.1. AGN Feedback in Cosmological Simulations

Hydrodynamical cosmological simulations are paramount for self-consistently modelling the highly non-linear formation of large-scale structure. They can simultaneously precisely solve for the gravitational and hydrodynamical aspects of structure formation. Yet, because these simulations have limited spatial and mass resolutions, one needs to implement simplified 'sub-grid' prescriptions for including crucial physical processes, such as cooling, star formation, and the feedback from supernovae and AGN, since these phenomena take place at scales that cannot be resolved by the simulations. For a complete review on numerical simulations of galaxy groups, we refer the reader to Oppenheimer et al. within this issue. Modern simulations of galaxy groups broadly fall into three categories:

1. High resolution zoom simulations (a few hundred pc spatial resolution and $\sim 10^5\ M_\odot$ particles) of a few groups or of a small volume. These types of simulations are typically used to develop new baryonic physics and study the details of its impact on the IGrM (e.g., ROMULUS [280]; NewHorizon [281]; FABLE [282]).
2. Moderate resolution simulations (spatial resolution of the order of $\sim$1 kpc and $\sim 10^6\ M_\odot$ particles) of volumes that are large enough (about 100 Mpc on a side) to contain a sizable sample of groups, but not many clusters (e.g., Horizon-AGN [283]; EAGLE [284]; Illustris (TNG) [285,286]; SIMBA [287,288]; MassiveBlack-II [289]).
3. Low resolution simulations (about 5 kpc spatial resolution and $\sim 10^9\ M_\odot$ particles) of much larger volumes ($\sim$300–1000 Mpc on a side with the most common value being around 500 Mpc) to contain a large sample of groups and clusters (e.g., IllustrisTNG-300 [286]; cosmo-OWLS [12]; BAHAMAS [41]; Magneticum [290]; Horizon Run 5 [291]).

These simulations have been run with codes that use different methods for solving the equations of hydrodynamics in a cosmological context. Namely, the Tree Particle Mesh (TreePM) and smoothed particle hydrodynamics (SPH) code GADGET [292] in various versions for the majority (EAGLE, MassiveBlack-II, cosmo-OWLS, BAHAMAS, Magneticum), the moving mesh codes AREPO [293,294], and GIZMO [295] for a smaller number (FABLE, Illustris(TNG), SIMBA), and, finally, the adaptive mesh refinement (AMR) code RAMSES [296] for an even smaller fraction of them (Horizon-AGN, Horizon Run 5, and NewHorizon). Note that ROMULUS was run with the Tree+SPH code CHANGA [297]. All of these simulations include a sophisticated modelization of the non-gravitational processes of galaxy formation, such as metal-dependent radiative cooling, star formation, chemical evolution, accretion onto supermassive black holes, and feedback processes from supernovae, asymptotic giant branch stars, as well as AGN. Some of them have even calibrated the free parameters of these models on observations (e.g., FABLE, EAGLE, Illustris(TNG), and BAHAMAS). Note that the value of these parameters are often at least informed by higher-resolution simulations.

In the majority of cases, cosmological simulations (type 2 and 3) implement some variation of the Booth and Schaye [298] AGN feedback model (hereafter BS09), which is itself largely based upon the Springel et al. [292] model (hereafter S05). In the BS09 model, halos are seeded with BH seeds in their center when their mass, as evaluated by an on-the-fly halo finder, first reaches $M_{\rm h,min} = 100 m_{DM}$, where m_{DM} is the mass of a dark matter particle (as in [21]). At that point, BH seeds are introduced at the bottom of the potential well, with masses $M_{seed} = 0.001 m_g$, where m_g is the mass of a gas particle. BH can then grow either by gas accretion or mergers. Specifically, BH accrete from the surrounding gas at a rate that is proportional to that given by the Eddington-limited Bondi–Hoyle–Lyttleton [299,300] formula,

$$\dot{M}_{acc} = \alpha \dot{M}_{\rm Bondi} = \alpha \frac{4\pi G^2 M_{\rm BH}^2 \rho}{(c_s^2 + v^2)^{3/2}}, \tag{18}$$

where v is the velocity of the BH relative to the ambient gas. The dimensionless 'boosting' factor α was introduced by S05 as a numerical correction factor that attempts to correct for the limitations of numerical simulations (see also Section 4.1). It was independent of density and had a constant value of $\sim$100 (e.g., [22,292,301–303]). BS09 introduced a density-dependent efficiency that varies as a power law of the density with a power-law index $\beta = 2$ when the density is above $n_H^* = 0.1$ cm^{-3} and $\beta = 1$ otherwise. Bondi–Hoyle accretion is spatially resolved when the local gas density $n_H^* < 0.1$ cm^{-3}, which corresponds to the threshold for the formation of a cold ($T \lesssim 10^4$ K) phase, and when the simulations resolve the Jeans length (see BS09 and e.g., [304] for a detailed discussion). The BH growth rate can then be determined from the mass accretion rate by assuming a given radiative efficiency ϵ_r, $\dot{M}_{\rm BH} = \dot{M}_{acc}(1 - \epsilon_r)$. The total radiative efficiency is always assumed to be 10 per cent, which is the mean value for the radiatively efficient [305] accretion on to a Schwarzschild BH. Bondi accretion—albeit very simple to implement in subgrid models—is far from a realistic representation of the feeding processes (such as CCA and multiphase precipitation), thus we advocate for fundamental updates of subgrid models in future works, as discussed in Section 4.1.

BH inject a fixed fraction of the rest–mass energy of the gas that they accrete into the surrounding medium. The feedback is only implemented thermally. In that case, the energy is deposited into the surrounding gas by increasing its internal energy, as opposed to kinetic feedback, which deposits energy by kicking the gas. The fraction of the accreted rest-mass energy that is injected is assumed to be independent of both the environment and accretion rate (i.e., no distinction between 'quasar mode' and 'radio mode' feedback as in the models of e.g., [302], which still injected energy thermally in both cases). The amount of energy returned by a BH to its surrounding medium in a time-step Δt is given by

$$E_{\rm feed} = \epsilon_f \epsilon_r \dot{M}_{\rm BH} c^2 \Delta t \quad (19)$$

where ϵ_f is the efficiency with which a BH couples the radiated energy into its surroundings (a free parameter) and c is the speed of light. In order to ensure that the thermal feedback from BHs is efficient, and that it is not immediately radiated away, BS09 introduced a minimum heating temperature $\Delta T_{\rm min}$. BHs store feedback energy until they have accumulated an energy $E_{\rm crit}$ that is large enough to increase the temperature of a number $n_{\rm heat}$ of their neighbours by an amount of $\Delta T_{\rm min}$, which is,

$$E_{\rm crit} = \frac{n_{\rm heat} m_g k_B \Delta T_{\rm min}}{(\gamma - 1) m_H}. \quad (20)$$

The internal energy of the heated gas is instantaneously increased by $E_{\rm crit}$. If $\Delta T_{\rm min}$ is set too low, the cooling time of the heated gas remains very short and the energy is efficiently radiated away. If $n_{\rm heat}\Delta T_{\rm min}$ is set too high, then the energy threshold and the time period between AGN heating events become very large. Thus, $n_{\rm heat}\Delta T_{\rm min}$ is connected to the AGN duty cycle. The energy is then isotropically deposited into the gas.

The recent increase in resolution (from category 3 to category 2) led to improvements of the AGN feedback modeling, as more physical processes could be taken into account. For instance, the EAGLE team modified the BS09 formula for $\dot{M}_{acc}$ differently from what had been previously done by BS09 to take the angular momentum of the gas accreted by the BH into account, such that the above accretion rate was multiplied by a factor $\alpha = \min(1, C_{visc}^{-1}(c_s/V_\phi)^3)$, where V_ϕ is the rotation speed of the gas around the BH [306] and C_{visc} is a free parameter that s related to the viscosity of the accretion disc. Improvements to the [302] model used by [22] have also been made as part of the Illustris, IllustrisTNG and FABLE projects for the same reasons. Specifically, the authors added a third mode of AGN feedback (i.e., the feedback is now thermal, mechanical, and radiative, as described in [307]) for Illustris. The kinetic AGN feedback model in the low accretion rate regime was updated for IllustrisTNG [308] (see [309] for an exhaustive discussion of the changes between Illustris and IllustrisTNG). In parallel, the FABLE team also modified the Illustris model to alleviate some of its shortcomings, such as the underestimation of the gas fractions of groups and clusters (see Section 5.2, and, in particular, the discussion of Figures 14 and 15 below). The parameters of the feedback model were calibrated on the gas mass fractions using a strategy similar to the one employed for BAHAMAS ([41]; see [282] for detailed discussion). In particular, being inspired by the BS09 model, they introduced a 25 Myr duty cycle for the AGN feedback to reduce artificial overcooling. We note that both Horizon-AGN [283] and NewHorizon [281] use two modes of feedback, as originally introduced by [302], but that the low accretion rate or 'radio' mode is kinetic instead of thermal (the detailed modelling can be found in [310]). Indeed, the observed and realistic astrophysical deposition of heating in hot halos is carried out most of the time via the AGN outflows and jets, as discussed in Section 4.2.

5.2. The Hot Gas Fraction and the AGN Feedback Model

The total baryon content and its partition between the various gas and stellar phases put fundamental constraints on galaxy formation models and, in particular, on the strength of AGN feedback, as stated in Section 3.1.4. Thus, here we present the gas mass fraction-M_{500} relation at $z = 0$ for two compilations of massive galaxies, groups and cluster simulations that include different sorts of baryonic physics: *(i)* historical simulations, which is, simulations run before 2014–2015 in Figure 14; and, *(ii)* modern simulations run as from 2014–2015 that calibrated the free parameters of their subgrid models to reproduce at least the galaxy stellar mass function at $z = 0$ in Figure 15. The various simulation sets will be compared with the compilation of observations that we presented in Section 3.1.4 and especially Figure 7, which is shown as a gray band on both figures.

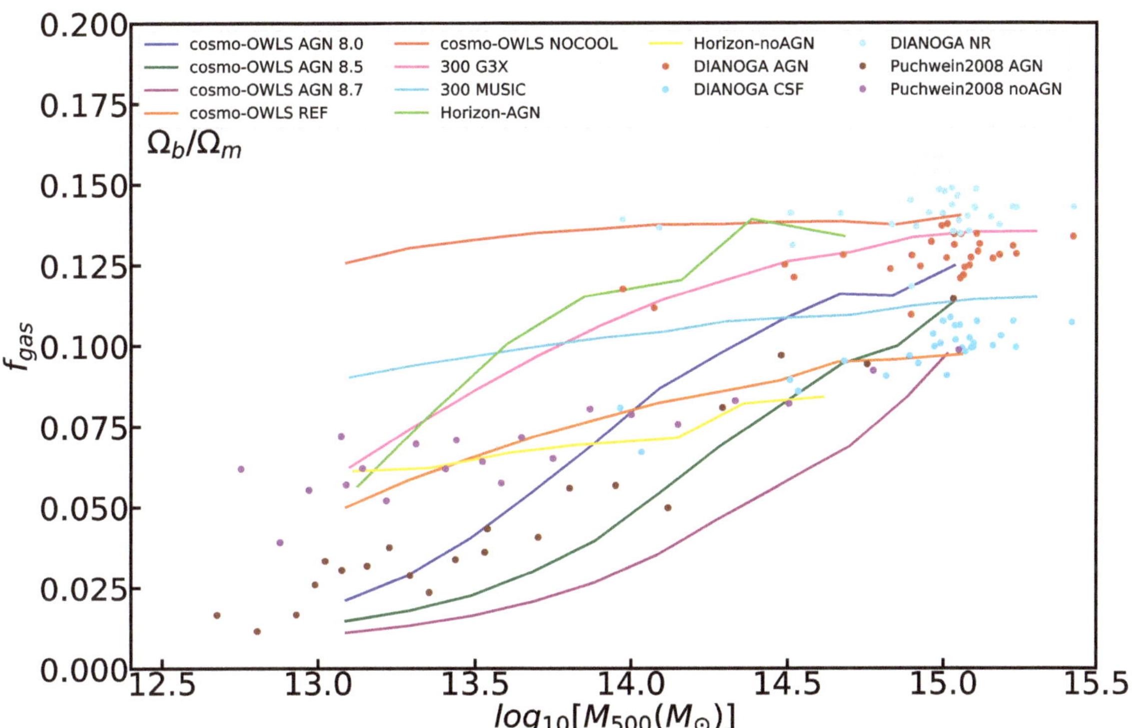

Figure 14. Compilation of historical simulation results for the gas fraction within R_{500} as a function of M_{500}. The red, orange, magenta, green and dark blue solid lines correspond to the different sub-grid models of cosmo-OWLS [12], the pink and cyan ones to the 300 clusters run with the GADGET3X [311] and MUSIC [312] codes, respectively, as part of The Three Hundred Project [313], the lime and gold ones correspond to Horizon-AGN and, its counterpart without AGN, Horizon-noAGN [283], the cyan, blue, and crimson ones to the various physical models of the DIANOGA suite [314] and, finally, the purple and brown symbols correspond to the simulations of [22] without and with AGN, respectively. The compilation of observations presented in Figure 7 is shown as a gray band.

In Figure 14, we present the results of a compilation of historical simulations for the gas fraction within R_{500} as a function of M_{500}. The NOCOOL model of cosmo-OWLS and the NR model of DIANOGA both correspond to classical non-radiative simulations, where one includes hydrodynamics, but do not allow the gas to cool through radiative processes. The REF model of cosmo-OWLS, the CSF model of DIANOGA, the 300 clusters run with the MUSIC code, as well as the noAGN models of [22] and the Horizon suite, all include prescriptions for radiative cooling, star formation, and stellar feedback, but not for AGN feedback. As first noted by [22,315], the inclusion of AGN feedback substantially lowers the gas fractions of both groups and clusters.The intensity and, thus, the duty cycle of the AGN feedback, as parameterized by $\Delta T_{\rm min}$ in BS09 (see Equation (20)), can be used to eject more or less gas from the potential well, as can be seen by comparing the models AGN 8.0, 8.5, and 8.7 of cosmo-OWLS. Here the number corresponds to the logarithm of the value of $\Delta T_{\rm min}$ chosen, i.e., 8.0 corresponds to $\Delta T_{\rm min} = 10^8$ K. The REF, CSF, and noAGN models also yield reasonable gas mass fractions, but the relation with mass is flatter than observed, because the SF efficiency does not strongly depend on halo mass. The low gas fractions in these models are achieved by overly efficient star formation (e.g., [12,315] and Figure 1). Note that, while the cosmo-OWLS models that include AGN feedback use the BS09 AGN feedback model summarized in Section 5.1, which is fully thermal, Horizon-AGN, DIANOGA and 300 G3X resort to a mixture of thermal and kinetic feedback, as originally developed by [310] for the former and [316] latter two.

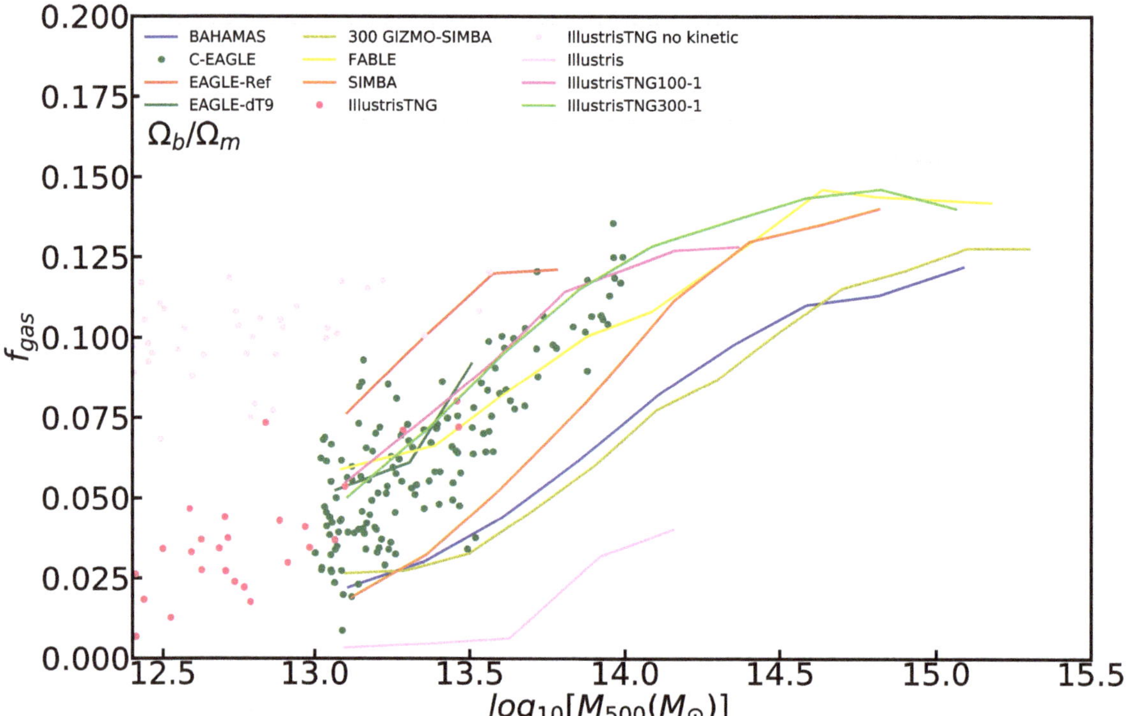

Figure 15. Compilation of gas fractions within R_{500} from modern simulations as a function of M_{500}. The blue line corresponds to BAHAMAS [41], the red and green lines to the Reference and dT9 models of EAGLE [284], the green symbols to C-EAGLE and Hydrangea [317,318], as it uses the same sub-grid model as EAGLE-dT9, the olive line to the 300 clusters run with the GIZMO-SIMBA code (Cui et al. in preparation) as part of The Three Hundred Project [313], the salmon line to Horizon-AGN [283], the gold one to FABLE [282], the orange one to SIMBA [287,288], and the pink and deep pink symbols, as well as the plum, orchid, and lime lines correspond to various models from the Illustris and IllustrisTNG suites [285,286]. The compilation of observations presented in Figure 7 is shown as a gray band.

In Figure 15, we present the results of a compilation of modern simulations for the gas fraction within R_{500} as a function of M_{500}. Despite the fact that most modern simulations have been calibrated to reproduce the local galaxy stellar mass function (see Figure 1), the predictions on the hot gas fraction are vastly different. For instance, Illustris (plum line) vastly underpredicts the observed hot gas fractions, whereas the reference EAGLE model (red line) clearly overpredicts them. Therefore, a setup that broadly reproduces the stellar content of galaxies in the Universe may simultaneously fail at reproducing the properties of the hot gas phase. Note that, in the case of BAHAMAS (blue line) and FABLE (gold line), the free parameters of the stellar and AGN feedback have been adjusted to reproduce both the $z = 0$ galaxy stellar mass function and gas content of groups and clusters (see discussions in [41,282]). IllustrisTNG and EAGLE-dT9 (and the associated simulations) are versions of Illustris and EAGLE in which the AGN feedback parameters were slightly adjusted to reduce the discrepancies with the gas content of massive groups and clusters. It is worth noting that SIMBA uses a fully kinetic AGN feedback model [287], while the simulations from the Illustris series and FABLE include a mix of thermal, kinetic/mechanical, and radiative feedback [282,285,309] in the vein of the one first developed by [302].

Generally speaking, we stress that the hot gas fraction of galaxy groups is an extremely sensitive probe of the feedback scheme implemented in cosmological simulations. Modern simulation suites have little predictive power on the baryon content of groups, even when the properties of the galaxy population are accurately reproduced (see Figure 1). Some of the simulations are actually *calibrated* on the gas mass fractions, i.e., the parameters governing the feedback model were tuned to produce reasonable gas fractions in the group regime. Major observational (Section 3.1.4) and theoretical advances (Section 4) are required to understand the ejection of baryons from halos by AGN feedback and inform the mainstream galaxy evolution models.

5.3. Co-Evolution between the IGrM and the Central AGN

SMBH masses are known to correlate with the properties of their host galaxy, in particular the integrated K-band luminosity L_K and the velocity dispersion of the stars in the bulge, σ_e (see [64] for a review), as discussed in Section 2.2. However, it is still unclear whether the optical scaling relations of SMBH are fundamental or derive from correlations with other key quantities. Recent findings have instead unveiled that the SMBH masses are more tightly correlated with the properties of the host X-ray gaseous halos, especially in the IGrM regime [76–78]. In Figure 16, we summarize our current knowledge of the relation between SMBH mass and X-ray temperature within the core of galaxy groups ($R \lesssim 0.15\,R_{500}$). It is important to note that the SMBH masses shall be directly observed via dynamical measurements to properly unveil intrinsic scaling relations. The largest existing study is provided by Gaspari et al. [77] with 85 systems with measured SMBH masses, most of which with temperatures ~0.5–1 keV that were typical of galaxy groups. A Bayesian fit to the relation finds slopes $M_{\rm BH} \propto T_{\rm x}^{2.1}$ (Figure 16, green) and $M_{\rm BH} \propto L_{\rm x}^{0.4}$. At the high-mass end, Bogdán et al. [76] measure a somewhat flatter slope, $M_{\rm BH} \propto T_{\rm x}^{1.7}$. Notably, the intrinsic scatter goes down to ~0.2 dex, with a very high correlation coefficient at or above the 0.9 level, when compared to ~0.5 dex for the K-band luminosity. The correlations hold, regardless of the large diversity of systems, from BGGs and ETGs to non-central lenticular/spiral galaxies. Varying the extraction radius by slightly enlarging or decreasing it (group outskirts or CGM) does not significantly vary such conclusions (see the companion Lovisari et al. review for the complementary R_{500} scalings, such as $M_{\rm BH} - L_{500}$ and $M_{\rm BH} - M_{\rm tot}$). We note that multi-variate X-ray correlations (a.k.a. 'fundamental planes') do not further improve the intrinsic scatter. In sum, by comparing the different X-ray/optical scaling relations, it has emerged that the extended plasma (collisional) atmospheres seem to play a more fundamental role than small-scale (collisionless) stellar properties in the co-evolution of SMBH and groups. This is further supported by zoom-in cosmological simulations [319–321]. On the other hand, the slope (and scatter) of the current cosmological simulations still remain too low when compared with the observations (dotted lines in Figure 16), indicating the need to model more realistic feeding and feedback physics (see Section 4) into the coarse subgrid numerical modules.

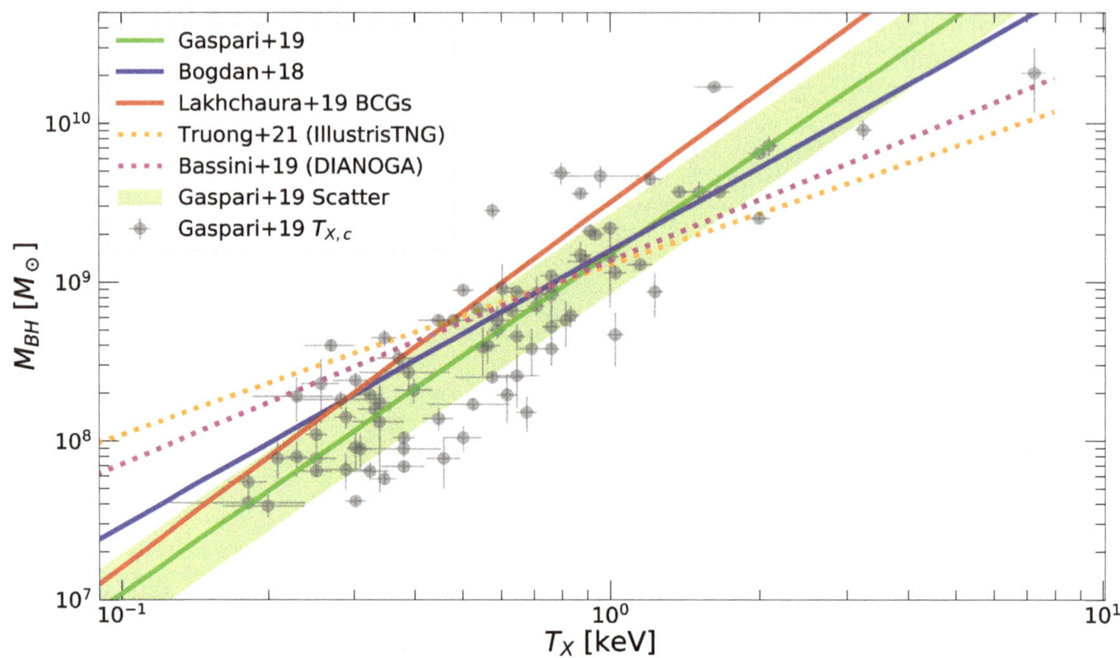

Figure 16. Relation between BH mass and IGrM X-ray temperature. The data points show dynamical measurements of BH mass plotted against the spectroscopic X-ray temperature of the host halo. The solid curves show the fitted observational relations from Gaspari et al. [77] (green; together with the quoted log-normal scatter), Bogdán et al. [76] (blue), and the "BCG" subsample of Lakhchaura et al. [78] (red). The dashed lines show the predictions of cosmological simulations with AGN feedback (Illustris TNG, Truong et al. [321]; DIANOGA, Bassini et al. [319]). The gray data points are taken from the sample of Gaspari et al. [77], which already included the smaller samples that were used by Bogdán et al. [76] and Lakhchaura et al. [78].

The above SMBH versus X-ray correlations are crucial for testing models of galaxy/group formation and evolution. Accretion/feeding models can be broadly divided into cold and hot accretion modes, as discussed in Section 4.1. Besides the cosmic dawn, hierarchical binary BH mergers are a present, but subdominant growth channel over most of cosmic time [77,319]. In hot accretion (usually Bondi or Advection Dominated Accretion Flow—ADAF; Bondi [249], Narayan and Fabian [250]), the larger the thermal entropy of the gas, the more strongly feeding is stifled, since the inflowing gas has to overcome the outward thermal pressure of the hot halo. This would induce negative correlations with the IGrM properties, which are ruled out by the strongly positive correlations that are shown in Figure 16. Conversely, cold-mode accretion (Gaspari et al. [245], Voit [246]; Section 4.1)—typically in chaotic form (due to the turbulent condensation out of the IGrM generating randomly colliding clouds)—is linearly and tightly correlated with the X-ray luminosity and gas mass. Figure 17 (left) shows the CCA final mass growth via theoretical/numerical predictions [77] when compared to direct measurements of SMBH masses. During the Gyr evolution, the turbulent IGrM locally condenses into extended warm filaments and cold molecular clouds via nonlinear thermal instability. The clouds inelastic collisions at the meso/kpc scale boost the micro accretion rate down to the Schwarzschild radius, hence triggering strong AGN feedback heating. Such recurrent SMBH growth drives the $M_{\rm BH}$ that is shown in Figure 17, with excellent agreement with the SMBH mass observed in our local universe.

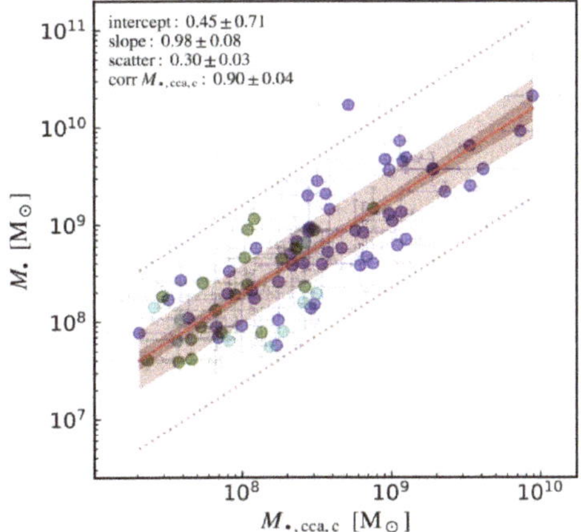

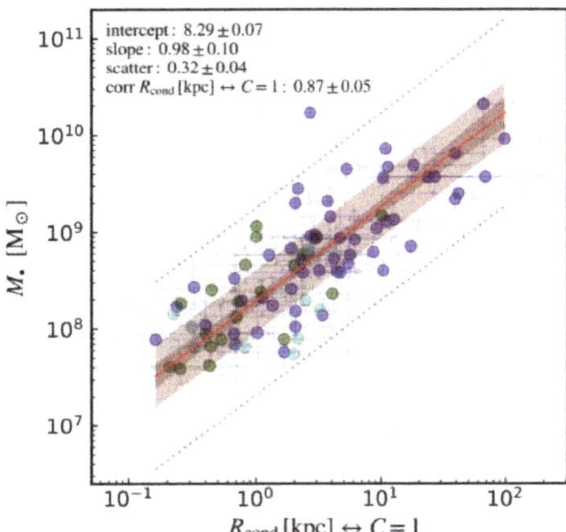

Figure 17. X-ray scaling relations are key to constrain different baryonic physics of the IGrM, here in terms of feeding models. ***Left***: Direct SMBH masses plotted against the mass derived from theoretical/numerical predictions of chaotic cold accretion (CCA) via the X-ray core properties—adapted from Gaspari et al. [77]. ***Right***: Hot-halo condensation radius as a function of BH mass and thus IGrM halo mass (the locus where $C \equiv t_{\rm cool}/t_{\rm eddy} = 1$; see Section 3.1.3 and Section 4.1). See Figure 12 for the description of the analysis, color coding, and sample. Figure reproduced from Gaspari et al. [77].

Scaling relations allow us to predict other baryonic physics of the IGrM, such as the extent of the IGrM multiphase condensation radius, which is shown in the right panel of Figure 17. As introduced in Section 4.1, such a radius is the locus at which the IGrM cooling time and the turbulent eddy-turnover time match ($C \equiv t_{\rm cool}/t_{\rm eddy} = 1$), both of which can be retrieved via the X-ray scaling relations as a function of $T_{\rm x}$ and $L_{\rm x}$ (Gaspari et al. [77]). Evidently, lower mass groups have condensation radii of less than a few kpc, while massive groups can reach $R_{\rm cond}$ of a few 10 kpc (e.g., David et al. [209], Olivares et al. [167], Lakhchaura et al. [206]). Overall, scaling relations between X-ray macro-scale properties translates into scaling relations of micro-scale properties ($M_{\rm BH}$), corroborating a tight co-evolution between multi-scale processes in the IGrM (as depicted in Figure 9).

5.4. Impact on Cosmological Probes

During the past few years, it has become clear that AGN feedback will play an important role as a leading source of systematic uncertainties for upcoming high-profile cosmology experiments. Indeed, the energy that is injected by the central AGN affects the global distribution of baryons (see Section 5.2), leading to local depletion or excesses of matter with respect to the expectations of models, including dark matter only. This effect is most important in galaxy groups, since these systems correspond to the peak of the local halo mass density and their baryonic properties are highly sensitive to feedback. As a result, the matter power spectrum at $z = 0$ can be substantially altered by baryonic physics and, in particular, AGN feedback (e.g., [322,323]). The shape of the power spectrum is strongly affected by baryonic processes on scales $k \gtrsim 10^{-1}\ h/{\rm Mpc}$. For $10^{-1} < k < 10\ h/{\rm Mpc}$, most of the simulations predict a *deficit* of power with respect to the N-body case, although the actual amplitude of the effect is highly uncertain [324]. The evacuation of baryons from the central regions of galaxy groups under the influence of feedback is responsible for the deficit of power on scales of ~ 1 Mpc, i.e., roughly the typical size of galaxy groups. On smaller scales ($k \gtrsim 10\ h/{\rm Mpc}$), cooling and condensation of baryons in the central regions lead to a rapidly increasing power.

Accurately predicting the shape of the matter power spectrum is crucial for the success of future cosmic shear experiments, such as *Euclid*, which aim at determining the growth of structures by measuring the matter power spectrum and its evolution [325]. Semboloni et al. [323] showed that neglecting baryonic effects would imply important systematics on the determination of cosmological parameters. Systematic effects can be mitigated by excluding the small scales ($k \gtrsim 10^{-1}$ h/Mpc) when fitting the measured power spectra, although that comes at the price of greatly increased uncertainties in the resulting cosmological parameters. Constraints on extended cosmologies, such as massive neutrinos, variable dark energy equation of state, or chameleon gravity, require sensitivity on smaller scales, and their effect is strongly degenerate with that of baryonic physics [326,327].

Chisari et al. [324] showed that numerical simulations have not yet converged on the actual impact of feedback on the power spectrum (see their Figure 3). For instance, the very strong feedback that was implemented in the original Illustris simulation, which was sufficient to completely evacuate the gas content from most groups (see Figure 15), leads to a very strong suppression of power (>30%) on scales of a few hundred kpc. Conversely, simulations implementing a more gentle feedback scheme (EAGLE, Horizon-AGN, MassiveBlackII) predict small corrections with respect to the fiducial DM-only case for $k \lesssim 10$ h/Mpc. These simulations also predict a high gas fraction in galaxy groups (see Section 5.2 and [328] for an extensive discussion). Recently, Schneider et al. [329] used a semi-analytic model to predict the impact of baryons on the matter power spectrum based on the observed gas properties of groups. The authors modified the mass profiles of halos in large N-body simulations to account for star formation and AGN feedback. In particular, the semi-analytic model of Schneider et al. [329] is highly sensitive to the parameter θ_{ej}, which governs the ejection of gas from the central regions of the halo by AGN feedback. In Figure 18, we show how the predicted matter power spectrum depends on θ_{ej}. With increasing feedback, a progressively larger fraction of the gas is ejected from the halo and, thus, the expected power gets more strongly suppressed. Calibrating their semi-analytic model on the observed gas fraction and gas density profiles of group-scale halos, Schneider et al. [329] provided a range of predictions matching the existing observational constraints. High-precision measurements of the gas density profiles in a representative sample of galaxy groups would allow us to precisely determine the expected shape of the power spectrum [327], thereby providing a key input for upcoming cosmology experiments.

In addition to the matter power spectrum, AGN feedback on the scales of galaxy groups also affects several other cosmological observables, such as the thermal SZ power spectrum (e.g., [330,331]). Indeed, AGN feedback affects the pressure profiles of halos and thus modifies the amplitude of the power spectrum on small scales ($\ell \gtrsim 1000$, [332]). Ramos-Ceja et al. [333] showed that tSZ models based on the universal pressure profile [334] overpredict the power measured by SPT and ACT on small scales. A strong feedback scenario and a low gas fraction on group scales are needed to fit the measured power. The effect of feedback also implies modifications to the cross-correlations between the tSZ and other observables (e.g., [335,336]). Finally, the choice of the feedback scheme affects the shape of the halo mass function (e.g., [337–339]) and the structure of dark-matter halos (e.g., [337,340]). We refer to the companion Oppenheimer et al. review for a more general discussion of the topic.

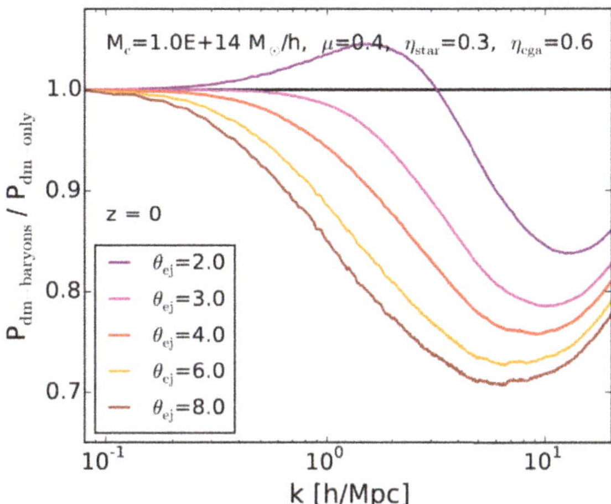

Figure 18. Modification of the local matter power spectrum with respect to pure N-body simulations in the presence of AGN feedback at the scale of galaxy groups in the semi-analytic model of Schneider et al. [329]. The various curves show how the power spectrum depends on the parameter θ_{ej} governing gas ejection from the central regions of groups under the influence of AGN feedback. A strong ejection of gas from the core of halos implies substantial modifications to the matter power spectrum on scales of ~1 Mpc. The figure reproduced from Schneider et al. [329].

6. Future Observatories

6.1. eROSITA

At present, our knowledge of the population of galaxy groups in the local Universe comes largely from the *ROSAT* All-Sky Survey (RASS). Groups that were identified in the RASS (or even the *Einstein* slew survey) form the basis of most studies of the mechanics of AGN feedback at this mass scale, but, unfortunately, groups are at the lower limit of sensitivity for these surveys. RASS is therefore biased toward the detection of relaxed, centrally-concentrated, cool-core systems, with the strength of the bias increasing as mass decreases from poor clusters to groups [185]. Searches tailored to the detection of more extended sources in RASS reveal a population of low surface brightness groups undetected by the original survey [341] confirming the expected bias, while XMM-*Newton* observations of optically-selected groups identify both low luminosity and disturbed systems previously not detected or not recognised as groups in RASS [106].

Spectrum Roentgen Gamma (SRG, launched in 2019) hosts the eROSITA instrument, a set of seven co-aligned soft X-ray telescopes covering the 0.2–10 keV band, with a field of view of 1° and ~15″ spatial resolution. SRG will spend four years surveying the whole sky once every six months, with eROSITA building up a map ~20× deeper than RASS in the 0.5–2 keV band [342]. This is sufficient to detect essentially every galaxy group with a virialized halo in the local universe [106]. More massive groups with luminosities $\sim 10^{42}$ erg s^{-1} should be detectable to $z \sim 0.1$ in the final eRASS:8 survey. Käfer et al. [343] performed detailed simulations to evaluate the sensitivity of eRASS:8 to galaxy groups. The authors used a wavelet decomposition algorithm sensitive to large-scale diffuse emission. In Figure 19, we show the corresponding sensitivity curves for two possible source detection setups: a decomposition over wavelets of scales 1–4′ that were optimized for relatively compact sources, and the other for scales in the range 1–16′ sensitive to the most extended nearby sources. Using these setups, the authors predict eRASS:8 will detect all the galaxy groups with $M_{500} > 10^{13} M_\odot$ out to $z = 0.05$. The most massive groups ($\sim 10^{14} M_\odot$) will be detected out to $z = 0.5$. While the survey observations will typically only provide luminosity and morphology information for individual halos, the group samples derived from them will be a solid base from which to investigate the impact of cooling and AGN feedback in groups, particularly when combined with radio surveys.

Pointed observations with eROSITA, possible once the survey phase is complete, may also prove useful for studies of groups. The combination of a large field of view and soft band effective area (roughly double that of the XMM-*Newton* EPIC-pn) is well-suited to observations of the outskirts of nearby groups, and the search for cavities or other structures that are associated with giant group-central radio galaxies. Thanks to its short focal length and very stable background [344], eROSITA is very sensitive to diffuse X-ray emission below 2 keV, which means that it is well suited to study the diffuse X-ray emission of galaxy groups.

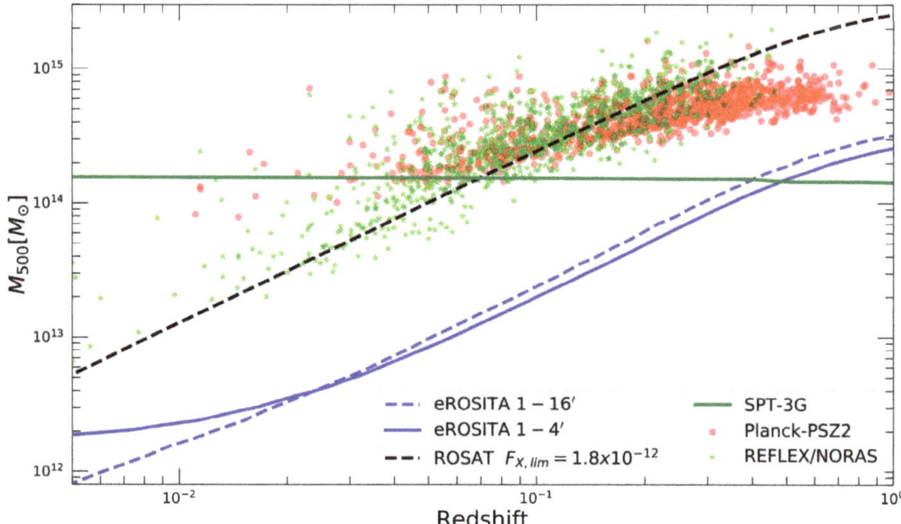

Figure 19. Expected sensitivity curve of the final eROSITA survey (eRASS:8) compared to the sensitivity of existing and upcoming X-ray and SZ surveys. The eROSITA sensitivity curve was computed from synthetic data using a wavelet decomposition algorithm [343] sensitive to scales of 1–4′ (solid blue curve) and 1–16′ (dashed blue curve). For comparison, the dashed black curve shows the sensitivity of the ROSAT all-sky survey assuming a fixed soft X-ray flux threshold of 1.8×10^{-12} erg s^{-1}, which is the typical sensitivity of the REFLEX and NORAS samples (green asterisks, [345]). The red points show the systems selected from the second *Planck* SZ catalogue [346]. The green curve shows the expected sensitivity of the SPT-3G experiment [347].

6.2. XRISM

The X-ray Imaging and Spectroscopy Mission (XRISM), expected to launch by April 2023, will open a new era of high spectral-resolution observations of galaxy groups. Its X-ray microcalorimeter (*Resolve*, [348]) will have a constant 7 eV energy resolution across its 0.3–12 keV band, resolving the forest of emission lines that characterizes emissions from the IGrM. This offers opportunities in a number of important areas, including measurements of bulk flows and turbulence in the hot gas. At present, grating spectra only provide upper limits on the turbulence of the IGrM [240,349], with possible hints of asymmetries that are associated with sloshing motions [350]. *Hitomi* demonstrated the capabilities of microcalorimeters, but it was only able to make a single turbulence measurement in the Perseus cluster [351]. XRISM should be able to measure turbulent velocities down to tens of km s^{-1}, providing a measure of the kinetic energy stored in the IGrM and allowing us to determine how much of the energy of AGN outbursts can be diffused out into the IGrM by these gas motions. The spectra from the *Resolve* microcalorimeter will also provide a detailed view of shock heating and cooling, with individual emission lines accurately tracing gas at different temperatures. Performance verification targets for the mission include NGC 5044 and NGC 4636, and, while the spatial resolution ($>1'$) and effective area of the observatory may limit its use to relatively bright nearby groups, its results are likely to be ground-breaking.

6.3. *Athena*

The Advanced Telescope for High Energy Astrophysics (Athena), which is expected to launch in the early 2030s, represents the next generation of major X-ray observatories, with $5''$ spatial resolution and an effective area of 1.4 m^2 at 1 keV (roughly 45$\times$ that of XRISM's *Resolve* instrument). It will carry a $40'\times40'$ active pixel detector (the wide field imager, WFI) providing CCD-like spectral imaging, and a $5'$-diameter microcalorimeter array (the X-ray Integral Field Unit, X-IFU) with $\leq5''$ pixels, capable of 2.5 eV spectral resolution (see, e.g., [352]). The combination of the very large collecting area with these two instruments will open up several new fields of study for galaxy groups. The WFI survey, performed over the first four years of operations, is expected to find >10,000 groups and clusters at $z \geq 0.5$, including ~20 groups with $M_{500} \geq 5\times10^{13}$ $M_\odot$ at $z \sim 2$, and measure their temperature to better than 25% accuracy [353]. This will provide a clear view of the evolution of AGN feedback in groups back to the era of peak star formation and black hole growth. The identification of cavities and spectral mapping will be possible for moderating redshift, showing us the impact of feedback over the past few gigayears.

X-IFU offers capabilities that are similar to that of XRISM's *Resolve*, but with greatly improved spatial resolution and the ability to examine even low luminosity systems in the local universe. It will allow the mapping of turbulence and bulk flows in the IGrM, tracing gas motions that are associated with mergers, sloshing, uplift behind rising radio bubbles, or AGN-driven outflows. By mapping the kinetic and thermal energy content of the IGrM on spatial scales similar to those at which energy is injected by radio galaxies, it will allow us to quantify how much energy is injected into the hot gas by outbursts, determine where and when energy is transferred out of the radio jets and lobes, and see how it is then transported out into the surrounding halo [354]. It will also provide a clear view of the location of the coolest gas and allow us to trace the process by which it cools out of the hot phase.

6.4. *Lynx*

The Lynx mission concept (https://www.lynxobservatory.com/report accessed on 25 April 2021, [355]) will, if approved, go beyond Athena, with sub-arcsecond resolution over a $22'\times22'$ field of view and an effective area of 2 m^2 at 1 keV, giving 50$\times$ the throughput of *Chandra*. As with Athena, an active pixel array (the high-definition X-ray imager, HDXI) would provide wide-field CCD-like spectral imaging, but with $0.3''$ pixels to take advantage of the exquisite spatial resolution. The Lynx X-ray Microcalorimeter (LXM) would bring 3 eV spectral resolution on $1''$ spatial scales over a $5'$ field of view, with sub-arrays offering $0.5''$ spatial resolution or 0.3 eV spectral resolution in $1'$ fields. Much of Lynx's proposed science relates to the detailed physics of accretion and galaxy evolution, and the X-ray universe at high redshift; the survey observations with the HDXI would be capable of detecting groups with masses as low as 2×10^{13} $M_\odot$ out to a $z \sim 3$. In low-redshift systems, LXM could examine the conditions within individual cooling filaments in group cores, and measure the velocities of the weak shocks and sound waves that are produced during AGN outbursts. As with Athena, the observatory would be sensitive enough to trace the IGrM out to the virial radius in a large sample of groups, but with fine spatial resolution, making the identification of structure easier and the rejection of background sources cleaner. It is notable that Lynx would be the first mission after *Chandra* to be able to provide the finely detailed images that have proved to be so useful in the study of AGN feedback, reaching scales that are comparable to those of optical and radio observations.

6.5. The Square Kilometer Array and Its Precursors

Radio astronomy has undergone something of a renaissance in recent years, with new and improved capabilities coming online, e.g., the upgraded Jansky Very Large Array (JVLA) and Giant Metrewave Radio Telescope (uGMRT), the Low-Frequency Array (LOFAR), and the Atacama Large Milimeter Array (ALMA). These are providing improved radio continuum surveys in the northern hemisphere and equatorial sky. However, new telescopes in southern Africa and Australia are opening up new opportunities, as they begin to survey the relatively poorly-explored southern sky. These include the Murchison Widefield Array (MWA), the Australian Square Kilometer Array Pathfinder (ASKAP), and MeerKAT in South Africa's Karoo region. The Galactic and Extragalactic All-sky MWA Survey (GLEAM, [356]) provides an early example, covering the entire southern sky below Declination +30° at frequencies 72–231 MHz. While its spatial resolution is modest ($\sim$100″), its high sensitivity at low frequency and provision of fluxes in multiple bands makes it a powerful tool for studying radio galaxies, particularly old, fading sources. Higher frequency ($\sim$1 GHz), higher spatial resolution (8-30″) surveys of continuum emission and HI are becoming available from ASKAP (e.g., RACS, WALLABY, [357,358]) and MeerKAT (e.g., MIGHTEE, [359]), and these observatories are beginning to produce interesting findings on, e.g., group dynamics and evolution [360,361] and the population of giant radio galaxies [197,201].

These telescopes are the precursors of the Square Kilometer Array (SKA), a set of next-generation telescopes that are expected to begin operations in the late 2020s, combining wide frequency coverage with unprecedented sensitivity. The SKA will be built in phases, with phase 1 consisting of two components: the SKA1-Low, covering the 50–350 MHz band, with baselines up to 65 km providing spatial resolution of $\sim$4″ at 300 MHz and sensitivity a factor of 5–10 better than LOFAR or GMRT; and the SKA1-Mid, covering 350 MHz to 15 GHz with resolution 0.4″ at 1.4 GHz and a sensitivity up to an order of magnitude better than JVLA [362]. The proposed phase 2 SKA would improve sensitivity by another order of magnitude. Much of the SKA's proposed science relates to the early universe, but it will be an extraordinary tool for studies of feedback in groups and clusters, tracing the entire AGN population to high redshift, not merely the radio-loud systems that dominate current samples [363]. SKA surveys are likely to be sensitive to sources down to 10^{22} W Hz^{-1} out to z = 3–4 [364]. This offers an opportunity to detect the radio counterparts of most galaxies that are identified in current optical surveys, including essentially all group and cluster-dominant galaxies, with sufficient resolution to allow AGN and star formation emission to be disentangled, and with the wide frequency coverage that is necessary to determine the state and age of jets and lobes. Given the very large numbers of groups and clusters that are likely to be detected in the southern sky by, e.g., eROSITA, such surveys will play an important role in identifying systems with active cooling and feedback. HI observations reaching low column densities may provide another window on cooling from the IGrM. SKA will also open up the study of diffuse radio structures and magnetic fields in groups (e.g., [365]), providing constraints on rates of energy transport and conduction in the IGrM, as well as information on gas motions [366], and perhaps even on turbulence and shocks.

6.6. Upcoming SZ Facilities

Upcoming surveys of the cosmic microwave background (CMB), such as CCAT-prime [367], Simons Observatory [368], and CMB-S4 [369], will likely also play a role in advancing our understanding of AGN feedback at group scale by providing complementary information to X-ray surveys. While the thermal SZ effect (hereafter tSZ) is a steep function of mass ($Y_{SZ} \propto M_{500}^{5/3}$), stacking of the tSZ effect over large samples (either X-ray or optically selected) can lead to a detection down to M_{500} $\sim$$10^{13} M_{\odot}$. As a pilot study, Planck Collaboration et al. [370] presented the stacked tSZ signal from a large sample of SDSS galaxies selected to be central to their halo, and found that the tSZ-to-mass scaling relation extends with no break all the way down to M_{500} $\sim$$10^{12.5} M_{\odot}$. However, the interpretation of the result is rendered difficult by the large *Planck* beam ($\sim 8'$), which dilutes the signal [371]. While detecting the tSZ signal from individual galaxy groups will be challenging for the new generation of CMB survey instruments, the angular resolution of the foreseen facilities ($\lesssim 1'$) will be sufficient to study the distribution of the stacked tSZ signal and determine the origin of the signal identified by Planck Collaboration et al. [370]. On the other hand, large single-dish facilities such as AtLAST [372] will be sensitive enough to detect the tSZ effect from galaxy groups [373].

Recently, the kinetic SZ effect (kSZ), i.e., the Doppler shift of the CMB spectrum induced by moving electron clouds, has emerged as a promising tool for studying the baryon content of galaxy groups [374,375]. The kSZ signal is independent of the gas temperature, which makes it, in principle, more suitable than the tSZ for the study of low-mass systems. The kSZ signal cannot be detected directly by stacking CMB observations, given that the average velocity of structures with respect to the CMB rest frame vanishes. However, the kSZ signal can be measured by cross-correlating CMB maps with spectroscopic galaxy surveys, thereby fixing each system's velocity; this technique is known as the *pairwise* kSZ. Several recent studies reported low-significance detections of the kSZ with this technique [376–378]. These early results may indicate that the flat gas density profiles that are inferred from X-ray data (see Section 3.1.4) extend far beyond the halo's virial radius, which, if confirmed, provides a detection of the gas expelled from the central regions of halos by AGN feedback. Cross-correlating the kSZ data from future CMB experiments with large spectroscopic surveys, like DESI, may yield a detection of the pairwise kSZ at high significance [379] and possibly out to high redshifts [380].

Author Contributions: D.E.: lead author, Sections 1, 2, 3.1.3, 3.1.4, 5.2, 5.3, 5.4, 6.1 and 6.6; M.G.: Sections 3.1.3, 4 and 5.3; F.G.: Sections 3.1.1 and 3.1.2; A.M.C.L.B.: Sections 2, 5.1 and 5.2; E.O.: Sections 3.2, 3.3 and 6. All authors have read and agreed to the published version of the manuscript.

Funding: M.G. acknowledges partial support by NASA Chandra GO8-19104X/GO9-20114X and HST GO-15890.020-A grants. A.M.C.L.B. is supported by a fellowship of PSL University at the Paris Observatory. E.O. acknowledges support from NASA through *XMM-Newton* award number 80NSSC19K1056 and *Chandra* award number GO8-19112A.

Institutional Review Board Statement: Not applicable.

Informed Consent Statement: Not applicable.

Data Availability Statement: The review is based on public data and/or published papers.

Acknowledgments: We deeply thank Yannick Bahé, Weigang Cui, Marco De Petris, Federico Sembolini, Yohan Dubois, Scott Kay, Alisson Pellissier, Ewald Puchwein, Elena Rasia, Dylan Robson, Marcel van Daalen, Mark Vogelsberger and Rainer Weinberger for providing data from their simulations for inclusion in Figures 14 and 15, even if some could not be included in the end. We also thank Paul Nulsen for providing cavity parameters for inclusion in Table A1, and the anonymous referees for useful comments. We thank Ming Sun, Mark Voit, Iurii Babyk, Trevor Ponman and Aurel Schneider for granting us permission to reprint some of their figures. E.O. thanks G. Schellenberger and K. Kolokythas for useful conversations. D.E. thanks A. Finoguenov for useful discussions. D.E. and A.M.C.L.B. thank Benjamin Oppenheimer for useful discussions. We are also grateful to Tony Mroczkowski, Joop Schaye and Ming Sun for sending comments on the accepted version while we were checking the proofs, which were taken into account for the published version.

Conflicts of Interest: The authors declare no conflict of interest.

Appendix A. List of the Properties of Detected AGN Cavities in Galaxy Groups

Table A1. Properties of the cavities detected in groups. Only quantities available from the literature are included, thus for some systems the listing will be incomplete. [1] Projected semi-major axis of the cavity. [2] Projected semi-minor axis of the cavity. [3] Projected distance from the cavity center to the core. [4] Ages are reported as t_s-t_{buoy}-t_{refill}. Where only a single value is reported, this is t_s. [5] Bolometric luminosity between 0.001 and 100 keV inside r_{cool}, where the cooling time is less than 7.7 Gyr. * Additional data provided by Nulsen, P., priv. comm.

Source	a [1] (kpc)	b [2] (kpc)	R [3] (kpc)	pV (10^{56} ergs)	Age [4] (Myr)	L_{cool} [5] (10^{42} ergs/s)	Ref.
HCG 62 N	5.0	4.3	8.4	$2.9^{+4.1}_{-1.5}$	18-15-31	1.8 ± 0.2	[92,126,381]
HCG 62 S	4.0	4.0	8.6	$2.1^{+3.7}_{-1.3}$	19-16-29		
3C88 E	23	23	28	95	60		[115]
3C449 S	13	13	39	14.6	70		[382]
IC1262 N	2.2	1.5	6.5	58.0	17-24-52	$3.3^{+0.2}_{-0.3}$	[383]
IC1262 S	4	2	6.1	50.1	12-21-42		
NGC 5813 in SW	0.95	0.95	1.3	0.11	1.2		[384]
NGC 5813 in NE	1.03	0.93	1.4	0.15	1.4		
NGC 5813 mid SW	3.9	3.9	7.7	1.53	7.2		
NGC 5813 mid-1 NE	2.9	2.2	4.9	0.93	4.6		
NGC 5813 mid-2 NE	2.8	2.4	9.3	0.41	8.8		
NGC 5813 out SW	5.2	3.0	22.2	0.6	20.8		
NGC 5813 out NE	8.0	4.4	18.0	2.6	17.0		
IC 4296 NW	80	80	230	920	220		[203]
NGC 741 W	8	8	16	12.2 ± 1.2	30		[385]
NGC 193	63.4	47.2	0.0	$22.7^{+17.3}_{-17.7}$	44.2-20.4-76.9	0.11 ± 0.01	[199]
NGC 507 E	21.7	8.7	22.1	308^{+494}_{-63}	48.4-229-38.9	1.37 ± 0.02	[199]
NGC 507 W	13.4	5.0	11.7	90^{+254}_{-30}	22.2-17.4-38.9		
NGC 1550 E	5.96	2.31	9.0	$6.70^{+10.7}_{-1.98}$	12.8-27.9-37.0	$2.79^{+0.03}_{-0.01}$	[193]
NGC 1550 W	4.09	1.88	14.65	$2.00^{+2.45}_{-0.68}$	19.8-52.6-33.0		
NGC 4261 E	20.94	15.51	24.82	$31.02^{+11.98}_{-6.15}$	40.5-36.6-105.3	$0.11^{+0.08}_{-0.01}$	[106,191]
NGC 4261 W	18.62	16.91	21.72	$32.78^{+6.30}_{-6.52}$	35.4-31.8-99.4		
NGC 4636 NE	2.67	1.11	3.25	$0.28^{+0.40}_{-0.06}$	8.8-20.2-46.8	0.18 ± 0.01	[199]
NGC 4636 SE	2.40	1.52	2.80	$0.47^{+0.28}_{-0.10}$	7.6-17.0-47.2		
NGC 4636 SW	2.78	1.88	4.62	$0.88^{+0.48}_{-0.31}$	11.7-31.2-62.1		
NGC 4636 NW	2.53	1.32	3.33	$0.38^{+0.36}_{-0.08}$	9.1-21.5-49.7		
NGC 4782	10.7	10.7	23.0	11.0	35-54-		[386]
NGC 5044 SW	6.54	2.84	8.60	$2.17^{+2.87}_{-0.49}$	21.1-14.4-35.5	4.72 ± 0.01	[106,191]
NGC 5044 NW	3.04	2.34	4.85	$0.62^{+0.22}_{-0.13}$	11.4-16.4-35.5		
NGC 5044 in SW	0.15	0.15	0.45	0.0007	1		[195,209]
NGC 5044 in NE	0.15	0.15	0.45	0.0007	1		[195,209]
NGC 5098 N	3.0	1.6	2.97	7.0	18		[387]
NGC 5098 S	3.0	1.6	2.97	7.0	18		
NGC 5846 N	0.74	0.58	0.64	$0.35^{+0.15}_{-0.13}$	1.7-1.2-4.6	0.27 ± 0.01	[199]
NGC 5846 S	0.74	0.58	0.68	$0.35^{+0.15}_{-0.13}$	1.8-1.4-4.8		
NGC 5903	16.0	13.0	24.6	2.3 ± 0.10	82.5	0.0047 ± 0.0005	[388]
NGC 6269 N	5.2	5.2	10.7	$22.4^{+7.2}_{-9.0}$	14.0-14.3-26.2	$0.78^{+0.04}_{-0.03}$	[199]
NGC 6269 S	5.5	5.5	12.3	$27.2^{+8.2}_{-10.5}$	16.1-17.1-28.9		
NGC 6338 in NE	4.60	3.22	3.96	$20.38^{+9.65}_{-4.83}$	7.3-5.6-19.3	$4.56^{+0.08}_{-0.06}$	[210]
NGC 6338 in SW	4.22	3.22	6.34	$10.20^{+3.85}_{-2.53}$	11.4-14.7-30.8		
NGC 6338 outer	6.49	4.01	18.21	$12.98^{+8.38}_{-2.88}$	25.8-33.5-32.7		
VII Zw 700 NE	3.96	2.43	5.54	$0.24^{+0.17}_{-0.11}$	44.0-41.7-95.5	0.54 ± 0.03	[210]

Table A1. *Cont.*

Source	a [1] (kpc)	b [2] (kpc)	R [3] (kpc)	pV (10^{56} ergs)	Age [4] (Myr)	L^{5}_{cool} (10^{42} ergs/s)	Ref.
VII Zw 700 SW	4.70	1.90	3.43	$0.43^{+0.65}_{-0.19}$	27.3-29.3-82.8		
NGC 6868 NW	11.7	11.7	38.7	1.48	88-107-119		[389]
NGC 6868 SE	8.14	8.14	25.3	1.0	55		
A 1991 N	16.8	5.5	12.4	496	18-28-68	60.4	[390]
A 1991 S	13.3	6.2	11.5	535	18-29-59		
A 3581 1	7.9	3.8	3.1				[105]
A 3581 2	3.4	3.7	3.1				[105]
NGC 533 1	2.2	1.3	1.2				[105]
NGC 533 2	3.1	1.6	1.6				[105]
NGC 4104	1.9	1.5	0.0				[105]
RXC J0352.9+1941 1	8.0	5.9	9.3				[107]
RXC J0352.9+1941 2	7.9	4.3	10.8				
RX J0419+0225	1.2	0.9	1.7				[107]
A 2550 1	18.9	9.3	10.3				[105]
A 2550 2	10.7	5.9	7.8				[105]
A2717 1	11.2	6.3	7.9				[105]
A2717 2	13.4	5.8	8.4				[105]
AS1101 1	21.0	14.7	24.2				[105]
AS1101 2	24.1	15.7	23.6				[105]
ESO 351-021	12.2	8.6	14.8				[105]
RX J1159+5531 1	7.7	3.9	7.5				[105]
RX J1159+5531 2	6.7	4.3	9.7				[105]
RX J1206-0744	27.6	21.4	29.1				[105]
NGC 2300	1.3	1.0	1.5				[105]
UGC 5088 1	7.3	5.4	8.4				[105]
UGC 5088 2	6.5	3.6	5.4				[105]
NGC 777 E	1.9	2.3	4.6			0.99	[104,113]
NGC 777 W	2.1	2.4	4.0				[113]
NGC 4235 E	2.4	4.6	11.8				[113]
NGC 4235 W	2.1	2.4	4.0				[113]
NGC 1553 1	4.1	3.5	4.6	$0.42^{+0.49}_{-0.22}$	13.5-10.7-33.0	1.72	[104] *
NGC 1553 2	3.5	2.7	3.3	$0.28^{+0.23}_{-0.13}$	9.8-7.4-25.2	1.72	[104] *
NGC 1600 1	0.87	0.82	1.21	$0.15^{+0.24}_{-0.09}$	2.2-1.6-4.2	0.12	[104] *
NGC 1600 2	0.83	0.72	1.42	$0.11^{+0.13}_{-0.06}$	2.5-2.1-4.3	0.12	[104] *
NGC 3608 1	3.2	2.2	6.0	$0.04^{+0.03}_{-0.01}$	18.7-15.9-26.4	0.008	[104] *
NGC 3608 2	2.5	1.7	5.3	$0.02^{+0.02}_{-0.01}$	16.3-14.8-21.5	0.008	[104] *
NGC 7626 1	5.6	3.1	14.4	$0.37^{+0.21}_{-0.14}$	33.1-36.5-39.9	0.12	[104] *
NGC 7626 2	1.4	0.7	3.0	$0.03^{+0.03}_{-0.01}$	6.8-7.0-8.7	0.12	[104] *
NGC 7626 3	1.6	1.1	3.8	$0.06^{+0.04}_{-0.02}$	8.7-8.8-11.5	0.12	[104] *
NGC 7626 4	4.7	4.0	16.2	$0.36^{+0.48}_{-0.21}$	37.2-42.5-43.2	0.12	[104] *
A 262 E	2.6	2.6	6.2	$1.7^{+3.2}_{-1.1}$	11-13-20		[92,391]

References

1. Springel, V. The cosmological simulation code GADGET-2. *Mon. Not. R. Astron. Soc.* **2005**, *364*, 1105–1134. [CrossRef]
2. Robotham, A.S.G.; Norberg, P.; Driver, S.P.; Baldry, I.K.; Bamford, S.P.; Hopkins, A.M.; Liske, J.; Loveday, J.; Merson, A.; Peacock, J.A.; et al. Galaxy and Mass Assembly (GAMA): The GAMA galaxy group catalogue (G^{3}Cv1). *Mon. Not. R. Astron. Soc.* **2011**, *416*, 2640–2668. [CrossRef]
3. Davidzon, I.; Ilbert, O.; Laigle, C.; Coupon, J.; McCracken, H.J.; Delvecchio, I.; Masters, D.; Capak, P.; Hsieh, B.C.; Le Fèvre, O.; et al. The COSMOS2015 galaxy stellar mass function. Thirteen billion years of stellar mass assembly in ten snapshots. *Astron. Astrophys.* **2017**, *605*, A70, [CrossRef]

4. Behroozi, P.S.; Wechsler, R.H.; Conroy, C. The Average Star Formation Histories of Galaxies in Dark Matter Halos from z = 0–8. *Astrophys. J.* **2013**, *770*, 57, [CrossRef]
5. Leauthaud, A.; Tinker, J.; Bundy, K.; Behroozi, P.S.; Massey, R.; Rhodes, J.; George, M.R.; Kneib, J.P.; Benson, A.; Wechsler, R.H.; et al. New Constraints on the Evolution of the Stellar-to-dark Matter Connection: A Combined Analysis of Galaxy-Galaxy Lensing, Clustering, and Stellar Mass Functions from z = 0.2 to z =1. *Astrophys. J.* **2012**, *744*, 159, [CrossRef]
6. Coupon, J.; Arnouts, S.; van Waerbeke, L.; Moutard, T.; Ilbert, O.; van Uitert, E.; Erben, T.; Garilli, B.; Guzzo, L.; Heymans, C.; et al. The galaxy-halo connection from a joint lensing, clustering and abundance analysis in the CFHTLenS/VIPERS field. *Mon. Not. R. Astron. Soc.* **2015**, *449*, 1352–1379. [CrossRef]
7. Silk, J.; Rees, M.J. Quasars and galaxy formation. *Astron. Astrophys.* **1998**, *331*, L1–L4,
8. McNamara, B.R.; Nulsen, P.E.J. Heating Hot Atmospheres with Active Galactic Nuclei. *Annu. Rev. Astron. Astrophys.* **2007**, *45*, 117–175. [CrossRef]
9. Fabian, A.C. Observational Evidence of Active Galactic Nuclei Feedback. *Annu. Rev. Astron. Astrophys.* **2012**, *50*, 455–489. [CrossRef]
10. Gitti, M.; Brighenti, F.; McNamara, B.R. Evidence for AGN Feedback in Galaxy Clusters and Groups. *Adv. Astron.* **2012**, *2012*, 1–24. [CrossRef]
11. Gaspari, M.; Tombesi, F.; Cappi, M. Linking macro-, meso- and microscales in multiphase AGN feeding and feedback. *Nat. Astron.* **2020**, *4*, 10–13. [CrossRef]
12. Le Brun, A.M.C.; McCarthy, I.G.; Schaye, J.; Ponman, T.J. Towards a realistic population of simulated galaxy groups and clusters. *Mon. Not. R. Astron. Soc.* **2014**, *441*, 1270–1290. [CrossRef]
13. Gastaldello, F.; Buote, D.A.; Temi, P.; Brighenti, F.; Mathews, W.G.; Ettori, S. X-ray Cavities, Filaments, and Cold Fronts in the Core of the Galaxy Group NGC 5044. *Astrophys. J.* **2009**, *693*, 43–55. [CrossRef]
14. Randall, S.W.; Nulsen, P.E.J.; Jones, C.; Forman, W.R.; Bulbul, E.; Clarke, T.E.; Kraft, R.; Blanton, E.L.; David, L.; Werner, N.; et al. A Very Deep Chandra Observation of the Galaxy Group NGC 5813: AGN Shocks, Feedback, and Outburst History. *Astrophys. J.* **2015**, *805*, 112, [CrossRef]
15. Larson, R.B. Effects of supernovae on the early evolution of galaxies. *Mon. Not. R. Astron. Soc.* **1974**, *169*, 229–246. [CrossRef]
16. Rees, M.J.; Ostriker, J.P. Cooling, dynamics and fragmentation of massive gas clouds: Clues to the masses and radii of galaxies and clusters. *Mon. Not. R. Astron. Soc.* **1977**, *179*, 541–559. [CrossRef]
17. White, S.D.M.; Rees, M.J. Core condensation in heavy halos: A two-stage theory for galaxy formation and clustering. *Mon. Not. R. Astron. Soc.* **1978**, *183*, 341–358. [CrossRef]
18. Granato, G.L.; De Zotti, G.; Silva, L.; Bressan, A.; Danese, L. A Physical Model for the Coevolution of QSOs and Their Spheroidal Hosts. *Astrophys. J.* **2004**, *600*, 580–594. [CrossRef]
19. Bower, R.G.; Benson, A.J.; Malbon, R.; Helly, J.C.; Frenk, C.S.; Baugh, C.M.; Cole, S.; Lacey, C.G. Breaking the hierarchy of galaxy formation. *Mon. Not. R. Astron. Soc.* **2006**, *370*, 645–655. [CrossRef]
20. Croton, D.J.; Springel, V.; White, S.D.M.; De Lucia, G.; Frenk, C.S.; Gao, L.; Jenkins, A.; Kauffmann, G.; Navarro, J.F.; Yoshida, N. The many lives of active galactic nuclei: Cooling flows, black holes and the luminosities and colours of galaxies. *Mon. Not. R. Astron. Soc.* **2006**, *365*, 11–28. [CrossRef]
21. Di Matteo, T.; Colberg, J.; Springel, V.; Hernquist, L.; Sijacki, D. Direct Cosmological Simulations of the Growth of Black Holes and Galaxies. *Astrophys. J.* **2008**, *676*, 33–53. [CrossRef]
22. Puchwein, E.; Sijacki, D.; Springel, V. Simulations of AGN Feedback in Galaxy Clusters and Groups: Impact on Gas Fractions and the L_X-T Scaling Relation. *Astrophys. J.* **2008**, *687*, L53–L56, [CrossRef]
23. Cowie, L.L.; Songaila, A.; Hu, E.M.; Cohen, J.G. New Insight on Galaxy Formation and Evolution From Keck Spectroscopy of the Hawaii Deep Fields. *Astron. J.* **1996**, *112*, 839, [CrossRef]
24. Shaver, P.A.; Wall, J.V.; Kellermann, K.I.; Jackson, C.A.; Hawkins, M.R.S. Decrease in the space density of quasars at high redshift. *Nature* **1996**, *384*, 439–441. [CrossRef]
25. Madau, P.; Ferguson, H.C.; Dickinson, M.E.; Giavalisco, M.; Steidel, C.C.; Fruchter, A. High-redshift galaxies in the Hubble Deep Field: Colour selection and star formation history to $z \sim 4$. *Mon. Not. R. Astron. Soc.* **1996**, *283*, 1388–1404. [CrossRef]
26. Benson, A.J.; Bower, R.G.; Frenk, C.S.; Lacey, C.G.; Baugh, C.M.; Cole, S. What Shapes the Luminosity Function of Galaxies? *Astrophys. J.* **2003**, *599*, 38–49. [CrossRef]
27. Bower, R.G.; McCarthy, I.G.; Benson, A.J. The flip side of galaxy formation: A combined model of galaxy formation and cluster heating. *Mon. Not. R. Astron. Soc.* **2008**, *390*, 1399–1410. [CrossRef]
28. Silk, J. On the fragmentation of cosmic gas clouds. I. The formation of galaxies and the first generation of stars. *Astrophys. J.* **1977**, *211*, 638–648. [CrossRef]
29. Cole, S. Modeling Galaxy Formation in Evolving Dark Matter Halos. *Astrophys. J.* **1991**, *367*, 45. [CrossRef]
30. White, S.D.M.; Frenk, C.S. Galaxy Formation through Hierarchical Clustering. *Astrophys. J.* **1991**, *379*, 52. [CrossRef]
31. Blanchard, A.; Valls-Gabaud, D.; Mamon, G.A. The origin of the galaxy luminosity function and the thermal evolution of the intergalactic medium. *Astron. Astrophys.* **1992**, *264*, 365–378.
32. Katz, N. Dissipational Galaxy Formation. II. Effects of Star Formation. *Astrophys. J.* **1992**, *391*, 502. [CrossRef]
33. Kauffmann, G.; Colberg, J.M.; Diaferio, A.; White, S.D.M. Clustering of galaxies in a hierarchical universe—I. Methods and results at z=0. *Mon. Not. R. Astron. Soc.* **1999**, *303*, 188–206. [CrossRef]

34. van Kampen, E.; Jimenez, R.; Peacock, J.A. Overmerging and mass-to-light ratios in phenomenological galaxy formation models. *Mon. Not. R. Astron. Soc.* **1999**, *310*, 43–56. [CrossRef]
35. Cole, S.; Lacey, C.G.; Baugh, C.M.; Frenk, C.S. Hierarchical galaxy formation. *Mon. Not. R. Astron. Soc.* **2000**, *319*, 168–204. [CrossRef]
36. Cole, S.; Norberg, P.; Baugh, C.M.; Frenk, C.S.; Bland-Hawthorn, J.; Bridges, T.; Cannon, R.; Colless, M.; Collins, C.; Couch, W.; et al. The 2dF galaxy redshift survey: Near-infrared galaxy luminosity functions. *Mon. Not. R. Astron. Soc.* **2001**, *326*, 255–273. [CrossRef]
37. Balogh, M.L.; Pearce, F.R.; Bower, R.G.; Kay, S.T. Revisiting the cosmic cooling crisis. *Mon. Not. R. Astron. Soc.* **2001**, *326*, 1228–1234. [CrossRef]
38. Lin, Y.T.; Mohr, J.J.; Stanford, S.A. Near-Infrared Properties of Galaxy Clusters: Luminosity as a Binding Mass Predictor and the State of Cluster Baryons. *Astrophys. J.* **2003**, *591*, 749–763. [CrossRef]
39. Yang, X.; Mo, H.J.; van den Bosch, F.C.; Jing, Y.P. A halo-based galaxy group finder: Calibration and application to the 2dFGRS. *Mon. Not. R. Astron. Soc.* **2005**, *356*, 1293–1307. [CrossRef]
40. Eke, V.R.; Baugh, C.M.; Cole, S.; Frenk, C.S.; Navarro, J.F. Galaxy groups in the 2dF Galaxy Redshift Survey: The number density of groups. *Mon. Not. R. Astron. Soc.* **2006**, *370*, 1147–1158. [CrossRef]
41. McCarthy, I.G.; Schaye, J.; Bird, S.; Le Brun, A.M.C. The BAHAMAS project: Calibrated hydrodynamical simulations for large-scale structure cosmology. *Mon. Not. R. Astron. Soc.* **2017**, *465*, 2936–2965. [CrossRef]
42. Tinker, J.; Kravtsov, A.V.; Klypin, A.; Abazajian, K.; Warren, M.; Yepes, G.; Gottlöber, S.; Holz, D.E. Toward a Halo Mass Function for Precision Cosmology: The Limits of Universality. *Astrophys. J.* **2008**, *688*, 709–728. [CrossRef]
43. Evrard, A.E.; Henry, J.P. Expectations for X-ray cluster observations by the ROSAT satellite. *Astrophys. J.* **1991**, *383*, 95–103. [CrossRef]
44. Kaiser, N. Evolution of Clusters of Galaxies. *Astrophys. J.* **1991**, *383*, 104. [CrossRef]
45. Balogh, M.L.; McCarthy, I.G.; Bower, R.G.; Eke, V.R. Testing cold dark matter with the hierarchical build-up of stellar light. *Mon. Not. R. Astron. Soc.* **2008**, *385*, 1003–1014. [CrossRef]
46. Bryan, G.L. Explaining the Entropy Excess in Clusters and Groups of Galaxies without Additional Heating. *Astrophys. J.* **2000**, *544*, L1–L5, [CrossRef]
47. Muanwong, O.; Thomas, P.A.; Kay, S.T.; Pearce, F.R. The effect of cooling and preheating on the X-ray properties of clusters of galaxies. *Mon. Not. R. Astron. Soc.* **2002**, *336*, 527–540. [CrossRef]
48. Voit, G.M.; Bryan, G.L. Regulation of the X-ray luminosity of clusters of galaxies by cooling and supernova feedback. *Nature* **2001**, *414*, 425–427. [CrossRef] [PubMed]
49. Voit, G.M.; Bryan, G.L.; Balogh, M.L.; Bower, R.G. Modified Entropy Models for the Intracluster Medium. *Astrophys. J.* **2002**, *576*, 601–624. [CrossRef]
50. Wu, X.P.; Xue, Y.J. The Effect of Radiative Cooling on the Scale Dependence of the Global Stellar and Gas Content of Groups and Clusters of Galaxies. *Astrophys. J.* **2002**, *572*, L19–L22, [CrossRef]
51. Menci, N.; Cavaliere, A. The history of cosmic baryons: X-ray emission versus star formation rate. *Mon. Not. R. Astron. Soc.* **2000**, *311*, 50–62. [CrossRef]
52. Bower, R.G.; Benson, A.J.; Lacey, C.G.; Baugh, C.M.; Cole, S.; Frenk, C.S. The impact of galaxy formation on the X-ray evolution of clusters. *Mon. Not. R. Astron. Soc.* **2001**, *325*, 497–508. [CrossRef]
53. Kay, S.T.; Thomas, P.A.; Theuns, T. The impact of galaxy formation on X-ray groups. *Mon. Not. R. Astron. Soc.* **2003**, *343*, 608–618. [CrossRef]
54. Valdarnini, R. Iron abundances and heating of the intracluster medium in hydrodynamical simulations of galaxy clusters. *Mon. Not. R. Astron. Soc.* **2003**, *339*, 1117–1134. [CrossRef]
55. Nagai, D.; Vikhlinin, A.; Kravtsov, A.V. Testing X-ray Measurements of Galaxy Clusters with Cosmological Simulations. *Astrophys. J.* **2007**, *655*, 98–108. [CrossRef]
56. Valageas, P.; Silk, J. The entropy history of the universe. *Astron. Astrophys.* **1999**, *350*, 725–742.
57. Kravtsov, A.V.; Yepes, G. On the supernova heating of the intergalactic medium. *Mon. Not. R. Astron. Soc.* **2000**, *318*, 227–238. [CrossRef]
58. Wu, K.K.S.; Fabian, A.C.; Nulsen, P.E.J. Non-gravitational heating in the hierarchical formation of X-ray clusters. *Mon. Not. R. Astron. Soc.* **2000**, *318*, 889–912. [CrossRef]
59. Martin, C.L. Properties of Galactic Outflows: Measurements of the Feedback from Star Formation. *Astrophys. J.* **1999**, *513*, 156–160. [CrossRef]
60. Fabian, A.C.; Voigt, L.M.; Morris, R.G. On conduction, cooling flows and galaxy formation. *Mon. Not. R. Astron. Soc.* **2002**, *335*, L71–L74, [CrossRef]
61. Dolag, K.; Jubelgas, M.; Springel, V.; Borgani, S.; Rasia, E. Thermal Conduction in Simulated Galaxy Clusters. *Astrophys. J.* **2004**, *606*, L97–L100, [CrossRef]
62. Pope, E.C.D.; Pavlovski, G.; Kaiser, C.R.; Fangohr, H. The effects of thermal conduction on the intracluster medium of the Virgo cluster. *Mon. Not. R. Astron. Soc.* **2005**, *364*, 13–28. [CrossRef]
63. Kauffmann, G.; Haehnelt, M. A unified model for the evolution of galaxies and quasars. *Mon. Not. R. Astron. Soc.* **2000**, *311*, 576–588. [CrossRef]

64. Kormendy, J.; Ho, L.C. Coevolution (Or Not) of Supermassive Black Holes and Host Galaxies. *Annu. Rev. Astron. Astrophys.* **2013**, *51*, 511–653. [CrossRef]
65. Magorrian, J.; Tremaine, S.; Richstone, D.; Bender, R.; Bower, G.; Dressler, A.; Faber, S.M.; Gebhardt, K.; Green, R.; Grillmair, C.; et al. The Demography of Massive Dark Objects in Galaxy Centers. *Astron. J.* **1998**, *115*, 2285–2305. [CrossRef]
66. Ferrarese, L.; Merritt, D. A Fundamental Relation between Supermassive Black Holes and Their Host Galaxies. *Astrophys. J.* **2000**, *539*, L9–L12, [CrossRef]
67. Gültekin, K.; Richstone, D.O.; Gebhardt, K.; Lauer, T.R.; Tremaine, S.; Aller, M.C.; Bender, R.; Dressler, A.; Faber, S.M.; Filippenko, A.V.; et al. The M-σ and M-L Relations in Galactic Bulges, and Determinations of Their Intrinsic Scatter. *Astrophys. J.* **2009**, *698*, 198–221. [CrossRef]
68. McConnell, N.J.; Ma, C.P.; Gebhardt, K.; Wright, S.A.; Murphy, J.D.; Lauer, T.R.; Graham, J.R.; Richstone, D.O. Two ten-billion-solar-mass black holes at the centres of giant elliptical galaxies. *Nature* **2011**, *480*, 215–218. [CrossRef]
69. McConnell, N.J.; Ma, C.P. Revisiting the Scaling Relations of Black Hole Masses and Host Galaxy Properties. *Astrophys. J.* **2013**, *764*, 184, [CrossRef]
70. van den Bosch, R.C.E. Unification of the fundamental plane and Super Massive Black Hole Masses. *Astrophys. J.* **2016**, *831*, 134, [CrossRef]
71. Fabian, A.C. The obscured growth of massive black holes. *Mon. Not. R. Astron. Soc.* **1999**, *308*, L39–L43, [CrossRef]
72. Cattaneo, A.; Haehnelt, M.G.; Rees, M.J. The distribution of supermassive black holes in the nuclei of nearby galaxies. *Mon. Not. R. Astron. Soc.* **1999**, *308*, 77–81. [CrossRef]
73. Saglia, R.P.; Opitsch, M.; Erwin, P.; Thomas, J.; Beifiori, A.; Fabricius, M.; Mazzalay, X.; Nowak, N.; Rusli, S.P.; Bender, R. The SINFONI Black Hole Survey: The Black Hole Fundamental Plane Revisited and the Paths of (Co)evolution of Supermassive Black Holes and Bulges. *Astrophys. J.* **2016**, *818*, 47, [CrossRef]
74. Ferrarese, L. Beyond the Bulge: A Fundamental Relation between Supermassive Black Holes and Dark Matter Halos. *Astrophys. J.* **2002**, *578*, 90–97. [CrossRef]
75. Booth, C.M.; Schaye, J. Dark matter haloes determine the masses of supermassive black holes. *Mon. Not. R. Astron. Soc.* **2010**, *405*, L1–L5, [CrossRef]
76. Bogdán, Á.; Lovisari, L.; Volonteri, M.; Dubois, Y. Correlation between the Total Gravitating Mass of Groups and Clusters and the Supermassive Black Hole Mass of Brightest Galaxies. *Astrophys. J.* **2018**, *852*, 131. [CrossRef]
77. Gaspari, M.; Eckert, D.; Ettori, S.; Tozzi, P.; Bassini, L.; Rasia, E.; Brighenti, F.; Sun, M.; Borgani, S.; Johnson, S.D.; et al. The X-ray Halo Scaling Relations of Supermassive Black Holes. *Astrophys. J.* **2019**, *884*, 169, [CrossRef]
78. Lakhchaura, K.; Truong, N.; Werner, N. Correlations between supermassive black holes, hot atmospheres, and the total masses of early-type galaxies. *Mon. Not. R. Astron. Soc.* **2019**, *488*, L134–L142, [CrossRef]
79. Fabian, A.C. Cooling Flows in Clusters of Galaxies. *Annu. Rev. Astron. Astrophys.* **1994**, *32*, 277–318. [CrossRef]
80. Kaastra, J.S.; Ferrigno, C.; Tamura, T.; Paerels, F.B.S.; Peterson, J.R.; Mittaz, J.P.D. XMM-Newton observations of the cluster of galaxies Sérsic 159-03. *Astron. Astrophys.* **2001**, *365*, L99–L103, [CrossRef]
81. Peterson, J.R.; Paerels, F.B.S.; Kaastra, J.S.; Arnaud, M.; Reiprich, T.H.; Fabian, A.C.; Mushotzky, R.F.; Jernigan, J.G.; Sakelliou, I. X-ray imaging-spectroscopy of Abell 1835. *Astron. Astrophys.* **2001**, *365*, L104–L109, [CrossRef]
82. Molendi, S.; Pizzolato, F. Is the Gas in Cooling Flows Multiphase? *Astrophys. J.* **2001**, *560*, 194–200. [CrossRef]
83. Peterson, J.R.; Fabian, A.C. X-ray spectroscopy of cooling clusters. *Phys. Rep.* **2006**, *427*, 1–39. [CrossRef]
84. Narayan, R.; Medvedev, M.V. Thermal Conduction in Clusters of Galaxies. *Astrophys. J.* **2001**, *562*, L129–L132, [CrossRef]
85. Motl, P.M.; Burns, J.O.; Loken, C.; Norman, M.L.; Bryan, G. Formation of Cool Cores in Galaxy Clusters via Hierarchical Mergers. *Astrophys. J.* **2004**, *606*, 635–653. [CrossRef]
86. ZuHone, J.A.; Markevitch, M.; Johnson, R.E. Stirring Up the Pot: Can Cooling Flows in Galaxy Clusters be Quenched by Gas Sloshing? *Astrophys. J.* **2010**, *717*, 908–928. [CrossRef]
87. Silk, J.; Djorgovski, S.; Wyse, R.F.G.; Bruzual, A. Self-regulated Cooling Flows in Elliptical Galaxies and in Cluster Cores: Is Excusively Low Mass Star Formation Really Necessary? *Astrophys. J.* **1986**, *307*, 415. [CrossRef]
88. Burns, J.O. The radio properties of cD galaxies in Abell clusters. I—An X-ray selected sample. *Astron. J.* **1990**, *99*, 14–30. [CrossRef]
89. Best, P.N.; von der Linden, A.; Kauffmann, G.; Heckman, T.M.; Kaiser, C.R. On the prevalence of radio-loud active galactic nuclei in brightest cluster galaxies: Implications for AGN heating of cooling flows. *Mon. Not. R. Astron. Soc.* **2007**, *379*, 894–908. [CrossRef]
90. Finoguenov, A.; Jones, C. Chandra Observation of M84, a Radio Lobe Elliptical Galaxy in the Virgo Cluster. *Astrophys. J.* **2001**, *547*, L107–L110, [CrossRef]
91. Vrtilek, J.M.; Grego, L.; David, L.P.; Ponman, T.J.; Forman, W.; Jones, C.; Harris, D.E. *A Sharper Picture of X-ray Bright Galaxy Groups: Chandra Imaging and Spectroscopy of HCG 62 and NGC 741*; American Physical Society: Albuquerque, NM, USA, 2002; p. B17.107.
92. Bîrzan, L.; Rafferty, D.A.; McNamara, B.R.; Wise, M.W.; Nulsen, P.E.J. A Systematic Study of Radio-induced X-ray Cavities in Clusters, Groups, and Galaxies. *Astrophys. J.* **2004**, *607*, 800–809. [CrossRef]
93. Forman, W.; Nulsen, P.; Heinz, S.; Owen, F.; Eilek, J.; Vikhlinin, A.; Markevitch, M.; Kraft, R.; Churazov, E.; Jones, C. Reflections of Active Galactic Nucleus Outbursts in the Gaseous Atmosphere of M87. *Astrophys. J.* **2005**, *635*, 894–906. [CrossRef]

94. Allen, S.W.; Dunn, R.J.H.; Fabian, A.C.; Taylor, G.B.; Reynolds, C.S. The relation between accretion rate and jet power in X-ray luminous elliptical galaxies. *Mon. Not. R. Astron. Soc.* **2006**, *372*, 21–30. [CrossRef]
95. Fabian, A.C.; Sanders, J.S.; Taylor, G.B.; Allen, S.W.; Crawford, C.S.; Johnstone, R.M.; Iwasawa, K. A very deep Chandra observation of the Perseus cluster: Shocks, ripples and conduction. *Mon. Not. R. Astron. Soc.* **2006**, *366*, 417–428. [CrossRef]
96. Dunn, R.J.H.; Fabian, A.C. Investigating AGN heating in a sample of nearby clusters. *Mon. Not. R. Astron. Soc.* **2006**, *373*, 959–971. [CrossRef]
97. Jetha, N.N.; Ponman, T.J.; Hardcastle, M.J.; Croston, J.H. Active galactic nuclei heating in the centres of galaxy groups: A statistical study. *Mon. Not. R. Astron. Soc.* **2007**, *376*, 193–204. [CrossRef]
98. Croston, J.H. Jet/Environment Interactions in Low-Power Radio Galaxies. In *Extragalactic Jets: Theory and Observation from Radio to Gamma Ray*; Rector, T.A., De Young, D.S., Eds.; Astronomical Society of the Pacific: Girdwood, AK, USA, 2008; Volume 386, p. 335.
99. Churazov, E.; Sunyaev, R.; Forman, W.; Böhringer, H. Cooling flows as a calorimeter of active galactic nucleus mechanical power. *Mon. Not. R. Astron. Soc.* **2002**, *332*, 729–734. [CrossRef]
100. Scheuer, P.A.G. Models of extragalactic radio sources with a continuous energy supply from a central object. *Mon. Not. R. Astron. Soc.* **1974**, *166*, 513–528. [CrossRef]
101. Nulsen, P.E.J.; McNamara, B.R.; Wise, M.W.; David, L.P. The Cluster-Scale AGN Outburst in Hydra A. *Astrophys. J.* **2005**, *628*, 629–636. [CrossRef]
102. McNamara, B.R.; Nulsen, P.E.J.; Wise, M.W.; Rafferty, D.A.; Carilli, C.; Sarazin, C.L.; Blanton, E.L. The heating of gas in a galaxy cluster by X-ray cavities and large-scale shock fronts. *Nature* **2005**, *433*, 45–47. [CrossRef]
103. Nulsen, P.E.J.; Jones, C.; Forman, W.R.; David, L.P.; McNamara, B.R.; Rafferty, D.A.; Bîrzan, L.; Wise, M.W. AGN Heating Through Cavities and Shocks. In *Heating versus Cooling in Galaxies and Clusters of Galaxies*; Springer-Verlag series "ESO Astrophysics Symposia."; Böhringer, H., Pratt, G.W., Finoguenov, A., Schuecker, P., Eds.; Springer: Heidelberg, Germany, 2007; p. 210. [CrossRef]
104. Cavagnolo, K.W.; McNamara, B.R.; Nulsen, P.E.J.; Carilli, C.L.; Jones, C.; Birzan, L. A relationship between AGN jet power and radio power. *Astrophys. J.* **2010**, *720*, 1066–1072. [CrossRef]
105. Dong, R.; Rasmussen, J.; Mulchaey, J.S. A Systematic Search for X-ray Cavities in the Hot Gas of Galaxy Groups. *Astrophys. J.* **2010**, *712*, 883–900. [CrossRef]
106. O'Sullivan, E.; Ponman, T.J.; Kolokythas, K.; Raychaudhury, S.; Babul, A.; Vrtilek, J.M.; David, L.P.; Giacintucci, S.; Gitti, M.; Haines, C.P. The Complete Local Volume Groups Sample—I. Sample selection and X-ray properties of the high-richness subsample. *Mon. Not. R. Astron. Soc.* **2017**, *472*, 1482–1505. [CrossRef]
107. Shin, J.; Woo, J.H.; Mulchaey, J.S. A Systematic Search for X-ray Cavities in Galaxy Clusters, Groups, and Elliptical Galaxies. *Astrophys. J.* **2016**, *227*, 31, [CrossRef]
108. Kim, D.W.; Anderson, C.; Burke, D.; D'Abrusco, R.; Fabbiano, G.; Fruscione, A.; Lauer, J.; McCollough, M.; Morgan, D.; Mossman, A.; et al. Chandra Early-type Galaxy Atlas. *Astrophys. J. Suppl. Ser.* **2019**, *241*, 36. [CrossRef]
109. McNamara, B.R.; Nulsen, P.E.J. Mechanical feedback from active galactic nuclei in galaxies, groups and clusters. *New J. Phys.* **2012**, *14*, 055023, [CrossRef]
110. Abdulla, Z.; Carlstrom, J.E.; Mantz, A.B.; Marrone, D.P.; Greer, C.H.; Lamb, J.W.; Leitch, E.M.; Muchovej, S.; O'Donnell, C.; Plagge, T.J.; et al. Constraints on the Thermal Contents of the X-ray Cavities of Cluster MS 0735.6+7421 with Sunyaev-Zel'dovich Effect Observations. *Astrophys. J.* **2019**, *871*, 195, [CrossRef]
111. Churazov, E.; Brüggen, M.; Kaiser, C.R.; Böhringer, H.; Forman, W. Evolution of Buoyant Bubbles in M87. *Astrophys. J.* **2001**, *554*, 261–273. [CrossRef]
112. Bîrzan, L.; McNamara, B.R.; Nulsen, P.E.J.; Carilli, C.L.; Wise, M.W. Radiative Efficiency and Content of Extragalactic Radio Sources: Toward a Universal Scaling Relation between Jet Power and Radio Power. *Astrophys. J.* **2008**, *686*, 859. [CrossRef]
113. Panagoulia, E.K.; Fabian, A.C.; Sanders, J.S.; Hlavacek-Larrondo, J. A volume-limited sample of X-ray galaxy groups and clusters—II. X-ray cavity dynamics. *Mon. Not. R. Astron. Soc.* **2014**, *444*, 1236–1259. [CrossRef]
114. Salvatier, J.; Wiecki, T.V.; Fonnesbeck, C. Probabilistic programming in Python using PyMC3. *PeerJ Comput. Sci.* **2016**, *2*, e55. [CrossRef]
115. Liu, W.; Sun, M.; Nulsen, P.; Clarke, T.; Sarazin, C.; Forman, W.; Gaspari, M.; Giacintucci, S.; Lal, D.V.; Edge, T. AGN feedback in galaxy group 3C 88: Cavities, shock, and jet reorientation. *Mon. Not. R. Astron. Soc. Lett.* **2019**, *484*, 3376–3392. [CrossRef]
116. Tang, X.; Churazov, E. Sound wave generation by a spherically symmetric outburst and AGN feedback in galaxy clusters. *Mon. Not. R. Astron. Soc.* **2017**, *468*, 3516–3532. [CrossRef]
117. Machacek, M.; Nulsen, P.E.J.; Jones, C.; Forman, W.R. Chandra Observations of Nuclear Outflows in the Elliptical Galaxy NGC 4552 in the Virgo Cluster. *Astrophys. J.* **2006**, *648*, 947–955. [CrossRef]
118. Forman, W.; Churazov, E.; Jones, C.; Heinz, S.; Kraft, R.; Vikhlinin, A. Partitioning the Outburst Energy of a Low Eddington Accretion Rate AGN at the Center of an Elliptical Galaxy: The Recent 12 Myr History of the Supermassive Black Hole in M87. *Astrophys. J.* **2017**, *844*, 122, [CrossRef]
119. Wise, M.W.; McNamara, B.R.; Nulsen, P.E.J.; Houck, J.C.; David, L.P. X-ray Supercavities in the Hydra A Cluster and the Outburst History of the Central Galaxy's Active Nucleus. *Astrophys. J.* **2007**, *659*, 1153–1158. [CrossRef]

120. Croston, J.H.; Kraft, R.P.; Hardcastle, M.J.; Birkinshaw, M.; Worrall, D.M.; Nulsen, P.E.J.; Penna, R.F.; Sivakoff, G.R.; Jordán, A.; Brassington, N.J.; et al. High-energy particle acceleration at the radio-lobe shock of Centaurus A. *Mon. Not. R. Astron. Soc.* **2009**, *395*, 1999–2012. [CrossRef]
121. Snios, B.; Nulsen, P.E.J.; Wise, M.W.; de Vries, M.; Birkinshaw, M.; Worrall, D.M.; Duffy, R.T.; Kraft, R.P.; McNamara, B.R.; Carilli, C.; et al. The Cocoon Shocks of Cygnus A: Pressures and Their Implications for the Jets and Lobes. *Astrophys. J.* **2018**, *855*, 71, [CrossRef]
122. Croston, J.H.; Hardcastle, M.J.; Mingo, B.; Evans, D.A.; Dicken, D.; Morganti, R.; Tadhunter, C.N. A Large-scale Shock Surrounding a Powerful Radio Galaxy? *Astrophys. J.* **2011**, *734*, L28, [CrossRef]
123. Blanton, E.L.; Randall, S.W.; Douglass, E.M.; Sarazin, C.L.; Clarke, T.E.; McNamara, B.R. Shocks and Bubbles in a Deep Chandra Observation of the Cooling Flow Cluster Abell 2052. *Astrophys. J.* **2009**, *697*, L95–L98, [CrossRef]
124. Kraft, R.P.; Birkinshaw, M.; Nulsen, P.E.J.; Worrall, D.M.; Croston, J.H.; Forman, W.R.; Hardcastle, M.J.; Jones, C.; Murray, S.S. An Active Galactic Nucleus Driven Shock in the Intracluster Medium around the Radio Galaxy 3C 310. *Astrophys. J.* **2012**, *749*, 19. [CrossRef]
125. Baldi, A.; Forman, W.; Jones, C.; Kraft, R.; Nulsen, P.; Churazov, E.; David, L.; Giacintucci, S. The Unusual X-ray Morphology of NGC 4636 Revealed by Deep Chandra Observations: Cavities and Shocks Created by Past Active Galactic Nucleus Outbursts. *Astrophys. J.* **2009**, *707*, 1034–1043. [CrossRef]
126. Gitti, M.; O'Sullivan, E.; Giacintucci, S.; David, L.P.; Vrtilek, J.; Raychaudhury, S.; Nulsen, P.E.J. Cavities and Shocks in the Galaxy Group HCG 62 as Revealed by Chandra, XMM-Newton, and Giant Metrewave Radio Telescope Data. *Astrophys. J.* **2010**, *714*, 758–771. [CrossRef]
127. Kaiser, N. Evolution and clustering of rich clusters. *Mon. Not. R. Astron. Soc.* **1986**, *222*, 323–345. [CrossRef]
128. Allen, S.W.; Fabian, A.C. The impact of cooling flows on the T_X-L_Bol relation for the most luminous clusters. *Mon. Not. R. Astron. Soc.* **1998**, *297*, L57–L62, [CrossRef]
129. Arnaud, M.; Evrard, A.E. The L_X-T relation and intracluster gas fractions of X-ray clusters. *Mon. Not. R. Astron. Soc.* **1999**, *305*, 631–640. [CrossRef]
130. Markevitch, M. The L_X-T Relation and Temperature Function for Nearby Clusters Revisited. *Astrophys. J.* **1998**, *504*, 27–34. [CrossRef]
131. Tozzi, P.; Scharf, C.; Norman, C. Detection of the Entropy of the Intergalactic Medium: Accretion Shocks in Clusters, Adiabatic Cores in Groups. *Astrophys. J.* **2000**, *542*, 106–119. [CrossRef]
132. Tozzi, P.; Norman, C. The Evolution of X-ray Clusters and the Entropy of the Intracluster Medium. *Astrophys. J.* **2001**, *546*, 63–84. [CrossRef]
133. Knight, P.A.; Ponman, T.J. The properties of the hot gas in galaxy groups and clusters from 1D hydrodynamical simulations—I. Cosmological infall models. *Mon. Not. R. Astron. Soc.* **1997**, *289*, 955–972. [CrossRef]
134. Ponman, T.J.; Cannon, D.B.; Navarro, J.F. The thermal imprint of galaxy formation on X-ray clusters. *Nature* **1999**, *397*, 135–137. [CrossRef]
135. Borgani, S.; Viel, M. The evolution of a pre-heated intergalactic medium. *Mon. Not. R. Astron. Soc.* **2009**, *392*, L26–L30, [CrossRef]
136. Eke, V.R.; Navarro, J.F.; Frenk, C.S. The Evolution of X-ray Clusters in a Low-Density Universe. *Astrophys. J.* **1998**, *503*, 569–592. [CrossRef]
137. Sun, M.; Voit, G.M.; Donahue, M.; Jones, C.; Forman, W.; Vikhlinin, A. Chandra Studies of the X-ray Gas Properties of Galaxy Groups. *Astrophys. J.* **2009**, *693*, 1142–1172. [CrossRef]
138. Vikhlinin, A.; Burenin, R.A.; Ebeling, H.; Forman, W.R.; Hornstrup, A.; Jones, C.; Kravtsov, A.V.; Murray, S.S.; Nagai, D.; Quintana, H.; et al. Chandra Cluster Cosmology Project. II. Samples and X-ray Data Reduction. *Astrophys. J.* **2009**, *692*, 1033–1059. [CrossRef]
139. Lloyd-Davies, E.J.; Ponman, T.J.; Cannon, D.B. The entropy and energy of intergalactic gas in galaxy clusters. *Mon. Not. R. Astron. Soc.* **2000**, *315*, 689–702. [CrossRef]
140. Ponman, T.J.; Sanderson, A.J.R.; Finoguenov, A. The Birmingham-CfA cluster scaling project—III. Entropy and similarity in galaxy systems. *Mon. Not. R. Astron. Soc.* **2003**, *343*, 331–342. [CrossRef]
141. Brighenti, F.; Mathews, W.G. Entropy Evolution in Galaxy Groups and Clusters: A Comparison of External and Internal Heating. *Astrophys. J.* **2001**, *553*, 103–120. [CrossRef]
142. Finoguenov, A.; Jones, C.; Böhringer, H.; Ponman, T.J. ASCA Observations of Groups at Radii of Low Overdensity: Implications for the Cosmic Preheating. *Astrophys. J.* **2002**, *578*, 74–89. [CrossRef]
143. Voit, G.M.; Kay, S.T.; Bryan, G.L. The baseline intracluster entropy profile from gravitational structure formation. *Mon. Not. R. Astron. Soc.* **2005**, *364*, 909–916. [CrossRef]
144. Humphrey, P.J.; Buote, D.A.; Brighenti, F.; Flohic, H.M.L.G.; Gastaldello, F.; Mathews, W.G. Tracing the Gas to the Virial Radius (R_{100}) in a Fossil Group. *Astrophys. J.* **2012**, *748*, 11, [CrossRef]
145. Simionescu, A.; Werner, N.; Mantz, A.; Allen, S.W.; Urban, O. Witnessing the growth of the nearest galaxy cluster: Thermodynamics of the Virgo Cluster outskirts. *Mon. Not. R. Astron. Soc.* **2017**, *469*, 1476–1495. [CrossRef]
146. Thölken, S.; Lovisari, L.; Reiprich, T.H.; Hasenbusch, J. X-ray analysis of the galaxy group UGC 03957 beyond R_{200} with Suzaku. *Astron. Astrophys.* **2016**, *592*, A37, [CrossRef]

147. Johnson, R.; Ponman, T.J.; Finoguenov, A. A statistical analysis of the Two-Dimensional XMM-Newton Group Survey: The impact of feedback on group properties. *Mon. Not. R. Astron. Soc.* **2009**, *395*, 1287–1308. [CrossRef]
148. Finoguenov, A.; Davis, D.S.; Zimer, M.; Mulchaey, J.S. The Two-dimensional XMM-Newton Group Survey: $z < 0.012$ Groups. *Astrophys. J.* **2006**, *646*, 143–160. [CrossRef]
149. Finoguenov, A.; Ponman, T.J.; Osmond, J.P.F.; Zimer, M. XMM-Newton study of $0.012 < z < 0.024$ groups—I. Overview of the IGM thermodynamics. *Mon. Not. R. Astron. Soc.* **2007**, *374*, 737–760. [CrossRef]
150. Mahdavi, A.; Finoguenov, A.; Böhringer, H.; Geller, M.J.; Henry, J.P. XMM-Newton and Gemini Observations of Eight RASSCALS Galaxy Groups. *Astrophys. J.* **2005**, *622*, 187–204. [CrossRef]
151. Sanderson, A.J.R.; O'Sullivan, E.; Ponman, T.J. A statistically selected Chandra sample of 20 galaxy clusters—II. Gas properties and cool core/non-cool core bimodality. *Mon. Not. R. Astron. Soc.* **2009**, *395*, 764–776. [CrossRef]
152. Cavagnolo, K.W.; Donahue, M.; Voit, G.M.; Sun, M. Intracluster Medium Entropy Profiles for a Chandra Archival Sample of Galaxy Clusters. *Astrophys. J.* **2009**, *182*, 12–32. [CrossRef]
153. Panagoulia, E.K.; Fabian, A.C.; Sanders, J.S. A volume-limited sample of X-ray galaxy groups and clusters—I. Radial entropy and cooling time profiles. *Mon. Not. R. Astron. Soc.* **2014**, *438*, 2341–2354. [CrossRef]
154. Hogan, M.T.; McNamara, B.R.; Pulido, F.A.; Nulsen, P.E.J.; Vantyghem, A.N.; Russell, H.R.; Edge, A.C.; Babyk, I.; Main, R.A.; McDonald, M. The Onset of Thermally Unstable Cooling from the Hot Atmospheres of Giant Galaxies in Clusters: Constraints on Feedback Models. *Astrophys. J.* **2017**, *851*, 66, [CrossRef]
155. Babyk, I.V.; McNamara, B.R.; Nulsen, P.E.J.; Russell, H.R.; Vantyghem, A.N.; Hogan, M.T.; Pulido, F.A. A Universal Entropy Profile for the Hot Atmospheres of Galaxies and Clusters within R_{2500}. *Astrophys. J.* **2018**, *862*, 39, [CrossRef]
156. Gaspari, M.; Ruszkowski, M.; Sharma, P. Cause and Effect of Feedback: Multiphase Gas in Cluster Cores Heated by AGN Jets. *Astrophys. J.* **2012**, *746*, 94, [CrossRef]
157. McCourt, M.; Sharma, P.; Quataert, E.; Parrish, I.J. Thermal instability in gravitationally stratified plasmas: Implications for multiphase structure in clusters and galaxy haloes. *Mon. Not. R. Astron. Soc.* **2012**, *419*, 3319–3337. [CrossRef]
158. Sharma, P.; McCourt, M.; Quataert, E.; Parrish, I.J. Thermal instability and the feedback regulation of hot haloes in clusters, groups and galaxies. *Mon. Not. R. Astron. Soc.* **2012**, *420*, 3174–3194. [CrossRef]
159. Voit, G.M.; Donahue, M.; Bryan, G.L.; McDonald, M. Regulation of star formation in giant galaxies by precipitation, feedback and conduction. *Nature* **2015**, *519*, 203–206. [CrossRef]
160. Voit, G.M.; Bryan, G.L.; O'Shea, B.W.; Donahue, M. Precipitation-regulated Star Formation in Galaxies. *Astrophys. J.* **2015**, *808*, L30, [CrossRef]
161. Field, G.B. Thermal Instability. *Astrophys. J.* **1965**, *142*, 531. [CrossRef]
162. Borgani, S.; Finoguenov, A.; Kay, S.T.; Ponman, T.J.; Springel, V.; Tozzi, P.; Voit, G.M. Entropy amplification from energy feedback in simulated galaxy groups and clusters. *Mon. Not. R. Astron. Soc.* **2005**, *361*, 233–243. [CrossRef]
163. Cavagnolo, K.W.; Donahue, M.; Voit, G.M.; Sun, M. An Entropy Threshold for Strong Hα and Radio Emission in the Cores of Galaxy Clusters. *Astrophys. J.* **2008**, *683*, L107–L110, [CrossRef]
164. Lau, E.T.; Gaspari, M.; Nagai, D.; Coppi, P. Physical Origins of Gas Motions in Galaxy Cluster Cores: Interpreting Hitomi Observations of the Perseus Cluster. *Astrophys. J.* **2017**, *849*, 54, [CrossRef]
165. Gaspari, M.; McDonald, M.; Hamer, S.L.; Brighenti, F.; Temi, P.; Gendron-Marsolais, M.; Hlavacek-Larrondo, J.; Edge, A.C.; Werner, N.; Tozzi, P.; et al. Shaken Snow Globes: Kinematic Tracers of the Multiphase Condensation Cascade in Massive Galaxies, Groups, and Clusters. *Astrophys. J.* **2018**, *854*, 167, [CrossRef]
166. Gaspari, M.; Temi, P.; Brighenti, F. Raining on black holes and massive galaxies: The top-down multiphase condensation model. *Mon. Not. R. Astron. Soc.* **2017**, *466*, 677–704. [CrossRef]
167. Olivares, V.; Salome, P.; Combes, F.; Hamer, S.; Guillard, P.; Lehnert, M.D.; Polles, F.L.; Beckmann, R.S.; Dubois, Y.; Donahue, M.; et al. Ubiquitous cold and massive filaments in cool core clusters. *Astron. Astrophys.* **2019**, *631*, A22, [CrossRef]
168. Juráňová, A.; Werner, N.; Gaspari, M.; Lakhchaura, K.; Nulsen, P.E.J.; Sun, M.; Canning, R.E.A.; Allen, S.W.; Simionescu, A.; Oonk, J.B.R.; et al. Cooling in the X-ray halo of the rotating, massive early-type galaxy NGC 7049. *Mon. Not. R. Astron. Soc.* **2019**, *484*, 2886–2895. [CrossRef]
169. Eckert, D.; Ghirardini, V.; Ettori, S.; Rasia, E.; Biffi, V.; Pointecouteau, E.; Rossetti, M.; Molendi, S.; Vazza, F.; Gastaldello, F.; et al. Non-thermal pressure support in X-COP galaxy clusters. *Astron. Astrophys.* **2019**, *621*, A40, [CrossRef]
170. Gonzalez, A.H.; Zaritsky, D.; Zabludoff, A.I. A Census of Baryons in Galaxy Clusters and Groups. *Astrophys. J.* **2007**, *666*, 147–155. [CrossRef]
171. Gastaldello, F.; Buote, D.A.; Humphrey, P.J.; Zappacosta, L.; Bullock, J.S.; Brighenti, F.; Mathews, W.G. Probing the Dark Matter and Gas Fraction in Relaxed Galaxy Groups with X-ray Observations from Chandra and XMM-Newton. *Astrophys. J.* **2007**, *669*, 158–183. [CrossRef]
172. Gonzalez, A.H.; Sivanandam, S.; Zabludoff, A.I.; Zaritsky, D. Galaxy Cluster Baryon Fractions Revisited. *Astrophys. J.* **2013**, *778*, 14, [CrossRef]
173. Pratt, G.W.; Croston, J.H.; Arnaud, M.; Böhringer, H. Galaxy cluster X-ray luminosity scaling relations from a representative local sample (REXCESS). *Astron. Astrophys.* **2009**, *498*, 361–378. [CrossRef]
174. Lovisari, L.; Reiprich, T.H.; Schellenberger, G. Scaling properties of a complete X-ray selected galaxy group sample. *Astron. Astrophys.* **2015**, *573*, A118, [CrossRef]

175. Eckert, D.; Ettori, S.; Coupon, J.; Gastaldello, F.; Pierre, M.; Melin, J.B.; Le Brun, A.M.C.; McCarthy, I.G.; Adami, C.; Chiappetti, L.; et al. The XXL Survey. XIII. Baryon content of the bright cluster sample. *Astron. Astrophys.* **2016**, *592*, A12, [CrossRef]
176. Ettori, S. The physics inside the scaling relations for X-ray galaxy clusters: Gas clumpiness, gas mass fraction and slope of the pressure profile. *Mon. Not. R. Astron. Soc.* **2015**, *446*, 2629–2639. [CrossRef]
177. Nugent, J.M.; Dai, X.; Sun, M. Suzaku Measurements of Hot Halo Emission at Outskirts for Two Poor Galaxy Groups: NGC 3402 and NGC 5129. *Astrophys. J.* **2020**, *899*, 160, [CrossRef]
178. Sun, M. Hot gas in galaxy groups: Recent observations. *New J. Phys.* **2012**, *14*, 045004, [CrossRef]
179. Croston, J.H.; Pratt, G.W.; Böhringer, H.; Arnaud, M.; Pointecouteau, E.; Ponman, T.J.; Sanderson, A.J.R.; Temple, R.F.; Bower, R.G.; Donahue, M. Galaxy-cluster gas-density distributions of the representative XMM-Newton cluster structure survey (REXCESS). *Astron. Astrophys.* **2008**, *487*, 431–443. [CrossRef]
180. Chiu, I.; Mohr, J.J.; McDonald, M.; Bocquet, S.; Desai, S.; Klein, M.; Israel, H.; Ashby, M.L.N.; Stanford, A.; Benson, B.A.; et al. Baryon content in a sample of 91 galaxy clusters selected by the South Pole Telescope at 0.2 <z < 1.25. *Mon. Not. R. Astron. Soc.* **2018**, *478*, 3072–3099. [CrossRef]
181. Sanderson, A.J.R.; O'Sullivan, E.; Ponman, T.J.; Gonzalez, A.H.; Sivanandam, S.; Zabludoff, A.I.; Zaritsky, D. The baryon budget on the galaxy group/cluster boundary. *Mon. Not. R. Astron. Soc.* **2013**, *429*, 3288–3304. [CrossRef]
182. Andreon, S. The stellar mass fraction and baryon content of galaxy clusters and groups. *Mon. Not. R. Astron. Soc.* **2010**, *407*, 263–276. [CrossRef]
183. Kravtsov, A.V.; Vikhlinin, A.A.; Meshcheryakov, A.V. Stellar Mass—Halo Mass Relation and Star Formation Efficiency in High-Mass Halos. *Astron. Lett.* **2018**, *44*, 8–34. [CrossRef]
184. Rasia, E.; Ettori, S.; Moscardini, L.; Mazzotta, P.; Borgani, S.; Dolag, K.; Tormen, G.; Cheng, L.M.; Diaferio, A. Systematics in the X-ray cluster mass estimators. *Mon. Not. R. Astron. Soc.* **2006**, *369*, 2013–2024. [CrossRef]
185. Eckert, D.; Molendi, S.; Paltani, S. The cool-core bias in X-ray galaxy cluster samples. I. Method and application to HIFLUGCS. *Astron. Astrophys.* **2011**, *526*, A79. [CrossRef]
186. Andreon, S.; Wang, J.; Trinchieri, G.; Moretti, A.; Serra, A.L. Variegate galaxy cluster gas content: Mean fraction, scatter, selection effects, and covariance with X-ray luminosity. *Astron. Astrophys.* **2017**, *606*, A24, [CrossRef]
187. Lin, Y.T.; Mohr, J.J. Radio Sources in Galaxy Clusters: Radial Distribution, and 1.4 GHz and K-band Bivariate Luminosity Function. *ApJS* **2007**, *170*, 71–94. [CrossRef]
188. Smolčić, V.; Finoguenov, A.; Zamorani, G.; Schinnerer, E.; Tanaka, M.; Giodini, S.; Scoville, N. On the occupation of X-ray-selected galaxy groups by radio active galactic nuclei since z = 1.3. *Mon. Not. R. Astron. Soc. Lett.* **2011**, *416*, L31–L35. [CrossRef]
189. Dunn, R.J.H.; Allen, S.W.; Taylor, G.B.; Shurkin, K.F.; Gentile, G.; Fabian, A.C.; Reynolds, C.S. The radio properties of a complete, X-ray selected sample of nearby, massive elliptical galaxies. *Mon. Not. R. Astron. Soc. Lett.* **2010**, *404*, 180. [CrossRef]
190. Kolokythas, K.; O'Sullivan, E.; Intema, H.; Raychaudhury, S.; Babul, A.; Giacintucci, S.; Gitti, M. The complete local volume groups sample—III. Characteristics of group central radio galaxies in the Local Universe. *Mon. Not. R. Astron. Soc. Lett.* **2019**, *489*, 2488–2504. [CrossRef]
191. Kolokythas, K.; O'Sullivan, E.; Raychaudhury, S.; Giacintucci, S.; Gitti, M.; Babul, A. The Complete Local Volume Groups Sample—II. A study of the central radio galaxies in the high-richness sample. *Mon. Not. R. Astron. Soc. Lett.* **2018**, *481*, 1550. [CrossRef]
192. Giacintucci, S.; O'Sullivan, E.; Vrtilek, J.M.; David, L.P.; Raychaudhury, S.; Venturi, T.; Athreya, R.M.; Clarke, T.E.; Murgia, M.; Mazzotta, P.; et al. A combined low-radio frequency/X-ray study of AGN feedback in groups of galaxies—I. GMRT observations at 235 MHz and 610 MHz. *Astrophys. J.* **2011**, *732*, 95. [CrossRef]
193. Kolokythas, K.; O'Sullivan, E.; Giacintucci, S.; Worrall, D.M.; Birkinshaw, M.; Raychaudhury, S.; Horellou, C.; Intema, H.; Loubser, I. Evidence of AGN feedback and sloshing in the X-ray luminous NGC 1550 galaxy group. *Mon. Not. R. Astron. Soc. Lett.* **2020**, *496*, 1471–1487. [CrossRef]
194. Bîrzan, L.; Rafferty, D.A.; Bruggen, M.; Botteon, A.; Brunettit, G.; Cuciti, V.; Edge, A.C.; Morganti, R.; Rottgering, H.J.A.; Shimwell, T. LOFAR Observations of X-ray Cavity Systems. *Mon. Not. R. Astron. Soc. Lett.* **2020**, *496*, 2613–2635. [CrossRef]
195. Schellenberger, G.; David, L.P.; Vrtilek, J.; O'Sullivan, E.; Giacintucci, S.; Forman, W.; Jones, C.; Venturi, T. A New Feedback Cycle in the Archetypal Cooling Flow Group NGC 5044. *Astrophys. J.* **2021**, *906*, 16. [CrossRef]
196. Ineson, J.; Croston, J.H.; Hardcastle, M.J.; Kraft, R.P.; Evans, D.A.; Jarvis, M. The link between accretion mode and environment in radio-loud active galaxies. *Mon. Not. R. Astron. Soc. Lett.* **2015**, *453*, 2682–2706. [CrossRef]
197. Pasini, T.; Brüggen, M.; de Gasperin, F.; Bîrzan, L.; O'Sullivan, E.; Finoguenov, A.; Jarvis, M.; Gitti, M.; Brighenti, F.; Whittam, I.H.; et al. The relation between the diffuse X-ray luminosity and the radio power of the central AGN in galaxy groups. *Mon. Not. R. Astron. Soc. Lett.* **2020**, *497*, 2163–2174. [CrossRef]
198. Bîrzan, L.; Rafferty, D.A.; Nulsen, P.E.J.; McNamara, B.R.; Röttgering, H.J.A.; Wise, M.W.; Mittal, R. The duty cycle of radio-mode feedback in complete samples of clusters. *Mon. Not. R. Astron. Soc. Lett.* **2012**, *427*, 3468. [CrossRef]
199. O'Sullivan, E.; Giacintucci, S.; David, L.P.; Gitti, M.; Vrtilek, J.M.; Raychaudhury, S.; Ponman, T.J. Heating the Hot Atmospheres of Galaxy Groups and Clusters with Cavities: The Relationship between Jet Power and Low-frequency Radio Emission. *Astrophys. J.* **2011**, *735*, 11. [CrossRef]

200. Giodini, S.; Smolčić, V.; Finoguenov, A.; Boehringer, H.; Bîrzan, L.; Zamorani, G.; Oklopčić, A.; Pierini, D.; Pratt, G.W.; Schinnerer, E.; et al. Radio Galaxy Feedback in X-ray-selected Groups from COSMOS: The Effect on the Intracluster Medium. *Astrophys. J.* **2010**, *714*, 218. [CrossRef]
201. Delhaize, J.; Heywood, I.; Prescott, M.; Jarvis, M.J.; Delvecchio, I.; Whittam, I.H.; White, S.V.; Hardcastle, M.J.; Hale, C.L.; Afonso, J.; et al. MIGHTEE: Are giant radio galaxies more common than we thought? *Mon. Not. R. Astron. Soc. Lett.* **2021**, *501*, 3833–3845. [CrossRef]
202. O'Sullivan, E.; Worrall, D.M.; Birkinshaw, M.; Trinchieri, G.; Wolter, A.; Zezas, A.; Giacintucci, S. Interaction between the intergalactic medium and central radio source in the NGC 4261 group of galaxies. *Mon. Not. R. Astron. Soc. Lett.* **2011**, *416*, 2916–2931. [CrossRef]
203. Grossová, R.; Werner, N.; Rajpurohit, K.; Mernier, F.; Lakhchaura, K.; Gabányi, K.; Canning, R.E.A.; Nulsen, P.; Massaro, F.; Sun, M.; et al. Powerful AGN jets and unbalanced cooling in the hot atmosphere of IC 4296. *Mon. Not. R. Astron. Soc. Lett.* **2019**, *488*, 1917–1925. [CrossRef]
204. Cantwell, T.M.; Bray, J.D.; Croston, J.H.; Scaife, A.M.M.; Mulcahy, D.D.; Best, P.N.; Brüggen, M.; Brunetti, G.; Callingham, J.R.; Clarke, A.O.; et al. Low-frequency observations of the giant radio galaxy NGC 6251. *Mon. Not. R. Astron. Soc. Lett.* **2020**, *495*, 143–159. [CrossRef]
205. Ineson, J.; Croston, J.H.; Hardcastle, M.J.; Mingo, B. A representative survey of the dynamics and energetics of FR II radio galaxies. *Mon. Not. R. Astron. Soc. Lett.* **2017**, *467*, 1586–1607. [CrossRef]
206. Lakhchaura, K.; Werner, N.; Sun, M.; Canning, R.E.A.; Gaspari, M.; Allen, S.W.; Connor, T.; Donahue, M.; Sarazin, C. Thermodynamic properties, multiphase gas, and AGN feedback in a large sample of giant ellipticals. *Mon. Not. R. Astron. Soc.* **2018**, *481*, 4472–4504. [CrossRef]
207. Pulido, F.A.; McNamara, B.R.; Edge, A.C.; Hogan, M.T.; Vantyghem, A.N.; Russell, H.R.; Nulsen, P.E.J.; Babyk, I.; Salomé, P. The Origin of Molecular Clouds in Central Galaxies. *Astrophys. J.* **2018**, *853*, 177. [CrossRef]
208. Werner, N.; Oonk, J.B.R.; Sun, M.; Nulsen, P.E.J.; Allen, S.W.; Canning, R.E.A.; Simionescu, A.; Hoffer, A.; Connor, T.; Donahue, M.; et al. The origin of cold gas in giant elliptical galaxies and its role in fuelling radio-mode AGN feedback. *Mon. Not. R. Astron. Soc. Lett.* **2014**, *439*, 2291–2306. [CrossRef]
209. David, L.P.; Vrtilek, J.; O'Sullivan, E.; Jones, C.; Forman, W.; Sun, M. The Presence of Thermally Unstable X-ray Filaments and the Production of Cold Gas in the NGC 5044 Group. *Astrophys. J.* **2017**, *842*, 84. [CrossRef]
210. O'Sullivan, E.; Schellenberger, G.; Burke, D.J.; Sun, M.; Vrtilek, J.M.; David, L.P.; Sarazin, C. Building a cluster: Shocks, cavities, and cooling filaments in the group-group merger NGC 6338. *Mon. Not. R. Astron. Soc. Lett.* **2019**, *488*, 2925–2946. [CrossRef]
211. Gomes, J.M.; Papaderos, P.; Kehrig, C.; Vílchez, J.M.; Lehnert, M.D.; Sánchez, S.F.; Ziegler, B.; Breda, I.; Dos Reis, S.N.; Iglesias-Páramo, J.; et al. Warm ionized gas in CALIFA early-type galaxies. 2D emission-line patterns and kinematics for 32 galaxies. *Astron. Astrophys.* **2016**, *588*, A68. [CrossRef]
212. Diniz, S.I.F.; Pastoriza, M.G.; Hernandez-Jimenez, J.A.; Riffel, R.; Ricci, T.V.; Steiner, J.E.; Riffel, R.A. Integral field spectroscopy of the inner kpc of the elliptical galaxy NGC 5044. *Mon. Not. R. Astron. Soc. Lett.* **2017**, *470*, 1703–1717. [CrossRef]
213. Schellenberger, G.; David, L.P.; Vrtilek, J.; O'Sullivan, E.; Lim, J.; Forman, W.; Sun, M.; Combes, F.; Salome, P.; Jones, C.; et al. Atacama Compact Array Measurements of the Molecular Mass in the NGC 5044 Cooling-flow Group. *Astrophys. J.* **2020**, *894*, 72. [CrossRef]
214. McDonald, M.; Gaspari, M.; McNamara, B.R.; Tremblay, G.R. Revisiting the Cooling Flow Problem in Galaxies, Groups, and Clusters of Galaxies. *Astrophys. J.* **2018**, *858*, 45, [CrossRef]
215. Kaneda, H.; Onaka, T.; Sakon, I.; Kitayama, T.; Okada, Y.; Suzuki, T. Properties of Polycyclic Aromatic Hydrocarbons in Local Elliptical Galaxies Revealed by the Infrared Spectrograph on Spitzer. *Astrophys. J.* **2008**, *684*, 270–281. [CrossRef]
216. Rose, T.; Edge, A.C.; Combes, F.; Gaspari, M.; Hamer, S.; Nesvadba, N.; Peck, A.B.; Sarazin, C.; Tremblay, G.R.; Baum, S.A.; et al. Constraining cold accretion on to supermassive black holes: Molecular gas in the cores of eight brightest cluster galaxies revealed by joint CO and CN absorption. *Mon. Not. R. Astron. Soc.* **2019**, *489*, 349–365. [CrossRef]
217. Babyk, I.V.; McNamara, B.R.; Tamhane, P.D.; Nulsen, P.E.J.; Russell, H.R.; Edge, A.C. Origins of Molecular Clouds in Early-type Galaxies. *Astrophys. J.* **2019**, *887*, 149. [CrossRef]
218. Davis, T.A.; Greene, J.E.; Ma, C.P.; Blakeslee, J.P.; Dawson, J.M.; Pandya, V.; Veale, M.; Zabel, N. The MASSIVE survey—XI. What drives the molecular gas properties of early-type galaxies. *Mon. Not. R. Astron. Soc. Lett.* **2019**, *486*, 1404–1423. [CrossRef]
219. O'Sullivan, E.; Combes, F.; Hamer, S.; Salomé, P.; Babul, A.; Raychaudhury, S. Cold gas in group-dominant elliptical galaxies. *Astron. Astrophys.* **2015**, *573*, A111. [CrossRef]
220. O'Sullivan, E.; Combes, F.; Salomé, P.; David, L.P.; Babul, A.; Vrtilek, J.M.; Lim, J.; Olivares, V.; Raychaudhury, S.; Schellenberger, G. Cold gas in a complete sample of group-dominant early-type galaxies. *Astron. Astrophys.* **2018**, *618*, A126. [CrossRef]
221. Temi, P.; Amblard, A.; Gitti, M.; Brighenti, F.; Gaspari, M.; Mathews, W.G.; David, L. ALMA Observations of Molecular Clouds in Three Group-centered Elliptical Galaxies: NGC 5846, NGC 4636, and NGC 5044. *Astrophys. J.* **2018**, *858*, 17. [CrossRef]
222. David, L.P.; Lim, J.; Forman, W.; Vrtilek, J.; Combes, F.; Salome, P.; Edge, A.; Hamer, S.; Jones, C.; Sun, M.; et al. Molecular Gas in the X-ray Bright Group NGC 5044 as Revealed by ALMA. *Astrophys. J.* **2014**, *792*, 94. [CrossRef]
223. Ruffa, I.; Laing, R.A.; Prandoni, I.; Paladino, R.; Parma, P.; Davis, T.A.; Bureau, M. The AGN fuelling/feedback cycle in nearby radio galaxies—III. 3D relative orientations of radio jets and CO discs and their interaction. *Mon. Not. R. Astron. Soc. Lett.* **2020**, *499*, 5719–5731. [CrossRef]

224. Boizelle, B.D.; Walsh, J.L.; Barth, A.J.; Buote, D.A.; Baker, A.J.; Darling, J.; Ho, L.C.; Cohn, J.; Kabasares, K.M. Black Hole Mass Measurements of Radio Galaxies NGC 315 and NGC 4261 Using ALMA CO Observations. *arXiv* **2020**, arXiv:2012.04669.
225. Shulevski, A.; Morganti, R.; Oosterloo, T.; Struve, C. Recurrent radio emission and gas supply: The radio galaxy B2 0258+35. *Astron. Astrophys.* **2012**, *545*, A91. [CrossRef]
226. Gaspari, M. Shaping the X-ray spectrum of galaxy clusters with AGN feedback and turbulence. *Mon. Not. R. Astron. Soc.* **2015**, *451*, L60–L64, [CrossRef]
227. Sutherland, R.S.; Dopita, M.A. Cooling functions for low-density astrophysical plasmas. *Astrophys. J.* **1993**, *88*, 253–327. [CrossRef]
228. Mernier, F.; de Plaa, J.; Kaastra, J.S.; Zhang, Y.Y.; Akamatsu, H.; Gu, L.; Kosec, P.; Mao, J.; Pinto, C.; Reiprich, T.H.; et al. Radial metal abundance profiles in the intra-cluster medium of cool-core galaxy clusters, groups, and ellipticals. *Astron. Astrophys.* **2017**, *603*, A80, [CrossRef]
229. Pope, E.C.D. Why and when is internally driven AGN feedback energetically favoured? *Mon. Not. R. Astron. Soc.* **2012**, *427*, 2–10. [CrossRef]
230. Dalgarno, A.; McCray, R.A. Heating and Ionization of HI Regions. *Annu. Rev. Astron. Astrophys.* **1972**, *10*, 375. [CrossRef]
231. Inoue, T.; Inutsuka, S.i. Two-Fluid Magnetohydrodynamic Simulations of Converging H I Flows in the Interstellar Medium. I. Methodology and Basic Results. *Astrophys. J.* **2008**, *687*, 303–310. [CrossRef]
232. Hamer, S.L.; Edge, A.C.; Swinbank, A.M.; Wilman, R.J.; Combes, F.; Salomé, P.; Fabian, A.C.; Crawford, C.S.; Russell, H.R.; Hlavacek-Larrondo, J.; et al. Optical emission line nebulae in galaxy cluster cores 1: The morphological, kinematic and spectral properties of the sample. *Mon. Not. R. Astron. Soc.* **2016**, *460*, 1758–1789. [CrossRef]
233. Juráňová, A.; Werner, N.; Nulsen, P.E.J.; Gaspari, M.; Lakhchaura, K.; Canning, R.E.A.; Donahue, M.; Hroch, F.; Voit, G.M. Hot gaseous atmospheres of rotating galaxies observed with XMM-Newton. *Mon. Not. R. Astron. Soc.* **2020**, *499*, 5163–5174. [CrossRef]
234. Vazza, F.; Brunetti, G.; Gheller, C.; Brunino, R.; Brüggen, M. Massive and refined. II. The statistical properties of turbulent motions in massive galaxy clusters with high spatial resolution. *Astron. Astrophys.* **2011**, *529*, A17, [CrossRef]
235. Valentini, M.; Brighenti, F. AGN-stimulated cooling of hot gas in elliptical galaxies. *Mon. Not. R. Astron. Soc.* **2015**, *448*, 1979–1998. [CrossRef]
236. Gaspari, M.; Brighenti, F.; Temi, P. Mechanical AGN feedback: Controlling the thermodynamical evolution of elliptical galaxies. *Mon. Not. R. Astron. Soc.* **2012**, *424*, 190–209. [CrossRef]
237. Hillel, S.; Soker, N. Gentle Heating by Mixing in Cooling Flow Clusters. *Astrophys. J.* **2017**, *845*, 91, [CrossRef]
238. Weinberger, R.; Springel, V.; Pakmor, R.; Nelson, D.; Genel, S.; Pillepich, A.; Vogelsberger, M.; Marinacci, F.; Naiman, J.; Torrey, P.; et al. Supermassive black holes and their feedback effects in the IllustrisTNG simulation. *Mon. Not. R. Astron. Soc.* **2018**, *479*, 4056–4072. [CrossRef]
239. Wittor, D.; Gaspari, M. Dissecting the turbulent weather driven by mechanical AGN feedback. *Mon. Not. R. Astron. Soc.* **2020**, *498*, 4983–5002. [CrossRef]
240. Sanders, J.S.; Fabian, A.C. Velocity width measurements of the coolest X-ray emitting material in the cores of clusters, groups and elliptical galaxies. *Mon. Not. R. Astron. Soc.* **2013**, *429*, 2727–2738. [CrossRef]
241. Ogorzalek, A.; Zhuravleva, I.; Allen, S.W.; Pinto, C.; Werner, N.; Mantz, A.B.; Canning, R.E.A.; Fabian, A.C.; Kaastra, J.S.; de Plaa, J. Improved measurements of turbulence in the hot gaseous atmospheres of nearby giant elliptical galaxies. *Mon. Not. R. Astron. Soc.* **2017**, *472*, 1659–1676. [CrossRef]
242. Collaboration, H.; Aharonian, F.; Akamatsu, H.; Akimoto, F.; Allen, S.W.; Angelini, L.; Audard, M.; Awaki, H.; Axelsson, M.; Bamba, A.; et al. Atmospheric gas dynamics in the Perseus cluster observed with Hitomi. *Publ. Astron. Soc. Jpn.* **2018**, *70*, 9, [CrossRef]
243. Gaspari, M.; Churazov, E. Constraining turbulence and conduction in the hot ICM through density perturbations. *Astron. Astrophys.* **2013**, *559*, A78, [CrossRef]
244. Zhuravleva, I.; Churazov, E.M.; Schekochihin, A.A.; Lau, E.T.; Nagai, D.; Gaspari, M.; Allen, S.W.; Nelson, K.; Parrish, I.J. The Relation between Gas Density and Velocity Power Spectra in Galaxy Clusters: Qualitative Treatment and Cosmological Simulations. *Astrophys. J.* **2014**, *788*, L13. [CrossRef]
245. Gaspari, M.; Ruszkowski, M.; Oh, S.P. Chaotic cold accretion on to black holes. *Mon. Not. R. Astron. Soc.* **2013**, *432*, 3401–3422. [CrossRef]
246. Voit, G.M. A Role for Turbulence in Circumgalactic Precipitation. *Astrophys. J.* **2018**, *868*, 102, [CrossRef]
247. Pizzolato, F.; Soker, N. On the Nature of Feedback Heating in Cooling Flow Clusters. *Astrophys. J.* **2005**, *632*, 821–830. [CrossRef]
248. McDonald, M.; Veilleux, S.; Mushotzky, R. The Effect of Environment on the Formation of Hα Filaments and Cool Cores in Galaxy Groups and Clusters. *Astrophys. J.* **2011**, *731*, 33, [CrossRef]
249. Bondi, H. On spherically symmetrical accretion. *Mon. Not. R. Astron. Soc.* **1952**, *112*, 195. [CrossRef]
250. Narayan, R.; Fabian, A.C. Bondi flow from a slowly rotating hot atmosphere. *Mon. Not. R. Astron. Soc.* **2011**, *415*, 3721–3730. [CrossRef]
251. Voit, G.M.; Donahue, M.; O'Shea, B.W.; Bryan, G.L.; Sun, M.; Werner, N. Supernova Sweeping and Black Hole Feedback in Elliptical Galaxies. *Astrophys. J.* **2015**, *803*, L21, [CrossRef]
252. McNamara, B.R.; Russell, H.R.; Nulsen, P.E.J.; Hogan, M.T.; Fabian, A.C.; Pulido, F.; Edge, A.C. A Mechanism for Stimulating AGN Feedback by Lifting Gas in Massive Galaxies. *Astrophys. J.* **2016**, *830*, 79, [CrossRef]

253. Voit, G.M. A Graphical Interpretation of Circumgalactic Precipitation. *Astrophys. J.* **2021**, *908*, L16, [CrossRef]
254. Singh, P.; Voit, G.M.; Nath, B.B. Constraints on precipitation-limited hot halos from massive galaxies to galaxy clusters. *Mon. Not. R. Astron. Soc.* **2020**, [CrossRef]
255. Gaspari, M.; Brighenti, F.; Temi, P. Chaotic cold accretion on to black holes in rotating atmospheres. *Astron. Astrophys.* **2015**, *579*, A62, [CrossRef]
256. Caon, N.; Macchetto, D.; Pastoriza, M. A Survey of the Interstellar Medium in Early-Type Galaxies. III. Stellar and Gas Kinematics. *Astrophys. J.* **2000**, *127*, 39–58. [CrossRef]
257. Diehl, S.; Statler, T.S. The Hot Interstellar Medium of Normal Elliptical Galaxies. I. A Chandra Gas Gallery and Comparison of X-ray and Optical Morphology. *Astrophys. J.* **2007**, *668*, 150–167. [CrossRef]
258. Ulrich, M.H.; Maraschi, L.; Urry, C.M. Variability of Active Galactic Nuclei. *Annu. Rev. Astron. Astrophys.* **1997**, *35*, 445–502. [CrossRef]
259. Peterson, B.M. Variability of Active Galactic Nuclei. In *Advanced Lectures on the Starburst-AGN*; World Scientific ; Aretxaga, I., Kunth, D., Mújica, R., Eds.; Tonantzintla: Puebla, Mexico, 2001; p. 3.
260. Hudson, D.S.; Mittal, R.; Reiprich, T.H.; Nulsen, P.E.J.; Andernach, H.; Sarazin, C.L. What is a cool-core cluster? a detailed analysis of the cores of the X-ray flux-limited HIFLUGCS cluster sample. *Astron. Astrophys.* **2010**, *513*, A37, [CrossRef]
261. Ghirardini, V.; Eckert, D.; Ettori, S.; Pointecouteau, E.; Molendi, S.; Gaspari, M.; Rossetti, M.; De Grandi, S.; Roncarelli, M.; Bourdin, H.; et al. Universal thermodynamic properties of the intracluster medium over two decades in radius in the X-COP sample. *Astron. Astrophys.* **2019**, *621*, A41, [CrossRef]
262. Gaspari, M.; Sądowski, A. Unifying the Micro and Macro Properties of AGN Feeding and Feedback. *Astrophys. J.* **2017**, *837*, 149, [CrossRef]
263. Gaspari, M.; Brighenti, F.; Temi, P.; Ettori, S. Can AGN Feedback Break the Self-similarity of Galaxies, Groups, and Clusters? *Astrophys. J.* **2014**, *783*, L10, [CrossRef]
264. Gaspari, M.; Brighenti, F.; D'Ercole, A.; Melioli, C. AGN feedback in galaxy groups: The delicate touch of self-regulated outflows. *Mon. Not. R. Astron. Soc.* **2011**, *415*, 1549–1568. [CrossRef]
265. Prasad, D.; Sharma, P.; Babul, A. Cool Core Cycles: Cold Gas and AGN Jet Feedback in Cluster Cores. *Astrophys. J.* **2015**, *811*, 108, [CrossRef]
266. Sądowski, A.; Gaspari, M. Kinetic and radiative power from optically thin accretion flows. *Mon. Not. R. Astron. Soc.* **2017**, *468*, 1398–1404. [CrossRef]
267. Fukumura, K.; Kazanas, D.; Contopoulos, I.; Behar, E. Magnetohydrodynamic Accretion Disk Winds as X-ray Absorbers in Active Galactic Nuclei. *Astrophys. J.* **2010**, *715*, 636–650. [CrossRef]
268. Tombesi, F.; Cappi, M.; Reeves, J.N.; Nemmen, R.S.; Braito, V.; Gaspari, M.; Reynolds, C.S. Unification of X-ray winds in Seyfert galaxies: From ultra-fast outflows to warm absorbers. *Mon. Not. R. Astron. Soc.* **2013**, *430*, 1102–1117. [CrossRef]
269. Tchekhovskoy, A.; Narayan, R.; McKinney, J.C. Efficient generation of jets from magnetically arrested accretion on a rapidly spinning black hole. *Mon. Not. R. Astron. Soc.* **2011**, *418*, L79–L83, [CrossRef]
270. Russell, H.R.; McNamara, B.R.; Edge, A.C.; Hogan, M.T.; Main, R.A.; Vantyghem, A.N. Radiative efficiency, variability and Bondi accretion on to massive black holes: The transition from radio AGN to quasars in brightest cluster galaxies. *Mon. Not. R. Astron. Soc.* **2013**, *432*, 530–553. [CrossRef]
271. Costa, T.; Sijacki, D.; Haehnelt, M.G. Feedback from active galactic nuclei: Energy- versus momentum-driving. *Mon. Not. R. Astron. Soc.* **2014**, *444*, 2355–2376. [CrossRef]
272. Giovannini, G. Observational Properties of Jets in Active Galactic Nuclei. *Astrophys. Space Sci.* **2004**, *293*, 1–13. [CrossRef]
273. Fiore, F.; Feruglio, C.; Shankar, F.; Bischetti, M.; Bongiorno, A.; Brusa, M.; Carniani, S.; Cicone, C.; Duras, F.; Lamastra, A.; et al. AGN wind scaling relations and the co-evolution of black holes and galaxies. *Astron. Astrophys.* **2017**, *601*, A143, [CrossRef]
274. Brighenti, F.; Mathews, W.G.; Temi, P. Hot Gaseous Atmospheres in Galaxy Groups and Clusters Are Both Heated and Cooled by X-ray Cavities. *Astrophys. J.* **2015**, *802*, 118, [CrossRef]
275. Gaspari, M.; Churazov, E.; Nagai, D.; Lau, E.T.; Zhuravleva, I. The relation between gas density and velocity power spectra in galaxy clusters: High-resolution hydrodynamic simulations and the role of conduction. *Astron. Astrophys.* **2014**, *569*, A67, [CrossRef]
276. Fabian, A.C.; Walker, S.A.; Russell, H.R.; Pinto, C.; Sanders, J.S.; Reynolds, C.S. Do sound waves transport the AGN energy in the Perseus cluster? *Mon. Not. R. Astron. Soc.* **2017**, *464*, L1–L5, [CrossRef]
277. Cielo, S.; Babul, A.; Antonuccio-Delogu, V.; Silk, J.; Volonteri, M. Feedback from reorienting AGN jets. I. Jet-ICM coupling, cavity properties and global energetics. *Astron. Astrophys.* **2018**, *617*, A58. [CrossRef]
278. Voit, G.M.; Bryan, G.L.; Prasad, D.; Frisbie, R.; Li, Y.; Donahue, M.; O'Shea, B.W.; Sun, M.; Werner, N. A Black Hole Feedback Valve in Massive Galaxies. *Astrophys. J.* **2020**, *899*, 70, [CrossRef]
279. Goulding, A.D.; Greene, J.E.; Ma, C.P.; Veale, M.; Bogdan, A.; Nyland, K.; Blakeslee, J.P.; McConnell, N.J.; Thomas, J. The MASSIVE Survey. IV. The X-ray Halos of the Most Massive Early-type Galaxies in the Nearby Universe. *Astrophys. J.* **2016**, *826*, 167, [CrossRef]
280. Tremmel, M.; Karcher, M.; Governato, F.; Volonteri, M.; Quinn, T.R.; Pontzen, A.; Anderson, L.; Bellovary, J. The Romulus cosmological simulations: A physical approach to the formation, dynamics and accretion models of SMBHs. *Mon. Not. R. Astron. Soc.* **2017**, *470*, 1121–1139. [CrossRef]

281. Dubois, Y.; Beckmann, R.; Bournaud, F.; Choi, H.; Devriendt, J.; Jackson, R.; Kaviraj, S.; Kimm, T.; Kraljic, K.; Laigle, C.; et al. Introducing the NewHorizon simulation: Galaxy properties with resolved internal dynamics across cosmic time. *arXiv* **2020**, arXiv:2009.10578.
282. Henden, N.A.; Puchwein, E.; Shen, S.; Sijacki, D. The FABLE simulations: A feedback model for galaxies, groups, and clusters. *Mon. Not. R. Astron. Soc.* **2018**, *479*, 5385–5412. [CrossRef]
283. Dubois, Y.; Pichon, C.; Welker, C.; Le Borgne, D.; Devriendt, J.; Laigle, C.; Codis, S.; Pogosyan, D.; Arnouts, S.; Benabed, K.; et al. Dancing in the dark: Galactic properties trace spin swings along the cosmic web. *Mon. Not. R. Astron. Soc.* **2014**, *444*, 1453–1468. [CrossRef]
284. Schaye, J.; Crain, R.A.; Bower, R.G.; Furlong, M.; Schaller, M.; Theuns, T.; Dalla Vecchia, C.; Frenk, C.S.; McCarthy, I.G.; Helly, J.C.; et al. The EAGLE project: Simulating the evolution and assembly of galaxies and their environments. *Mon. Not. R. Astron. Soc.* **2015**, *446*, 521–554. [CrossRef]
285. Vogelsberger, M.; Genel, S.; Springel, V.; Torrey, P.; Sijacki, D.; Xu, D.; Snyder, G.; Nelson, D.; Hernquist, L. Introducing the Illustris Project: Simulating the coevolution of dark and visible matter in the Universe. *Mon. Not. R. Astron. Soc.* **2014**, *444*, 1518–1547. [CrossRef]
286. Springel, V.; Pakmor, R.; Pillepich, A.; Weinberger, R.; Nelson, D.; Hernquist, L.; Vogelsberger, M.; Genel, S.; Torrey, P.; Marinacci, F.; et al. First results from the IllustrisTNG simulations: Matter and galaxy clustering. *Mon. Not. R. Astron. Soc.* **2018**, *475*, 676–698. [CrossRef]
287. Davé, R.; Anglés-Alcázar, D.; Narayanan, D.; Li, Q.; Rafieferantsoa, M.H.; Appleby, S. SIMBA: Cosmological simulations with black hole growth and feedback. *Mon. Not. R. Astron. Soc.* **2019**, *486*, 2827–2849. [CrossRef]
288. Robson, D.; Davé, R. X-ray emission from hot gas in galaxy groups and clusters in SIMBA. *Mon. Not. R. Astron. Soc.* **2020**, *498*, 3061–3076. [CrossRef]
289. Khandai, N.; Di Matteo, T.; Croft, R.; Wilkins, S.; Feng, Y.; Tucker, E.; DeGraf, C.; Liu, M.S. The MassiveBlack-II simulation: The evolution of haloes and galaxies to $z \sim 0$. *Mon. Not. R. Astron. Soc.* **2015**, *450*, 1349–1374. [CrossRef]
290. Hirschmann, M.; Dolag, K.; Saro, A.; Bachmann, L.; Borgani, S.; Burkert, A. Cosmological simulations of black hole growth: AGN luminosities and downsizing. *Mon. Not. R. Astron. Soc.* **2014**, *442*, 2304–2324. [CrossRef]
291. Lee, J.; Shin, J.; Snaith, O.N.; Kim, Y.; Few, C.G.; Devriendt, J.; Dubois, Y.; Cox, L.M.; Hong, S.E.; Kwon, O.K.; et al. The Horizon Run 5 Cosmological Hydrodynamical Simulation: Probing Galaxy Formation from Kilo- to Gigaparsec Scales. *Astrophys. J.* **2021**, *908*, 11, [CrossRef]
292. Springel, V.; Di Matteo, T.; Hernquist, L. Modelling feedback from stars and black holes in galaxy mergers. *Mon. Not. R. Astron. Soc.* **2005**, *361*, 776–794. [CrossRef]
293. Springel, V. E pur si muove: Galilean-invariant cosmological hydrodynamical simulations on a moving mesh. *Mon. Not. R. Astron. Soc.* **2010**, *401*, 791–851. [CrossRef]
294. Weinberger, R.; Springel, V.; Pakmor, R. The AREPO Public Code Release. *Astrophys. J.* **2020**, *248*, 32, [CrossRef]
295. Hopkins, P.F. A new class of accurate, mesh-free hydrodynamic simulation methods. *Mon. Not. R. Astron. Soc.* **2015**, *450*, 53–110. [CrossRef]
296. Teyssier, R. Cosmological hydrodynamics with adaptive mesh refinement. A new high resolution code called RAMSES. *Astron. Astrophys.* **2002**, *385*, 337–364. [CrossRef]
297. Menon, H.; Wesolowski, L.; Zheng, G.; Jetley, P.; Kale, L.; Quinn, T.; Governato, F. Adaptive techniques for clustered N-body cosmological simulations. *Comput. Astrophys. Cosmol.* **2015**, *2*, 1, [CrossRef]
298. Booth, C.M.; Schaye, J. Cosmological simulations of the growth of supermassive black holes and feedback from active galactic nuclei: Method and tests. *Mon. Not. R. Astron. Soc.* **2009**, *398*, 53–74. [CrossRef]
299. Hoyle, F.; Lyttleton, R.A. The effect of interstellar matter on climatic variation. *Proc. Camb. Philos. Soc.* **1939**, *35*, 405. [CrossRef]
300. Bondi, H.; Hoyle, F. On the mechanism of accretion by stars. *Mon. Not. R. Astron. Soc.* **1944**, *104*, 273. [CrossRef]
301. Di Matteo, T.; Springel, V.; Hernquist, L. Energy input from quasars regulates the growth and activity of black holes and their host galaxies. *Nature* **2005**, *433*, 604–607. [CrossRef]
302. Sijacki, D.; Springel, V.; Di Matteo, T.; Hernquist, L. A unified model for AGN feedback in cosmological simulations of structure formation. *Mon. Not. R. Astron. Soc.* **2007**, *380*, 877–900. [CrossRef]
303. Bhattacharya, S.; Di Matteo, T.; Kosowsky, A. Effects of quasar feedback in galaxy groups. *Mon. Not. R. Astron. Soc.* **2008**, *389*, 34–44. [CrossRef]
304. Schaye, J.; Dalla Vecchia, C.; Booth, C.M.; Wiersma, R.P.C.; Theuns, T.; Haas, M.R.; Bertone, S.; Duffy, A.R.; McCarthy, I.G.; van de Voort, F. The physics driving the cosmic star formation history. *Mon. Not. R. Astron. Soc.* **2010**, *402*, 1536–1560. [CrossRef]
305. Shakura, N.I.; Sunyaev, R.A. Reprint of 1973A&A....24..337S. Black holes in binary systems. Observational appearance. *Astron. Astrophys.* **1973**, *500*, 33–51.
306. Rosas-Guevara, Y.M.; Bower, R.G.; Schaye, J.; Furlong, M.; Frenk, C.S.; Booth, C.M.; Crain, R.A.; Dalla Vecchia, C.; Schaller, M.; Theuns, T. The impact of angular momentum on black hole accretion rates in simulations of galaxy formation. *Mon. Not. R. Astron. Soc.* **2015**, *454*, 1038–1057. [CrossRef]
307. Vogelsberger, M.; Genel, S.; Sijacki, D.; Torrey, P.; Springel, V.; Hernquist, L. A model for cosmological simulations of galaxy formation physics. *Mon. Not. R. Astron. Soc.* **2013**, *436*, 3031–3067. [CrossRef]

308. Weinberger, R.; Springel, V.; Hernquist, L.; Pillepich, A.; Marinacci, F.; Pakmor, R.; Nelson, D.; Genel, S.; Vogelsberger, M.; Naiman, J.; et al. Simulating galaxy formation with black hole driven thermal and kinetic feedback. *Mon. Not. R. Astron. Soc.* **2017**, *465*, 3291–3308. [CrossRef]
309. Pillepich, A.; Springel, V.; Nelson, D.; Genel, S.; Naiman, J.; Pakmor, R.; Hernquist, L.; Torrey, P.; Vogelsberger, M.; Weinberger, R.; et al. Simulating galaxy formation with the IllustrisTNG model. *Mon. Not. R. Astron. Soc.* **2018**, *473*, 4077–4106. [CrossRef]
310. Dubois, Y.; Devriendt, J.; Slyz, A.; Teyssier, R. Self-regulated growth of supermassive black holes by a dual jet-heating active galactic nucleus feedback mechanism: Methods, tests and implications for cosmological simulations. *Mon. Not. R. Astron. Soc.* **2012**, *420*, 2662–2683. [CrossRef]
311. Beck, A.M.; Murante, G.; Arth, A.; Remus, R.S.; Teklu, A.F.; Donnert, J.M.F.; Planelles, S.; Beck, M.C.; Förster, P.; Imgrund, M.; et al. An improved SPH scheme for cosmological simulations. *Mon. Not. R. Astron. Soc.* **2016**, *455*, 2110–2130. [CrossRef]
312. Sembolini, F.; Yepes, G.; De Petris, M.; Gottlöber, S.; Lamagna, L.; Comis, B. The MUSIC of galaxy clusters—I. Baryon properties and scaling relations of the thermal Sunyaev-Zel'dovich effect. *Mon. Not. R. Astron. Soc.* **2013**, *429*, 323–343. [CrossRef]
313. Cui, W.; Knebe, A.; Yepes, G.; Pearce, F.; Power, C.; Dave, R.; Arth, A.; Borgani, S.; Dolag, K.; Elahi, P.; et al. The Three Hundred project: A large catalogue of theoretically modelled galaxy clusters for cosmological and astrophysical applications. *Mon. Not. R. Astron. Soc.* **2018**, *480*, 2898–2915. [CrossRef]
314. Rasia, E.; Borgani, S.; Murante, G.; Planelles, S.; Beck, A.M.; Biffi, V.; Ragone-Figueroa, C.; Granato, G.L.; Steinborn, L.K.; Dolag, K. Cool Core Clusters from Cosmological Simulations. *Astrophys. J.* **2015**, *813*, L17, [CrossRef]
315. McCarthy, I.G.; Schaye, J.; Ponman, T.J.; Bower, R.G.; Booth, C.M.; Dalla Vecchia, C.; Crain, R.A.; Springel, V.; Theuns, T.; Wiersma, R.P.C. The case for AGN feedback in galaxy groups. *Mon. Not. R. Astron. Soc.* **2010**, *406*, 822–839. [CrossRef]
316. Steinborn, L.K.; Dolag, K.; Hirschmann, M.; Prieto, M.A.; Remus, R.S. A refined sub-grid model for black hole accretion and AGN feedback in large cosmological simulations. *Mon. Not. R. Astron. Soc.* **2015**, *448*, 1504–1525. [CrossRef]
317. Barnes, D.J.; Kay, S.T.; Bahé, Y.M.; Dalla Vecchia, C.; McCarthy, I.G.; Schaye, J.; Bower, R.G.; Jenkins, A.; Thomas, P.A.; Schaller, M.; et al. The Cluster-EAGLE project: Global properties of simulated clusters with resolved galaxies. *Mon. Not. R. Astron. Soc.* **2017**, *471*, 1088–1106. [CrossRef]
318. Bahé, Y.M.; Barnes, D.J.; Dalla Vecchia, C.; Kay, S.T.; White, S.D.M.; McCarthy, I.G.; Schaye, J.; Bower, R.G.; Crain, R.A.; Theuns, T.; et al. The Hydrangea simulations: Galaxy formation in and around massive clusters. *Mon. Not. R. Astron. Soc.* **2017**, *470*, 4186–4208. [CrossRef]
319. Bassini, L.; Rasia, E.; Borgani, S.; Ragone-Figueroa, C.; Biffi, V.; Dolag, K.; Gaspari, M.; Granato, G.L.; Murante, G.; Taffoni, G.; et al. Black hole mass and cluster mass correlation in cosmological hydro-dynamical simulations. *arXiv* **2019**, arXiv:1903.03142.
320. Martín-Navarro, I.; Burchett, J.N.; Mezcua, M. Black hole feedback and the evolution of massive early-type galaxies. *Mon. Not. R. Astron. Soc.* **2020**, *491*, 1311–1319. [CrossRef]
321. Truong, N.; Pillepich, A.; Werner, N. Correlations between supermassive black holes and hot gas atmospheres in IllustrisTNG and X-ray observations. *Mon. Not. R. Astron. Soc.* **2021**, *501*, 2210–2230. [CrossRef]
322. van Daalen, M.P.; Schaye, J.; Booth, C.M.; Dalla Vecchia, C. The effects of galaxy formation on the matter power spectrum: A challenge for precision cosmology. *Mon. Not. R. Astron. Soc.* **2011**, *415*, 3649–3665. [CrossRef]
323. Semboloni, E.; Hoekstra, H.; Schaye, J.; van Daalen, M.P.; McCarthy, I.G. Quantifying the effect of baryon physics on weak lensing tomography. *Mon. Not. R. Astron. Soc.* **2011**, *417*, 2020–2035. [CrossRef]
324. Chisari, N.E.; Mead, A.J.; Joudaki, S.; Ferreira, P.G.; Schneider, A.; Mohr, J.; Tröster, T.; Alonso, D.; McCarthy, I.G.; Martin-Alvarez, S.; et al. Modelling baryonic feedback for survey cosmology. *Open J. Astrophys.* **2019**, *2*, 4, [CrossRef]
325. Amendola, L.; Appleby, S.; Avgoustidis, A.; Bacon, D.; Baker, T.; Baldi, M.; Bartolo, N.; Blanchard, A.; Bonvin, C.; Borgani, S.; et al. Cosmology and fundamental physics with the Euclid satellite. *Living Rev. Relativ.* **2018**, *21*, 2, [CrossRef] [PubMed]
326. Schneider, A.; Stoira, N.; Refregier, A.; Weiss, A.J.; Knabenhans, M.; Stadel, J.; Teyssier, R. Baryonic effects for weak lensing. Part I. Power spectrum and covariance matrix. *J. Cosmol. Astropart. Phys.* **2020**, *2020*, 19, [CrossRef]
327. Schneider, A.; Refregier, A.; Grandis, S.; Eckert, D.; Stoira, N.; Kacprzak, T.; Knabenhans, M.; Stadel, J.; Teyssier, R. Baryonic effects for weak lensing. Part II. Combination with X-ray data and extended cosmologies. *J. Cosmol. Astropart. Phys.* **2020**, *2020*, 20, [CrossRef]
328. van Daalen, M.P.; McCarthy, I.G.; Schaye, J. Exploring the effects of galaxy formation on matter clustering through a library of simulation power spectra. *Mon. Not. R. Astron. Soc.* **2020**, *491*, 2424–2446. [CrossRef]
329. Schneider, A.; Teyssier, R.; Stadel, J.; Chisari, N.E.; Le Brun, A.M.C.; Amara, A.; Refregier, A. Quantifying baryon effects on the matter power spectrum and the weak lensing shear correlation. *J. Cosmol. Astropart. Phys.* **2019**, *2019*, 020, [CrossRef]
330. Battaglia, N.; Bond, J.R.; Pfrommer, C.; Sievers, J.L. On the Cluster Physics of Sunyaev-Zel'dovich and X-ray Surveys. II. Deconstructing the Thermal SZ Power Spectrum. *Astrophys. J.* **2012**, *758*, 75, [CrossRef]
331. McCarthy, I.G.; Le Brun, A.M.C.; Schaye, J.; Holder, G.P. The thermal Sunyaev-Zel'dovich effect power spectrum in light of Planck. *Mon. Not. R. Astron. Soc.* **2014**, *440*, 3645–3657. [CrossRef]
332. McCarthy, I.G.; Bird, S.; Schaye, J.; Harnois-Deraps, J.; Font, A.S.; van Waerbeke, L. The BAHAMAS project: The CMB-large-scale structure tension and the roles of massive neutrinos and galaxy formation. *Mon. Not. R. Astron. Soc.* **2018**, *476*, 2999–3030. [CrossRef]
333. Ramos-Ceja, M.E.; Basu, K.; Pacaud, F.; Bertoldi, F. Constraining the intracluster pressure profile from the thermal SZ power spectrum. *Astron. Astrophys.* **2015**, *583*, A111, [CrossRef]

334. Arnaud, M.; Pratt, G.W.; Piffaretti, R.; Böhringer, H.; Croston, J.H.; Pointecouteau, E. The universal galaxy cluster pressure profile from a representative sample of nearby systems (REXCESS) and the Y_{SZ}—M_{500} relation. *Astron. Astrophys.* **2010**, *517*, A92, [CrossRef]
335. Battaglia, N.; Hill, J.C.; Murray, N. Deconstructing Thermal Sunyaev-Zel'dovich—Gravitational Lensing Cross-correlations: Implications for the Intracluster Medium. *Astrophys. J.* **2015**, *812*, 154, [CrossRef]
336. Hojjati, A.; McCarthy, I.G.; Harnois-Deraps, J.; Ma, Y.Z.; Van Waerbeke, L.; Hinshaw, G.; Le Brun, A.M.C. Dissecting the thermal Sunyaev-Zeldovich-gravitational lensing cross-correlation with hydrodynamical simulations. *J. Cosmol. Astropart. Phys.* **2015**, *2015*, 047, [CrossRef]
337. Velliscig, M.; van Daalen, M.P.; Schaye, J.; McCarthy, I.G.; Cacciato, M.; Le Brun, A.M.C.; Dalla Vecchia, C. The impact of galaxy formation on the total mass, mass profile and abundance of haloes. *Mon. Not. R. Astron. Soc.* **2014**, *442*, 2641–2658. [CrossRef]
338. Cui, W.; Borgani, S.; Murante, G. The effect of active galactic nuclei feedback on the halo mass function. *Mon. Not. R. Astron. Soc.* **2014**, *441*, 1769–1782. [CrossRef]
339. Bocquet, S.; Saro, A.; Dolag, K.; Mohr, J.J. Halo mass function: Baryon impact, fitting formulae, and implications for cluster cosmology. *Mon. Not. R. Astron. Soc.* **2016**, *456*, 2361–2373. [CrossRef]
340. Schaller, M.; Frenk, C.S.; Bower, R.G.; Theuns, T.; Jenkins, A.; Schaye, J.; Crain, R.A.; Furlong, M.; Dalla Vecchia, C.; McCarthy, I.G. Baryon effects on the internal structure of ΛCDM haloes in the EAGLE simulations. *Mon. Not. R. Astron. Soc.* **2015**, *451*, 1247–1267. [CrossRef]
341. Xu, W.; Ramos-Ceja, M.E.; Pacaud, F.; Reiprich, T.H.; Erben, T. A new X-ray-selected sample of very extended galaxy groups from the ROSAT All-Sky Survey. *Astron. Astrophys.* **2018**, *619*, A162, [CrossRef]
342. Merloni, A.; Predehl, P.; Becker, W.; Böhringer, H.; Boller, T.; Brunner, H.; Brusa, M.; Dennerl, K.; Freyberg, M.; Friedrich, P.; et al. eROSITA Science Book: Mapping the Structure of the Energetic Universe. *arXiv* **2012**, arXiv:1209.3114.
343. Käfer, F.; Finoguenov, A.; Eckert, D.; Clerc, N.; Ramos-Ceja, M.E.; Sanders, J.S.; Ghirardini, V. Toward the low-scatter selection of X-ray clusters. Galaxy cluster detection with eROSITA through cluster outskirts. *Astron. Astrophys.* **2020**, *634*, A8, [CrossRef]
344. Freyberg, M.; Perinati, E.; Pacaud, F.; Eraerds, T.; Churazov, E.; Dennerl, K.; Predehl, P.; Merloni, A.; Meidinger, N.; Bulbul, E.; et al. SRG/eROSITA in-flight background at L2. In Proceedings of the SPIE Digital Library, SPIE Astronomical Telescopes + Instrumentation, Online Conference, 13–18 December 2020; Society of Photo-Optical Instrumentation Engineers (SPIE) Conference Series; 2020; Volume 11444, p. 114441O. [CrossRef]
345. Böhringer, H.; Chon, G.; Collins, C.A.; Guzzo, L.; Nowak, N.; Bobrovskyi, S. The extended ROSAT-ESO flux limited X-ray galaxy cluster survey (REFLEX II) II. Construction and properties of the survey. *Astron. Astrophys.* **2013**, *555*, A30, [CrossRef]
346. Planck Collaboration XXVII. Planck 2015 results. XXVII. The second Planck catalogue of Sunyaev-Zeldovich sources. *Astron. Astrophys.* **2016**, *594*, A27, [CrossRef]
347. Benson, B.A.; Ade, P.A.R.; Ahmed, Z.; Allen, S.W.; Arnold, K.; Austermann, J.E.; Bender, A.N.; Bleem, L.E.; Carlstrom, J.E.; Chang, C.L.; et al. SPT-3G: A next-generation cosmic microwave background polarization experiment on the South Pole telescope. In Proceedings of the SPIE Astronomical Telescopes + Instrumentation, Montréal, QC, Canada, 22–27 June 2014; Volume 9153, p. 91531P. [CrossRef]
348. XRISM Science Team. Science with the X-ray Imaging and Spectroscopy Mission (XRISM). *arXiv* **2020**, arXiv:2003.04962.
349. Pinto, C.; Sanders, J.S.; Werner, N.; de Plaa, J.; Fabian, A.C.; Zhang, Y.Y.; Kaastra, J.S.; Finoguenov, A.; Ahoranta, J. Chemical Enrichment RGS cluster Sample (CHEERS): Constraints on turbulence. *Astron. Astrophys.* **2015**, *575*, A38, [CrossRef]
350. Ahoranta, J.; Finoguenov, A.; Pinto, C.; Sanders, J.; Kaastra, J.; de Plaa, J.; Fabian, A. Observations of asymmetric velocity fields and gas cooling in the NGC 4636 galaxy group X-ray halo. *Astron. Astrophys.* **2016**, *592*, A145. [CrossRef]
351. Hitomi Collaboration. The quiescent intracluster medium in the core of the Perseus cluster. *Nature* **2016**, *535*, 117–121. [CrossRef] [PubMed]
352. Barret, D.; Decourchelle, A.; Fabian, A.; Guainazzi, M.; Nandra, K.; Smith, R.; den Herder, J.W. The Athena space X-ray observatory and the astrophysics of hot plasma. *Astron. Nachrichten* **2020**, *341*, 224–235. [CrossRef]
353. Zhang, C.; Ramos-Ceja, M.E.; Pacaud, F.; Reiprich, T.H. High-redshift galaxy groups as seen by ATHENA/WFI. *Astron. Astrophys.* **2020**, *642*, A17. [CrossRef]
354. Croston, J.H.; Sanders, J.S.; Heinz, S.; Hardcastle, M.J.; Zhuravleva, I.; Bîrzan, L.; Bower, R.G.; Brüggen, M.; Churazov, E.; Edge, A.C.; et al. The Hot and Energetic Universe: AGN feedback in galaxy clusters and groups. *arXiv* **2013**, arXiv:1306.2323.
355. Gaskin, J.A.; Swartz, D.A.; Vikhlinin, A.; Özel, F.; Gelmis, K.E.; Arenberg, J.W.; Bandler, S.R.; Bautz, M.W.; Civitani, M.M.; Dominguez, A.; et al. Lynx X-ray Observatory: An overview. *J. Astron. Telesc. Instruments, Syst.* **2019**, *5*, 021001. [CrossRef]
356. Hurley-Walker, N.; Callingham, J.R.; Hancock, P.J.; Franzen, T.M.O.; Hindson, L.; Kapińska, A.D.; Morgan, J.; Offringa, A.R.; Wayth, R.B.; Wu, C.; et al. GaLactic and Extragalactic All-sky Murchison Widefield Array (GLEAM) survey—I. A low-frequency extragalactic catalogue. *Mon. Not. R. Astron. Soc. Lett.* **2017**, *464*, 1146–1167. [CrossRef]
357. McConnell, D.; Hale, C.L.; Lenc, E.; Banfield, J.K.; Heald, G.; Hotan, A.W.; Leung, J.K.; Moss, V.A.; Murphy, T.; O'Brien, A.; et al. The Rapid ASKAP Continuum Survey I: Design and first results. *Publ. Astron. Soc. Aust.* **2020**, *37*, e048. [CrossRef]
358. Koribalski, B.S.; Staveley-Smith, L.; Westmeier, T.; Serra, P.; Spekkens, K.; Wong, O.I.; Lee-Waddell, K.; Lagos, C.D.P.; Obreschkow, D.; Ryan-Weber, E.V.; et al. WALLABY—An SKA Pathfinder HI survey. *Astrophys. Space Sci.* **2020**, *365*, 118. [CrossRef]

359. Jarvis, M.; Taylor, R.; Agudo, I.; Allison, J.R.; Deane, R.P.; Frank, B.; Gupta, N.; Heywood, I.; Maddox, N.; McAlpine, K.; et al. The MeerKAT International GHz Tiered Extragalactic Exploration (MIGHTEE) Survey. In Proceedings of the 2016 MeerKAT Science: On the Pathway to the SKA, MeerKAT, Stellenbosch, South Africa, 25–27 May 2016; p. 6.
360. Oosterloo, T.A.; Zhang, M.L.; Lucero, D.M.; Carignan, C. Galaxy interactions in loose galaxy groups: KAT-7 and VLA HI Observations of the IC 1459 group. *arXiv* **2018**, arXiv:1803.08263.
361. For, B.Q.; Staveley-Smith, L.; Westmeier, T.; Whiting, M.; Oh, S.H.; Koribalski, B.; Wang, J.; Wong, O.I.; Bekiaris, G.; Cortese, L.; et al. WALLABY early science—V. ASKAP H I imaging of the Lyon Group of Galaxies 351. *Mon. Not. R. Astron. Soc. Lett.* **2019**, *489*, 5723–5741. [CrossRef]
362. Braun, R.; Bonaldi, A.; Bourke, T.; Keane, E.; Wagg, J. Anticipated Performance of the Square Kilometre Array—Phase 1 (SKA1). *arXiv* **2019**, arXiv:1912.12699.
363. Prandoni, I.; Seymour, N. Revealing the Physics and Evolution of Galaxies and Galaxy Clusters with SKA Continuum Surveys. In Proceedings of theAdvancing Astrophysics with the Square Kilometre Array (AASKA14), Giardini Naxos, Italy, 8–13 June 2015; p. 67,
364. McAlpine, K.; Prandoni, I.; Jarvis, M.; Seymour, N.; Padovani, P.; Best, P.; Simpson, C.; Guidetti, D.; Murphy, E.; Huynh, M.; et al. The SKA view of the Interplay between SF and AGN Activity and its role in Galaxy Evolution. In Proceedings of the Advancing Astrophysics with the Square Kilometre Array (AASKA14), Giardini Naxos, Italy, 8–13 June 2015; p. 83.
365. Heald, G.; Mao, S.; Vacca, V.; Akahori, T.; Damas-Segovia, A.; Gaensler, B.; Hoeft, M.; Agudo, I.; Basu, A.; Beck, R.; et al. Magnetism Science with the Square Kilometre Array. *Galaxies* **2020**, *8*, 53, [CrossRef]
366. Kale, R.; Dwarakanath, K.S.; Vir Lal, D.; Bagchi, J.; Paul, S.; Malu, S.; Datta, A.; Parekh, V.; Sharma, P.; Pandey-Pommier, M. Clusters of Galaxies and the Cosmic Web with Square Kilometre Array. *J. Astrophys. Astron.* **2016**, *37*, 31, [CrossRef]
367. Stacey, G.J.; Aravena, M.; Basu, K.; Battaglia, N.; Beringue, B.; Bertoldi, F.; Bond, J.R.; Breysse, P.; Bustos, R.; Chapman, S.; et al. CCAT-Prime: Science with an ultra-widefield submillimeter observatory on Cerro Chajnantor. In Proceedings of the SPIE Astronomical Telescopes + Instrumentation, Austin, TX, USA, 12–14 June 2018; Volume 10700, p. 107001M. [CrossRef]
368. Lee, A.; Abitbol, M.H.; Adachi, S.; Ade, P.; Aguirre, J.; Ahmed, Z.; Aiola, S.; Ali, A.; Alonso, D.; Alvarez, M.A.; et al. The Simons Observatory. *Bull. Am. Astron. Soc.* **2019**, *51*, 147.
369. Abazajian, K.N.; Adshead, P.; Ahmed, Z.; Allen, S.W.; Alonso, D.; Arnold, K.S.; Baccigalupi, C.; Bartlett, J.G.; Battaglia, N.; Benson, B.A.; et al. CMB-S4 Science Book, First Edition. *arXiv* **2016**, arXiv:1610.02743.
370. Planck Collaboration; Ade, P.A.R.; Aghanim, N.; Arnaud, M.; Ashdown, M.; Atrio-Barandela, F.; Aumont, J.; Baccigalupi, C.; Balbi, A.; Banday, A.J.; et al. Planck intermediate results. XI. The gas content of dark matter halos: The Sunyaev-Zeldovich-stellar mass relation for locally brightest galaxies. *Astron. Astrophys.* **2013**, *557*, A52, [CrossRef]
371. Le Brun, A.M.C.; McCarthy, I.G.; Melin, J.B. Testing Sunyaev-Zel'dovich measurements of the hot gas content of dark matter haloes using synthetic skies. *Mon. Not. R. Astron. Soc.* **2015**, *451*, 3868–3881. [CrossRef]
372. Klaassen, P.D.; Mroczkowski, T.K.; Cicone, C.; Hatziminaoglou, E.; Sartori, S.; De Breuck, C.; Bryan, S.; Dicker, S.R.; Duran, C.; Groppi, C.; et al. The Atacama Large Aperture Submillimeter Telescope (AtLAST).In Proceedings of the SPIE Astronomical Telescopes + Instrumentation, Online Conference, 13–18 December 2020; Society of Photo-Optical Instrumentation Engineers (SPIE) Conference Series; Volume 11445, p. 114452F, [CrossRef]
373. Mroczkowski, T.; Nagai, D.; Basu, K.; Chluba, J.; Sayers, J.; Adam, R.; Churazov, E.; Crites, A.; Di Mascolo, L.; Eckert, D.; et al. Astrophysics with the Spatially and Spectrally Resolved Sunyaev-Zeldovich Effects. A Millimetre/Submillimetre Probe of the Warm and Hot Universe. *Space Sci. Rev.* **2019**, *215*, 17, [CrossRef]
374. Hill, J.C.; Ferraro, S.; Battaglia, N.; Liu, J.; Spergel, D.N. Kinematic Sunyaev-Zel'dovich Effect with Projected Fields: A Novel Probe of the Baryon Distribution with Planck, WMAP, and WISE Data. *Phys. Rev. Lett.* **2016**, *117*, 051301, [CrossRef] [PubMed]
375. Ferraro, S.; Hill, J.C.; Battaglia, N.; Liu, J.; Spergel, D.N. Kinematic Sunyaev-Zel'dovich effect with projected fields. II. Prospects, challenges, and comparison with simulations. *Phys. Rev. D* **2016**, *94*, 123526, [CrossRef]
376. De Bernardis, F.; Aiola, S.; Vavagiakis, E.M.; Battaglia, N.; Niemack, M.D.; Beall, J.; Becker, D.T.; Bond, J.R.; Calabrese, E.; Cho, H.; et al. Detection of the pairwise kinematic Sunyaev-Zel'dovich effect with BOSS DR11 and the Atacama Cosmology Telescope. *J. Cosmol. Astropart. Phys.* **2017**, *2017*, 008, [CrossRef]
377. Calafut, V.; Gallardo, P.A.; Vavagiakis, E.M.; Amodeo, S.; Aiola, S.; Austermann, J.E.; Battaglia, N.; Battistelli, E.S.; Beall, J.A.; Bean, R.; et al. The Atacama Cosmology Telescope: Detection of the Pairwise Kinematic Sunyaev-Zel'dovich Effect with SDSS DR15 Galaxies. *arXiv* **2021**, arXiv:2101.08374.
378. Planck Collaboration; Ade, P.A.R.; Aghanim, N.; Arnaud, M.; Ashdown, M.; Aubourg, E.; Aumont, J.; Baccigalupi, C.; Banday, A.J.; Barreiro, R.B.; et al. Planck intermediate results. XXXVII. Evidence of unbound gas from the kinetic Sunyaev-Zeldovich effect. *Astron. Astrophys.* **2016**, *586*, A140, [CrossRef]
379. Battaglia, N.; Ferraro, S.; Schaan, E.; Spergel, D.N. Future constraints on halo thermodynamics from combined Sunyaev-Zel'dovich measurements. *J. Cosmol. Astropart. Phys.* **2017**, *2017*, 040, [CrossRef]
380. Chaves-Montero, J.; Hernandez-Monteagudo, C.; Angulo, R.E.; Emberson, J.D. Measuring the evolution of intergalactic gas from z=0 to 5 using the kinematic Sunyaev-Zel'dovich effect. *arXiv* **2019**, arXiv:1911.10690.
381. Rafferty, D.A.; McNamara, B.R.; Nulsen, P.E.J.; Wise, M.W. The Feedback-regulated Growth of Black Holes and Bulges through Gas Accretion and Starbursts in Cluster Central Dominant Galaxies. *Astrophys. J.* **2006**, *652*, 216–231. [CrossRef]

382. Lal, D.V.; Kraft, R.P.; Randall, S.W.; Forman, W.R.; Nulsen, P.E.J.; Roediger, E.; ZuHone, J.A.; Hardcastle, M.J.; Jones, C.; Croston, J.H. Gas Sloshing and Radio Galaxy Dynamics in the Core of the 3C 449 Group. *Astrophys. J.* **2013**, *764*, 83, [CrossRef]
383. Pandge, M.B.; Sonkamble, S.S.; Parekh, V.; Dabhade, P.; Parmar, A.; Patil, M.K.; Raychaudhury, S. AGN Feedback in Galaxy Groups: A Detailed Study of X-ray Features and Diffuse Radio Emission in IC 1262. *Astrophys. J.* **2019**, *870*, 62, [CrossRef]
384. Randall, S.W.; Forman, W.R.; Giacintucci, S.; Nulsen, P.E.J.; Sun, M.; Jones, C.; Churazov, E.; David, L.P.; Kraft, R.; Donahue, M.; et al. Shocks and Cavities from Multiple Outbursts in the Galaxy Group NGC 5813: A Window to Active Galactic Nucleus Feedback. *Astrophys. J.* **2011**, *726*, 86. [CrossRef]
385. Schellenberger, G.; Vrtilek, J.M.; David, L.; O'Sullivan, E.; Giacintucci, S.; Johnston-Hollitt, M.; Duchesne, S.W.; Raychaudhury, S. NGC 741—Mergers and AGN Feedback on a Galaxy-group Scale. *Astrophys. J.* **2017**, *845*, 84, [CrossRef]
386. Machacek, M.E.; Kraft, R.P.; Jones, C.; Forman, W.R.; Hardcastle, M.J. X-ray Constraints on Galaxy-Gas-Jet Interactions in the Dumbbell Galaxies NGC 4782 and NGC 4783 in the LGG 316 Galaxy Group. *Astrophys. J.* **2007**, *664*, 804–819. [CrossRef]
387. Randall, S.W.; Jones, C.; Markevitch, M.; Blanton, E.L.; Nulsen, P.E.J.; Forman, W.R. Gas Sloshing and Bubbles in the Galaxy Group NGC 5098. *Astrophys. J.* **2009**, *700*, 1404–1414. [CrossRef]
388. O'Sullivan, E.; Kolokythas, K.; Kantharia, N.G.; Raychaudhury, S.; David, L.P.; Vrtilek, J.M. The origin of the X-ray, radio and H I structures in the NGC 5903 galaxy group. *Mon. Not. R. Astron. Soc. Lett.* **2018**, *473*, 5248. [CrossRef]
389. Machacek, M.E.; O'Sullivan, E.; Randall, S.W.; Jones, C.; Forman, W.R. The Mysterious Merger of NGC 6868 and NGC 6861 in the Telescopium Group. *Astrophys. J.* **2010**, *711*, 1316, [CrossRef]
390. Pandge, M.B.; Vagshette, N.D.; Sonkamble, S.S.; Patil, M.K. Investigation of X-ray cavities in the cooling flow system Abell 1991. *Astrophys. Space Sci.* **2013**, *345*, 183–193. [CrossRef]
391. Blanton, E.L.; Sarazin, C.L.; McNamara, B.R.; Clarke, T.E. Chandra Observation of the Central Region of the Cooling Flow Cluster A262: A Radio Source That Is a Shadow of Its Former Self? *Astrophys. J.* **2004**, *612*, 817–824. [CrossRef]

MDPI

Review

Simulating Groups and the IntraGroup Medium: The Surprisingly Complex and Rich Middle Ground between Clusters and Galaxies

Benjamin D. Oppenheimer [1,2,*], Arif Babul [3], Yannick Bahé [4], Iryna S. Butsky [5] and Ian G. McCarthy [6]

1 CASA, Department of Astrophysical and Planetary Sciences, University of Colorado, 389 UCB, Boulder, CO 80309, USA
2 Harvard Smithsonian Center for Astrophysics, 60 Garden Street, Cambridge, MA 02138, USA
3 Department of Physics and Astronomy, University of Victoria, Victoria, BC V8W 2Y2, Canada; babul@uvic.ca
4 Leiden Observatory, Leiden University, P.O. Box 9513, 2300 RA Leiden, The Netherlands; bahe@strw.leidenuniv.nl
5 Astronomy Department, University of Washington, Seattle, WA 98195, USA; ibutsky@uw.edu
6 Astrophysics Research Institute, Liverpool John Moores University, 146 Brownlow Hill, Liverpool L53RF, UK; i.g.mccarthy@ljmu.ac.uk
* Correspondence: benjamin.oppenheimer@colorado.edu

Abstract: Galaxy groups are more than an intermediate scale between clusters and halos hosting individual galaxies, they are crucial laboratories capable of testing a range of astrophysics from how galaxies form and evolve to large scale structure (LSS) statistics for cosmology. Cosmological hydrodynamic simulations of groups on various scales offer an unparalleled testing ground for astrophysical theories. Widely used cosmological simulations with $\sim$(100 Mpc)^3 volumes contain statistical samples of groups that provide important tests of galaxy evolution influenced by environmental processes. Larger volumes capable of reproducing LSS while following the redistribution of baryons by cooling and feedback are the essential tools necessary to constrain cosmological parameters. Higher resolution simulations can currently model satellite interactions, the processing of cool ($T \approx 10^{4-5}$ K) multi-phase gas, and non-thermal physics including turbulence, magnetic fields and cosmic ray transport. We review simulation results regarding the gas and stellar contents of groups, cooling flows and the relation to the central galaxy, the formation and processing of multi-phase gas, satellite interactions with the intragroup medium, and the impact of groups for cosmological parameter estimation. Cosmological simulations provide evolutionarily consistent predictions of these observationally difficult-to-define objects, and have untapped potential to accurately model their gaseous, stellar and dark matter distributions.

Keywords: black holes; galaxy groups; galaxy surveys; intragroup medium/plasma; hydrodynamical and cosmological simulations; active galactic nuclei; X-ray observations; UV observations; cosmological parameters

Citation: Oppenheimer, B.D.; Babul, A.; Bahé, Y.; Butsky, I.S.; McCarthy, I.G. Simulating Groups and the IntraGroup Medium: The Surprisingly Complex and Rich Middle Ground between Clusters and Galaxies. *Universe* **2021**, *7*, 209. https://doi.org/10.3390/universe7070209

Academic Editors: Lorenzo Lovisari and Stefano Ettori

Received: 9 April 2021
Accepted: 2 June 2021
Published: 24 June 2021

Publisher's Note: MDPI stays neutral with regard to jurisdictional claims in published maps and institutional affiliations.

1. Introduction

Galaxy groups are versatile laboratories to study a range of astrophysics spanning non-gravitational, baryonic processes associated with galaxy formation to large-scale structure statistics constraining cosmology. Their intermediate scale between galactic halos and clusters offers a unique set of theoretical challenges that are often overlooked relative to adjacent mass bins, but this scale offers crucial constraints for how a large proportion of galaxies evolved to their present state. The perspective of this review, focusing on halos with masses $M_{\rm halo} \approx 10^{13}$–$10^{14}$ M$_\odot$, differs from the companion reviews because a cosmological simulation tracks the evolution of all gas, stars and dark matter, not just the X-ray emitting intragroup medium (IGrM). Therefore we consider all phases of gas—from the extended, hot IGrM that stretches beyond the virial radius to the cold interstellar medium (ISM) within individual group galaxies—alongside the stars and dark matter in central and satellite group galaxies, as well as the surrounding cosmological large scale structure (LSS). Simulations have enabled breakthroughs in our understanding of all these components, but their limited resolution and incomplete physics models still represent a major obstacle on the path to a truly complete understanding of the observed gas, galaxies and LSS statistics in and around groups.

The IGrM is mainly very hot ($T \sim 10^7$ K), but does not primarily radiate via Bremsstrahlung radiation like the intracluster medium (ICM; see Figure 1 of the companion review by Lovisari et al. [1]). Cooling via line emission offers a greater opportunity to form multi-phase, cool ($T < 10^5$ K) gas, and potentially provide additional fuel to galaxies. Nevertheless, IGrM observations are currently dominated by hot X-ray probes from especially *Chandra* [2–4] and *XMM-Newton* (e.g., Lovisari et al. [5]). X-ray-derived profiles of IGrM properties, especially its entropy, provide rigorous tests for simulations that often make diverging predictions (e.g., Mitchell et al. [6], Le Brun et al. [7]). Yet these simulations also produce lower mass groups that can be tested against the Complete Local Volume Groups Sample (CLoGS) survey [8], and sometimes higher mass objects for comparison with observed clusters (e.g., [9,10]). With large-scale simulations projects containing multiple volumes and/or zoom-in simulations of massive objects, simulations of the IGrM can be compared and contrasted to lower mass galaxy halos and higher mass clusters.

Similar to a cluster, most groups contain a dominant "brightest group galaxy" (BGG) near the halo center, usually where the X-ray-traced IGrM peaks in brightness. In contrast to brightest cluster galaxies (BCGs), BGGs are observed to less likely be quenched early-type galaxies (ETGs; e.g., [11,12]), and more likely to have disc-like morphologies (e.g., [13]). Understanding the—likely intimate—connection between the BGG and IGrM involves studying how the galactic baryon cycle (the interplay of gas accretion, outflows, and recycling that is understood as fundamental for the co-evolution of field galaxies and their circumgalactic medium; [14,15]) transitions to the cluster version of precipitation, jet-driven active galactic nuclei (AGN) feedback, and chaotic cold accretion (e.g., [16,17]).

The IGrM also includes cool and warm ($T \sim 10^5$–10^6 K) gas phases. Although subdominant to the hot IGrM component by mass as simulations clearly demonstrate (e.g., [18,19]), these phases are thought to represent the link between the IGrM and individual group galaxies: from them, gas can accrete onto the BGG and fuel further star formation (e.g., [20,21]), whereas gas stripped from satellite galaxies (e.g., [22–24]) or ejected from them through superwind feedback (e.g., [25]) is also initially less hot than the virialized IGrM halo. Furthermore, observations of quasar absorption lines in the UV indicate a substantial reservoir of H I and metals in the IGrM at temperatures of 10^4–$10^{5.5}$ K [26,27], and 21-cm emission shows extended IGrM structures, at least for more compact, spiral-rich groups [28]. The theoretical modeling of cool/warm gas is crucial to understand for how groups diverge from clusters owing to their lower temperatures promoting more cooling.

Lower IGrM pressures and galaxy velocities process satellite galaxies differently than in clusters: simulations find both the atomic hydrogen and star forming gas is removed less rapidly after infall [29], providing an explanation for observed group galaxies being less H I deficient than in clusters [30] and having lower quenched fractions (e.g., [11,12]). However, the lower velocity dispersion of group satellites makes dynamical friction more efficient, so that mergers—in particular between satellites and the BGG—are more common [31]. Finally, although groups are (typically) dynamically older than clusters, their galaxies are more likely to have been accreted directly from the field, rather than via an intermediate "pre-processing" phase in a lower-mass halo (e.g., [32]). The total time that $z = 0$ galaxies have spent as satellites is therefore $\approx$50% shorter for groups than massive clusters (ca. 4 vs. 6 Gyr; [33], see also Donnari et al. [34]).

Relative to clusters, the $\sim$10$\times$ higher number density of groups (e.g., [35]) combined with their shallower potential wells—which make it easier for AGN feedback to eject baryons—gives groups particular significance for cosmological parameter estimates that rely on the total mass distribution, for example, lensing measurements, cosmic shear, and redshift space distortions. The distribution of baryons in halos corresponding to groups (10^{13-14} $M_\odot$) is known to be significantly affected by feedback processes. Therefore, the self-consistent modeling of the entire baryonic (gas+stellar) and dark matter distributions of groups, in way that realistically captures the effects of galaxy formation, is a necessary tool for precision cosmology.

In this review, we focus on how groups process their baryons in a cosmological context, and therefore discuss predominantly cosmological simulations that contain groups. Entire separate reviews could be written about idealized simulations and analytical models at the group scale, but these are not able to confront the intersection of galaxy formation and cosmology. We will, however, refer to idealized simulations and other methods simulating important physics when they are relevant to groups, especially in regards to how cosmological simulations can improve. Ultimately, our understanding of how gas and galaxies evolve together to create the observed distribution of groups would not be as far along without these tools.

Throughout this review, we use a theorist's definition of a group as a system of galaxies and gas hosted by a halo within a certain mass range. Following common convention, we define these masses as M_Δ, the sum of all matter species (i.e., dark matter and baryons) within a spherical aperture r_Δ inside which the mean density equals Δ times the critical density of the universe; we adopt the value $\Delta = 500$ as commonly used in X-ray studies of the IGrM/ICM and thus define a group halo as one with $M_{500(c)} = 10^{13}$–10^{14} $M_\odot$ ($r_{500(c)} \approx$ 340–600 kpc) [1]. This definition places groups between lower-mass galactic halos ($M_{500} \lesssim 10^{13}\,M_\odot$) and more massive clusters ($M_{500} \gtrsim 10^{14}\,M_\odot$). For the benefit of readers more used to masses within other overdensity thresholds, we note that in this halo mass range masses calculated within spheres of average density equal to 200 times the critical, or 200 times the mean, density (M_{200} and M_{200m}, respectively), or to the virial overdensity $\Delta_{\rm vir} \approx 18\pi^2 + 82(\Omega(z) - 1) - 39(\Omega(z) - 1)^2$ [36] that is based on the analytic solution of spherical top-hat collapse ($M_{\rm vir}$), are offset from M_{500} by +0.16 dex (M_{200}), +0.31 dex (M_{200m}), and +0.25 dex ($M_{\rm vir}$), respectively (and the corresponding radii from r_{500} by factors of 1.5, 2.6, and 2.1) [2]. A typical group with mass of $M_{500} = 10^{13.5}$ $M_\odot$ has a radius $R_{500} = 480$ kpc, a virial temperature $T_X \sim 1$ keV, and a velocity dispersion $\sigma \approx 440$ km s^{-1} at $z = 0$.

This review is organized into sections as follows: We begin with an overview of the cosmological simulations we discuss in Section 2. The methods used in simulation of groups are discussed in a series of subsections throughout Section 3. Section 4 comprises the results of current simulations creating groups, and is divided into five main subsections: the baryonic content of groups (Section 4.1), the connection between the central galaxy and the IGrM (Section 4.2), the multiphase IGrM (Section 4.3), satellite galaxies in groups (Section 4.4), and the impact of galaxy group astrophysics on LSS cosmology (Section 4.5). We discuss future directions in Section 5 and make a short final statement in Section 6.

2. Overview of Simulations That Model Groups

Cosmological simulations use a variety of hydrodynamics schemes, span many orders of magnitude in mass and spatial resolution, and model volumes of vastly different sizes and contents. This rich diversity in modeling approaches reflects the wide range of astrophysical processes and objects that different simulations attempt to model. In Table 1, we list recent simulations that include galaxy groups and are therefore of particular interest to our review.

Table 1. Modern cosmological hydrodynamic simulations with groups that run to $z = 0$.

Simulation	Simulation Code	Hydrodynamic Scheme	Baryon Resolution ($M_\odot$)	Volume (Mpc^3)	AGN Feedback Scheme	$f_{gas,500}$ at $M_{500} = 10^{13.5} M_\odot$	$f_{*,500}$ at $M_{500} = 10^{13.5} M_\odot$
cosmo-OWLS [a]	GADGET-3	Classical SPH	1.2×10^9	2.1×10^8	Thermal	0.05	
Illustris [b]	AREPO	Moving Mesh	1.3×10^6	1.2×10^6	Dual	0.01	0.04
EAGLE [c]	GADGET-3	Modern SPH	1.8×10^6	1.0×10^6	Thermal	0.11	0.01
Liang et al. [37]	GADGET-2	Classical SPH	9.0×10^7	2.9×10^6	None	0.09	0.06
Horizon-AGN [d]	RAMSES	AMR	1×10^7	2.9×10^6	Dual	0.09	
BAHAMAS [e]	GADGET-3	Classical SPH	1.2×10^9	2.1×10^8	Thermal	0.04	0.02
C-EAGLE/Hydrangea [f]	GADGET-3	Modern SPH	1.8×10^6	30 zooms	Thermal	0.08	0.02
FABLE [g]	AREPO	Moving Mesh	9.4×10^6	2.0×10^5 + 6 zooms	Dual	0.07	0.02
The Three Hundred [h]	GADGET-3	Modern SPH	3.5×10^8	324 zooms	Dual	0.10	0.02
IllustrisTNG100 [i]	AREPO	Moving Mesh	1.4×10^6	1.4×10^6	Dual	0.08	0.02
IllustrisTNG300 [j]	AREPO	Moving Mesh	1.1×10^7	2.8×10^7	Dual	0.08	
ROMULUS [k]	CHANGA	Modern SPH	2.1×10^5	1.5×10^4 + 3 zooms	Thermal	0.11	0.04
SIMBA [l]	GIZMO	Meshless Finite Mass	1.8×10^7	2.9×10^6	Dual	0.04	0.02
IllustrisTNG50 [m]	AREPO	Moving Mesh	8.5×10^4	1.4×10^5	Dual	0.09	
Magneticum-Box2/hr [n]	GADGET-3	Modern SPH	1.4×10^8	1.3×10^8	Dual		

[a] Le Brun et al. [7]; [b] Vogelsberger et al. [38]; [c] Schaye et al. [39]; [d] Dubois et al. [40]; [e] McCarthy et al. [41]; [f] Bahé et al. [42], Barnes et al. [43]; [g] Henden et al. [44]; [h] Cui et al. [45]; [i] Pillepich et al. [46]; [j] Nelson et al. [47]; [k] Tremmel et al. [48,49]; [l] Davé et al. [50]; [m] Nelson et al. [51]; [n] http://www.magneticum.org (accessed on 10 June 2021).

This list encompasses simulations with three broad classes of hydrodynamics schemes, as indicated in the third column of Table 1. Adaptive Mesh Refinement (AMR; [52]) uses a fixed (Eulerian) grid whose cells are locally and dynamically (de-/)refined, typically depending on the local gas density. Smoothed Particle Hydrodynamics (SPH; [53,54]), on the other hand, is a Lagrangian scheme in which gas *mass* is discretized into a finite number of particles that move under the influence of gravity and hydrodynamic forces; the latter are calculated by smoothing over neighboring particles. The third class is a hybrid of these two: in the "moving mesh" approach [55], an unstructured grid is used that moves with the local gas flow, whereas in "meshless finitie mass" simulations ([56], see also [57]), mass is discretized into particles like in SPH, but with explicit accounting of mass flows between neighboring particles [3]. We note that, as we discuss in Section 3.1, early SPH implementations suffered from systematic problems that have motivated improved "modern" formulations of SPH (see e.g., [58–61]); most SPH simulations that we discuss use one of these modern variants.

Many simulations listed in Table 1 evolve periodic, cosmological cubes with side length ≈100 Mpc, large enough to contain at least several dozen groups. In addition to modeling hydrodynamics and gravity, they all contain "subgrid" prescriptions for astrophysical processes that originate on unresolved scales, such as gas cooling, the formation of stars and black holes, and feedback associated with it (see Section 3). The "galaxy formation model" formed by these prescriptions is nowadays often calibrated to reproduce a particular set of galaxy properties. The Illustris [38], EAGLE (Evolution and Assembly of GaLaxies and their Environments; [39]) [4], and Horizon-AGN [40] simulations are representative examples of this approach. Their ability to (broadly) reproduce galactic stellar mass functions, galaxy colors and star formation rates (SFRs), and even galaxy morphologies signified a transformational advance over previous generations of simulations. Particularly relevant for groups, almost all the simulations listed in Table 1 model the accretion of gas onto SMBHs, and the resulting AGN feedback. In some recent simulations, including the IllustrisTNG project [46,47] [5] and SIMBA [50], these models are explicitly calibrated against gaseous properties of group-scale halos.

Simulations using the same or slightly modified codes as in some of the above-mentioned projects target volumes $\gg 10^6$ Mpc^3, with the aim to model clusters and/or LSS. For example, the C-EAGLE/Hydrangea simulations [42,43] extend EAGLE with 30 "zoom-in" simulations centered on clusters with $M_{200} = 10^{14.0}$–$10^{15.4}$ $M_\odot$, 24 of which (the Hydrangea suite) with high-resolution regions that extend to 10 r_{200} at $z = 0$ [42]. The zoom technique allows these simulations to reach the same resolution (≈2×10^6 $M_\odot$ for baryons) as the 100 Mpc EAGLE "Reference" run, but with slightly adjusted AGN feedback parameters for more realistic gas fractions in groups as we detail in Section 3.8. The only simulation that reaches a comparable resolution in a full $\gg 10^6$ Mpc^3 volume is the TNG300 run of the IllustrisTNG suite: it evolves a ($\approx$300 Mpc)^3 volume with a baryon mass resolution of $\approx 1.1 \times 10^7$ M$_\odot$, that is, $\approx$6 or 8 times lower resolution than (C-)EAGLE or the $\approx$100 Mpc TNG100 simulation, respectively. We note that, irrespective of this resolution difference, all TNG simulations use the same subgrid model and parameters. The degree of numerical convergence between these different resolution levels is discussed in for example, Pillepich et al. [67] and Donnari et al. [68]; we refer to Schaye et al. [39] for the opposite "weak convergence" approach of re-calibrating parameters for different resolution levels.

The cosmo-OWLS [7] and BAHAMAS (BAryons and HAloes of MAssive Systems) [41] simulations run larger periodic volumes at lower resolution. cosmo-OWLS varied different aspects of the subgrid models, switching different physics models on and off as well as changing the efficiencies of stellar and AGN feedback. No attempt was however made to calibrate the simulations to match observations, even though the default model reproduces different aspects of groups relatively well. BAHAMAS, on the other hand, explicitly calibrated the stellar and AGN feedback to reproduce the gas fractions of galaxy groups and the galaxy stellar mass function in order to ensure a realistic treatment of the effects of baryons on the matter power spectrum. The Magneticum simulations [69] are a series of volumes that mainly concentrate on LSS and cluster astrophysics. We list their Box2/hr simulation in Table 1, described in Castro et al. [70], which is capable of resolving a statistical sample of groups in a volume 500 Mpc on a side [6]. Lower resolution, Gpc-scale Magneticum volumes employ a strategy of calibrating subgrid modules to a *Planck* cosmology, and then exclusively varying cosmological parameters to explore the impact on baryonic properties. The Three Hundred project [45] simulated 324 clusters with $M_{\rm vir}(z=0) \gtrsim 1.2 \times 10^{15}$ M$_\odot$ out to a radius of 15 h^{-1}Mpc; similar to Hydrangea, the simulations therefore also contain many groups in the periphery of the central clusters.

We also list the Liang et al. [37] simulations and the Henden et al. [44] FABLE (Feedback Acting on Baryons in Large-scale Environments) simulations, both of which were run at lower resolution than their contemporary counterparts but with the aim to reproduce properties of groups and clusters. The former is an example of a simulation suite that does not include AGN feedback. In contrast, FABLE calibrated their subgrid models to reproduce the galactic stellar mass functions and, specifically, the gas and stellar contents of $M_{500} \approx 10^{13-15}$ M$_\odot$ halos. They simulated one smaller volume, supplemented by 6 zoom simulations extending up to a $M_{500} = 10^{15}$ M$_\odot$ cluster.

Finally, we will also discuss two higher-resolution simulations that include groups: the ROMULUS suite [48,49,72,73] and the IllustrisTNG50 (TNG50; [51,74]) simulation. ROMULUS includes a small-volume run, ROMULUS25, and three group/cluster zooms called ROMULUSG1, ROMULUSG2 and ROMULUSC, all at a baryon mass resolution of 2.1×10^5 M$_\odot$. TNG50 is a larger volume, $\approx$50 Mpc on a side, which contains 20 halos with $M_{500} > 10^{13}$ M$_\odot$ and reaches an even higher mass resolution of 8.5×10^4 M$_\odot$ for baryons.

Many other simulations have modeled groups from cosmological initial conditions, often with the zoom-in approach. These include the three 10^{13} M$_\odot$ group zooms by Feldmann et al. [75], which focus on the evolution and morphology of the central group galaxy; the 10 EAGLE-CGM zooms of $\sim 10^{13}$ M$_\odot$ groups by Oppenheimer et al. [76] to understand how the circumgalactic medium (CGM) around passive galaxies differs from star-forming galaxies at 10$\times$ lower halo mass, and the Joshi et al. [77] 2–3 $\times 10^{13}$ M$_\odot$ zoom to study environmental processing of satellite galaxies.

These simulations span a factor of 10,000 in both mass resolution and volume, and were run to study objects on a wide range of scales—from $M_\star < 10^8$ M$_\odot$ dwarf galaxies to the $\gg$Mpc scale LSS. However, they all overlap at the mass scale of our fiducial intermediate-mass group with $M_{500} = 10^{13.5}$ M$_\odot$. In Table 1, we therefore list the approximate total gas and stellar mass fractions inside R_{500}, defined as

$$f_{\rm gas,500} \equiv \frac{M_{\rm gas}(<R_{500})}{M_{\rm tot}(<R_{500})} \quad (1)$$

and

$$f_{\star,500} \equiv \frac{M_{\star}(< R_{500})}{M_{\text{tot}}(< R_{500})}. \quad (2)$$

For a more detailed discussion of these fractions, and their dependence on halo mass, we refer the interested reader to Figures 14 and 15 in the companion review by Eckert et al. [78].

3. Computational Methods Relevant for Simulations of Groups

The cosmological simulations we discuss are N-body+hydro simulations beginning from cosmological initial conditions at $z \gtrsim 100$. They all contain cooling, star formation, and stellar feedback, which are necessary to form realistic galaxies. All but the Liang et al. [37] simulation include the SMBH seeding, SMBH accretion, and AGN feedback, which are necessary to reproduce key properties of groups. All Magneticum simulations include a passive magnetic field [79] and IllustrisTNG uses a magnetic hydrodynamic (MHD) solver that self-consistently follows the magnetic field, which is initially seeded in the initial conditions [80]. Separate code modules are written for these different aspects and while our discussion here is not an exhaustive list of code modules in every simulation, we focus on specific methods, prescriptions and subgrid models that impact the group scale, while providing some additional context.

3.1. Hydrodynamics

Historically, the choice of hydrodynamic scheme has had significant effects on the appearance of simulated clusters, in particular the entropy profiles near cluster centers. Grid-based, Eulerian mesh codes tend to produce higher entropy, flatter cores than particle-based schemes as first shown by Frenk et al. [81]. This behavior is believed to be a result of numerical diffusion and over-mixing in mesh codes and/or the absence of heat diffusion in "classical" SPH simulations [82]. Mitchell et al. [6] explored these phenomena using idealized cluster merger simulations in both FLASH (grid-based) and GADGET-2 (SPH), finding that the former creates flatter high-entropy cores by mixing entropy through vortices and turbulent eddies. They argued that the suppression of mixing and fluid instabilities (e.g., Kelvin-Helmholtz) in SPH simulations appear to preserve the entropy low of individual gas particles resulting in a power-law entropy distribution reminiscent of a cool core cluster.

Recent numerical improvements in modern SPH and AMR codes have greatly improved their ability to recreate a wider range of entropy profiles. For example, adding artificial conduction to GADGET-3 simulations allowed for the creation of cored profiles in SPH simulations by mimicking thermal diffusion, resulting in the production of both cool core (CC) and non-cool core (NCC) entropy profiles across a sample of massive cluster simulations [83]. Hahn et al. [84] ran a set of RAMSES AMR simulations which also produced the dichotomy of CC/NCC clusters. The nIFTy simulation code comparison project compared one 10^{15} $M_{\odot}$ cluster [85], finding that modern SPH methods could create entropy cores just as grid-based and moving mesh methods do. These updated SPH implementations, such as ANARCHY [59] used by EAGLE and the Beck et al. [60] scheme used by Magneticum, often include a combination of pressure-entropy formulations of SPH, higher-order SPH kernels, and new treatments for thermal conduction and artificial viscosity. Cluster simulation comparison projects using modern SPH schemes (like the nIFTy project and The Three Hundred project [45]) now find that hydrodynamic scheme makes less of a difference than the inclusion of subgrid models on the appearance of entropy cores. While the appearance of entropy cores does not necessarily depend on the AGN feedback scheme at the cluster scale, the lower potentials of galaxy groups may yield a different answer.

The majority of the simulations explored here use either AREPO moving mesh or GADGET-2/GADGET-3 SPH, either classical or modern as listed in Table 1. The CHANGA code has a number of SPH updates putting it in the modern category.

3.2. Gas Cooling and Heating

All simulations we discuss include radiative cooling and photoionization heating by the meta-galactic UV/X-ray background (UVB, for e.g., [86]). A module accesses cooling/photo-heating rate tables usually as a function of density, temperature and redshift. These lookup tables are usually calculated from CLOUDY [87] models element-by-element, where the redshift dependence accounts for the evolving UVB. Most of the GADGET-3 simulations use Wiersma et al. [88] rates calculated for 11 elements, SIMBA uses the GRACKLE library [89] that tabulates self-shielding from the UVB for dense gas, and the AREPO-based simulations track additional photo-heating from local AGN that suppresses cooling in nearby gas [90].

For IGrM temperatures and densities that emit in the X-ray, cooling is driven by the balance of recombination and collisional ionization, with photo-ionization/heating usually playing an insignificant role. Often ionization equilibrium is assumed, but some simulations follow non-equilibrium rates of primordial elements (e.g., SIMBA). Cen and Fang [91] integrated non-equilibrium tracking of high oxygen ions (O V – O IX) in uniform mesh simulations, following how these ion species could deviate from equilibrium conditions when the recombination time became significant compared to the cooling time. Oppenheimer et al. [76] ran zoom SPH simulations following all ions from 11 metal species and including collisional [92], photo-ionization [93], Auger ionization and charge exchange [94] processes in low-mass groups. While non-equilibrium processes involving metals are not likely to significantly alter dynamics of cooling at IGrM temperatures, they can have an impact on the observational diagnostics, such as O VII and O VIII line emission [92,94]. Further exploration of IGrM simulations integrating non-equilibrium ionization and cooling is necessary in light of the coming launch of *XRISM* that will allow the measurement of line emission in the IGrM and ICM.

In contrast to simulations with radiative cooling, non-radiative (or adiabatic) simulations do not contain gas cooling (or any galaxy or SMBH formation). Frenk et al. [81], Lewis et al. [95], and Voit et al. [96] ran such simulations, finding that a baseline $K \propto R^{1.1}$ entropy profile is created in the absence of cooling, feedback, and other processes related to galaxy formation. [7].

We also note that the ROMULUS simulations do not include metal-line cooling at $T > 10^4$ K [48].

3.3. Star Formation, Stellar Evolution, and Nucleosynthetic Production

The formation of stars, their evolution and the release of elements are included in subgrid models for all the simulations we discuss. The multi-phase ISM is never resolved in separate phases, but relies on applying subgrid models that encompass multiple phases in a single gas parcel. A density threshold, above which the multi-phase ISM forms, is often the only criterion used for determining if gas can form stars. Different subgrid ISM models (e.g., [99–101]) use different motivations for their choices of unresolved phases, but all are calibrated to reproduce the Kennicutt [102] SFR surface density as a function of gas surface density relation. Star particles are spawned from gas parcels probabilistically and represent a population of individual equal-age stars, since they are at minimum $\sim 10^5\,M_\odot$. Stellar death is modeled as a time-dependent return of a star particle's mass to surrounding gas as a function of star particle age, based on expectations from stellar evolution. The initial mass function (IMF) determines the proportion of stars that die and can induce a systematic shift in the stellar mass formed per amount of stellar luminosity emitted. Fortunately, the simulations used in the comparisons of M_* in Section 4.2 use either a Kroupa [103] IMF (ROMULUS) or a Chabrier [104] IMF (every other simulation), which are fairly similar in their shapes.

The release of elements from stars is the source for the enrichment of the IGrM, about which an entire review by Gastaldello et al.[105] is written in this Special Issue "The Physical Properties of the Groups of Galaxies". In their Section 3.2, they include a discussion of chemical evolution models in cosmological hydrodynamic simulations. We note here that the simulations we discuss frequently all use chemical evolution models that include the elemental yields taken from stellar evolution models of Type II Supernovae (SNe), Type Ia SNe, and AGB stars. We note that yields have a great deal of uncertainty associated with them as explained in Wiersma et al. [106]. Uncertainties arise in the production of iron from Type Ia SNe since the Type Ia rate is not well constrained, in the transition stellar mass from Type II SNe to AGB stars, in the shape of the IMF, and in the calculations of the elemental yields themselves. These uncertainties can directly affect the rates of gas cooling (Section 3.2), which can be dominated by metal-line emission at many IGrM temperatures (i.e., between $T \approx 10^{4.5}$–$10^{6.5}$ K for a $Z = 0.3\ \mathrm{Z}_\odot$ plasma, [92,94]).

3.4. Metal Spreading

The distribution of metals in the IGrM depends on (1) how metals are released from star particles into the surrounding gas; and (2) how metals diffuse between gas elements. This is a separate issue from stellar feedback that applies mechanical feedback energy to gas elements, which we discuss in the next section.

Metal enrichment to large distances can occur without mechanical feedback if the number of enrichment neighbors extends over a large volume, which may unrealistically enrich diffuse environments. In cases where the number of neighbors is large and/or the resolution is low, simulated star particles may artificially enrich diffuse environments like the IGrM without superwinds. This effect is most obvious in the uniform mesh simulations of Cen and Fang [91], where O VII and O VIII absorption statistics of diffuse gas remain at similar levels with and without superwinds. Modern simulations typically release metals over several dozen neighboring gas elements, based on either the smoothing kernel in the case of SPH simulations (e.g., [39]), or mesh cells in the case of moving mesh simulations (e.g., [46]), which appear capable of enriching diffuse environments in simulations without invoking superwind feedback (e.g., [107]). Tornatore et al. [108] tested the effect of the number of neighboring SPH particles on ICM properties, finding only moderate effects on the ICM metallicity. On the other hand, the Davé et al. [109] simulations spread metals over very few (i.e., 3) neighboring SPH particles, resulting in much lower IGrM oxygen metallicity arising from Type II SNe without mechanical feedback in contrast to their stellar feedback runs that have $\sim$1 dex higher oxygen levels that better agree with observational constraints [110]. However, the Davé et al. [109] IGrM iron enrichment remains similar with and without stellar feedback, indicating enrichment from a different source—Type Ia SNe from older intracluster stars. Higher IGrM/ICM enrichment levels can also be achieved through increased numerical resolution as reviewed in Section 2.1.2 of Biffi et al. [111], owing primarily to the ability to resolve metal-enrichment from smaller halos at high-z.

While metal diffusion occurs naturally in mesh-based codes via advection between gas cells, metal diffusion between gas elements must be explicitly modeled in modern SPH implementations. The EAGLE simulations use SPH kernel-smoothed metallicities to calculate cooling rates [106] to estimate the dynamical effects of metal spreading in a manner that is consistent with the SPH formalism. The ROMULUS simulations apply the Shen et al. [112] metal diffusion algorithm to mimic turbulent diffusion based on the velocity shear between particles. Variations of metal mixing implementations are explored using GIZMO meshless finite mass codes in the FIRE-2 simulations [113] and the Rennehan [114] simulations, which found that the mixing algorithm significantly affects the resultant CGM and warm-hot IGM metallicity distributions.

In summary, the algorithm that describes how stars release metals into surrounding gas, the simulation resolution, and the metal mixing algorithm (if any is applied), are important considerations when simulating IGrM metallicities as discussed in the companion review by Gastaldello et al.[105].

3.5. Stellar Feedback

Stellar feedback, directly associated with stellar mass loss and enrichment, is discussed independently and is usually modeled by separate subgrid prescriptions. This owes to its uncertainty and range of possible outcomes that dramatically change the properties and appearances of galaxies in cosmological simulations (e.g., [62,115–119]). Stellar feedback has been identified as a key component of the overall solution to the overcooling problem whereby gaseous baryons are too efficiently converted into stars (e.g., [120]). Additionally, this feedback provides a pathway to enriching the high-redshift intergalactic medium (IGM; e.g., [121–126]), which is observed to have metal absorption far from the locations of galaxies (e.g., [127–131]). The enriched high-z IGM certainly contributes to the IGrM metal content of simulated $z \sim 0$ groups that show enrichment levels consistent with observations (e.g., [109,132,133]).

Early cosmological simulations took the expected energy from Type II SNe and imparted velocity kicks to gas articles in star-forming regions [99]. Oppenheimer and Davé [134] applied a momentum-driven wind scaling [135] to their simulations to mimic the acceleration of dust-driven winds by stellar UV radiation pressure, resulting in greater mass-loading for lower-mass galaxies. The large range of stellar feedback prescriptions explored in the OWLS (OverWhelmingly Large Simulations; [116]) project, including varying the wind mass-loading (as a proportion of the SFR), wind velocity ($v_{\rm wind}$), and attempting thermally heated wind prescriptions, demonstrated very different outcomes for galaxies and gas.

Many modern cosmological simulations treat stellar feedback explicitly as a tunable subgrid model to reproduce key observations of galaxies including the galactic stellar mass function, galaxy sizes, and their central SMBH masses. Different physical motivations are used to justify parameter choices, but the resultant models are often far from the physical mechanisms of feedback. EAGLE tunes their feedback to heat SPH particles to $10^{7.5}$ K [39], justifying the choice of temperature to prevent catastrophic cooling and allow feedback to efficiently apply mechanical work [136]. IllustrisTNG uses a group finder to estimate the dark matter halo velocity dispersion ($\sigma_{\rm DM}$) based upon Oppenheimer and Davé [137], and scale their kinetic wind velocities in proportion to $\sigma_{\rm DM}$ and the mass loading (i.e., the proportion of mass launched relative to SFR) in inverse proportion to $\sigma_{\rm DM}$ to a power that was calibrated [46]. SIMBA applies the wind velocity and mass loading scalings derived by Muratov et al. [138] from FIRE zoom simulations [139] that followed multiple sources of stellar feedback (radiation pressure, SNe, stellar winds, and photo-ionization from stars) in their kinetic wind model [50].

The complexity of these modern subgrid stellar feedback prescriptions can be quantified in the amount of energy returned per unit of stellar mass formed. Davies et al. [140] calculated that EAGLE dumps 1.74×10^{49} erg $M_\odot^{-1}$ via thermal heating, and IllustrisTNG dumps 1.08×10^{49} erg $M_\odot^{-1}$ with 90% going to a velocity kick and 10% going to thermal heating. This is the approximate energy injection rate expected if all SNe energy input into winds, that is, the expectation if there is $\sim$1 SN with 10^{51} ergs per 100 $M_\odot$ formed. SIMBA limits their wind velocity to the total supernova energy available, which applies only to high-z small galaxies. The ROMULUS simulations apply the Stinson et al. [141] 'blastwave' feedback model at an efficiency of $\approx$0.75, with cooling temporarily disabled to prevent radiative losses.

It should also be noted that metal mass loading is treated differently than total mass loading in some simulations. Illustris and the TNG simulations apply *reduced* metal-loading that is 2.5$\times$ lower than mass-loading, because Vogelsberger et al. [90] argued that winds punch through low-density cavities in the ISM, reducing the metallicity of the ejected material. Alternatively, SIMBA applies higher metal loading that can be up to a factor of twice as high but is more typically 10%–20% higher.

While stellar feedback can reduce the efficiency of galaxy formation at lower masses, this mechanism becomes inefficient at preventing star formation in halos greater than 10^{12} $M_\odot$ (e.g., [115,142,143]), which is one of the key motivations for applying SMBH feedback.

3.6. Black Hole Seeding

Subgrid implementations of black hole (BH) seeding (Section 3.6), accretion (Section 3.7), and AGN feedback (Section 3.8) in simulations are introduced in Section 5.1 of the companion review by Eckert et al. [78], which we expand upon here with discussions of specific implementations applied in the simulations listed in Table 1. Most of these use a version of the Booth and Schaye [144] BH seeding module, which applies a friends-of-friends group finder to identify halos, originally performed by Di Matteo et al. [145], then seeding $10^{-3} M_{\rm gas}$ sink particles in halos resolved with 100 dark matter particles (as in OWLS, Cosmo-OWLS, and BAHAMAS).

Illustris and EAGLE add $10^5\,h^{-1}M_\odot$ BH seeds in $10^{10}\,h^{-1}M_\odot$ halos, while IllustrisTNG uses larger seeds, $8 \times 10^5\,h^{-1}M_\odot$ to avoid the need for boosted Bondi accretion (see next Section). SIMBA seeds $10^4\,h^{-1}M_\odot$ BHs when $M_* > 10^{9.5}\,M_\odot$, skipping the attempt to follow BH accretion in low-mass halos. In general, dynamical friction is insufficiently resolved, therefore the default of these models is to continually re-position BHs to the local potential minimum (with the exception of Magneticum; see [146]).

The ROMULUS simulations instead seed SMBHs based on local gas properties: they can form whenever and wherever (i) gas density is 15 times the threshold for star formation; (ii) the local metallicity is low ($Z < 3 \times 10^{-4}$); and (iii) the temperature is just below the limit for atomic cooling. Consequently, SMBHs appear in ROMULUS at a higher redshift and in lower mass halos (10^8–$10^9\,M_\odot$) than in the other models discussed above. Moreover, SMBHs are not pinned to the local potential minimum; instead, the effects of unresolved dynamical friction are captured with a subgrid model [147]. This degree of freedom alters the SMBHs' growth and feedback trajectories [48,49].

3.7. Black Hole Growth

Once seeded, black holes primarily grow through the accretion of gas, but can also grow through SMBH-SMBH mergers. For accretion, the most common implementation uses the Bondi-Hoyle accretion rate limited to the Eddington luminosity. As discussed in Section 5.1 of the companion review by Eckert et al. [78], initial implementations boosted the Bondi-Hoyle accretion rate by a large constant factor ($\approx$100) to compensate for the lack of dense gas in early low-resolution simulations [148]. Instead, Booth and Schaye [144] introduced a density-dependent boost to the Bondi-Hoyle accretion rate that only affects black holes surrounded by gas with an (unresolved) dense phase.

The Booth and Schaye [144] boost formula is used by many simulations, including cosmo-OWLS, BAHAMAS, Horizon-AGN, ROMULUS, while Illustris and FABLE use the older constant Springel et al. [148] boost. Magneticum uses a temperature-dependent boost [149] that increases for cooler gas to approximate turbulent-driven chaotic cold accretion rates [150]. (C-)EAGLE and IllustrisTNG forgo the need for boosted Bondi at their higher resolutions. The former modify their accretion rates by an additional viscous timescale to account for the angular momentum of infalling gas [151], with the intention to significantly *reduce* SMBH growth in low-mass halos. In this model, a longer viscous timescale due to high angular momentum of gas around the black hole delays accretion and translates into a higher $M_\star$ threshold where the SMBH accretion rate approaches the $\propto M_{\rm BH}^2$ Bondi limit ([62], but see [143]). A similar motivation is used in ROMULUS to modify the Bondi rate to account for gas rotation in addition to the relative velocity between the gas and the SMBH [48].

SIMBA calculates black hole accretion rates from cool gas with the 'torque-limited accretion' model [152,153], in an attempt to diverge from the self-regulated nature of Bondi accretion [144]. This model results in a much shallower dependence on $M_{\rm BH}$ than the Bondi formula [154], while also recognizing that there are preferred orientations for accretion in addition to feedback. Accretion from hot gas is modeled with the standard Bondi-Hoyle formula, but the torque-limited mode from cool gas generally dominates SMBH growth in SIMBA.

3.8. AGN Feedback

Groups provide some of the best evidence that AGN feedback significantly transforms the distribution of baryons, both in terms of the bulk transport of gas outward and the reduction in star formation [44,132]. Figure 1 demonstrates how adding AGN feedback using the Booth and Schaye [144] AGN prescription both reduces the gas fraction within R_{500} and the luminosities of galaxies [132].

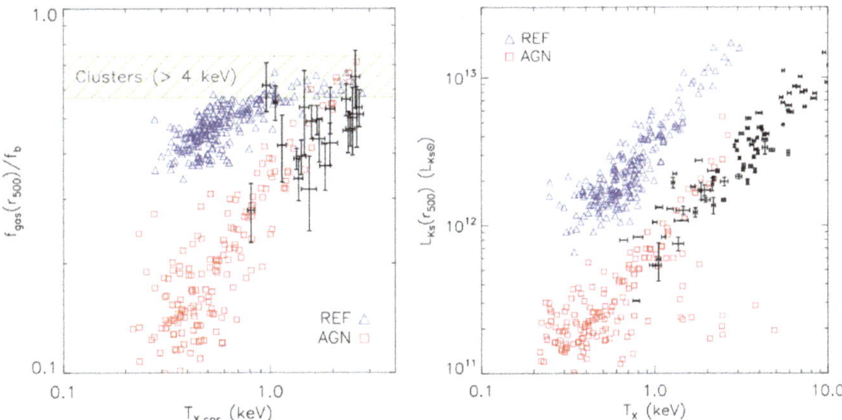

Figure 1. Figure panels adapted with permission from McCarthy et al. [132] showing OWLS simulations, with only stellar feedback (REF, blue triangles) and with additional AGN feedback (AGN, red squares). (**Left**) The AGN feedback provide a better fit to the Sun et al. [2] observationally derived f_{500} data (black points) as a function of X-ray temperature; (**Right**) AGN feedback reduces the integrated K-band luminosity within R_{500} to observable values compiled by Lin and Mohr [155].

Simulations usually apply an AGN feedback model that uses an AGN feedback efficiency term, ε, such that the feedback power is defined as $\dot{E}_{\rm AGN} = \varepsilon \dot{M}_{\rm BH} c^2$. Physically, ε is a product of two efficiencies, the radiative efficiency (ε_r), and the feedback efficiency (ε_f). The radiative efficiency is the energy radiated away from the accretion onto the black hole, and is often assumed to be 10% based on the accretion onto a Schwarzschild BH [156]. The feedback efficiency is the fraction of the radiated energy that is imparted to the gas either thermally and/or kinetically.

A number of simulations that we list in Table 2 use a single-mode thermal model assuming $\varepsilon_f = 0.15$ (i.e., resulting in $\varepsilon = 0.015$), based on the value calibrated by Booth and Schaye [144] to reproduce the SMBH mass-halo mass relation derived from observations. The EAGLE Reference simulations apply a temperature increase of $\Delta T = 10^{8.5}$ K to a neighboring SPH particle, once enough SMBH feedback energy has been accumulated to heat at least one SPH particle. The C-EAGLE/Hydrangea simulations run at the same resolution use a $\Delta T = 10^{9.0}$ K, which leads to less frequent, more bursty feedback that becomes more effective when heating gas above the virial temperatures of massive clusters [158]. The multiple cosmo-OWLS simulations varied ΔT while keeping $\varepsilon = 0.015$ [7]. Their spatial resolution, nearly 1 dex lower than EAGLE, makes their heating temperature not directly comparable to the EAGLE ΔT, since heating a single particle involves injecting a much larger amount of energy. In both cases, however, larger ΔT leads to a higher efficiency in transporting baryons beyond the virial radius. In BAHAMAS, feedback energy is accumulated until multiple gas particles can be heated simultaneously; McCarthy et al. [41] found that heating 20 SPH particles ($\sim 2 \times 10^{10}$ $M_\odot$) to $\Delta T = 10^{7.8}$ K, which was chosen as a calibration to prevent over-efficient assembly of intermediate-mass galaxies (see their Figure 3).

ROMULUS, with its higher resolution, uses a much lower efficiency ($\varepsilon = 0.002$), which was selected from a parameter search performed by Tremmel et al. [48] using galaxies in 10^{11-12} $M_\odot$ halos and not group/cluster-mass objects. Unlike stellar feedback, AGN feedback in ROMULUS is *not* subject to cooling shutoff.

Sijacki et al. [159] introduced the dual AGN model, with a "quasar" mode injecting thermal energy at high accretion rates, which is often defined as a relative fraction of the Eddington accretion rate, $f_{\rm Edd} \equiv \dot{m}_{\rm Edd}/\dot{m}_{\rm BH}$ where $\dot{m}_{\rm Edd}$ is the Eddington accretion limit, and a "radio" mode injecting mechanical energy in the form of bubbles at lower accretion rates. The radio mode (at low Eddington ratios) has more energy to impart into the gas since it bypasses the (inefficient) conversion from thermal to kinetic energy. Illustris switches to this mode when $f_{\rm Edd} < 0.05$, and releases built-up bubble events with sizes $\sim$100 kpc when the BH grows by 15% ($\delta M_{\rm BH} = 0.15$). The aggressiveness of this scheme over-evacuates the IGrM [160]. FABLE applied the same algorithm with similar energy efficiencies, but using smaller bubble events and a lower $f_{\rm Edd} = 0.01$ transition to radio mode that were the result of tuning their AGN feedback to reproduce IGrM gas fractions and galaxy stellar masses. Finally, more recent Magneticum simulations use the Steinborn et al. [149] thermal AGN feedback model that smoothly transitions between radio-mode and quasar-mode based on the fractional Eddington accretion rate.

Table 2. AGN feedback modules used in simulations.

Simulation	Mode	Injection	Energy Dump	Efficiency (ε)	Frequency	Loading Factor
cosmo-OWLS	–	Thermal	$10^{8.0}$ K	0.015	Build-up	1 particle
Illustris	Quasar	Thermal	–	0.01	Continuous	
–	Radio	Bubble	–	0.07	$\delta M_{BH} = 0.15$	
EAGLE	–	Thermal	$10^{8.5}$ K	0.015	Build-up	1 particle
Horizon-AGN	Quasar	Thermal	10^7 K	0.015	Build-up	
–	Jet	Kinetic	10^4 km s^{-1}	0.10	Continuous	
BAHAMAS	–	Thermal	$10^{7.8}$ K	0.015	Build-up	20 particles
C-EAGLE/Hydrangea	–	Thermal	$10^{9.0}$ K	0.015	Build-up	1 particle
ROMULUS	–	Thermal	–	0.002	Continuous	
FABLE	Quasar	Thermal	–	0.01	$\delta t = 25$ Myr	
–	Radio	Bubble	–	0.08	$\delta M_{BH} = 0.01$	
IllustrisTNG	"High"	Thermal	–	0.02	Continuous	
–	"Low"	Kinetic "Pulse"	–	≤ 0.2	Build-Up	Weinberger et al. [157] Equation (13)
SIMBA	"Radiative"	Kinetic	1000 km s^{-1}	0.003 [a]	Continuous	
–	"Jet"	Kinetic "Jet"	8000 km s^{-1}	0.03 [a]	Continuous	

[a] Values for a $10^9 M_\odot$ SMBH.

Several simulations attempt to simulate jets or jet-like feedback at low Eddington rates. Horizon-AGN uses the Dubois et al. [161] bipolar kinetic jet aligned with the spin of the SMBH. IllustrisTNG integrates the Weinberger et al. [157] model, where randomly-oriented, directional kinetic "pulses" predominantly take over once a SMBH grows above a certain mass based on their Equation (5). In practice, this choice yields significant implications as TNG100 SMBHs growing above a threshold mass of $M_{\rm SMBH} \approx 10^{8.1}\,M_\odot$ at late times leads to a sharp transition of galaxy properties (reduced SFRs, quenched galaxies, redder colors) and gaseous halo properties (lower $f_{\rm gas}$) [162,163]. Low Eddington ratio feedback in SIMBA uses a bipolar kinetic jet perpendicular to the angular momentum of the local disc that has a constant momentum input rate (20$\times$ the radiative luminosity of the accretion disc divided by the speed of light). This is meant to complement the preferred orientation of the Davé et al. [50] torque-limited accretion model. This model also decouples for 10^{-4} of a Hubble time, which can transport launched jet particles over $\sim$10 kpc.

3.9. Transport Processes and Magnetic Fields

It is fair to say that the effects of viscosity, thermal conduction, and turbulence (which we collectively refer to as 'transport processes') on the IGrM have received significantly less attention in comparison to, for example, the overall gravitational and hydrodynamical evolution of groups and the impact of processes associated with galaxy formation (such as radiative cooling and feedback processes). Indeed, from the cosmological simulations standpoint, the vast majority of existing simulations completely neglect the roles of viscosity and conduction (treating the IGrM as an inviscid and non-conducting fluid), while the effects of turbulence are generally only captured on relatively large, well-resolved scales.

However, there are no a priori compelling physical reasons for neglecting these processes, as the IGrM and the ICM are both plasmas where, generally speaking, one expects such transport processes to be active [164]. To include their effects in cosmological simulations is non-trivial, though, as the hydro solvers that normally evaluate the standard (inviscid) hydrodynamic equations must be replaced with more complex solvers capable of evaluating the full Navier Stokes equations in a stable fashion, at least if one wishes to model the effects of viscosity (e.g., [165]). Furthermore, since the transport of heat by viscosity and conduction preferentially occurs along magnetic field lines (and is strongly suppressed perpendicular to the field lines), it really only makes sense to include their effects in the context of MHD simulations that self-consistently follow the evolution of the magnetic fields. While there is a growing interest in the inclusion of anisotropic thermal conduction, viscosity and magnetic fields in simulations (e.g., [55,166]), we are presently unaware of any large-scale cosmological simulations that include both and have evaluated their impacts on the plasma in groups and clusters [8]. In terms of modeling turbulence, modern hydrodynamical solvers accurately follow the cascade of energy, momentum, and mass down to the effective resolution scale of the simulations but generally not below this. Thus, there are ongoing efforts to model the turbulent cascade to smaller scales using physically-motivated subgrid models (e.g., [112,168]). Whether turbulence plays a large role or if viscosity is able to strongly damp the cascade depends on the dimensionless Reynolds number but, at present, this is a very poorly constrained quantity for the IGrM and ICM.

What impact might we expect from including anisotropic conduction, viscous heating and/or small-scale turbulence in simulations of the IGrM? After the launch of *Chandra* there was much excitement with the discovery of large bubbles of relativistic plasma that appear, at least in some cases, to remain structurally intact even after buoyantly rising to relatively large distances from their inflation points (e.g., [169]). This is difficult to understand from an unmagnetized, inviscid (and therefore highly turbulent) fluid standpoint, as the bubbles should have been rapidly destroyed/mixed via Kelvin-Helmholtz and Rayleigh-Taylor instabilities. High-resolution idealized simulations of viscous and/or magnetized clusters, however, demonstrated that the bubbles were much more structurally stable and long-lived when these processes were included (e.g., [170,171]). Thus, how and where bubbles transmit their energy to the IGrM will likely be impacted by the inclusion of transport processes and magnetic fields. In addition, damping of sound waves (e.g., produced during the inflation of bubbles) via the viscous friction has been proposed as another way in which the central AGN might couple its energy to the gas over a large volume (e.g., [172,173]). The evolution of satellite galaxies (i.e., when and how fast they are stripped) may also be strongly affected by the inclusion of conduction, viscosity, magnetic fields and/or small-scale turbulence (see, e.g., Figure 8 of [165] for a dramatic demonstration of viscous stripping in clusters). While, theoretically, we expect thermal conduction and viscous dissipation to be considerably more important in massive clusters compared to groups, as the transport coefficients have strong temperature dependencies, much depends on the geometry of the magnetic field lines and, at present, the impact of these transport processes on the evolution of the IGrM and the satellites in groups remains an important unknown. The advent of new codes such as AREPO and GIZMO that are capable of accurately incorporating their effects is a promising step in the right direction and we expect to see important progress in answering these questions in the coming years.

3.10. Cosmic Rays

Although cosmic ray (CR) physics has not yet been included in cosmological simulations of groups or clusters, recent advancements in CR hydrodynamics and results from both idealized simulations of massive galaxies and cosmological simulations of lower mass galaxies demonstrate that cosmic rays may be an important source of pressure and energy in galaxy groups. In this subsection we briefly describe existing approaches to modeling CR physics and their expected effects in galaxy groups.

In hydrodynamics galaxy simulations, cosmic rays are typically modeled as a relativistic fluid of GeV protons, separate from the thermal gas [174–176]. This CR fluid advects with the gas and can provide non-thermal pressure support, inject momentum, or heat the gas. Additionally, cosmic rays can move relative to the gas through diffusion or streaming, both of which are approximations of the bulk flow of CR energy density along magnetic field lines (see [177,178] for a comprehensive review). Unfortunately, there is no empirical or theoretical consensus on the CR pressure in galaxy groups or on the correct model for CR hydrodynamics. For this reason, quantitative predictions of galaxy, outflow, and halo properties from simulations can vary by orders of magnitude depending on the model parameters [179–181]. However, there are several *qualitative* ways in which cosmic rays alter galaxy and halo properties that are consistently demonstrated in simulations.

(1) Cosmic ray transport (either streaming or diffusion) can drive galactic outflows [182–186]. Cosmic rays are injected into the ISM during stellar feedback events (typically ~10% of the total supernova energy). CR transport redistributes CR pressure out of the galaxy, creating a non-thermal pressure gradient that exerts a force opposing gravity. If the force exerted by the CR pressure gradient is sufficiently strong, it will trigger galactic outflows. Relative to thermally driven winds, CR driven winds are cooler, smoother, and more mass-loaded [187]. Additionally, since cosmic rays do not suffer radiative losses, CR-driven winds may continue accelerating gas at large distances from the galactic disc. However, since the gravitational force is stronger in more massive galaxies, CR-driven winds may become inefficient in galaxy groups [188].

(2) Streaming cosmic rays impart energy to heat the surrounding gas. In massive galaxies, this CR heating rate can efficiently balance radiative cooling, preventing a cooling catastrophe (e.g., [189–192]). CR heating may also be a key aspect in the self-regulated AGN feedback cycle. As cosmic rays lose energy, gas cools more efficiently, fueling AGN feedback which re-injects CR energy into the IGrM [193].

(3) CR pressure qualitatively alters the structure of multiphase gas in galactic halos (e.g., [180,194–196]). Non-thermal pressure support enables cool gas to exist at lower densities than expected from purely thermal equilibrium. Figure 2 demonstrates how the density contrast between cold and hot gas in a two-phase medium diminishes with increasing CR pressure support. In the extreme case of a CR pressure-dominated galaxy halo, cool and hot gas can exist at the same densities. However, CR pressure is unlikely to be the dominant source of pressure in the halos of massive galaxies. Therefore, the likely effect of cosmic rays in the IGrM is a modest decrease of cool cloud and cool filament densities [197,198].

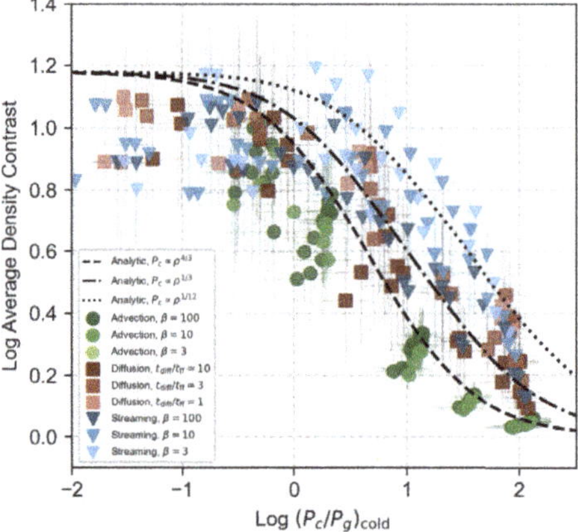

Figure 2. The average density contrast between cold and hot gas phases as a function of cosmic ray pressure support in the cold gas [199]. Each point represents a time-averaged measurement from an idealized simulation of thermal instability with varying initial conditions and cosmic ray physics. The black lines show different analytical predictions for various degrees of coupling between cosmic rays and gas. With increasing cosmic ray pressure, the density of cool gas decreases. However, the detailed quantitative predictions are sensitive to the invoked cosmic ray transport model.

4. Results of Simulations at the Group Scale

4.1. The Baryonic Content of Group Halos

The intermediate scale of group halos between galactic halos and clusters provides a unique lever arm on the nature of superwind feedback. Unlike most galactic halos, the temperature, density, and metallicity of their gaseous baryons can be probed via soft X-ray emission without stacking, while their shallower gravitational potential wells relative to clusters allow energetic feedback to remove a significant fraction of baryons. Twenty years ago, IGrM observations from missions including *ROSAT* provided only singular data points per group, which were then compiled into the $L_X - T_X$ relation. The steep $L_X \propto T_X^{4.9}$ relation observed for groups [200] relative to clusters [201] indicated a significant deviation from the $L_X \propto T_X^2$ relationship expected for virialized gaseous halos retaining all their baryons in hydrostatic equilibrium (e.g., [202], see companion review by Lovisari et al. [1]).

With the launch of more powerful X-ray missions, such as *Chandra* and *XMM-Newton*, hot gas profiles around groups could be resolved. This revealed that the IGrM had far fewer gaseous baryons than the cosmic ratio, $f_b \equiv \Omega_b/\Omega_M \simeq 0.16$, inside R_{2500} and a still low proportion inside R_{500} [2,5]. Altogether these results indicated that many of the baryons are removed from the hot phase. Earlier theoretical work argued that more efficient line cooling at cooler group temperatures, as opposed to clusters where Bremsstrahlung cooling dominates ($\Lambda_{cool} \propto T^{0.5}$), could more efficiently build the stellar component of galaxies [203,204]. While Gonzalez et al. [205] measured that the integrated group stellar masses from the BGG, the intragroup light (IGrL), and satellite galaxies could account for the missing gaseous baryon resulting in a groups retaining all baryons inside R_{500}, this was later challenged. Balogh et al. [206] argued that groups could not be far more efficient than clusters in converting their gas to stars due to the hierarchical requirement that clusters are assembled from progenitor groups. More recent compilations of group stellar masses summed from BGGs, satellites and IGrL [207] find lower stellar mass fractions, strongly suggesting that groups are missing baryons.

Cosmological volume simulations with periodic volumes 100–150 comoving Mpc on a side contain populations of groups that can be statistically compared to observations. Initial simulations by Davé et al. [109], including only stellar superwind feedback, were able to enrich the IGrM to observed levels [200] and additionally add entropy to group halos as observed [208]; however the continued late-time star formation in these groups is a telltale sign that these simulations fail to solve the cooling flow problem [209]. Liang et al. [37] used GADGET-2 SPH simulations similar to Davé et al. [109] to successfully fit a range of X-ray observations, including IGrM masses inside R_{500}. Nevertheless, the total stellar masses exceeded observations by at least a factor of two, and the total baryon content (gas+stars) of groups exceeded 80% of the cosmic fraction. These simulations demonstrated that it is possible to reproduce a wide range of IGrM properties, while assembling the wrong galaxies, which in these cases had too much late-time star formation.

McCarthy et al. [132] made the case for AGN feedback by using simulations from the OWLS [116] suite of simulations by comparing to *Chandra* observations from Sun et al. [2] that resolved X-ray emission profiles out to R_{500} in group-scale objects. Comparing OWLS simulations without and with Booth and Schaye [144] AGN feedback, McCarthy et al. [132] demonstrated that the latter could much better reproduce $f_{gas}(R_{500})$, the $L_X - T_X$ relation, as well as the stellar K-band luminosity of the BGG and all stars within R_{500} for the approximately 200 group-sized objects in 100 h^{-1}Mpc boxes with gas element resolution of 1.2×10^8 $M_\odot$ (see Figure 1). We refer the reader to the Eckert et al. [78] companion review (their Section 5.2) for a discussion of a broader range of simulations and their baryon fractions within R_{500}.

4.1.1. Gaseous and Stellar Masses in Recent Simulations

One can consider as "contemporary" intermediate resolution simulations those that resolve gas resolution elements at $\sim 10^6$ M$_\odot$ mass resolution in $\sim 10^6$ Mpc^3 volumes. The first simulations to satisfy these criteria were Illustris [38] and EAGLE [39]. Both simulations follow SMBH growth and AGN feedback, and are tuned to match the $z = 0$ stellar mass function, some other galaxy characteristics such as galaxy sizes, and SMBH demographics. However, they both were not tuned to reproduce the properties of X-ray emitting halos, and fail to reproduce gaseous properties of groups and poor clusters. Genel et al. [160] showed that Illustris severely underpredicts the IGrM mass within R_{500}, by as much as a factor of 10$\times$ compared to observations by Giodini et al. [210]. EAGLE produces group halos with much higher gas fractions, exceeding 80% of $f_{\rm b}$ at $M_{200} > 10^{13.5}$ M$_\odot$ [211]. We plot gas, stellar, and baryon fractions inside R_{200} as a function of M_{200} in Figure 3 for EAGLE and other contemporary simulations [9]. Schaye et al. [39] found the IGrM masses derived from virtual X-ray observations to be too high by a factor of two and the $L_X - T_X$ relationship to be too luminous for a given T_X. This reveals that simulations tuned to fit galaxies can deviate significantly for gaseous properties on group scales, which is why accurately simulating the IGrM can provide orthogonal constraints on the processes governing galaxy formation and evolution.

BAHAMAS explicitly tuned their AGN feedback prescriptions to reproduce properties of groups/clusters and massive galaxies as explained in Section 3.8. The FABLE simulations ([44], not shown in Figure 3) also explicitly tuned their feedback to reproduce massive halos, at $\sim$100$\times$ higher mass resolution than BAHAMAS but in a 1000$\times$ smaller volume that is augmented by a series of zooms extending up to cluster masses. These simulations have $f_{\rm gas,500}$ for a $M_{500} = 10^{13.5}$ M$_\odot$ between 0.04 and 0.07 in Table 1, which reflects the uncertainty in the Sun et al. [2] and Lovisari et al. [5] observations (see Figure 5 of the companion review by Eckert et al. [78]).

The C-EAGLE/Hydrangea zooms [42,43] use the EAGLE model with a higher AGN heating temperature ($\Delta T = 10^{9.0}$ K). As shown by Schaye et al. [39], this change lowers (i.e., improves) $f_{\rm gas,500}$ in $M_{500} \lesssim 10^{13.5}$ M$_\odot$ groups, but more massive objects remain too baryon rich, especially in the regime of rich groups/poor clusters [43]. Similarly, the IllustrisTNG AGN feedback was calibrated to $f_{\rm gas,\,500}$ of groups at TNG100 resolution [157], though the simulations still predict $f_{\rm bar,\,200} \gtrsim 0.12$ for halos with $M_{200} > 10^{13.5}$ M$_\odot$, substantially higher than the explicitly calibrated BAHAMAS simulation (see the right-hand panel of Figure 3). SIMBA [50] predicts $f_{\rm gas,200} \approx 0.05$ at $M_{500} = 10^{13.5}$ M$_\odot$ and agrees well with BAHAMAS in this metric within the overlapping halo mass range. Interestingly, the companion review by Eckert et al., however, shows that $f_{\rm gas,500}$ in SIMBA increases more rapidly with halo mass in SIMBA compared to BAHAMAS (their Figure 15), so that more baryons are contained near the center of poor clusters with $M_{500} = 10^{14-14.5}$ M$_\odot$. We will return to the importance of the radial range for assessing the baryon content of simulated groups below.

Higher resolution simulations ($m_{\rm bar} \sim 10^5$ M$_\odot$) currently contain at best a handful of group halos, but still yield constraining results. The ROMULUS suite [48,49] predicts $f_{\rm gas,200} = 0.10 - 0.12$ between $M_{200} = 10^{12.5-14.0}$ M$_\odot$ and a higher $f_{*,200}$ than all the aforementioned simulations (orange lines in Figure 3). These halos therefore retain nearly all their baryons, in strong conflict with the observational evidence outlined above. It is unlikely that this shift is a direct consequence of the higher resolution, since the higher-resolution simulation of the IllustrisTNG family, TNG50 [51], predicts $f_{\rm gas,500}$ rising from 0.06 to 0.12 over the mass range $M_{500} = 10^{13.0-14.0}$ M$_\odot$ (not shown), in excellent convergence with TNG100 (we find a similar result for the lower-resolution version, TNG300).

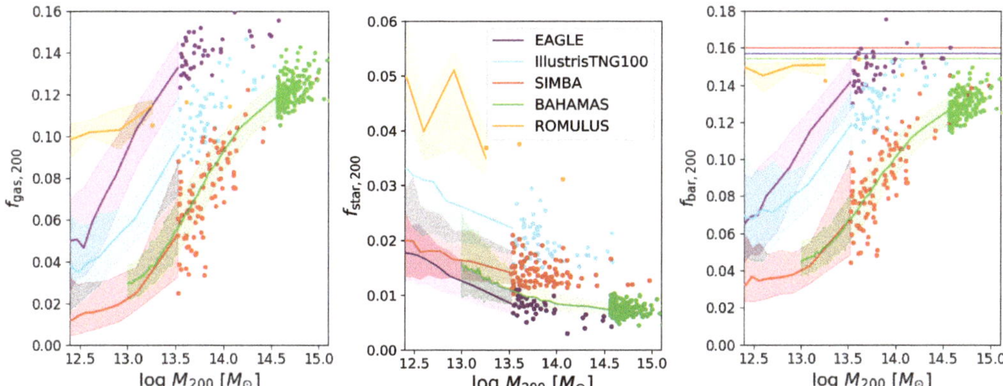

Figure 3. Gas, stellar and combined baryon fractions inside R_{200} from a variety of widely-used, contemporary cosmological hydrodynamic simulations. Medians and 1-σ spreads are shown with lines and shading up to a mass where individual objects are plotted as points. The cosmic baryon fraction $f_b = \Omega_b/\Omega_M$ is plotted in the right-hand panel for the given simulation cosmology. IGrM measurements rarely extend to R_{200}, therefore this plot does not include observations. It is notable that stellar contents do vary by a factor of more than $3\times$ between different simulations in the group regime, and more than double between EAGLE and IllustrisTNG. ROMULUS retains almost all of its baryons within R_{200} while BAHAMAS and SIMBA eject more than half their baryons at $M_{200} \lesssim 10^{13.5}$ $M_\odot$. For reference, M_{500} is typically 0.16 dex lower than M_{200} for this halo mass range.

The determination of IGrM properties within R_{500} is physically motivated, but measurements within fixed apertures are observationally more straightforward. In Figure 4, we therefore show the integrated IGrM masses as a function of halo mass as predicted by EAGLE, SIMBA, and the three IllustrisTNG boxes out to fixed radii of 100, 200, and 400 kpc in the 3 subpanels. For comparison, we obtain the equivalent masses from the observations of Sun et al. [2] and Lovisari et al. [5] by integrating their measured electron density profiles out to the same radii. While the relative differences between the simulations shown in Figure 4 are consistent across the three radial cuts—remarkably close agreement between the three TNG runs, with EAGLE and SIMBA offset by $\approx +0.1$ and -0.5 dex, respectively, at $M_{500} = 10^{13.5}$ $M_\odot$—the comparison to the observations reveals additional details in each panel. Close to the group center ($r \leq 100$ kpc), EAGLE shows promising agreement with the observations, owing to its comparatively less aggressive AGN feedback. SIMBA, on the other hand, despite having calibrated its AGN model to match the IGrM fraction inside R_{500}, is evacuating the central region too efficiently. Within the larger 400 kpc aperture (right-hand panel), the IGrM masses of EAGLE are higher than observed, while SIMBA is at least marginally consistent with the observations at the high-mass end ($M_{500} \gtrsim 10^{14}$ $M_\odot$). The IGrM masses of IllustrisTNG are consistent with the observations across radii, albeit with a tendency of being too low in the center ($r < 100$ kpc) of low-mass groups ($M_{500} \lesssim 10^{13.5}$ $M_\odot$).

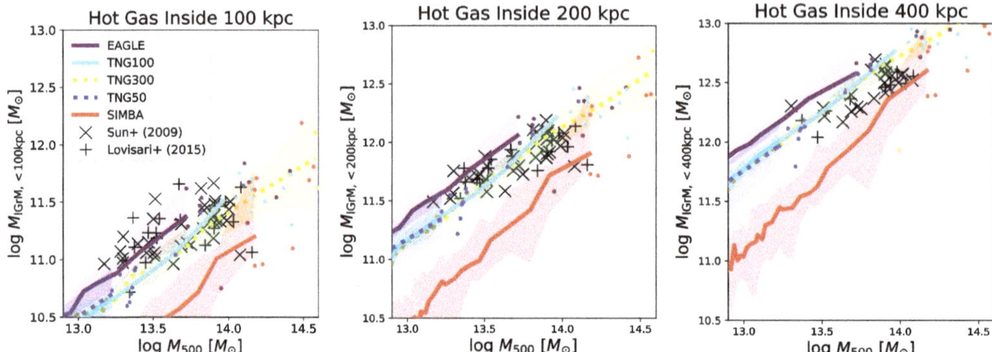

Figure 4. IGrM masses (gas at $T \geq 10^6$ K) within three fixed 3D apertures plotted for EAGLE (purple), SIMBA (red), and all three IllustrisTNG volumes (dark blue, cyan, and yellow). Medians and 1σ scatter are indicated by lines (dotted for TNG50 and TNG300, solid for others) and shaded regions, respectively. Different panels correspond to limiting radii of 100 kpc (**left**), 200 kpc (**middle**), and 400 kpc (**right**). Observed IGrM masses of individual groups within these radii, obtained from the (3D) profiles of Sun et al. [2] and Lovisari et al. [5], are shown as black symbols. While differences between simulations are approximately consistent across the three limiting radii, the agreement with observations differs noticeably between the group center (left-hand panel) and outskirts (right-hand panel).

4.1.2. Gaseous Profiles in Recent Simulations

We have seen above that even simulations that were calibrated to match gas and stellar fractions within the virial radius can make discrepant predictions about the IGrM content within other radii. A more challenging test of the simulations is therefore provided by their predicted IGrM profiles, which are plotted for $z \approx 0$ groups of $M_{500} = 10^{13.5} - 10^{14.0}$ $M_\odot$ from EAGLE, TNG100, SIMBA, and ROMULUS [10] in Figure 5. In the top left panel, we show stacked profiles of electron density n_e, calculated as $n_e = \rho_{IGrM}/(\mu_e m_H)$ with ρ_{IGrM} the density of the hot IGrM gas (here defined as $T > 10^6$ K), $\mu_e = 1.14$ the mean molecular weight per free electron, and m_H the proton mass. Stacked IGrM temperature profiles are shown in the top-right panel. In both cases, we compare to observed profiles of individual groups in the same mass range from thin dashed lines color-coded by mass (Sun et al. [2]); for density we also plot the stacked observed profile of black dash-dotted lines (Lovisari et al. [5]).

All simulations (except ROMULUS, which may be affected by its small sample size) agree closely with each other for the temperature profile beyond $\approx$0.1 R_{500}, and are broadly consistent with the observations of Sun et al. [2]. This similarity is interesting in the context of AGN feedback prescriptions that heat the gas differently resulting in similar temperature profiles throughout most of the IGrM, indicating that virialization primarily sets the IGrM temperature. Significantly more variety is seen in the density profiles (top-left); all simulations predict a comparable and realistic density around R_{500}, the increase towards smaller radii is substantially stronger for ROMULUS, and weaker for SIMBA, than observed. TNG100 and EAGLE, on the other hand, agree quite closely with each other and the observations, albeit with a slight ($\lesssim$0.2 dex) excess of gas towards the outskirts—especially for EAGLE—and a more substantial deficit ($\approx$0.2–0.5 dex) near the center that is stronger for TNG100.

Of particular interest are two physically motivated combinations of density and temperature: the IGrM pressure $P \equiv n_e T$ and its entropy $K \equiv T/(n_e^{2/3})$. These not only highlight additional discrepancies between simulations and observational data, but reveal imprints of the subgrid prescriptions, most notably AGN feedback schemes. We show their radial profiles, normalized to their analytically expected values within R_{500}: for pressure, $P_{500} \equiv k_B T_{500} n_{e,500}$ and for entropy, $K_{500} \equiv k_B T_{500} (n_{e,500})^{-2/3}$, where T_{500} is taken as the virial temperature $k_B T_{500} \equiv G M_{500} \mu m_H / R_{500}$ with mean molecular weight $\mu = 0.59$ and the electron density $n_{e,500}$ as the ideal value of $n_{e,500} \equiv 500 f_b \rho_{crit}/(\mu_e m_H)$ in the bottom row of Figure 5.

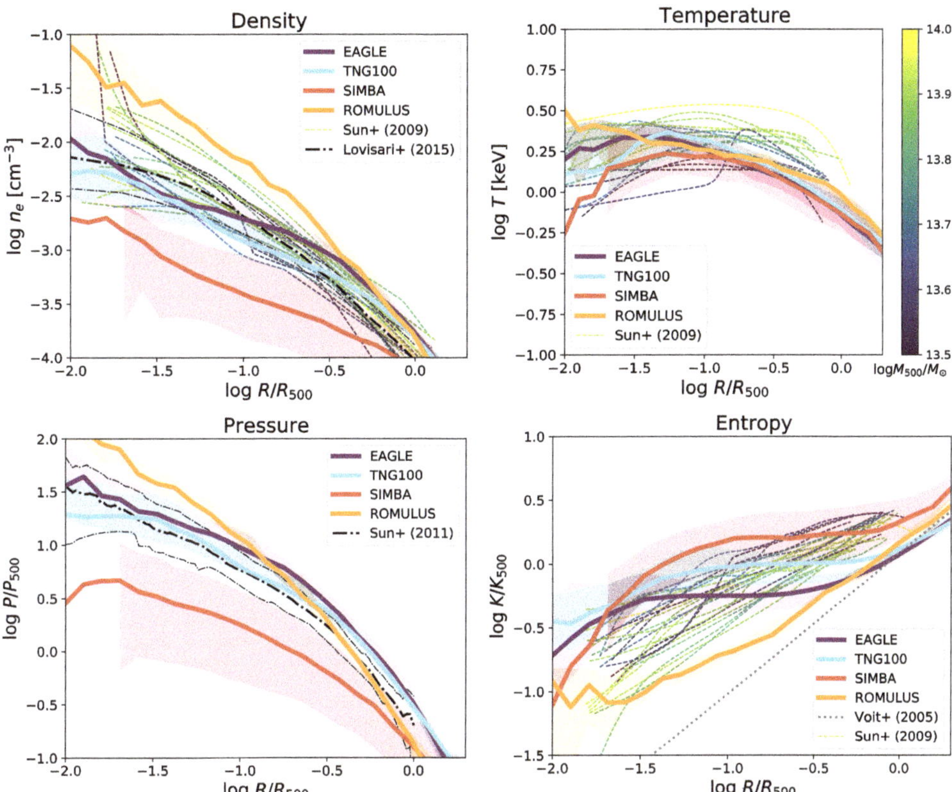

Figure 5. Mass-weighted 3D radial profiles of IGrM density (**top left**), temperature (**top right**), pressure (**bottom left**) and entropy (**bottom right**) in groups with $M_{500} = 10^{13.5} - 10^{14.0}$ M$_\odot$ from the EAGLE ($N = 18$ groups, purple), TNG100 ($N = 35$, cyan), SIMBA ($N = 80$, red), and ROMULUS ($N = 1$, orange) simulations. For the first three, we show profiles at $z = 0$, while for ROMULUS we have stacked the profiles obtained from the single group in this mass range at five snapshots with $z \leq 0.36$. Running medians within each simulation are plotted as solid lines, with shaded bands representing the 1σ scatter. For comparison, profiles of individual groups in the same mass range derived from X-ray observations [2] are shown as thin dashed lines in three panels, colored by (temperature-derived) halo mass; for pressure, we instead show the SZ-based profiles of the same groups by Sun et al. [212] as black dash-dotted lines (thick and thin ones for the median and 1σ scatter, respectively). In the top-left panel, the stacked density profile from X-ray observations of Lovisari et al. [5] is shown in the same fashion, while the grey dotted line in the bottom right panel represents the "base line" $K \propto r^{-1.1}$ entropy profile seen in non-radiative simulations (e.g., Lewis et al. [95], Voit et al. [96]). With the exception of temperature at large radii, there is little agreement between simulations. The entropy profiles in particular are also clearly different from what is observed in all four cases, even for simulations such as EAGLE and TNG100 that approximately reproduce the observed pressure profile. In general, stronger AGN feedback prescriptions raise entropies and reduce densities and pressures.

At large radii, all simulations follow a power-law entropy profile with an index close to 1.1, as expected from accretion and associated shocks [213] and as found in non-radiative simulations (e.g., [95,96]). Within $\approx$0.3 R_{500}, the median profiles of three simulations (SIMBA, TNG and EAGLE) flatten to an extended entropy core, while ROMULUS entropy profiles keep decreasing and only flatten in the very center. The magnitude of these simulated entropy cores appears to scale broadly with the aggressiveness of the AGN feedback. A factor of 100 difference in the energy coupling efficiency ε of AGN feedback progressively raises the entropy levels from ROMULUS ($\varepsilon = 0.002$) to TNG ($\varepsilon = 0.2$ for their pulse mode). SIMBA, however, exemplifies that the situation is more complex; it has the highest entropy core levels ($\gtrsim K_{500}$ beyond 0.03 R_{500}) despite an AGN coupling efficiency comparable to EAGLE (see Table 2). It is plausible that the high entropy core of SIMBA is, at least in part, due to its decoupled kinetic AGN feedback scheme [50], which can efficiently inject entropy at large radii. At the same time, it still allows transport of low-entropy gas towards the center, leading to the strong drop in central entropy (which, to a lesser extent, is also seen for EAGLE).

Although the core entropy levels in TNG100 and EAGLE are lower than for SIMBA, their AGN feedback schemes—randomly-oriented, pulsed feedback [157] and highly energetic thermal injection [39], respectively—are still creating extended cores. The only simulation without a clear high-entropy core is ROMULUS, which injects AGN energy thermally but with a much lower temperature increase. As we have seen above, this low entropy injection correlates with higher group baryon fractions than in the other simulations.

A comparison of these predictions to the Sun et al. [2] profiles clearly reveals that *none of the scaled entropy profiles from any of the simulations resemble the observations.* The latter do not have large extended plateaus; they typically scale with radius as $R^{0.7}$ for $R < R_{500}$ (see also Figure 10 of Sun et al. [2] and Figure 4 of O'Sullivan et al. [8] for the CLoGS group sample). This is despite reasonable agreement in terms of the pressure profiles (especially for EAGLE and TNG100), which indicates that the simulated IGrM remains approximately in pressure equilibrium throughout the (significant) AGN energy and entropy injection [43]. The FABLE simulations (not shown [44]) show more promising agreement of IGrM profiles (including entropy), although Henden et al. [44] discuss that even their explicitly calibrated (bubble) feedback might be too energetic at late times, since the $z = 0$ FABLE groups fall within the scatter, but mostly below the median, of the Sun et al. [2] density profiles (see their Figure 11). It will be interesting to see whether future simulations can overcome this shortcoming with more sophisticated AGN feedback models and higher resolution, or whether observational selection biases (i.e., preferential inclusion of cool-core systems with dense, bright centers in X-ray selected samples as discussed by, e.g., Henden et al. [44]) are responsible for at least part of the discrepancy (see Section 5).

For deeper insight into the predicted IGrM entropy profiles, it is instructive to consider them in context with their more massive cluster counterparts. This comparison is shown in the left panel of Figure 6, where we plot scaled $z \approx 0$ entropy profiles in analogy to the bottom-left panel of Figure 5 but over a wide range of halo mass, $M_{500} = 10^{13.0-15.0}$. For clarity, only IllustrisTNG (TNG100 and TNG300 combined, with TNG300 dominating because of its larger volume; solid lines) and ROMULUS (dashed lines) are shown, with halos median-stacked within 0.5 dex bins in M_{500}. These are compared to three observational samples that together span a similar mass range: the Sun et al. [2] and (lower-mass) CLoGS [8] groups as well as clusters from the ACCEPT survey [9].

There are two features of this comparison that are particularly worth highlighting. First, the median entropy profiles of IllustrisTNG feature prominent entropy cores that are significantly higher than observed across the selected mass range, from poor groups (purple) to massive clusters (yellow). The same is true for EAGLE and SIMBA clusters (not shown). On cluster scales, these high-entropy cores corresponding to non-cool-core (NCC) systems have previously been highlighted by C-EAGLE (Barnes et al. [43]), IllustrisTNG (Barnes et al. [214]), and SIMBA (Robson and Davé [215]); in the case of IllustrisTNG, Barnes et al. [214] found a rapid decline in the fraction of cool-core (CC) systems at $z < 1$ and identified the cause to be the AGN feedback implementation, which appears too efficient at removing baryons from the inner $0.01 R_{500}$.

Second, there is a clear trend towards lower (normalized) entropy with increasing halo mass at fixed radius, in both the observations and the IllustrisTNG simulations [11]. We speculate that these features may be a signature of quantized feedback dumping a fixed amount of entropy per AGN feedback event; further investigation of this topic is clearly warranted.

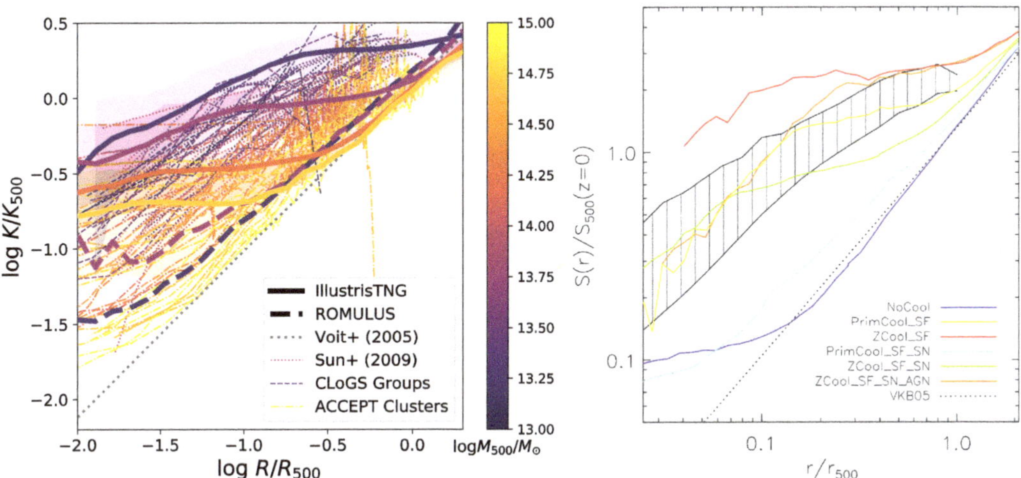

Figure 6. (**Left**) Normalized mass-weighted 3D entropy profiles at $z \sim 0$ for groups and clusters with $M_{500} = 10^{13}$–10^{15} M$_\odot$ in TNG100 and TNG300 (combined, thick solid lines) and ROMULUS (thick dashed lines). We show median-stacked profiles of halos in 0.5 dex bins, with 264 (7), 110 (5), 91, and 15 halos per bin for TNG (ROMULUS) in order of increasing mass, color-coded by mass from low-mass groups (purple) to massive clusters (yellow). For TNG, we also indicate their 1σ scatter with shaded bands. These predictions are compared to individual halos from three observed samples, all colored analogously by halo mass: low-mass CLoGS groups (dashed lines [8]), massive groups (dotted lines[2]), and a random 20% subset of the ACCEPT clusters (dash-dotted lines [9]). The gray dotted line indicates the $K \propto r^{1.1}$ scaling found in non-radiative simulations [96]. In agreement with observations, TNG (but not ROMULUS) predicts higher normalized entropy at lower mass, but with higher-entropy cores than observed. In contrast, ROMULUS groups have steeper and lower entropy profiles that are more typical of observed massive clusters. (**Right**) Normalized mass-weighted 3D entropy profiles of $M_{500} = 10^{13.25}$–$10^{14.25}$ M$_\odot$ groups predicted by OWLS simulations with different subgrid models [116], reproduced with permission from McCarthy et al. [158] (note that their symbol for astrophysical entropy is S, rather than K). Simulations include a run without any cooling or feedback (blue); without feedback but with star formation and cooling without (yellow) or with (red) the contribution from metal lines; with the addition of SNe feedback (cyan and green, respectively); and with AGN feedback added in addition to metal-line cooling, star formation, and SNe feedback (brown). The black-hatched band indicates the 1σ scatter of the Sun et al. [2] observations, the black dotted line a $r^{1.1}$ power law. While AGN feedback has a clear effect on the IGrM, the entropy profiles are also sensitively affected by cooling, star formation, and SNe feedback.

Interestingly, such a trend is not seen in ROMULUS, although the small number of groups and absence of massive clusters prevents strong conclusions here. If anything, the lower-mass groups (blue dashed line) have entropy profiles that are even closer to observed CC clusters, plausibly also because of a late-time merger in the more massive ROMULUSC group [49]. As discussed further below, it is plausible that this CC-like behaviour of ROMULUS groups is due to the absence of metal-line cooling—which may limit the removal of low-entropy gas from its IGrM—as well as the highly collimated nature of their AGN-driven outflows.

There are additional reasons why the ubiquitous conversion from CC to NCC halos in most simulations is problematic. McCarthy et al. [216] showed that the cost, in energy, is much greater than the typical jet power of an AGN outburst [217], with most of the energy being expended to lift the gas rather than raise its entropy. This suggests that the AGN outbursts in the simulations are likely injecting much more energy than observed radio AGNs in galaxy clusters. Observations also show little evidence of AGN feedback affecting CC-to-NCC transformations. The entropy profiles of NCC as well as CC CLoGS groups (with and without observed jet activity) are all, to first order, similar in shape and normalization [2,8]. Moreover, none of the clusters showing evidence of having experienced an *extreme* AGN feedback event in the recent past—e.g., MS 0735.6+7421 [218] and Hydra

A [217,219]—have core entropies as high as the median value in the simulations. We therefore speculate that the action of the AGN feedback models in SIMBA, TNG, and (C-)EAGLE may be too aggressive in the low-z Universe (see also the discussion in Barnes et al. [43]). Although we are comparing individual observed profiles to medians from simulations, our examination of simulated individual profiles (not shown) indicates that the median profiles are fairly representative, so that this is unlikely a source of significant bias.

Given the mismatched entropy profiles in contemporary simulations, it is helpful to refer to McCarthy et al. [158], who demonstrated using multiple OWLS simulations that adding cooling to non-radiative simulations can alter the entropy slopes as much as feedback (Figure 6, right panel). While non-radiative simulations fall on the baseline $K \propto R^{-1.1}$ relationship, adding only cooling without feedback "cools out" low-entropy gas, raising the entropy of the remaining IGrM. Metal-line cooling, which is comparatively more important for groups than clusters, raises entropy even more. Adding SNe feedback without metal-line cooling actually lowers the entropy nearly to the $R^{1.1}$ line, which may explain the ROMULUS entropy slopes that have weak AGN feedback that is more similar to SNe feedback and no metal-line cooling. While McCarthy et al. [158] confirm that adding AGN feedback indeed increase the entropy (cf. orange vs. green lines), the difference is smaller than changing the cooling physics. Furthermore, they showed that the AGN-induced entropy increase is, for the largest part, also an indirect effect: in their simulation, it is *not* (primarily) caused by heating gas that remains in the IGrM, but by the selective ejection of low-entropy gas from the less tightly bound $z \approx 2$–4 progenitor halos.

We end our discussion of IGrM properties by pointing out that even radial profiles convey only a simplified view of the processes shaping the gaseous halos of groups. This is exemplified particularly clearly by high-resolution simulations such as ROMULUSC, for which we show 2D maps of the gas temperature and density near the center of its most massive group at $z = 0.53$ in Figure 7 (adapted from Figure 10 of Tremmel et al. [49]). A biconical jet is clearly seen in the temperature map, which emerges naturally by collimation of their (intrinsically isotropic) AGN feedback by a central gas disc. This jet evacuates bubbles in the vicinity of the group center—not unlike X-ray cavities observed in real clusters (e.g., Hlavacek-Larrondo et al. [220])—but does not strongly affect gas away from its axis, thus preserving a steep $R^{1.1}$ entropy slope down to $0.04R_{500}$ in a spherically averaged sense (Figure 6).

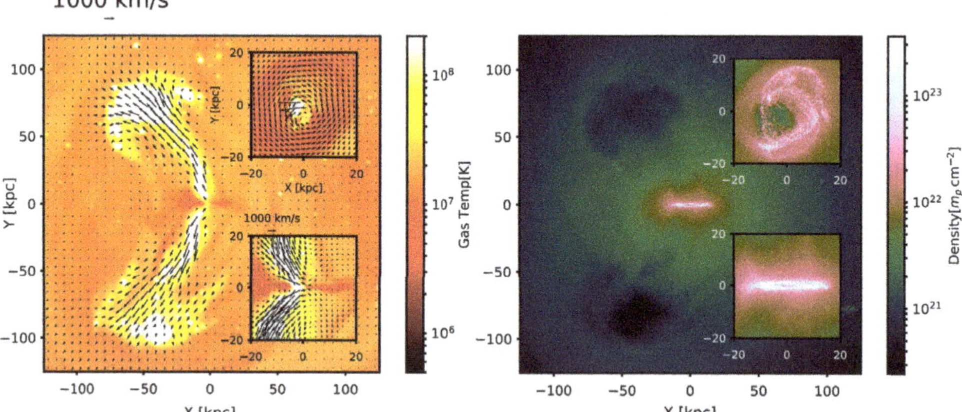

Figure 7. Wind-blown bubbles driven by AGN feedback from the central galaxy in the ROMULUSC simulation at $z = 0.53$, adapted with permission from Tremmel et al. [49]. This $M_{500} = 10^{13.6}$ $M_\odot$ group produces a cool core that is elusive in other simulations, although the central gas densities are high compared to observations. This AGN feedback is a thermal dump model without *explicit* collimation, and is the most obvious example of bubble-driven feedback at the group scale found in the cosmological simulations we explore.

4.2. Brightest Groups Galaxies

For more than two decades, attempts to model the formation and evolution of galaxy groups and to identify the role of different physical processes that shape these systems have focused primarily on the properties of the hot IGrM/ICM (e.g., [202,213,216,221–223], and references therein). Even with the advent of cosmological simulations, this tendency has largely continued [7,37,39,44,109,132,204,215,224]. However, observational studies find that a number of BGG properties (e.g., stellar mass, size, morphology, the nature of the surface brightness and stellar velocity dispersion profiles, whether the BGG is a fast or slow rotator, etc.) are also correlated with the properties of their host system [155,225–231]. Such correlations are not entirely unexpected given that BGGs are typically found at the bottom of their host halo's gravitational potential well. As such, the BGGs are thought to have experienced numerous mergers and close tidal encounters with other group galaxies over cosmic time. Simulations suggest that such interactions result not only in the growth of the BGGs' total stellar mass but also, induce structural and kinematic transformations [232–237]. Mergers can also potentially transport in cool gas and fuel in-situ star formation in disc-like structures. Additionally, as explicitly demonstrated by Lewis et al. [95] using the first hydrodynamic cosmological simulation of a rich galaxy group/poor cluster that allowed for radiative cooling, any gas cooling out of the IGrM/ICM flows towards the group center and ends up in the central galaxy. Consequently, the study of the BGGs, their properties, and the existence of any correlations between the latter and the properties of the host groups offer a complementary window onto the physical processes underlying the evolution of galaxy groups. Just as importantly, in the context of the present review, detailed comparisons of the observed BGG properties and the characteristics of those in numerical simulations offers an equally powerful opportunity for testing the efficacy of latest generation of cosmological simulations. In recent years, a handful of papers have assessed the characteristics of BGGs forming in hydrodynamic simulations (e.g., [7,39,43,48–50,238]) although this was done as part of a broad survey of the general properties of simulated galaxies. There are however notable recent exceptions, like Davison et al. [239], Henden et al. [240], Jackson et al. [241], Pillepich et al. [67], Katsianis et al. [242], Remus et al. [243], Tacchella et al. [244] and Jung et al. [73], that have treated BGGs as a distinct class, investigating both the structural and the kinematic properties of the simulated BGGs and comparing these to observations.

4.2.1. Central Galaxy Stellar Masses

The first BGG (and BCG) property we consider is their stellar mass−halo mass (SMHM) relationship. This is shown in Figure 8, where we plot $M_\star/M_{200}$ against M_{200}. $M_\star$ is the mass of the stars associated with a BGG/BCG. Stellar observations show that once the light from individual resolved galaxies other than the central BGG/BCG is excluded, the resulting surface brightness can be separated into two components: that due to light from a localized concentration of stars usually identified with the BGG/BCG proper and that due to light from a diffuse, often extended, component: the intragroup/intracluster light (IGrL/ICL) (e.g., see [245]). The $M_\star$ we quote here is the sum of the mass in the two components. The SMHM offers a window onto not just the efficiency with which the gas in the central galaxies is turned into stars but also the tug-of-war between heating and cooling in the group cores as well as the role of mergers in the build-up of the central plus IGrL stellar mass.

For the purposes of clarity, we plot the simulation results across the three panels in Figure 8 but for reference and comparison, we show the same set of nine representative observationally determined SMHMs [12] in all three panel. All of the observed results are based on the projected stellar light distribution and correspond to the central+IGrL stellar mass within a cylinder aligned along the line of sight with some aperture of radius R (i.e., $M_{\star,R,2D}$). This mass estimate explicitly *excludes* the contribution from resolved satellite galaxies. The dashed, dotted and solid curves are results from Girelli et al. [247], Yang et al. [248,249], Moster et al. [250], van Uitert et al. [251], Erfanianfar et al. [252]. There are two sets of Kravtsov et al. [207] datapoints: the filled stars are based on $M_{\star,50,2D}$ and the open stars are based on "total" aperture masses, derived by extrapolating and integrating the ICL profiles to large radii in the sky ($M_{\star,\mathrm{tot},2D}$). The blue crosses with error bars are total aperture masses from Loubser et al. [231] and Kolokythas et al. [253]: points with $\log(M_{200}/\mathrm{M}_{\odot}) < 14.0$ are for BGGs from the high richness subset of CLoGS groups [8] while the BCG results are from the Multi Epoch Nearby Cluster Survey and the Canadian Cluster Comparison Project (MENeaCS and CCCP, respectively [254–259]). There is a considerable spread in the published observationally derived SMHM results. This spread is due to a variety of factors, including (i) the groups and clusters are sourced from deep targeted observations as well as large surveys (e.g., SDSS, COSMOS and GAMA) of varying depth; (ii) the use of apertures of different sizes; (iii) the use of different strategies to estimate the background around large massive galaxies in crowded environments [207], as well as to model and extrapolate their observed light profiles of interest; (iv) the use of different estimates s of $M_{\star}/L$ ratio to convert measured light distribution into stellar mass; and (v) the use of different approaches for estimating the halo mass, ranging from the use of X-ray observations under the assumption of hydrostatic equilibrium and abundance matching, to weak gravitational lensing estimates. The observational results shown in Figure 8, taken together, show how $M_{\star}/M_{200}$ scales with M_{200}. We interpret the spread in the SMHM determinations as a measure of the uncertainty when comparing to the simulation results.

As for the simulations, we first consider the results from the ROMULUS simulations (i.e., ROMULUS25 [48], ROMULUSC [49,72,73], ROMULUSG1 and ROMULUSG2 [73]). These are plotted in the first (top left) panel. Following Liang et al. [37], Robson and Davé [215], and Jung et al. [73], we identify systems with $\log(M_{200}/\mathrm{M}_{\odot}) \geq 12.5$ as groups/clusters; these are shown as filled yellow circles. The lower mass systems are plotted as open yellow circles. The $M_{\star}$ for ROMULUS systems is the projected stellar mass within an $R = 50$ pkpc aperture (i.e., $M_{\star,50,2D}$). The $M_{\star}/M_{200}$ for the ROMULUS groups is in very good agreement with the observations; however, the overall trend suggests a drop with increasing M_{200} that is not as steep as the observed trend.

In the same panel, we also show the results for TNG100 (cyan line/band; [67]) and FABLE (magenta line/band; [240]) simulations. The available TNG stellar masses are $M_{\star,30,3D}$, that is, stellar mass within a *sphere* of radius $R = 30$ pkpc. We have corrected these masses to stellar mass within a 50 pkpc sphere using mass profiles from Pillepich et al. [67]. We suggest that it is preferable to compare this corrected stellar mass to the observed aperture masses. The FABLE results are $M_{\star,50,2D}$, like ROMULUS. The thick solid line is the median SMHM relationship and the shaded region encompasses 95% of the systems. Both TNG and FABLE results are in very good agreement with the observations on the group and low-mass cluster scales (i.e., $\log(M_{200}/\mathrm{M}_{\odot}) < 14.0$ for TNG and $\log(M_{200}/\mathrm{M}_{\odot}) < 14.3$ for FABLE). However, neither the TNG nor the FABLE median curve decreases as steeply as the observed relationship on the cluster scales. We also note that in the case of TNG, the median curve may become shallower still if one were to use a cylindrical volume.

In the second (top right) panel, we show the results from SIMBA (red line/band; [50]) and The Three Hundred project (brown line/band; [45]). Both results are $M_{\star,50,2D}$. Considering the SIMBA results first, we find that on the group scale, that is, $\log(M_{200}) = [12.5, 13.8]$, the distribution of $M_{\star}/M_{200}$ points for individual SIMBA BGGs (not explicitly shown) is consistent with the observations. The median curve, however does not have as steep a slope as the SMBHs based on observations. Consequently, the SIMBA central galaxies on the cluster scale (BCGs) have larger stellar mass than their observed counterparts and the discrepancy grows with increasing halo mass. Overall, SIMBA is in modest agreement with the observations when compared with observational results across the entire span, from low mass groups to massive clusters. The Three Hundred SMHM does not extend to low mass groups but over the mass range covered ($\log(M_{200}) = [13, 15]$), the relationship is in excellent agreement with the observations.

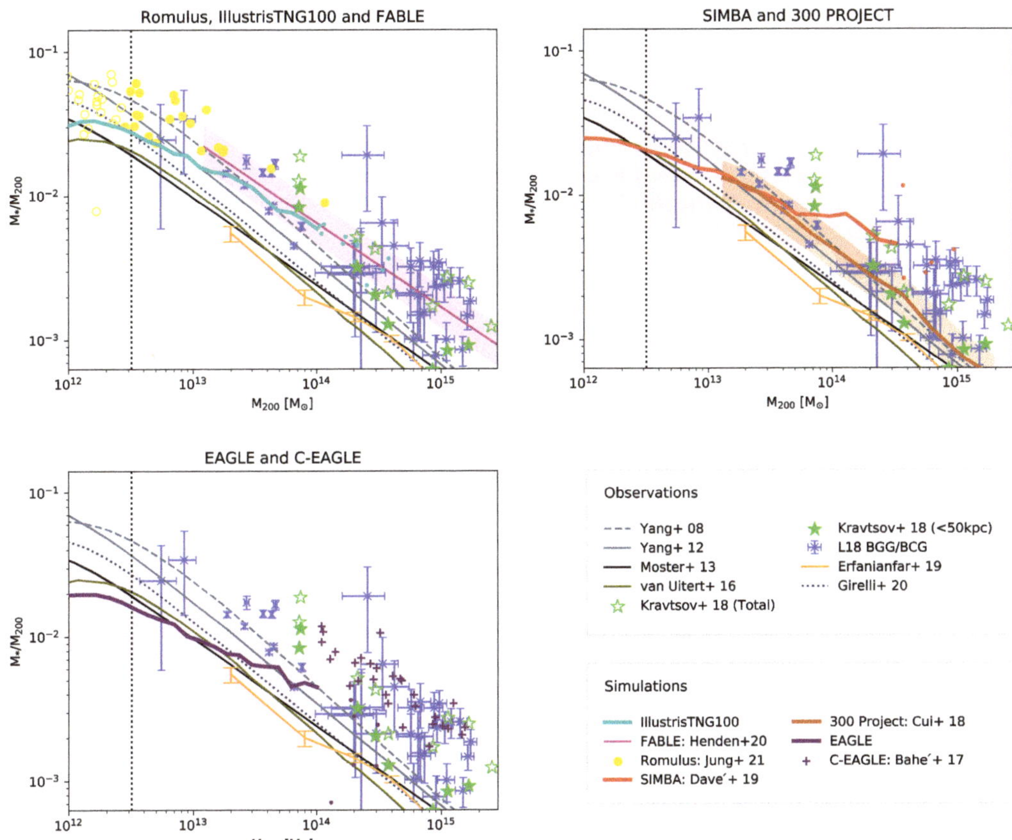

Figure 8. Figure adapted from Jung et al. [73]: The $z = 0$ stellar mass−halo mass relationship ($M_\star/M_{200}$) versus M_{200} for central galaxies in observed and simulated galaxy groups and clusters. The top left panel shows results for TNG100 (turquoise line/band and dots; [67]), FABLE (magenta line/band; [240]), and ROMULUS suite of simulations (open and filled yellow points; [49,72,73]). The thick solid lines are the median SMHM; the shaded region spans 95% of the systems. The top right panel shows results for SIMBA (red line/band and dots; [50]) and The Three Hundred project (brown line/band; [45]). The bottom left panel shows results for the EAGLE (purple line/band and dots; [39]) and C-EAGLE (purple crosses; [42]). We identify systems with $\log(M_{200}/\mathrm{M}_\odot) \geq 12.5$ (to the right of the dotted vertical line) as groups/clusters. The ROMULUS, FABLE, SIMBA, C-EAGLE and The Three Hundred project stellar masses are $M_{\star,50,2D}$; EAGLE and TNG stellar masses are $M_{\star,30,3D}$ corrected to $M_{\star,50,3D}$. For comparison, we also plot the same nine observationally determined SMHMs in all three panels. These are described in the text. There is considerable spread between the nine relationships but jointly, they show how $M_\star/M_{200}$ scales with M_{200}. We interpret the spread as a measure of the uncertainty.

The third (bottom left) panel shows the results from C-EAGLE (purple crosses Bahé et al. [42]) and the EAGLE reference run (purple line/band; [39]). The C-EAGLE stellar masses are $M_{\star,50,2D}$. The results for the lowest mass C-EAGLE systems are consistent with the observations but in massive clusters, as Bahé et al. [42] notes, the C-EAGLE stellar masses exceed the $M_{\star,50,2D}$ (filled stars) from Kravtsov et al. [207] by up to 0.6 dex. As for EAGLE, the reported $M_\star$ is $M_{\star,30,3D}$; we used a mean stellar mass-dependent correction factor, derived from plots in Schaye et al. [39] and McCarthy et al. [41], to map $M_{\star,30,3D}$ to $M_{\star,50,3D}$. The correction factor is negligible for galaxies with $\log(M_{\star,30,3D}/\mathrm{M}_\odot) < 10.7$ [39]; on the other hand, the stellar mass in the most massive BGG/BCGs approximately doubles. Over the mass range $\log(M_{200}) = [13, 14]$, the distribution of individual $M_\star/M_{200}$ EAGLE points (not explicitly shown) is consistent with the observed SMHM results; however, like the SIMBA results, the EAGLE median curve is shallower than the observed trend. Consequently, for $\log(M_{200}/\mathrm{M}_{\odot}) < 13$, more than half of EAGLE BGGs fall below Moster et al. [250] line, which is the lowest of the observationally-derived SMHM curves. Overall, the EAGLE SMHM is in modest agreement with the observations across the entire span, from low mass groups to massive clusters. Finally, we point out that the EAGLE median curve in Figure 8 is flatter than that in Schaye et al. [39] because of the mass-dependent correction, and we expect that the corresponding results based on $R = 50$ pkpc aperture mass will be shallower still since our correction maps to a sphere, not a cylinder.

4.2.2. Central Galaxy Star Formation Rates

Figure 9 shows the second trend that we consider here: the star formation rate−stellar mass relationship (hereafter, SFR$-M_{\star}$). The SFR is based on star formation within a $R = 30$ pkpc sphere centered on the BCG (very little star formation is observed beyond 30 pkpc) while the M_{*} is the same as that in Figure 8. We first consider the four sets of observational data points that appear in all three panels: (i) results based on combined data from XMM–LSS, COSMOS, and AEGIS surveys for BGGs and BCGs hosted by X-ray bright galaxy groups and clusters (green pluses; [229]); (ii) results for BCGs in Mittal et al. [260] sample of CC clusters (grey crosses); (iii) results for the high richness subset of the CLoGS sample (blue crosses; [8,231,253]), which consists of groups containing at least 4 optically bright ($\log(L_{\mathrm{B}}/L_{\odot}) \geq 10.2$) galaxies, of which the central galaxy is an ETG; and (iv) results for BCGs from the COSMOS survey (magenta squares; [261]).

The Gozaliasl et al. [229], Loubser et al. [231], Cooke et al. [261] and Mittal et al. [260] data collectively illustrate the approximate dichotomy in the population of BGGs and BCGs with respect to their star formation properties. On the cluster scale, this phenomenon has been widely discussed (e.g., [255,263], and references there in) and linked to the CC/NCC dichomotomy in the X-ray properties, including the X-ray luminosity at fixed halo mass, the core entropy value and the core hot gas cooling time as highlighted by Cavagnolo et al. [9], McCarthy et al. [216,264] and others. Essentially, not all BCGs are "red and dead"; there exists a population of star-forming BCGs that reside in strong cool core clusters [255]. *These BCGs' SFRs are such that they are distributed within approximately* ± 0.75 *dex of the extension of the observed* $z < 0.5$ *star forming main sequence* (SFMS) of Whitaker et al. [262]. The star-forming BCGs comprise ∼25%–30% of all BCGs; this is also the fraction of strong CC clusters [256,265]. Gozaliasl et al. [229] find that this dichotomy is also present on the group scale and the fraction of star-forming BCGs is comparable to that of star-forming BCGs; that is, 20%–25%. The majority of the observed central galaxies in Figure 9 have SFRs less than 0.75 dex *below* the SFMS. We refer to these as "quenched" BGGs/BCGs.

As for the simulations, we restrict ourselves to groups and clusters whose BGG/BCG satisfy $\log(M_{\star}/\mathrm{M}_{\odot}) \lesssim 12.0$; there are too few systems with higher stellar masses. We classify the simulation BGGs/BCGs in the same way as the observed systems: We designate all simulation BGGs/BCGs with SFRs less than 0.75 dex *below* the observed $z < 0.5$ SFMS [262] as "quenched" systems, and the rest of the galaxies as star-forming. The numbers at the bottom of the panels specify the fraction of quenched galaxies in 0.5 dex M_{*} bins for TNG100, SIMBA, and EAGLE simulations. Many of the quenched galaxies have low/unresolved SFRs. In order to display these galaxies in the panels, we assign them an SFR of $\log(\mathrm{SFR}/\mathrm{M}_{\odot}\mathrm{yr}^{-1}) = -4$. To the extent that it is of interest, we find that the number of galaxies with very low/unresolved SFRs relative to the total number varies considerably from simulation to simulation. However, we do not differentiate between quenched galaxies with low/unresolved SFRs and those with low but measurable SFRs, and there is there is no observational basis for doing so. Measuring low SFRs using observationally accessible diagnostics is extremely challenging [266–271], and strictly speaking, very low published SFRs ought to be treated as upper limits.

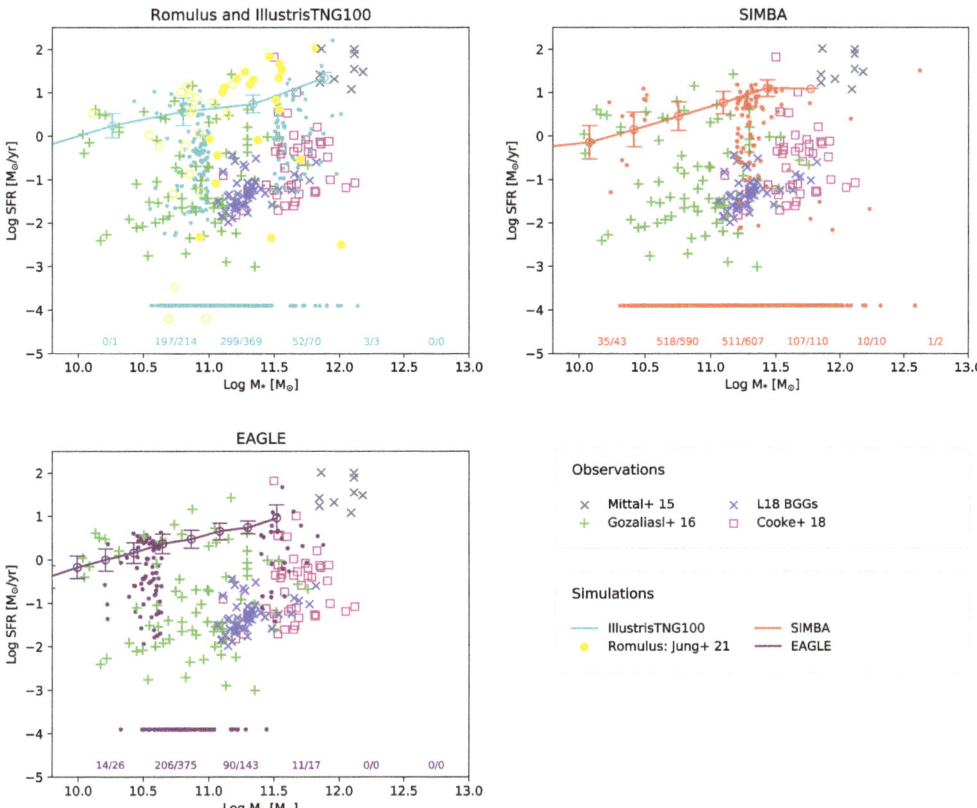

Figure 9. Figure adapted from Jung et al. [73]: Comparison of the $z = 0$ star formation rate (SFR)-$M_\star$ relationship for observed and simulated BGGs and BCGs. Arranged as Figure 8, each panel plots the same four sets of observational results (see text for details). The top left panel shows the results for ROMULUS (yellow filled and open circles; filled circles represent systems we identify as groups/clusters) and TNG100 simulations (cyan); the top right panel shows the results for SIMBA (red); and the bottom left panel shows the results for EAGLE (purple). The simulation SFRs are extracted from a sphere of radius 30 pkpc encompassing the central BGG/BCG while the stellar masses as the same as in Figure 8. The line connecting the open circles with error bars shows the star forming main sequence (SFMS) for TNG100, SIMBA, and EAGLE from Davé et al. [154]. The shaded regions, and the handful of individual points to the left and right, show where the simulation BGGs/BCGs with *measurable SFR* lie on this plot. BGGs/BCGs with unresolved/too low SFRs are assigned $\log(\mathrm{SFR}/\mathrm{M}_\odot \mathrm{yr}^{-1}) = -4$ and plotted accordingly. We identify all simulation BGGs/BCGs with SFRs $\gtrsim 0.75$ dex *below* the observed $z < 0.5$ SFMS of Whitaker et al. [262] as "quenched". For TNG100, SIMBA, and EAGLE, we specify the fraction of quenched galaxies in 0.5 dex M_* bins at the bottom of the panels. We do not differentiate between quenched systems with low but measurable SFRs and those with very low/unresolved SFRs. Measuring very low SFRs using observationally accessible diagnostics is extremely challenging; strictly speaking, the corresponding published SFRs ought to be treated as upper limits.

Considering the ROMULUS results (yellow points) in the top left panel first, we find that this model gives rise to both star-forming and quenched BGGs; however, the corresponding fractions are inverse of the observed fractions: ∼60% of the ROMULUS BGGs are star-forming and ∼40% are quenched. The top left panel also shows the entire TNG100 population of BGG/BCGs. Only 17% of these are star-forming. This is lower than the observed fraction. The SIMBA results in the top right panel are similar to TNG100 and also on the low side: only about 14% of all the SIMBA BGGs/BCGs are star-forming. The fraction of EAGLE BGGs (lower left panel) that are star-forming is ∼40%, which is higher than the observed fraction though not as high as the ROMULUS fraction. We will return to these results in Section 4.2.4.

4.2.3. Central Galaxy Morphologies

The final trend we consider is the stellar mass-morphology relationship for the central galaxies. Qualitatively, the observed morphologies of central galaxies span the full continuum, from disk dominated to pure spheroids. Weinmann et al. [228] found that approximately 50% of the BGGs in low-mass SDSS groups are late-type galaxies (see also [272]); this fraction drops to $\sim$10% in rich groups/poor clusters. Conversely, the fraction of early-type central galaxies rises from $\sim$30% in low-mass groups to 70% in high-mass systems. The morphological mix of the ROMULUS BGGs, which are from mainly low-mass groups, is consistent with Weinmann et al. [228] results in that nearly a half are disky and star forming [73]. Similarly, a visual inspection of the images of BCGs from the 20 most massive TNG100 systems with halos masses between 9×10^{13} M$_\odot$ and 4×10^{14} M$_\odot$ ([67]) also suggests that their morphological types are also compatible with Weinmann et al. [228] results for low-mass clusters: A significant fraction are ellipsoidal or S0-like and appear to be red.

More quantitatively, a galaxy's morphology is commonly characterized by the fraction of its total stellar mass that is in the spheroidal component; that is, its spheroidal-to-total ratio (S/T). This ratio has been computed using photometric observations, by decomposing the 2D projected light profile into disc and spheroidal components, but it can also be derived using stellar kinematics to identify the spheroidal component. The two approaches do not also give similar results. Simulation studies have shown that photometrically derived S/Ts tend to be significantly lower than the kinematic S/Ts [273], with Bottrell et al. [274] finding that the former leads to a higher likelihood of a galaxy being classified as "disky even when stellar kinematics show no ordered rotation". In effect, the Scannapieco et al. [273] and Bottrell et al. [274] studies show that it is not sufficient to only consider the BGG/BCG properties derived from photometric data. There is much to be learnt from examining the galaxies' kinematic properties.

In Figure 10, we show the kinematic S/T ratios for the central galaxies from ROMULUS [73], TNG100 [244] and EAGLE [238], as a function of galaxy stellar mass. We use comparable apertures as Tacchella et al. [244] to compute ROMULUS S/T ratios: For BGGs with $\log(M_*/\mathrm{M}_\odot) \leq 10.9$, S/Ts are computed using star particles within spheres of radius R = 15–20 pkpc; for more massive BGGs, we use spheres of radius R = 25–30 pkpc. We also use the same criterion to define the spheroidal component as Tacchella et al. [244]: the mass of the spheroid is the sum of the mass of stellar particles with $\epsilon_J < 0.7$ and the 15% of the stellar particles with $\epsilon_J > 0.7$, where $\epsilon_J = J_z/J_{\rm circ}(E)$, J_z is a z-component of the specific angular momentum of a stellar particle, z is the net spin axis of a galaxy, and $J_{\rm circ}(E)$ is a specific angular momentum of the stellar particle on a circular orbit with the same orbital energy.

The EAGLE criterion for determining S/T ratios differs slightly but we have confirmed that the two criteria give comparable results. Figure 10 also shows the S/T for central galaxies in the GAMA groups sample derived using photometric data (orange points; [13]), as well as the kinematic S/T for BGG/BCGs from the CALIFA survey (red points; [275]). We use the same prescription as Tacchella et al. [244] to compute the CALIFA S/Ts. The simulation results should be comparable to the CALIFA results.

Overall, there is a broad agreement between the three simulation results and for $M_* \gtrsim 10^{11}$ M$_\odot$, all three simulation results are consistent with the CALIFA results. In these galaxies, the spheroidal component dominates: S/T rises from approximate 0.7 to 0.9 with stellar mass. Towards lower M_*, however, the median TNG100 and EAGLE curves diverge from the CALIFA results. At present, it is unclear how much weight this divergence ought to be given. Firstly, there is a considerable spread in the simulation results and the CALIFA results are well within the shaded region; secondly, the CALIFA sample is known to be increasingly incomplete towards at lower masses. It is possible that the trend in the CALIFA results reflects this incompleteness.

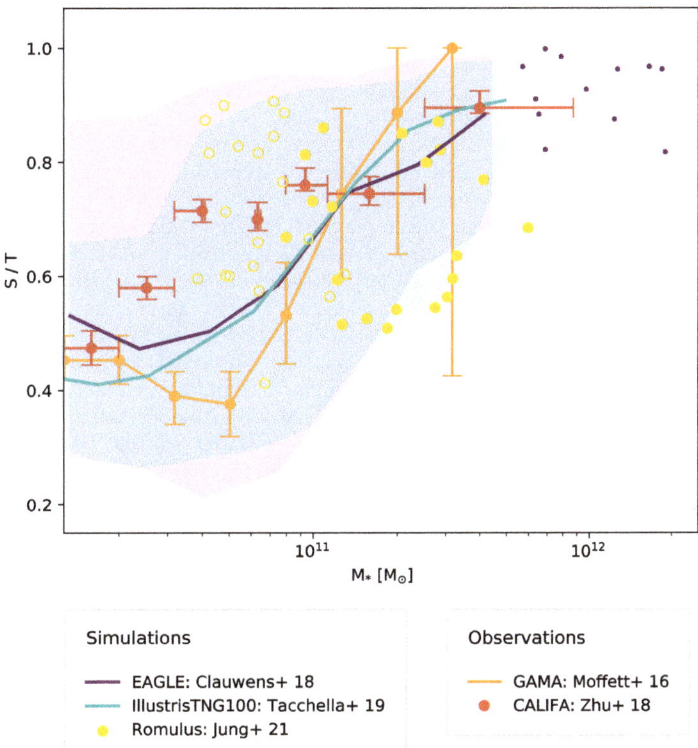

Figure 10. Figure adapted from Jung et al. [73]: The $z = 0$ spheroidal-to-total (S/T) for central galaxies in simulated galaxy groups and clusters from three simulations (ROMULUS, TNG100, and EAGLE), juxtaposed against results from two observational studies. The yellow filled and open circles show the ROMULUS kinematically derived results. (As in previous two figures, filled circles correspond to systems we identify as groups and clusters.) The cyan and purple curves show the median kinematically derived S/T results for TNG100 [244] and EAGLE [238], respectively. The purple shaded region indicates the 10–90% range while the cyan shaded region spans the 16%–84% range. The ROMULUS and TNG100 S/T are computed using the same criterion while the EAGLE results are based on a slightly different criterion. The two criteria, however, give very similar results. Of the two observational results shown, the CALIFA data are kinematically derived [275] while the GAMA results are based on photometry [13]. The two approaches are not equivalent. We discuss this further in the text. The simulation results should, in the first instance, be compared to the CALIFA results.

Finally, we note that S/T is not the only kinematic measure that is correlated with either the central galaxies' stellar mass or the host systems' properties. Other measures less commonly discussed in the galaxy formation literature include (i) the shape of the velocity dispersion profiles and whether they rise or fall with radius; (ii) the "anisotropy parameter" $V_{\rm rot}/\sigma_0$, which characterizes the global dynamical importance of rotation and random motions of stars in a galaxy; and (iii) whether the galaxy is a fast or slow rotator. Jung et al. [73] investigates some of these measures for ROMULUS galaxies and compare them to observations and we refer interested readers to that paper for further details and discussion.

4.2.4. The Link between the BGG and IGrM

The properties reported in Figures 8–10 are the observable byproducts of a myriad of physical processes that the BGGs and BCGs are subject to over the course of cosmic time. These properties are also sensitive to the details of the subgrid prescriptions used to model those processes that cannot be directly resolved in the simulations. All three of the properties discussed in this subsection are related to each other and more importantly, they are all correlated with the the current or the recent state of the IGrM/ICM in the group and cluster cores. Here, we briefly elaborate on this relationship, focusing on the SFR-$M_{\star}$ results.

As highlighted throughout this review, among the most important present-day challenges in simulating galaxy groups and clusters is (i) preventing strong radiative cooling flows from forming in the IGrM/ICM and (ii) ensuring that the resultant population of groups and clusters span the spectrum from CC to NCC systems. Nearly two decades ago, Babul et al. [213], Valageas and Silk [276], Nath and Roychowdhury [277] advocated for AGN feedback as the key mechanism for addressing these challenges. The current generation of simulation models of galaxy formation/evolution (see Table 2) all allow for SMBH seeding and growth as well as AGN feedback although the manner in which these are implemented vary from one simulation to another.

To examine how well the current models fare, we first consider the ROMULUS simulations. As noted previously, the fact that nearly 60% of the ROMULUS BGGs are star-forming (Figure 9) means that while AGN feedback in ROMULUS may temper cooling, it does not entirely halt it and consequently, a majority of the systems sustain some degree of cooling flow in their cores. A closer examination of the ROMULUS entropy profiles, such as those shown in Figure 6, offer some insights. For the most part, ROMULUS groups behave like CC clusters, not CC *groups*. Their entropy profiles are $\sim$2–3$\times$ lower than the Sun et al. [2] profiles at large radii, and the profiles themselves are steeper ($R^{1.1}$ versus $R^{0.7}$). We suspect this is the result of a combination of AGN feedback being too weak and metal-line cooling being absent. ROMULUS simulations do not produce stable NCC systems although once in a while, they do produce weak NCC systems [72] and during this phase, the BGG is quenched [73]. This phase typically arises when the BGG suffers a sizable merger that elevates the core entropy; however, the phase is temporary, lasting $\lesssim 2$ Gyrs.

Turning to the TNG and SIMBA simulations, we find that 78% and 91% of the BCGs, respectively, with $\log(M_{*}/\mathrm{M}_{\odot}) \geq 11.3$ are quenched. Such high fractions are not surprising given that the vast majority of the $z = 0$ TNG and SIMBA clusters are NCC systems with high entropy cores (see, for example, the orange and yellow curves in Figure 6). Neither TNG nor SIMBA produce clusters with typical CC cluster entropy profiles, and in both cases, the main reason appears to be that feedback is both too efficient, resulting in larger and higher entropy central cores than observed (c.f. [214,215]).

The main takeaway, therefore, is that while all of the simulation models appear to find reasonable agreement with a subset of the available observations of galaxy groups and clusters, none of them correctly describe galaxy groups and clusters in a comprehensive manner. It is very likely that the problems with the entropy profiles and quenched/star-forming BGGs/BCGs are due to either a problem with the physical models that inform the subgrid modeling of the various aspects of SMBH formation and evolution, and/or due to the implementation these models in the simulations. All simulation models have ad hoc features embedded within their subgrid prescriptions and when the models run into difficulties, quite reasonably a common response is to attempt to patch one or more of the subgrid schemes in the hope that that fixes the problem without causing any breakage elsewhere. In the case of AGN feedback, we assert that minimal patches have generally not led to a realistic population of simulated galaxy groups and clusters. This should be seen as a clear invitation to revisit the treatment of SMBHs and AGNs in the simulations. We offer some thought on the subject in Section 5.3.

4.3. The Multiphase IGrM

Observations of the IGrM are dominated by the X-ray emission from hot, dense gas surrounding the BGG. Since X-ray emission scales with the square of gas density, diffuse warm and cool phases of the IGrM are prohibitively difficult to observe directly in emission. While several groups have pioneered efforts to characterize the warm and cool IGrM/ICM with UV absorption specifically targeting group environments [27,278,279], the Virgo [280,281] and Coma [282] clusters, other clusters [26,283], cluster outskirts [284,285], as well as low-mass group halos hosting ETGs [286], such studies are limited by the available number of background sources with existing instrument sensitivities. For this reason, simulations of groups offer a unique insight into the multiphase structure of the IGrM and the physical origins of its different components. Because warm and cool gas is expected to form structures that are significantly smaller than the volume-filling hot phase, simulations that aim to accurately model the multiphase IGrM require extremely high resolution. Recent advancements in the resolution of group-scale simulations (like RomulusC and TNG50) have allowed for unprecedented insights into the nature of the multiphase IGrM. Figure 11 demonstrates the multiphase gas structure in the ROMULUSC simulation at z = 0.31. The top row shows 5 × 5 Mpc (9.5 × 9.5 R_{500}) projections of gas density, temperature, and metallicity. The bottom row shows synthetic X-ray emission and UV absorption maps, roughly calibrated to the sensitivity of existing instruments. This figure clearly demonstrates that X-ray emission and UV absorption studies probe highly complementary regions of the IGrM. Additionally, it highlights the rich multiphase structure of the IGrM that exists out to the edges of the group outskirts.

We quantify the fractional breakdown of multiphase gas throughout the extended IGrM for $z = 0$ snapshots of ROMULUSC, EAGLE, SIMBA, and TNG100, in Figure 12. Gas is divided into "hot" ($T \geq 10^6 K$), "warm" ($10^5 K \leq T < 10^6 K$), "cool" ($10^4 K \leq T < 10^5 K$), and "cold" ($T \leq 10^4 K$). The IGrM is clearly dominated by the hot phase all the way out to and beyond $4R_{500}$, and the cooler phases are not dramatically different between the 4 simulations. SIMBA has the least amount of multi-phase gas, and by total mass even less inside R_{500}, since the IGrM is most evacuated in this simulation. ROMULUSC has the most warm-hot gas at large radii, but far less than at $z = 0.31$ where Butsky et al. [19] showed more multi-phase gas that was disrupted by a 1:8 merger [72]. The properties of this hot inner region are intimately tied to the feedback from the central BGG. In particular, the temperature and entropy of the IGrM near the BGG depends on the recent AGN activity. After periods of relatively low AGN activity, the entropy profile of the inner 0.04 R_{500} is low and susceptible to thermal instability, which fuels star formation and AGN activity. After vigorous AGN activity, the inner 0.2 R_{500} develops hotter temperatures and a steeper entropy profile [72]. The metallicity of the hot phase is around 0.3 $Z_\odot$ and remains relatively constant throughout the inner region [19]. The prevalence and uniformity of metals in the hot phase implies that the hot phase was primarily enriched over time by diffuse gas in the group outskirts and mergers with group satellites (e.g., [111,287]).

The statistical samples of EAGLE, SIMBA, and TNG100 show more similar patterns of cool gas inside R_{500} including a fractional increase inside $0.25R_{500}$, which we detail below. Further out, warm gas begins to constitute a substantial fraction of the IGrM mass and beyond $\sim 2.5R_{500}$, but unlike the smooth, large volume-filling hot phase, the cool gas distribution is patchy and traces gas that is stripped from satellite galaxies as they move through the IGrM. In the most extreme cases, the satellite galaxies and their stripped tails are referred to as "jellyfish galaxies" as seen in ROMULUSC O VI and H I maps in Figure 11. In ROMULUSC, gas stripped from satellite galaxies tends to be cooler and more metal-enriched than the ambient IGrM gas in the group outskirts, leading to a wider distribution of metallicities traced by cool and warm gas. ROMULUSC satellite galaxy CGMs (i.e., gas within $\lesssim$ 150 kpc of satellites) show decreasing covering fractions of O VI, C VI and H I at lower IGrM radii with a significant decline inside $3R_{500}$ [19]. Eventually, the stripped halo gas mixes with the rest of the IGrM. Constraining this mixing rate will be important for understanding how galaxies lose their gas and how the group environment is important for galaxy evolution.

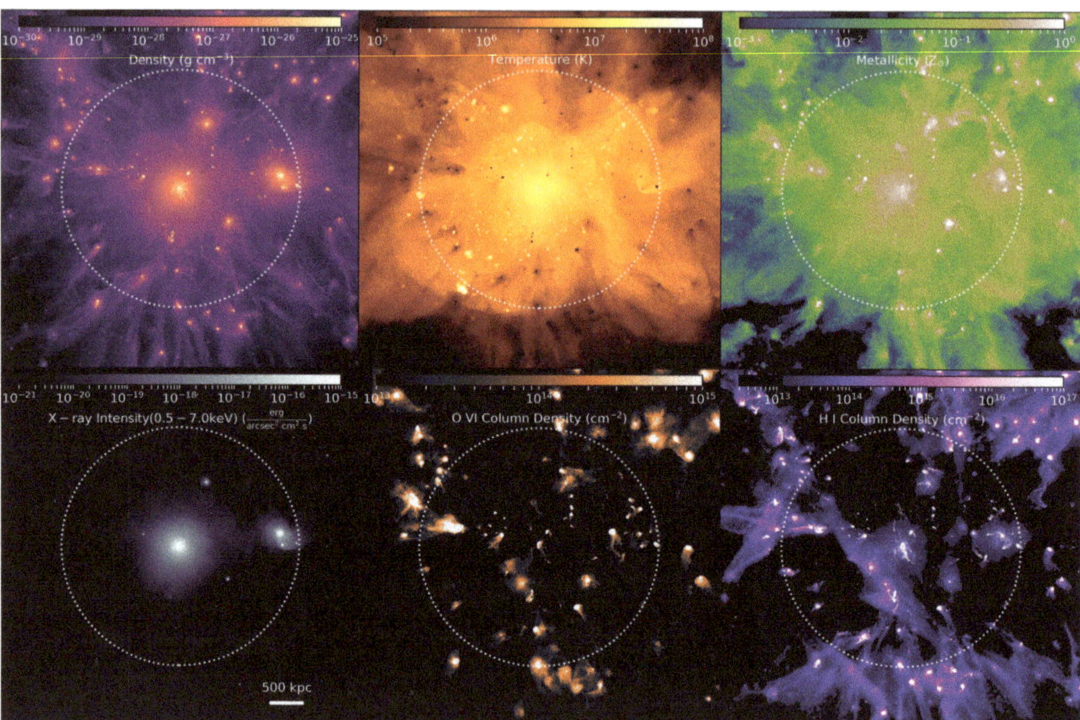

Figure 11. The ROMULUSC simulation at z = 0.31 [19]. The top row shows the projected density, temperature, and metallicity. The bottom row shows the predicted X-ray emission as well as the O VI and H I column densities. Each image spans 5 Mpc across and the white dashed circle has a radius of $3R_{500}$. X-ray emission probes the inner hot, dense region of the IGrM. UV absorption of ions like O VI and H I provide a highly complementary view of the IGrM, tracing the filamentary structure of cool and warm gas as it is stripped from its host galaxies.

In addition to being stripped from satellite galaxies, cool gas can also form through thermal instabilities, which is prevalent at a $\sim 10^{-2}$ fraction in the inner $0.25R_{500}$ of most simulations of $\sim 10^{14}\,M_\odot$ groups. When the local cooling time (t_{cool}) of gas is roughly less than 10× the gravitational freefall time (t_{ff}) [288], small perturbations can seed a runaway cooling effect through which cool gas condenses out of the background medium and precipitates onto the central BGG. This process of local thermal instability in a globally stable atmosphere is self regulating and maintains a global cooling to freefall time ratio (t_{cool}/t_{ff}) $\simeq$ 10–20 throughout the IGrM (e.g., [289,290]). Any gas with $t_{cool}/t_{ff} \leq 10$ will form cool filaments, lowering the density (and cooling time) of the remaining hot IGrM. This cool gas is accreted onto the central BGG and can trigger stellar or AGN feedback, driving the baryonic feedback cycle and maintaining global thermal stability in the IGrM. This process of thermal instability promoting the formation of clumps of cool gas was traced by Nelson et al. [291] using TNG50, which we check show very similar fractional values as TNG100 for the two most massive group halos in that 50^3 Mpc^3 box.

Finally, we show the mass trend of fractional IGrM phases for changing group halo mass bins in the subpanels on the bottom right of Figure 12 for TNG100. The uptick in cool gas inside $0.25R_{500}$ goes from 2%, to nearly 20%, to over 30% as one progresses down the mass scale from $M_{500} = 10^{13.9}$ to $10^{13.1}\,M_\odot$. Thus it should not be surprising that ions like H I, Si III, and C IV in halo gas around relatively isolated ETGs that may occupy $\sim 10^{13}\,M_\odot$

halos [286,292]. However, surveys targeting more massive groups show very little cool (H I) or warm (O VI) gas inside $R_{\rm vir}$ [27].

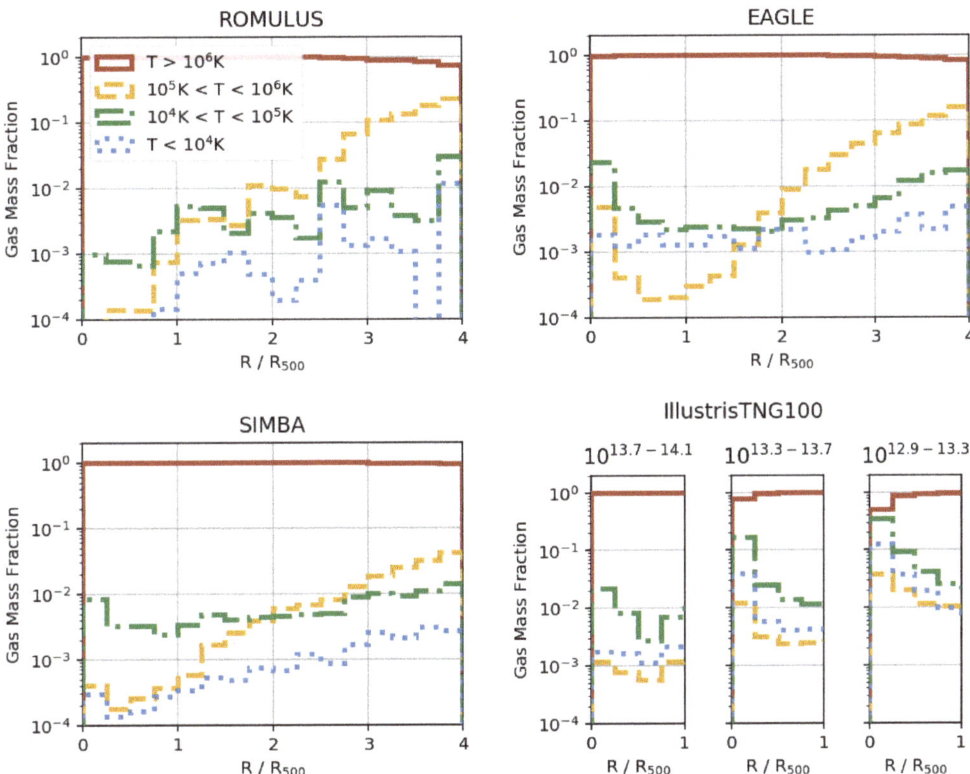

Figure 12. The total mass fraction of hot ($T \geq 10^6K$), warm ($10^5K \leq T < 10^6K$), cool ($10^4K \leq T < 10^5K$), and cold ($T \leq 10^4K$) gas as a function of distance from the group center at $z = 0$ for halos at $M_{500} \approx 10^{13.9}\,\mathrm{M}_\odot$ for ROMULUS (ROMULUSC in this case), EAGLE ($N = 7$ groups), SIMBA ($N = 39$), and the inner regions of TNG100 for this bin (left subpanel, $N = 13$) and two lower mass halo bins (middle, $M_{500} = 10^{13.3-13.7}\,\mathrm{M}_\odot$, $N = 41$; & right, $M_{500} = 10^{12.9-13.3}\,\mathrm{M}_\odot$, $N = 105$). Although hot gas dominates the mass fraction within $4R_{500}$ of the BGG in all cases, cool gas never constitute less than 10^{-3} of the total gas mass within R_{500}.

4.4. Satellite Galaxies in Groups

Group satellites have been studied extensively in observational surveys, not least because groups are more than an order of magnitude more common than massive clusters [35]; even medium-size surveys such as GAMA therefore include more than a thousand of them [293]. Although each individual group contains fewer satellite galaxies than a rich cluster, collectively they still host $\gtrsim$2 times as many satellites (see Figure 13). These observations generally place group satellites between clusters and the field with, for example, quenched fractions of ≈60% [11] and H I mass fraction ratios ($M_{\rm H\,I}/M_\star$) of ≈0.15 [30] at $M_\star = 2 \times 10^{10}\,\mathrm{M}_\odot$, (extended) X-ray detection fractions of bright ETGs as high as ≈90% [294], and elliptical galaxy fractions of ≈50% [295].

In simulations, stripping of the (extended) warm-hot gas halos of group satellites is robustly predicted (e.g., Butsky et al. [19], Bahé et al. [296], Zinger et al. [297]), even for galaxies that are still (well) beyond the virial radius of the group. Ram pressure is strong enough to explain this gas loss, in particular for galaxies that (temporarily) move through a denser part of the IGrM or with a higher velocity than typical at a given radius [298]. As discussed above, this gas loss is simultaneously predicted to be an important route of IGrM enrichment [19].

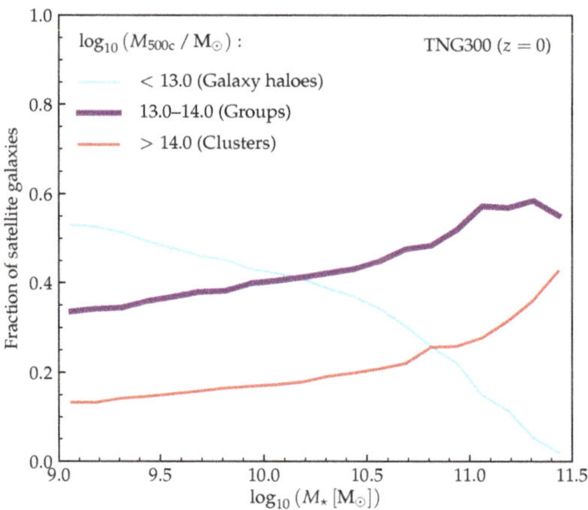

Figure 13. Fraction of satellite galaxies in the IllustrisTNG300 simulation that are hosted by galaxy-scale halos ($M_{500c} < 10^{13}\,M_\odot$, light blue), groups ($M_{500c} = 10^{13}$–$10^{14}\,M_\odot$, thick purple), and clusters ($M_{500c} > 10^{14}\,M_\odot$, light red), respectively, as a function of their stellar mass. Groups host around a third of low-mass satellites, and more than half of those with $M_\star \approx 10^{11}\,M_\odot$; their contribution exceeds that of clusters at all stellar masses shown here.

To our knowledge, no simulation that stretches up to group scales is currently able to self-consistently model the evolution of atomic and molecular hydrogen (see e.g., Hopkins et al. [113], Applebaum et al. [299] for examples of such simulations on smaller scales). Some authors have, however, modelled H I in post-processing with theoretically and/or empirically motivated relations (e.g., Blitz and Rosolowsky [300], Rahmati et al. [301], Gnedin and Draine [302]) to derive H I and H_2 masses of group satellites. For EAGLE, Marasco et al. [303] demonstrated agreement with ALFALFA observations and found that H I loss on group scales was driven by a complex mixture of tidal stripping, ram pressure, and satellite–satellite encounters. Stevens et al. [304] created detailed H I mock observations of the TNG100 simulation, showing that this step was critical in achieving a match to the observed H I mass fractions of Brown et al. [30]. They also found, however, that the H I loss in TNG100 parallels the decline in SFR, contrary to the observations that show lower H I mass fractions even at fixed sSFR [30].

In Figure 14, we compare the H I deficiency of group galaxies ΔH I $= \log_{10}(M_{\rm H\,I,sat}/M_{\rm H\,I,cen})$ (where $M_{\rm H\,I,sat}$ and $M_{\rm H\,I,cen}$ are the total H I masses of group satellite and central galaxies in a narrow range of $M_\star$) as predicted by the Illustris, IllustrisTNG, and Hydrangea simulations. For the first two, we take the H I masses computed by Diemer et al. [305] on a cell-by-cell basis with the Gnedin and Draine [302] H I/H_2 partition; for Hydrangea the H I have been computed in analogy to Bahé et al. [306] with the empirical Blitz and Rosolowsky [300] H I/H_2 partition. Despite the variety of simulations, resolutions, and H I models, the predictions are remarkably uniform: all simulations (except for Illustris) predict an H I deficiency of ΔH I ≈ 0.9 at the low-mass end ($M_\star \approx 10^{9.5}\,M_\odot$), and a less extreme difference at high masses (ΔH I ≈ 0.2).

Stevens et al. [307] investigated the molecular (H_2) masses of satellite galaxies in TNG100 through a similar mock imaging approach as Stevens et al. [304]. Despite the lack of a directly modelled cold ISM phase in these (and other) simulations, they obtained an H_2 mass fraction for group satellites that is $\approx$0.6 dex lower than in the field, consistent with data from the xCOLD GASS survey [308].

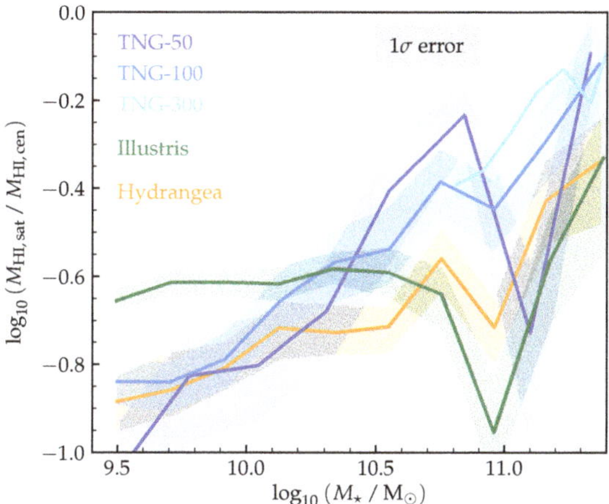

Figure 14. Atomic hydrogen (H I) deficiency of group satellites compared to centrals of the same stellar mass, as predicted by the Illustris (green), IllustrisTNG (shades of blue), and Hydrangea (orange) simulations (computed following Diemer et al. [305] and Bahé et al. [306], respectively). Solid lines show the difference between mean H I masses in each bin, shaded bands the corresponding 1σ uncertainties obtained from bootstrapping. All simulations predict H I-deficient group satellites, with the difference generally largest (almost 1 dex) for the least massive galaxies. Note that for IllustrisTNG300 (light blue), H I masses are only computed for all galaxies with $M_\star > 5 \times 10^{10}\,\mathrm{M}_\odot$.

The impact of this gas loss on star formation is arguably the most well-studied aspect of simulation works on group galaxies. Unanimously, simulations predict that the quenched (or red) fractions of group galaxies are significantly higher than for equal-mass field galaxies (e.g., Donnari et al. [34], Bahé et al. [42], Tremmel et al. [49]). In simulation suites that include groups, as well as more massive clusters (Hydrangea/C-EAGLE, IllustrisTNG), clear trends with halo mass are seen, at least for $M_\star \lesssim 10^{11}\,\mathrm{M}_\odot$ (see Figure 15): quenched fractions in groups are below those for clusters by up to a factor of ≈ 2 [34,42]. When only considering satellites that were directly accreted onto their $z = 0$ host and not quenched already, however, Donnari et al. [34] report that this trend reverses for $M_\star \gtrsim 3 \times 10^{10}\,\mathrm{M}_\odot$: at the massive end, the quenched fraction of this subset of satellites is highest for groups (up to 80%) and lowest in massive clusters (40%). Donnari et al. [34] interpret this "host rank inversion" as a result of AGN feedback: while this still operates efficiently for massive satellites in groups, it is suppressed in more massive clusters and therefore does not quench massive galaxies as efficiently.

A second indicator of changes to the baryon cycle in group galaxies, their ISM metallicity, was investigated by Genel [309], Bahé et al. [310], and Gupta et al. [311] with the Illustris, EAGLE, and TNG100 simulations, respectively. All three studies found elevated metallicities of satellites compared to the field, a difference that is more pronounced for lower $M_\star$ and higher halo mass, in qualitative agreement with observations [312]. For EAGLE, Bahé et al. [310] showed that this enhancement also agrees quantitatively with the observations, and is also predicted for stellar metallicities. Together, these studies identified three mechanisms that contribute to the elevated metallicities. Firstly, ram pressure stripping removes predominantly gas at larger radii, where the metallicity is lower [309,310]. Secondly, suppressed inflows of pristine gas within satellites [313] prevent the dilution of the ISM [310]. Finally, Gupta et al. [311] showed that the gas that is still replenishing the satellite ISM has higher metallicity than for isolated galaxies. Both TNG and EAGLE predict that this enhancement is not restricted to satellites within the virial radius of their group, but already affects galaxies during their infall [310,311].

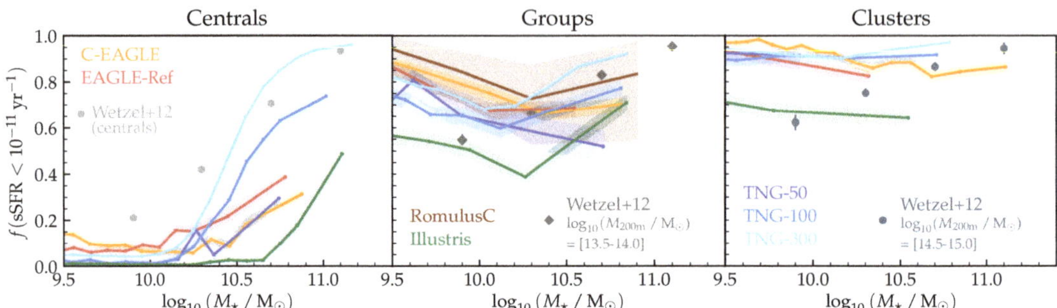

Figure 15. Quenched fractions f_q of central galaxies (**left**) and of satellites in groups ($M_{500} = 10^{13} - 10^{14}\,\mathrm{M}_\odot$, (**middle**)) and clusters ($M_{500} > 10^{14}\,\mathrm{M}_\odot$, (**right**)) as predicted by different simulations. For (C-)EAGLE, Illustris, and IllustrisTNG, these are computed directly from the simulation outputs; the quenched fractions for RomulusC are taken from Tremmel et al. [49]. Especially on the low stellar mass end ($M_\star \lesssim 2 \times 10^{10}\,\mathrm{M}_\odot$), different simulations agree closely for centrals ($f_q \approx 0$) and clusters ($f_q \approx 1$), whereas predictions on group scales show a much larger diversity (f_q between 0.6 and 1). Observational data from Wetzel et al. [11] are shown for approximate guidance, but neither the satellite selection nor the host mass ranges are matched to their analysis.

While ram pressure is acting on group galaxies, it distorts their gas into long "jellyfish" tails. In TNG100, these have been studied by Yun et al. [23] through visual inspection. Depending on group mass, these authors found that ≈25–45% of gas-bearing satellites show evidence of such tails in their (total) gas density maps, only moderately lower than the equivalent fraction for clusters (≈65%). Even when considering only the (observable) H I component in TNG100, Watts et al. [314] find statistically significant asymmetries that visually resemble observed H I tails (e.g., Chung et al. [315]), with a slightly higher occurrence (21 vs. 28%) amongst group galaxies compared to those in lower-mass halos.

While the stripping of gas is the clearest predicted effect of ram pressure, simulations have also begun revealing second-order effects due to the compressive effect on the leading edge of group satellites. In RomulusC, for instance, Ricarte et al. [316] demonstrated a correlation between ram pressure and black hole accretion rates as well as star formation rates, evidence for which has also been seen in recent observations [317]. A compression-induced enhancement of star formation has also been described in the EAGLE simulations by Troncoso-Iribarren et al. [318]. A caveat applicable to both simulations, however, is that stars are formed directly from the tenuous ISM phase rather than from dense molecular gas that may have a different susceptibility to the effect of ram pressure, at least in detail.

The high resolution of contemporary galaxy group simulations has also enabled studies of their (stellar) morphology. Feldmann et al. [319] studied the transformation from disc to elliptical galaxy morphologies in one zoom-in simulation, and identified major mergers prior to accretion as the key driver of this change. More recently, Joshi et al. [320] presented a detailed analysis of galaxy morphology in groups and low-mass clusters from the TNG50 and TNG100 simulations. They found that up to 95% of (satellite) disc galaxies are transformed into non-discs by $z = 0$, with redistribution of stars by tidal shocks during pericentric passages as the dominant mechanism behind the transformation.

Even more fundamentally, simulations have investigated the tidal stripping of dark matter and stars from group satellites. With the caveat that this stripping may be artificially enhanced by numerical artefacts (van den Bosch and Ogiya [321], but see Bahé et al. [31]), the clear prediction is that dark matter stripping far outweighs that of stars: for example, Joshi et al. [77] found in an individual zoom-in simulation of a galaxy group that stellar stripping at a >10% level typically only occurs after the loss of ≈80% of their dark matter halo. This relatively minor role of stellar stripping, also borne out by EAGLE [310] and IllustrisTNG [322], leads to a stellar-to-halo mass relation with a much higher peak ratio ($\approx$0.15) for group satellites than centrals ($\approx$0.02). The most extreme form of satellite stripping, their complete disruption, is even predicted to be somewhat more common in groups than clusters ($\approx$65 vs. 50% of all accreted satellites with a total mass of $\sim 10^{12}\,\mathrm{M}_{\odot}$), due to the higher efficiency of dynamical friction driving satellites towards their dense centers [31].

Finally, we note that simulations are also increasingly demonstrating the importance of groups for the evolution of *cluster* satellites: around 50% of all $z = 0$ satellites in massive clusters of the TNG300 simulation were quenched in a group before being accreted onto their final host [34]; Pallero et al. [323] came to a similar conclusion with the Hydrangea/C-EAGLE simulations. Similarly, Jung et al. [324] found that almost half of all $M_{\star} \gtrsim 10^{9}\,\mathrm{M}_{\odot}$ galaxies accreted onto clusters as (group) satellites are already gas poor at the time of cluster infall, compared to only 6% of central galaxies. Even where groups and their galaxies have not (yet) joined a cluster, they therefore contribute to the large-scale environmental dependence of galaxy properties far beyond the "edge" of the cluster [325].

4.5. Simulating the Impact of Galaxy Group Astrophysics on Large-Scale Structure Cosmology

Measurements of the growth of large-scale structure (LSS) can provide powerful tests of our cosmological framework [326–330]. Importantly, they are independent of, and complementary to, constraints from analyses of fluctuations in the cosmic microwave background (CMB) and geometric probes, such as Type Ia SNe and baryon acoustic oscillations (BAOs). Generally speaking, the different LSS tests (e.g., Sunyaev-Zel'dovich power spectrum, cosmic shear, group and cluster number counts, redshift space distortions, etc.) are just different ways of characterising the 'lumpiness' of the matter distribution on different scales. On very large scales, perturbation theory is sufficiently accurate to calculate this distribution reliably. However, most existing LSS tests probe well into the non-linear regime. The standard approach is therefore either to calibrate the 'halo model' (e.g., HMcode package; [331]) using large N-body cosmological simulations, or to use such simulations to correct linear theory empirically (e.g., HALOFIT package; [332]).

If the matter in the universe were composed entirely of dark matter, these approaches would likely be sufficient. However, baryons contribute a significant fraction of the matter density and work by a number of different groups has shown, using cosmological hydrodynamical simulations, that feedback processes associated with galaxy formation can have a significant effect on the matter distribution on scales of up to a few tens of megaparsecs [66,333–335]. Therefore, while such effects are typically ignored or treated in a simple way as a first step when modelling LSS data, it is of critical importance to understand their impact, as they can introduce significant biases in the inferred cosmological parameters in upcoming surveys if no action is taken (e.g., [70,336,337]).

Galaxy groups (taken here to be bound systems with total masses of $\sim 10^{13-14}\,\mathrm{M}_{\odot}$) play a particularly important role in LSS cosmology. This is simply because a sizeable fraction of the galaxies, baryons, and overall matter in the Universe resides in groups. Consequently, LSS cosmology tests that probe more 'typical' environments (such as cosmic shear, galaxy-galaxy lensing, galaxy clustering, redshift space distortions, CMB lensing, etc.), as opposed to tests that sample only the most massive systems (such as the SZ effect power spectrum and current cluster count surveys), will be sensitive to the abundance of galaxy groups and the spatial and kinematical distributions of matter within and around them.

A quantitative demonstration of the importance of the galaxy groups on LSS cosmology can be provided by examining the contribution by halo mass to the total matter power spectrum, $P(k)$. Note that, at present, virtually all current LSS tests probe cosmology through its effects on the matter power spectrum. In the left panel of Figure 16 we show the contribution to the dimensionless matter power spectrum by halos of different mass, as calculated in Mead et al. [338] using the HMCODE halo model. Note that the dimensionless matter power spectrum, $\Delta^2(k)$, is related to $P(k)$ via a multiplicative factor

$4\pi(k/2\pi)^3$ and k is the wavenumber related to the comoving size scale (λ) by $k = 2\pi/\lambda$. The curves show the resulting power spectrum when integrated up to different choices for the maximum halo mass. The results demonstrate that halos corresponding to galaxy groups contribute the majority of the power on the scales relevant for most LSS probes (typically $k \lesssim 10\ h\mathrm{Mpc}^{-1}$). This conclusion is consistent with previous simulation-based findings presented in van Daalen and Schaye [339].

A clear ramification of groups contributing a large fraction of the signal to current LSS tests of cosmology is that theoretical models/simulations must be able to predict the abundance of groups and the matter distribution within them to a very high level of precision on average. For example, the upcoming Rubin Observatory (formerly LSST), Euclid, and Roman Space Telescope surveys are expected to measure the matter power spectrum to better than a few percent accuracy over a very wide range of scales, implying that the theoretical uncertainties in predicting $P(k)$ should be smaller than this to avoid biasing cosmological parameter constraints (e.g., Huterer and Takada [340], Hearin et al. [341]). As already noted, because baryons contribute a non-negligible fraction of the matter density, this means an accurate theoretical description of the baryons within groups (and their back reaction on the dark matter) is also required.

Given the complexity of the physical processes involved in setting the thermodynamic properties of the IGrM and the difficulty in simulating the full range of scales at play (see Section 3), the prospects for accurately (to typically percent level) describing the impact of baryons and group astrophysics on LSS would at first sight seem daunting, if not altogether hopeless at present. Indeed, previous simulation work has shown that variations of the parameters associated with the efficiencies of feedback processes even within plausible bounds can lead to relatively large differences in the predicted properties of groups (e.g., [7,132,133,342]). Variations in resolution and method of solving the hydrodynamic equations may also produce important changes (e.g., [84]), though they are arguably of secondary importance compared to changes in the subgrid modelling associated with feedback (e.g., [85]). A consequence of these large simulation-to-simulation variations in the predicted properties of groups are relatively large study-to-study variations in the predicted impact of baryons on the matter power spectrum (see the simulation comparisons in [343,344]).

The study-to-study variation in the predicted properties of groups and the impact of baryons on $P(k)$ is not unexpected. It is a consequence of not being able to derive the efficiencies for the relevant feedback processes from first principles (see discussion in [39]). This problem is made particularly challenging in the context of simulations with finite resolution and approximations for other (coupled) physical phenomena. As we cannot derive the efficiencies from first principles, the feedback in simulations must be *calibrated* in order to ensure they reproduce particular quantities, after which the realism of the simulations can be tested against other, independent quantities. For LSS cosmology, the main problem we are trying to solve is to accurately model the impact of baryons on $P(k)$. One way this could be achieved is to directly measure $P(k)$ from observations (e.g., via cosmic shear or galaxy-galaxy lensing and galaxy clustering) and compare this with the $P(k)$ predicted in the absence of baryon physics to measure the impact of baryons. However, such an approach is generally a non-starter, as to predict $P(k)$ in the absence of baryons requires that we assume a cosmology and therefore the process of deriving the impact of baryons on $P(k)$ becomes explicitly dependent on cosmology and we would have adopted a circular line of reasoning in our aim to constrain cosmology with LSS measurements.

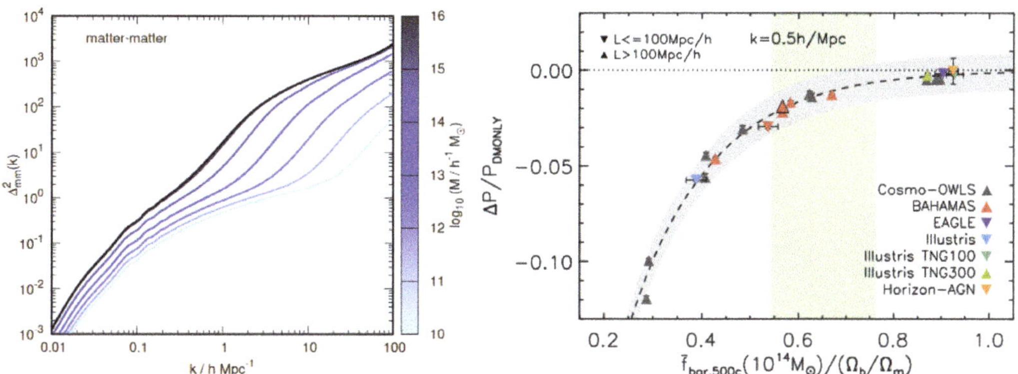

Figure 16. (**Left**) The contribution to the dimensionless matter power spectrum for halos of different mass. The curves show the resulting power spectrum when integrated up to different choices of the maximum halo mass. Here we can see that halos corresponding to galaxy groups (10^{13-14} M$_\odot$) contribute a very large fraction of the power over the range of wavenumbers probed by LSS measurements (typically $k \lesssim 10\,h$Mpc^{-1}). This figure was reproduced with permission from Mead et al. [338]. (**Right**) The effect of galaxy formation on the matter power spectrum at the scale $k = 0.5\,h$Mpc^{-1} as a function of the mean normalized baryon fraction in $\sim 10^{14}$ M$_\odot$ halos. Shown are simulations from cosmo-OWLS and BAHAMAS (grey and red), EAGLE (purple), Illustris (blue), TNG100 and TNG300 (cyan and green) and Horizon-AGN (orange). Baryon fractions were calculated within r_{500c} for halos in the mass range $M_{500c} = [6 \times 10^{13}, 2 \times 10^{14}]$ M$_\odot$. The dashed curve shows that at this k, a simple exponential function of the baryon fraction fits the predictions for the suppression of power of all simulations to within 1% (grey band). The vertical green band shows a range of mean group-scale baryon fractions roughly consistent with observations. These results demonstrate that the differences in the simulation predictions for the impact of baryons on $P(k)$ can be understood to high accuracy based on differences in the baryon fraction on the group scale and that calibration to observed baryon fractions is a promising tool for constraining feedback processes in the simulations. This figure was reproduced with permission from van Daalen et al. [344].

An alternative approach is to modify the gravity-only predictions with simple, physically-motivated prescriptions for baryon physics that have some number of associated free parameters and to *jointly* constrain the cosmological and feedback parameters through comparisons to LSS observables. Examples of this approach include the halo model approach HMCODE of Mead et al. [331,338] and the 'baryonification' approaches (which directly modifies the outputs of gravity-only simulations) of Schneider and Teyssier [334], Schneider et al. [345] and Aricò et al. [346]. The advantages of these approaches include: (i) they are considerably cheaper than running full cosmological hydrodynamical simulations; (ii) the baryon prescriptions are generally flexible and easy to adjust/improve; (iii) they can be straightforwardly incorporated within existing pipelines based on gravity-only simulations or the halo model. The disadvantage of these approaches are that the modeling of baryon physics and its back reaction on dark matter is simplistic and generally not self-consistent and that, unless the impact of baryons is very different than the cosmological variations being explored, one expects there to be important degeneracies between the cosmological and feedback 'nuisance' parameters and a degrading of cosmological constraining power (due to marginalization over uncertain baryon physics).

What about using cosmological hydrodynamical simulations directly for LSS cosmology predictions? The approach of the BAHAMAS program [41,347] (see also the recent FABLE simulations; Henden et al. [44]), is to explicitly calibrate the feedback efficiencies so that they reproduce the observed baryon fractions of galaxy groups. Except for an explicit dependence on the universal fraction, Ω_b/Ω_m, the baryon fractions of groups should be insensitive to changes in cosmology [348] and therefore they represent a highly useful metric on which to calibrate. Furthermore, since the growth of matter fluctuations is fundamentally a gravitational process, by ensuring the simulations have the correct baryon fractions on the scale of groups and gravitational forces are computed self-consistently, the impact of baryons on $P(k)$ should be strongly constrained by this calibration approach. Indeed, van Daalen et al. [344] have recently shown that one can understand the differences in the predicted impact of baryons on $P(k)$ from different simulations *at the percent level* in terms of the differences in baryon fraction in the various simulations at a mass scale of $\sim 10^{14}$ $M_{\odot}$ (see right panel of Figure 16). It is worth noting here that the simulations analysed in that study varied in resolution by more than a factor of 1000 in mass, used different hydro solvers, and assumed different baseline cosmologies.

The implication of this recent development is that, through calibration of feedback efficiencies on the observed baryon fractions of galaxy groups, we strongly limit the ways in which feedback can affect $P(k)$ (in other words we have a very strong *prior* on the impact of baryons). This, in turn, means much stronger (and more robust) cosmological constraints from LSS. However, before claiming victory, a number of important issues require further attention. Firstly, as the calibration is reliant on observations of galaxy groups, the uncertainties in the observed baryon fractions need to be properly included in any cosmological analysis. It goes without saying that the selection function of the group calibration data set must be also reasonably well understood and accounted for (otherwise we risk miscalibrating the feedback). In addition, while the results of van Daalen et al. [344] look very promising, we need to explicitly verify that different kinds of simulations (e.g., that vary how feedback, star formation, and so forth are implemented, resolution, how the hydro equations are solved, etc.) that are calibrated in the same way to the same precision actually produce the same $P(k)$. In other words, we need to check whether differences in the details and evolution of the simulations affect the end state (e.g., $P(k)$ at $z = 0$) if some aspect of that end state has an imposed boundary condition (the baryon fractions at $z = 0$). We also have precious few observational constraints on the baryon fractions of groups beyond $z \sim 0.3$ and it is therefore unclear to what extent the simulations calibrated at $z = 0$ remain 'well behaved' at significantly earlier times. Finally, the level of degeneracy between the various cosmological parameters and the parameters governing feedback efficiency needs to be fully quantified. As already noted, there is an explicit dependence on the universal baryon fraction, Ω_b/Ω_m, but there may well be other less obvious degeneracies that require understanding in order to obtain percent level constraints on parameters such as the dark energy equation of state. Thus, while a promising start has been made in addressing this complex challenge, a number of important steps remain in order to fully account for the impact of group astrophysics on high-precision LSS cosmology.

Finally, we note that, in this section, we have discussed the impact of group astrophysics on LSS cosmology, with a focus on the non-linear matter power spectrum, $P(k)$, which is the basis of many precision LSS probes of cosmology including cosmic shear, CMB lensing, galaxy clustering, and so on. We have not specifically discussed the impact of group astrophysics on attempts to use the abundances (number counts) of galaxy group themselves to constrain cosmology. Of course, one advantage that groups have over clusters in this regard is that they are much more numerous, potentially allowing for stronger cosmological constraints than what might be obtained by clusters alone (e.g., [349–352]). The challenge is that they are more difficult to detect and to model, as already discussed. These issues are particularly pertinent for group counts, as they affect the cosmological observable in a much more direct way than, for example, cosmic shear. Nevertheless, as our observational picture and ability to model galaxy groups improves, we expect galaxy group counts to play an increasingly important probe of cosmology and one that is complementary to constraints coming from cosmic shear, CMB lensing, and other LSS probes.

5. Future Directions

5.1. Using Simulations to Make Predictions for Existing and Future Missions and Telescopes

Cosmological simulations are widely used to make predictions for future missions, often in such a capacity that a mission's approval or rejection may hinge on these predictions. It is therefore incumbent upon simulators to generate mock observations that do not suffer from numerical effects, poorly implemented modules, or insufficient resolution. However, simulation predictions that are later refuted are not necessarily the result of numerical problems: instead, such discrepancies may also reveal new physical processes. This applies in particular to the complex interplay of non-gravitational physics and dynamics in groups.

A near-term example relevant for groups covers predictions of X-ray line emission from the EAGLE simulation that should be observed by several future missions, including *XRISM* to be launched in 2022. Figure 17 from Wijers et al. (in prep) demonstrates that *XRISM* should be able to detect O VII and Mg XII emission tracing the approximate temperatures of virialized IGrM gas. This prediction follows on from the "virial temperature thermometer" model of Oppenheimer et al. [76] and Wijers et al. [353] that specific metal ions should trace the volume-filling virialized halo gas corresponding to the temperature of the ion's peak collisional ionization fraction. While (UV-band) O VI traces outer virialized galactic halo gas at 3×10^5 K, O VII and O VIII in the X-ray spectrum are predicted to trace $\geq 10^6$ K IGrM in groups. Figure 8 of [353] predict that O VIII absorption is strongest in poor groups with $M_{500} \sim 10^{13}$ $M_\odot$ and Fe XVII in intermediate groups with $M_{500} \sim 10^{13.5} M_\odot$.

XRISM should measure the significant metal and (for an assumed metallicity), baryon contents of the IGrM out to 100 kpc from the central galaxy as predicted by the EAGLE simulations in Figure 17. Micro-calorimeters on *Athena* and *Lynx* should be able to resolve the interior metal emission of the IGrM at superb (<10 eV) spectral resolution and signal to noise ratio. If, however, *XRISM* does *not* detect O VIII or Mg XII at the levels predicted in this figure, this might point to one of the following scenarios. Firstly, the IGrM might not be as metal-enriched as EAGLE predicts, for example, because the nucleosynthetic yields assumed by the simulation are too high, or due to a higher-than-predicted fraction of metals being retained within galaxies. Secondly, EAGLE might over-predict the IGrM baryon content, and hence the gas density and emission line luminosity [13]. A third possibility is that metals in real groups are distributed more (or in principle also less) homogeneously than in EAGLE, which would affect the radially averaged cooling rates, and hence line emission luminosities—metal-line emission is highly sensitive to the distribution of gas density and metallicity distributions on both the macro and micro IGrM scales. Finally, non-equilibrium ionization [94] and/or dual temperature electron-ion plasmas [354] could alter simulation predictions that almost always assume ionization equilibrium and equipartition between electrons and ions.

When simulators generate mock observations, it is important to present their predictions in a manner that accurately showcases the capabilities of existing and proposed instruments. Often it is advisable to work with instrumentalists and observers. An example of an attempt to fairly compare capabilities of existing and future missions appears in Figure 18 for a TNG100 $M_{500} = 10^{13.6}$ $M_\odot$ halo at $z = 0.05$. Each panel shows a mock 100ksec X-ray image based on the instrument capability at launch (Cycle 0 for *Chandra*), but the leftmost panels show mock observations without any noise. We apply a forward modeling technique using the packages pyXSIM [14] [357] and SOXS [15]; the SIXTE simulation software [358] is used to create the *eROSITA* mock.

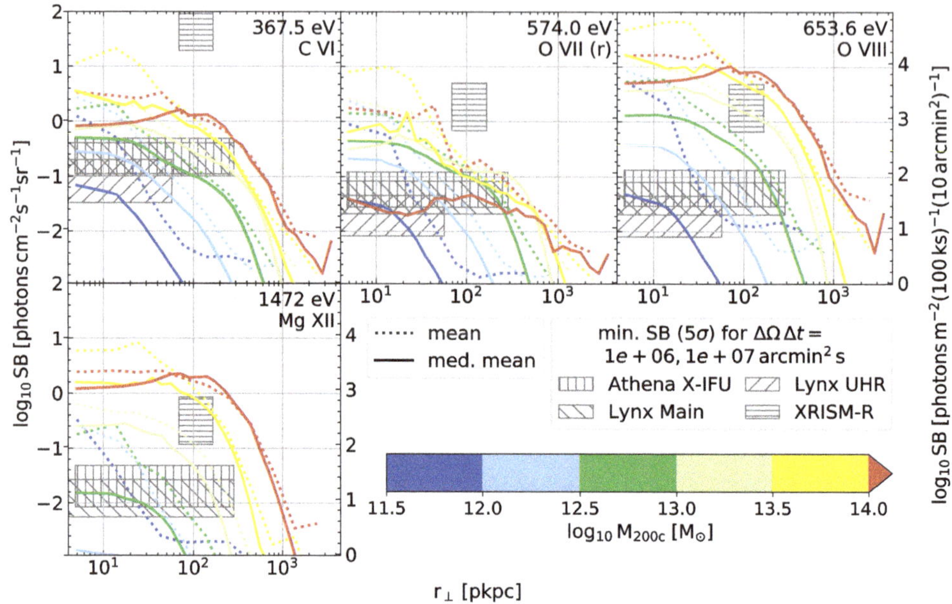

Figure 17. X-ray line emission predictions from the EAGLE simulation via Nastasha Wijers (in prep.) demonstrating the detectability by *XRISM*, *Athena*, & *Lynx* space-borne micro-calorimeters as a function of impact parameter from the central galaxy. Lines are colored by halo mass with light green and yellow corresponding approximately to low-mass and high-mass groups. Faded vertical stripes indicate R_{200} for each halo mass. High-mass groups are detectable by *XRISM* in O VIII and Mg XII at 100 kpc, while low-mass groups are detectable only in O VIII. *Athena* and *Lynx* should measure IGrM line emission for all group halos with high spatial resolution.

We choose surface brightness units (counts s^{-1} $arcmin^{-2}$) to show the relative throughput of the detectors. At launch, *Chandra* was able to detect groups out to R_{500} in $\lesssim$ 100 ksec, which appears consistent with the longest exposed groups of Sun et al. [2]. *eROSITA* can detect extended emission out to a good fraction of R_{500}, although it will only approach such exposures at the ecliptic poles during its eRASS:8 4-year survey [359], and will require targeted follow-up on most groups in the sky to achieve such depth. Biffi et al. [360] simulated *eROSITA* observations of clusters/groups between z = 0.1–2.0 from the Magneticum Box2/hr run, inputting AGN to determine if the underlying ICM/IGrM emission can be separated from AGN contamination. Oppenheimer et al. [361] simulated the stacking of galactic halos assuming 2 ksec exposures aimed at 50 $z = 0.01$ galaxies with an average halo mass of $M_{500} = 10^{12.5}$ $M_\odot$ to show that the eRASS:8 survey should be able to resolve the stacked profile out beyond 100 kpc. Their forward modeling analysis used EAGLE and TNG100 galaxies as inputs, added in noise and attempted to subtract it, and excised mock point sources from the cosmic X-ray background. However, Biffi et al. [360] and Oppenheimer et al. [361] still did not attempt to mock the scanning mode of the 8 individual all-sky surveys, instead assuming single pointed exposures with the object placed at the center, which likely under-estimates sources of systematic errors. Future data collected from *eROSITA* will need to be compared to simulations applying the same scanning exposures used by the eRASS:8 survey.

Continuing on the lower panels of Figure 18, *Athena* will be able to distinguish IGrM structure associated with the central galaxy and satellites. Finally, *Lynx* should be able to clearly resolve azimuthal dependence, a variety of satellite interactions, and sharp shock fronts bow shocks associated with infalling satellite as seen to the left of the simulated central.

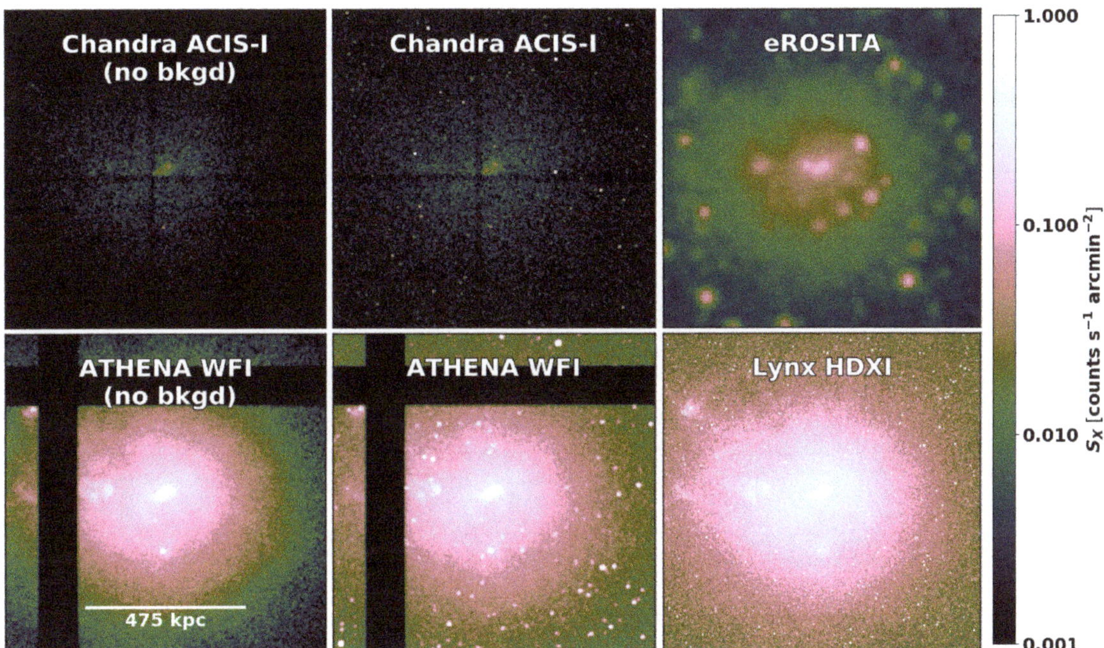

Figure 18. Simulated 100 ksec surface brightness maps of the same TNG100 $M_{500} = 10^{13.6}$ M$_\odot$ group placed at $z = 0.05$ by CCD detectors on *Chandra* (Cycle 0 capability), *eROSITA*, *Athena*, & *Lynx*. The two left panels show *Chandra* and *Athena* mocks without noise, while all the other panels add noise from the instrument, the Milky Way foreground, and a randomly generated cosmic X-ray background (CXB). The field of view covers out to R_{500}. All detectors use the color scale on the right. A. Simionescu collaborated on the production of this figure. Note: the CXB is differently randomly generated in each panel.

5.2. Observations and Simulations in Support of Each Other

It is clear that the increasing wealth of observational data on galaxy groups has led to continual improvement in cosmological simulations, as we gain increased knowledge about the physical processes at play and constraints on their 'efficiencies', both of which help inform the simulations. However, an important caveat to bear in mind is that observations themselves are subject to considerable uncertainties which should not be ignored when comparing with simulations, particularly if one intends to calibrate aspects of the simulations on said observations. While good strides have been made in generating realistic synthetic X-ray, thermal Sunyaev-Zel'dovich (tSZ), optical, and so forth, 'observations' of simulated groups to enable like-with-like comparisons (e.g., [7,357,362–364]), to date much less attention has been devoted to ensuring a consistent method of *selection*. While the use of mock catalogs is standard practice in galaxy surveys to quantify the selection function (e.g., [365–367]), the use of realistic mocks to quantify the selection function of groups selected on the basis of their hot gas properties (particularly X-ray and tSZ) has lagged behind. Instead very simplistic models (e.g., spherical beta models) are still regularly employed in characterizing the selection function. There is also often a key difference in the way observers speak about the selection function (which they normally cast in terms of observable quantities such as flux, surface brightness, or signal-to-noise ratio) and what a theorist or simulator would regard as the selection function (which is almost always with respect to halo mass). Ultimately, what is required is an iterative process involving simulations and observations, whereby mock surveys of the simulations are used to inform a consistent definition of the selection function. A like-with-like comparison is then made between the simulations and observations, shortcomings of the simulations are identified, new simulations are produced, and the cycle repeats. Each cycle yields not only an improved simulation and physical picture (hopefully), but also informs our knowledge of how biased observational methods are in selecting groups and estimating their physical parameters. Of course, the realism of the simulations must also be tested against independent observations which are not part of the calibration process (e.g., evolution of galaxy groups, environmental effects on galaxies, etc.).

While the heterogeneous nature of most pointed X-ray observations with *Chandra* and *XMM-Newton* do not naturally lend themselves to the development of simple selection functions, upcoming *eROSITA* observations should be much more tractable in this regard, given the homogeneity of the survey. Likewise, wide-field tSZ observations should benefit from mock surveys based on large hydrodynamical simulations, to complement existing work based on simple spatial templates (e.g., [368]), which are not expected to hold deep into the group regime [369].

5.3. *Timely Research Topics for Simulations of Groups*

Arguably, simulations of 10^{13}–10^{14} $M_\odot$ halos have so far received less attention than neighboring halo mass ranges, which might be due to the current difficulty in the observational identification and characterization of groups, as well as the complexity in their theoretical modeling. On the observational side, breakthroughs are imminent on multiple fronts: deep and highly complete spectroscopic galaxy redshift surveys such as the 4MOST Wide-Field Vista Extragalactic Survey (WAVES) will deliver robust group catalogues; IGrM probes with linear dependence on gas density—such as UV and X-ray absorption, kinetic Sunyaev-Zeldovich (kSZ), and fast radio bursts [370,371]—instead of the ρ^2 scaling of X-ray emission will map the diffuse IGrM out to large radii. Below, we therefore list a selection of timely open research topics in the field of group simulations.

- **The relationship between the central SMBH and properties of the IGrM/ICM/CGM:** The strong correlations between the mass of the central SMBH in a group ($M_{\rm SMBH}$) and the temperature and X-ray luminosity (T_X and L_X) of the IGrM/ICM [372,373], provide fundamental tests of AGN feedback in simulations (see also Section 4 of the companion review by Lovisari et al. [1]). On the one hand, the $M_{\rm SMBH}$ scaling with $M_{\rm halo}$, assuming T_X measures $M_{\rm halo}$, is a natural expectation of the SMBH growth being controlled by the binding energy of the halo [374]. The EAGLE simulation prediction that L_X scales inversely with $M_{\rm SMBH}$ *at fixed halo mass* and most strongly for the CGM (i.e., galactic halo masses below the group scale; Davies et al. [162] suggests the opposite, inverse trend, which is also seen in TNG100 [140]. As the latter paper explains, the CGM L_X is reduced in response to the integrated SMBH feedback lifting baryons out of galaxy halos, lowering the density, and significantly increasing cooling times. However, by group masses, the Gaspari et al. [372] correlations appear reproduced by TNG100 with $T_X(< R_{500})$ showing surprisingly little scatter for quenched galaxies with $M_{\rm SMBH} > 10^{8.2}$ $M_\odot$ [375]. This link between $M_{\rm BH}$ and halo-wide T_X, which is the best X-ray observational proxy for halo mass, indicates that simulations predict a fundamental relationship between $M_{\rm BH}$ and $M_{\rm halo}$ transmitted through the virialization of halo gas. The nature of this relationship contains both the virial temperature being set by hierarchical growth of group/cluster-scale halos (as discussed by [375]), and the mechanisms of gas accretion and AGN feedback determining SMBH growth, which span scales from the SMBH radius to $R_{\rm vir}$ (as discussed in Section 4.1 of the companion review by [78]. and in [376]). Bassini et al. [377] explored GADGET-3 cluster zoom simulations, finding that they were able to reproduce the observed $T_{500} - M_{\rm SMBH}$ and other correlations in the group and cluster regime. We emphasize the need for more simulations to explore the rich and diverse astrophysics contained in the relationship between group properties and their central SMBHs.
- **Cooling flows or cold rain?:** One of the key phenomenon that links the IGrM/ICM to the BGG/BCG and ultimately, to the SMBH hosted by the central galaxy is the flow of gas from the former to the latter two, particularly in CC clusters. During the cooling phase, the conventional view is that the gas typically flows inwards subsonically and *en*

masse, meaning that if gas is multiphased, then all phases move inwards in a comoving fashion. This is how cooling flows were originally conceived ([378]; see also reviews by [379,380]); this is what pre-AGN feedback simulations found (see Figures 17 and 19 of [95]); and recently, this is how the inflow is thought to behave during times when the central AGN is quiescent (or in quadrants about the cluster center where cooling is dominant). Recent very high resolution simulations of idealized galaxy groups and clusters [150,381–386] find that when the ratio $t_{\rm cool}/t_{\rm ff}$ in a cooling group/cluster core drops below some threshold (nominally $\sim$10), local density perturbations can become thermally unstable [288,290], leading to the formation of cold dense clouds. These clouds then separate from the rest of the CGM and stochastically rain down upon central galaxy and its SMBH. This "cold rain" fuels both star formation events as well as AGN outbursts.

There are number of problems with the conventional Bondi accretion model indicating that it is untenable (see, e.g., [384,385,387,388]), which the cold rain model appears to resolve. We argue that this warrants further investigation of the cold rain model within the context cosmological hydrodynamics simulations of the formation/evolution of massive galaxies, groups, and clusters. Two potential directions of study stand out: Firstly, current insights about the cold rain phenomenon come from idealized simulations that neither have satellite galaxies moving through the IGrM/ICM and inducing perturbations in their wakes, nor do they allow for interactions, like ram pressure stripping of these satellite galaxies. How these complications alter the thermal instability/cold rain picture remains unexplored. Secondly, it is not currently feasible to directly model the cold rain phenomenon in cosmological hydrodynamics simulations because that would require being able to resolve spatial scales approaching $\sim$1 pc (see [389] for further details). This however means that there is an opportunity for developing innovative subgrid models that can capture the most important elements of the cold rain model. There is precedence for the second option in that the torque-limited accretion subgrid model of Anglés-Alcázar et al. [153] was created to encompass idealized simulations of gas-rich accretion discs from 10 kpc to 0.1 pc [390].

- **New models for AGN feedback:** Non-spherical, jet-like feedback appears necessary to impart energy to the IGrM/ICM while not over-evacuating the inner region, as discussed in Section 4 of the Lovisari et al. [1] companion review where they show in their Figure 7 results from an idealized simulation by Gaspari et al. [391] demonstrating a self-regulated jet capable of preserving the cool core (see also [382–386]). Collimated jet feedback is currently inadequately modeled in cosmological simulations; however, thermal blast feedback should not be dismissed as a potential mode operating at late times until it is confirmed that cored NCC underluminous groups do not exist. Meece et al. [392] found a hybrid kinetic jet with thermal heating in idealized hydro simulations could best achieve self-regulation and produce a cool core, whereas a thermal-only jet results in a cored profile that rapidly radiates energy away leading to a cooling catastrophe. Their kinetic-only model also achieves self-regulation, but appears too steady compared to observed AGN duty cycles.

  The failure of cosmological simulations to reproduce the observed thermal structure of the IGrM (Section 4.1.2) may be related to cosmological simulation's inability to model narrow, high momentum flux, jet outflows. Narrow beams are better able to drill their way out through a highly pressurized IGrM/ICM that can easily stall an isotropic outflow, preventing the deposition of energy where it is needed. Nonetheless, narrow outflows have their own challenges, which become apparent through running idealized simulations. Firstly, as shown by Vernaleo and Reynolds [393] and confirmed by Cielo et al. [394], jets that fire in a fixed direction tend to deposit their energy at increasing larger distances and ultimately, end up doing so beyond the group/cluster core. As a result, such jets only delay the onset of catastrophic cooling, not prevent it. The second and equally vexing problem concerns the coupling between the narrow jets and the IGrM/ICM: how do narrow, bipolar jets manage to heat gas in the group/cluster cores in a near-isotropic fashion? This motivated Babul et al. [395] to argue for tilting jets, which change direction every so often as evidenced by observations detailed in this paper. Cielo et al. [394]—and most recently, Su et al. [396]—found that not only is tilting necessary, but the angle between jet events must also be reasonably large, and furthermore heated jets work better than cold jets. In effect, the desired outflows are those that have the appropriate energy/momentum flux, create near-spherical cocoons because these optimize energy transfer in transverse directions relatively to the jets, and in a time-averaged sense, distribute their energy in a near-isotropic fashion within the group/cluster core.

- **X-ray detectability of new classes of groups:** The largest number of diffuse object detections by *eROSITA* will be groups (e.g., [352]). The complete eRASS:8 survey should detect groups with $M_{500} > 10^{13}$ M$_\odot$ out to $z = 0.05$ [397]. While simulations of EAGLE and TNG100 eRASS:8 stacking show galactic-scale halos at $M_{500} < 10^{13}$ M$_\odot$ will not be individually detected [361], *eROSITA* should observe groups in the local volume covered by CLoGS. Simulations will provide necessary guidance in the interpretation of any prospective cored NCC under-luminous groups and/or coalescing groups that have yet to virialize. It may well be that significantly under-luminous groups potentially exist as Pearson et al. [398] cannot detect with *Chandra* two of their 10 optically selected groups, which do not show signs of being unvirialized. The near future holds promise to detect new potential classes of groups—poor, under-luminous, and coalescing—in X-rays.
- **Multi-phase gas stripping from group satellites:** A shortcoming in common to all simulations that we have discussed in this review is the lack of a cold ($T \ll 10^4$ K) and dense molecular phase in the ISM of galaxies. Both observations (e.g., [399]) and idealized hydrodynamic simulations (e.g., [400]) clearly indicate that ram pressure has a different effect on the dense molecular phase from which stars are formed than the more tenuous, warmer components traced by H I and H II. Although post-processing can be used to estimate the molecular content of group satellites (albeit with strong assumptions; [307]), it cannot capture the different dynamical evolution of the two phases. Simulations with direct modeling of molecular gas—as is now often done in high-resolution zooms of individual galaxies (e.g., [113,299])—would therefore reveal a fundamentally new aspect of the interaction between the IGrM and satellite galaxies. Recent advances in subgrid cooling models [401] make such large-scale cold ISM simulations possible, but the high resolution required to resolve giant molecular clouds at least marginally ($\lesssim 10^4$ M$_\odot$) makes them unfeasible on cluster scales for the foreseeable future. Galaxy groups, on the other hand, would be perfectly suited to exploring this additional facet of the baryon cycle in a full cosmological setting.
- **The Sunyaev-Zel'dovich Effect:** SZ stacking is already measuring the pressure and density profiles of groups from large radii inward. Cross-correlating large spectroscopic surveys (e.g., BOSS [402]) with high-resolution maps of the CMB from the Atacama Cosmology Telescope (ACT) has measured the extended pressure and density profiles of groups via the tSZ and kSZ effect respectively. Amodeo et al. [403] detected elevated gas pressure profiles outside R_{200} of $z = 0.55$ $M_{200} = 10^{13.5}$ M$_\odot$ groups indicating that feedback energy equivalent to double the gaseous halo binding energy needs to coupled directly to the IGrM, which is significantly above predictions from TNG100 simulations and even more so the EAGLE simulations [140]. Schaan et al. [404] showed that $z = 0.31$ $M_{200} = 10^{13.7}$ M$_\odot$ groups are far more devoid of baryons in kSZ measurements than a Navarro et al. [405] (NFW) profile. Lim et al. [406] tested groups in Illustris, EAGLE, TNG300, and Magneticum simulations against Planck Collaboration et al. [407] stacks, finding that the $M_{500} \sim 10^{13.0-13.5}$ M$_\odot$ scale provides a very promising scale to constrain the nature of AGN feedback. The measurements of pressure, density, and, through division, temperature

profiles of groups will dramatically increase in the 2020's as the Rubin Telescope comes on line and the Roman and Euclid Telescopes are launched, providing spectroscopic surveys to cross-correlate further CMB observations from the ACT, the Simons Observatory, the Large Millimeter Telescope, and CMB-S4. These future SZ surveys will provide standard calibrations against which simulated groups are compared.

6. Final Statement

Cosmological hydrodynamic simulations provide an aggregate picture of groups that is not yet available observationally. Some of the latest state-of-the-art simulation projects can reproduce key stellar observations of galaxies, while others are able to match essential gaseous properties of clusters and large scale structure. We approach the breakthrough when a single high-resolution cosmological simulation suite can match gaseous and stellar properties of both galaxies and clusters. The simulated groups in the intermediate range provide true testable predictions for upcoming observational datasets that include comprehensive galaxy surveys, all-sky X-ray maps, and deep radio surveys. Future observed group datasets synthesizing galaxy catalogues down to dwarf galaxies, X-ray emission extending beyond R_{500} and to masses below $M_{500} < 10^{13.5}\ \mathrm{M}_{\odot}$, thermal and kinetic Sunyaev-Zel'dovich measurements, and UV absorption compilations plus 21-cm maps accounting for warm and cool gas component will provide creative new stress tests of the non-gravitational, baryonic physics in cosmological simulations.

Author Contributions: B.D.O.: lead author Sections 1, 2, 3.1–3.8, 4.1, 5.1, 5.3 and 6; A.B.: lead author Section 4.2, major contributions to Sections 4.1 and 5.3; Y.B.: lead author Section 4.4, major contributions to Sections 1–3 and 4.1; I.S.B.: lead author Sections 3.10 and 4.3; I.G.M.: lead author Sections 3.9, 4.5 and 5.2. All authors have read and agreed to the published version of the manuscript.

Funding: This research received no external funding.

Data Availability Statement: Much of the simulation data used here are publicly available, including for the EAGLE [408], IllustrisTNG [409], and SIMBA (http://simba.roe.ac.uk/ (accessed on 10 June 2021)) datasets. Figures uniquely generated for this review by the authors include Figures 3–5, 6 (left panel), and Figures 12–18.

Acknowledgments: We are grateful to the three anonymous referees who provided thorough reports that substantially improved the scope and clarity of this review. We thank the following astrophysicists for thought-provoking conversations, observational and simulation data and figures, and essential guidance in the production of this review: Sarah Appleby, Michelle Cluver, Weiguang Cui, Romeel Davé, Dominique Eckert, Seoyoung Lyla Jung, Amandine Le Brun, Ilani Loubser, Lorenzo Lovisari, Daisuke Nagai, Ewan O'Sullivan, Douglas Rennehan, Vida Saeedzadeh, Zhiwei Shao, Prateek Sharma, Aurora Simionescu, Ming Sun, Michael Tremmel, Nastasha Wijers, and Mark Voit. This research was supported in part by the KITP National Science Foundation under Grant No. NSF PHY-1748958, and by the Netherlands Organization for Scientific Research (NWO) through Veni grant number 639.041.751. This review is reliant on the public data release and accessibility generously provided by the EAGLE, IllustrisTNG, and SIMBA teams. A number of calculations and figures presented in this review were done using (i) high performance computing facilities at Liverpool John Moores University, partly funded by the Royal Society and LJMU's Faculty of Engineering and Technology, and (ii) advanced research computing resources provided by Compute/Calcul Canada. I.G.M. acknowledges support by the European Research Council (ERC) under the European Union's Horizon 2020 research and innovation programme (grant agreement No. 769130). A.B. acknowledges research support from Natural Sciences and Engineering Research Council of Canada (NSERC) and Compute Canada.

Conflicts of Interest: The authors declare no conflict of interest.

Notes

1. In the remainder of this review, we omit the 'c' suffix that identifies the overdensity as measured with respect to the critical, rather than for example, mean, density of the universe
2. All conversions are median differences obtained from the IllustrisTNG300 simulation

3 We note that the ability of particles to move through the simulation volume is not particular to this approach, and is also an integral feature of the SPH approach

4 See also Crain et al. [62]. for calibration and McAlpine et al. [63] for public release

5 See also Marinacci et al. [64], Naiman et al. [65], Springel et al. [66]

6 We note the work of Ragagnin et al. [71] who measured the "fossil-ness" of Magneticum groups in the even larger 900 Mpc Box2b/hr volume, which apparently ran to $z = 0$

7 The $\sim R^{1.1}$ scaling of the gas entropy profile is set by gravitational infall-related processes, including accretion shocks and subsequent thermalization. As shown by Lewis et al. [95], it echoes the pseudo-phase space density profile of the dark matter: $\rho(R)/\sigma^3(R) \propto R^{-1.8}$ [97,98]. The entropy scaling is $\propto [\sigma^3/\rho(R)]^{2/3}$

8 IllustrisTNG does include anisotropic thermal conduction (and magnetic fields), though the impact on the evolution on hot gas in groups has not yet, to our knowledge, been examined in detail. However, Barnes et al. [167] have examined the impact of anisotropic thermal conduction on more massive clusters, concluded that it has the effect of making cool cores more prevalent

9 The companion review by Eckert et al. [78], Figures 14 and 15, plots gas fractions, $f_{\rm gas,500}$, as a function of M_{500}, which are more evacuated for a given halo than $f_{\rm gas,200}$ as plotted here. The gas, stellar, and baryon fractions within R_{200} are less well constrained due to group X-ray measurements not extending out to R_{200} yet, and therefore represent predictions for future observations

10 For EAGLE, TNG100, and SIMBA, all profiles shown are at $z = 0$; for ROMULUS we combine five snapshots at $z \leq 0.36$ because the simulation contains only a single halo in this mass range

11 At first sight, this offset may be surprising given the explicit normalization of our profiles by the integrated K_{500}. The reason for this apparent contradiction lies in the definition of K_{500}, which is calculated under the assumption that the baryon fraction within R_{500} is equal to the cosmic average f_b. As shown in Figure 3, the actual gas fraction, and hence electron density, is lower by a factor of up to $\approx$3 at $M_{500} = 10^{13}$ M$_\odot$ even within R_{200}, so that $<K> > K_{500}$

12 readers interested in seeing a more complete set of available SMHMs results are referred to Figure 10 of Coupon et al. [246] or Figure 9 of Girelli et al. [247]

13 As discussed in Section 4.1.1, there is already evidence for this, although at least the central regions of EAGLE groups appear to reproduce current observations (see Figure 5)

14 http://hea-www.cfa.harvard.edu/~jzuhone/pyxsim/ (accessed on 10 June 2021) pyXSIM is an implementation of the PHOX algorithm [355,356]

15 http://hea-www.cfa.harvard.edu/~jzuhone/soxs/ (accessed on 10 June 2021)

References

1. Lovisari, L.; Ettori, S.; Gaspari, M.; Giles, P.A. Scaling Properties of Galaxy Groups. *Universe* **2021**, *7*, 139. [CrossRef]
2. Sun, M.; Voit, G.M.; Donahue, M.; Jones, C.; Forman, W.; Vikhlinin, A. Chandra Studies of the X-Ray Gas Properties of Galaxy Groups. *Astrophys. J.* **2009**, *693*, 1142–1172. [CrossRef]
3. Eckmiller, H.J.; Hudson, D.S.; Reiprich, T.H. Testing the low-mass end of X-ray scaling relations with a sample of Chandra galaxy groups. *Astron. Astrophys.* **2011**, *535*, A105. [CrossRef]
4. Bharadwaj, V.; Reiprich, T.H.; Lovisari, L.; Eckmiller, H.J. Extending the L_X—T relation from clusters to groups. Impact of cool core nature, AGN feedback, and selection effects. *Astron. Astrophys.* **2015**, *573*, A75. [CrossRef]
5. Lovisari, L.; Reiprich, T.H.; Schellenberger, G. Scaling properties of a complete X-ray selected galaxy group sample. *Astron. Astrophys.* **2015**, *573*, A118. [CrossRef]
6. Mitchell, N.L.; McCarthy, I.G.; Bower, R.G.; Theuns, T.; Crain, R.A. On the origin of cores in simulated galaxy clusters. *Mon. Not. R. Astron. Soc.* **2009**, *395*, 180–196. [CrossRef]
7. Le Brun, A.M.C.; McCarthy, I.G.; Schaye, J.; Ponman, T.J. Towards a realistic population of simulated galaxy groups and clusters. *Mon. Not. R. Astron. Soc.* **2014**, *441*, 1270–1290. [CrossRef]
8. O'Sullivan, E.; Ponman, T.J.; Kolokythas, K.; Raychaudhury, S.; Babul, A.; Vrtilek, J.M.; David, L.P.; Giacintucci, S.; Gitti, M.; Haines, C.P. The Complete Local Volume Groups Sample—I. Sample selection and X-ray properties of the high-richness subsample. *Mon. Not. R. Astron. Soc.* **2017**, *472*, 1482–1505. [CrossRef]
9. Cavagnolo, K.W.; Donahue, M.; Voit, G.M.; Sun, M. Intracluster Medium Entropy Profiles for a Chandra Archival Sample of Galaxy Clusters. *Astrophys. J. Suppl.* **2009**, *182*, 12–32. [CrossRef]
10. Pratt, G.W.; Croston, J.H.; Arnaud, M.; Böhringer, H. Galaxy cluster X-ray luminosity scaling relations from a representative local sample (REXCESS). *Astron. Astrophys.* **2009**, *498*, 361–378. [CrossRef]
11. Wetzel, A.R.; Tinker, J.L.; Conroy, C. Galaxy evolution in groups and clusters: Star formation rates, red sequence fractions and the persistent bimodality. *Mon. Not. R. Astron. Soc.* **2012**, *424*, 232–243. [CrossRef]
12. Davies, L.J.M.; Robotham, A.S.G.; Lagos, C.d.P.; Driver, S.P.; Stevens, A.R.H.; Bahé, Y.M.; Alpaslan, M.; Bremer, M.N.; Brown, M.J.I.; Brough, S.; et al. Galaxy and Mass Assembly (GAMA): Environmental quenching of centrals and satellites in groups. *Mon. Not. R. Astron. Soc.* **2019**, *483*, 5444–5458. [CrossRef]

13. Moffett, A.J.; Ingarfield, S.A.; Driver, S.P.; Robotham, A.S.G.; Kelvin, L.S.; Lange, R.; Meštrić, U.; Alpaslan, M.; Baldry, I.K.; Bland-Hawthorn, J.; et al. Galaxy And Mass Assembly (GAMA): The stellar mass budget by galaxy type. *Mon. Not. R. Astron. Soc.* **2016**, *457*, 1308–1319. [CrossRef]
14. Somerville, R.S.; Davé, R. Physical Models of Galaxy Formation in a Cosmological Framework. *Annu. Rev. Astron. Astrophys.* **2015**, *53*, 51–113. [CrossRef]
15. Tumlinson, J.; Peeples, M.S.; Werk, J.K. The Circumgalactic Medium. *Annu. Rev. Astron. Astrophys.* **2017**, *55*, 389–432. [CrossRef]
16. Voit, G.M.; Donahue, M.; Bryan, G.L.; McDonald, M. Regulation of star formation in giant galaxies by precipitation, feedback and conduction. *Nature* **2015**, *519*, 203–206. [CrossRef]
17. Gaspari, M.; Temi, P.; Brighenti, F. Raining on black holes and massive galaxies: The top-down multiphase condensation model. *Mon. Not. R. Astron. Soc.* **2017**, *466*, 677–704. [CrossRef]
18. Emerick, A.; Bryan, G.; Putman, M.E. Warm gas in and around simulated galaxy clusters as probed by absorption lines. *Mon. Not. R. Astron. Soc.* **2015**, *453*, 4051–4069. [CrossRef]
19. Butsky, I.S.; Burchett, J.N.; Nagai, D.; Tremmel, M.; Quinn, T.R.; Werk, J.K. Ultraviolet signatures of the multiphase intracluster and circumgalactic media in the ROMULUSC simulation. *Mon. Not. R. Astron. Soc.* **2019**, *490*, 4292–4306. [CrossRef]
20. McDonald, M.; Veilleux, S.; Mushotzky, R. The Effect of Environment on the Formation of Hα Filaments and Cool Cores in Galaxy Groups and Clusters. *Astrophys. J.* **2011**, *731*, 33. [CrossRef]
21. Tremblay, G.R.; Combes, F.; Oonk, J.B.R.; Russell, H.R.; McDonald, M.A.; Gaspari, M.; Husemann, B.; Nulsen, P.E.J.; McNamara, B.R.; Hamer, S.L.; et al. A Galaxy-scale Fountain of Cold Molecular Gas Pumped by a Black Hole. *Astrophys. J.* **2018**, *865*, 13. [CrossRef]
22. Tonnesen, S.; Bryan, G.L.; Chen, R. How to Light it Up: Simulating Ram-pressure Stripped X-ray Bright Tails. *Astrophys. J.* **2011**, *731*, 98. [CrossRef]
23. Yun, K.; Pillepich, A.; Zinger, E.; Nelson, D.; Donnari, M.; Joshi, G.; Rodriguez-Gomez, V.; Genel, S.; Weinberger, R.; Vogelsberger, M.; et al. Jellyfish galaxies with the IllustrisTNG simulations—I. Gas-stripping phenomena in the full cosmological context. *Mon. Not. R. Astron. Soc.* **2019**, *483*, 1042–1066. [CrossRef]
24. Campitiello, M.G.; Ignesti, A.; Gitti, M.; Brighenti, F.; Radovich, M.; Wolter, A.; Tomičić, N.; Bellhouse, C.; Poggianti, B.M.; Moretti, A.; et al. GASP XXXIV: Unfolding the Thermal Side of Ram Pressure Stripping in the Jellyfish Galaxy JO201. *Astrophys. J.* **2021**, *911*, 144. [CrossRef]
25. McGee, S.L.; Bower, R.G.; Balogh, M.L. Overconsumption, outflows and the quenching of satellite galaxies. *Mon. Not. R. Astron. Soc.* **2014**, *442*, L105–L109. [CrossRef]
26. Burchett, J.N.; Tripp, T.M.; Wang, Q.D.; Willmer, C.N.A.; Bowen, D.V.; Jenkins, E.B. Warm-hot gas in X-ray bright galaxy clusters and the H I-deficient circumgalactic medium in dense environments. *Mon. Not. R. Astron. Soc.* **2018**, *475*, 2067–2085. [CrossRef]
27. Stocke, J.T.; Keeney, B.A.; Danforth, C.W.; Oppenheimer, B.D.; Pratt, C.T.; Berlind, A.A.; Impey, C.; Jannuzi, B. The Ultraviolet Detection of Diffuse Gas in Galaxy Groups. *Astrophys. J. Suppl.* **2019**, *240*, 15. [CrossRef]
28. Borthakur, S.; Yun, M.S.; Verdes-Montenegro, L. Detection of Diffuse Neutral Intragroup Medium in Hickson Compact Groups. *Astrophys. J.* **2010**, *710*, 385–407. [CrossRef]
29. Oman, K.A.; Bahé, Y.M.; Healy, J.; Hess, K.M.; Hudson, M.J.; Verheijen, M.A.W. A homogeneous measurement of the delay between the onsets of gas stripping and star formation quenching in satellite galaxies of groups and clusters. *Mon. Not. R. Astron. Soc.* **2021**, *501*, 5073–5095. [CrossRef]
30. Brown, T.; Catinella, B.; Cortese, L.; Lagos, C.d.P.; Davé, R.; Kilborn, V.; Haynes, M.P.; Giovanelli, R.; Rafieferantsoa, M. Cold gas stripping in satellite galaxies: From pairs to clusters. *Mon. Not. R. Astron. Soc.* **2017**, *466*, 1275–1289. [CrossRef]
31. Bahé, Y.M.; Schaye, J.; Barnes, D.J.; Dalla Vecchia, C.; Kay, S.T.; Bower, R.G.; Hoekstra, H.; McGee, S.L.; Theuns, T. Disruption of satellite galaxies in simulated groups and clusters: The roles of accretion time, baryons, and pre-processing. *Mon. Not. R. Astron. Soc.* **2019**, *485*, 2287–2311. [CrossRef]
32. Fujita, Y. Pre-Processing of Galaxies before Entering a Cluster. *Publ. Astron. Soc. Jpn.* **2004**, *56*, 29–43. [CrossRef]
33. Wetzel, A.R.; Tinker, J.L.; Conroy, C.; van den Bosch, F.C. Galaxy evolution in groups and clusters: Satellite star formation histories and quenching time-scales in a hierarchical Universe. *Mon. Not. R. Astron. Soc.* **2013**, *432*, 336–358. [CrossRef]
34. Donnari, M.; Pillepich, A.; Joshi, G.D.; Nelson, D.; Genel, S.; Marinacci, F.; Rodriguez-Gomez, V.; Pakmor, R.; Torrey, P.; Vogelsberger, M.; et al. Quenched fractions in the IllustrisTNG simulations: The roles of AGN feedback, environment, and pre-processing. *Mon. Not. R. Astron. Soc.* **2021**, *500*, 4004–4024. [CrossRef]
35. Jenkins, A.; Frenk, C.S.; White, S.D.M.; Colberg, J.M.; Cole, S.; Evrard, A.E.; Couchman, H.M.P.; Yoshida, N. The mass function of dark matter haloes. *Mon. Not. R. Astron. Soc.* **2001**, *321*, 372–384. [CrossRef]
36. Bryan, G.L.; Norman, M.L. Statistical Properties of X-ray Clusters: Analytic and Numerical Comparisons. *Astrophys. J.* **1998**, *495*, 80–99. [CrossRef]
37. Liang, L.; Durier, F.; Babul, A.; Davé, R.; Oppenheimer, B.D.; Katz, N.; Fardal, M.; Quinn, T. The growth and enrichment of intragroup gas. *Mon. Not. R. Astron. Soc.* **2016**, *456*, 4266–4290. [CrossRef]
38. Vogelsberger, M.; Genel, S.; Springel, V.; Torrey, P.; Sijacki, D.; Xu, D.; Snyder, G.; Nelson, D.; Hernquist, L. Introducing the Illustris Project: Simulating the coevolution of dark and visible matter in the Universe. *Mon. Not. R. Astron. Soc.* **2014**, *444*, 1518–1547. [CrossRef]

39. Schaye, J.; Crain, R.A.; Bower, R.G.; Furlong, M.; Schaller, M.; Theuns, T.; Dalla Vecchia, C.; Frenk, C.S.; McCarthy, I.G.; Helly, J.C.; et al. The EAGLE project: Simulating the evolution and assembly of galaxies and their environments. *Mon. Not. R. Astron. Soc.* **2015**, *446*, 521–554. [CrossRef]
40. Dubois, Y.; Peirani, S.; Pichon, C.; Devriendt, J.; Gavazzi, R.; Welker, C.; Volonteri, M. The HORIZON-AGN simulation: Morphological diversity of galaxies promoted by AGN feedback. *Mon. Not. R. Astron. Soc.* **2016**, *463*, 3948–3964. [CrossRef]
41. McCarthy, I.G.; Schaye, J.; Bird, S.; Le Brun, A.M.C. The BAHAMAS project: Calibrated hydrodynamical simulations for large-scale structure cosmology. *Mon. Not. R. Astron. Soc.* **2017**, *465*, 2936–2965. [CrossRef]
42. Bahé, Y.M.; Barnes, D.J.; Dalla Vecchia, C.; Kay, S.T.; White, S.D.M.; McCarthy, I.G.; Schaye, J.; Bower, R.G.; Crain, R.A.; Theuns, T.; et al. The Hydrangea simulations: Galaxy formation in and around massive clusters. *Mon. Not. R. Astron. Soc.* **2017**, *470*, 4186–4208. [CrossRef]
43. Barnes, D.J.; Kay, S.T.; Bahé, Y.M.; Dalla Vecchia, C.; McCarthy, I.G.; Schaye, J.; Bower, R.G.; Jenkins, A.; Thomas, P.A.; Schaller, M.; et al. The Cluster-EAGLE project: Global properties of simulated clusters with resolved galaxies. *Mon. Not. R. Astron. Soc.* **2017**, *471*, 1088–1106. [CrossRef]
44. Henden, N.A.; Puchwein, E.; Shen, S.; Sijacki, D. The FABLE simulations: A feedback model for galaxies, groups, and clusters. *Mon. Not. R. Astron. Soc.* **2018**, *479*, 5385–5412. [CrossRef]
45. Cui, W.; Knebe, A.; Yepes, G.; Pearce, F.; Power, C.; Dave, R.; Arth, A.; Borgani, S.; Dolag, K.; Elahi, P.; et al. The Three Hundred project: A large catalogue of theoretically modelled galaxy clusters for cosmological and astrophysical applications. *Mon. Not. R. Astron. Soc.* **2018**, *480*, 2898–2915. [CrossRef]
46. Pillepich, A.; Springel, V.; Nelson, D.; Genel, S.; Naiman, J.; Pakmor, R.; Hernquist, L.; Torrey, P.; Vogelsberger, M.; Weinberger, R.; et al. Simulating galaxy formation with the IllustrisTNG model. *Mon. Not. R. Astron. Soc.* **2018**, *473*, 4077–4106. [CrossRef]
47. Nelson, D.; Pillepich, A.; Springel, V.; Weinberger, R.; Hernquist, L.; Pakmor, R.; Genel, S.; Torrey, P.; Vogelsberger, M.; Kauffmann, G.; Marinacci, F.; Naiman, J. First results from the IllustrisTNG simulations: The galaxy colour bimodality. *Mon. Not. R. Astron. Soc.* **2018**, *475*, 624–647. [CrossRef]
48. Tremmel, M.; Karcher, M.; Governato, F.; Volonteri, M.; Quinn, T.R.; Pontzen, A.; Anderson, L.; Bellovary, J. The Romulus cosmological simulations: A physical approach to the formation, dynamics and accretion models of SMBHs. *Mon. Not. R. Astron. Soc.* **2017**, *470*, 1121–1139. [CrossRef]
49. Tremmel, M.; Quinn, T.R.; Ricarte, A.; Babul, A.; Chadayammuri, U.; Natarajan, P.; Nagai, D.; Pontzen, A.; Volonteri, M. Introducing ROMULUSC: A cosmological simulation of a galaxy cluster with an unprecedented resolution. *Mon. Not. R. Astron. Soc.* **2019**, *483*, 3336–3362. [CrossRef]
50. Davé, R.; Anglés-Alcázar, D.; Narayanan, D.; Li, Q.; Rafieferantsoa, M.H.; Appleby, S. SIMBA: Cosmological simulations with black hole growth and feedback. *Mon. Not. R. Astron. Soc.* **2019**, *486*, 2827–2849. [CrossRef]
51. Nelson, D.; Pillepich, A.; Springel, V.; Pakmor, R.; Weinberger, R.; Genel, S.; Torrey, P.; Vogelsberger, M.; Marinacci, F.; Hernquist, L. First results from the TNG50 simulation: Galactic outflows driven by supernovae and black hole feedback. *Mon. Not. R. Astron. Soc.* **2019**, *490*, 3234–3261. [CrossRef]
52. Teyssier, R. Grid-Based Hydrodynamics in Astrophysical Fluid Flows. *Annu. Rev. Astron. Astrophys.* **2015**, *53*, 325–364. [CrossRef]
53. Springel, V. Smoothed Particle Hydrodynamics in Astrophysics. *Annu. Rev. Astron. Astrophys.* **2010**, *48*, 391–430. [CrossRef]
54. Price, D.J. Smoothed particle hydrodynamics and magnetohydrodynamics. *J. Comput. Phys.* **2012**, *231*, 759–794. [CrossRef]
55. Springel, V. E pur si muove: Galilean-invariant cosmological hydrodynamical simulations on a moving mesh. *Mon. Not. R. Astron. Soc.* **2010**, *401*, 791–851. [CrossRef]
56. Hopkins, P.F. A new class of accurate, mesh-free hydrodynamic simulation methods. *Mon. Not. R. Astron. Soc.* **2015**, *450*, 53–110. [CrossRef]
57. Koshizuka, S.; Oka, Y. Moving-particle semi-implicit method for fragmentation of incompressible fluid. *Nucl. Sci. Eng.* **1996**, *123*, 421–434. [CrossRef]
58. Hu, C.Y.; Naab, T.; Walch, S.; Moster, B.P.; Oser, L. SPHGal: Smoothed particle hydrodynamics with improved accuracy for galaxy simulations. *Mon. Not. R. Astron. Soc.* **2014**, *443*, 1173–1191. [CrossRef]
59. Schaller, M.; Dalla Vecchia, C.; Schaye, J.; Bower, R.G.; Theuns, T.; Crain, R.A.; Furlong, M.; McCarthy, I.G. The EAGLE simulations of galaxy formation: The importance of the hydrodynamics scheme. *Mon. Not. R. Astron. Soc.* **2015**, *454*, 2277–2291. [CrossRef]
60. Beck, A.M.; Murante, G.; Arth, A.; Remus, R.S.; Teklu, A.F.; Donnert, J.M.F.; Planelles, S.; Beck, M.C.; Förster, P.; Imgrund, M.; et al. An improved SPH scheme for cosmological simulations. *Mon. Not. R. Astron. Soc.* **2016**, *455*, 2110–2130. [CrossRef]
61. Borrow, J.; Schaller, M.; Bower, R.G.; Schaye, J. Sphenix: Smoothed Particle Hydrodynamics for the next generation of galaxy formation simulations. *arXiv* **2020**, arXiv:2012.03974.
62. Crain, R.A.; Schaye, J.; Bower, R.G.; Furlong, M.; Schaller, M.; Theuns, T.; Dalla Vecchia, C.; Frenk, C.S.; McCarthy, I.G. The EAGLE simulations of galaxy formation: Calibration of subgrid physics and model variations. *Mon. Not. R. Astron. Soc.* **2015**, *450*, 1937–1961. [CrossRef]
63. McAlpine, S.; Helly, J.C.; Schaller, M.; Trayford, J.W.; Qu, Y.; Furlong, M.; Bower, R.G.; Crain, R.A.; Schaye, J.; Theuns, T.; et al. The EAGLE simulations of galaxy formation: Public release of halo and galaxy catalogues. *Astron. Comput.* **2016**, *15*, 72–89. [CrossRef]

64. Marinacci, F.; Vogelsberger, M.; Pakmor, R.; Torrey, P.; Springel, V.; Hernquist, L.; Nelson, D.; Weinberger, R.; Pillepich, A.; Naiman, J.; et al. First results from the IllustrisTNG simulations: Radio haloes and magnetic fields. *Mon. Not. R. Astron. Soc.* **2018**, *480*, 5113–5139. [CrossRef]
65. Naiman, J.P.; Pillepich, A.; Springel, V.; Ramirez-Ruiz, E.; Torrey, P.; Vogelsberger, M.; Pakmor, R.; Nelson, D.; Marinacci, F.; Hernquist, L.; et al. First results from the IllustrisTNG simulations: A tale of two elements—Chemical evolution of magnesium and europium. *Mon. Not. R. Astron. Soc.* **2018**, *477*, 1206–1224. [CrossRef]
66. Springel, V.; Pakmor, R.; Pillepich, A.; Weinberger, R.; Nelson, D.; Hernquist, L.; Vogelsberger, M.; Genel, S.; Torrey, P.; Marinacci, F.; et al. First results from the IllustrisTNG simulations: Matter and galaxy clustering. *Mon. Not. R. Astron. Soc.* **2018**, *475*, 676–698. [CrossRef]
67. Pillepich, A.; Nelson, D.; Hernquist, L.; Springel, V.; Pakmor, R.; Torrey, P.; Weinberger, R.; Genel, S.; Naiman, J.P.; Marinacci, F.; et al. First results from the IllustrisTNG simulations: The stellar mass content of groups and clusters of galaxies. *Mon. Not. R. Astron. Soc.* **2018**, *475*, 648–675. [CrossRef]
68. Donnari, M.; Pillepich, A.; Nelson, D.; Marinacci, F.; Vogelsberger, M.; Hernquist, L. Quenched fractions in the IllustrisTNG simulations: Comparison with observations and other theoretical models. *arXiv* **2020**, arXiv:2008.00004.
69. Dolag, K.; Komatsu, E.; Sunyaev, R. SZ effects in the Magneticum Pathfinder simulation: Comparison with the Planck, SPT, and ACT results. *Mon. Not. R. Astron. Soc.* **2016**, *463*, 1797–1811. [CrossRef]
70. Castro, T.; Borgani, S.; Dolag, K.; Marra, V.; Quartin, M.; Saro, A.; Sefusatti, E. On the impact of baryons on the halo mass function, bias, and cluster cosmology. *Mon. Not. R. Astron. Soc.* **2021**, *500*, 2316–2335. [CrossRef]
71. Ragagnin, A.; Dolag, K.; Moscardini, L.; Biviano, A.; D'Onofrio, M. Dependency of halo concentration on mass, redshift and fossilness in Magneticum hydrodynamic simulations. *Mon. Not. R. Astron. Soc.* **2019**, *486*, 4001–4012. [CrossRef]
72. Chadayammuri, U.; Tremmel, M.; Nagai, D.; Babul, A.; Quinn, T. Fountains and storms: The effects of AGN feedback and mergers on the evolution of the intracluster medium in the ROMULUSC simulation. *Mon. Not. R. Astron. Soc.* **2021**, *504*, 3922–3937. [CrossRef]
73. Jung, S.L.; Rennehan, D.; Saeedzadeh, V.; Babul, A.; Tremmel, M.; Quinn, T.R.; S.I., L.; Sukyoung, K.Y. Brightest Group Galaxies in Romulus Simulations: A link between kinematics and morphology. In preparation.
74. Pillepich, A.; Nelson, D.; Springel, V.; Pakmor, R.; Torrey, P.; Weinberger, R.; Vogelsberger, M.; Marinacci, F.; Genel, S.; van der Wel, A.; et al. First results from the TNG50 simulation: The evolution of stellar and gaseous discs across cosmic time. *Mon. Not. R. Astron. Soc.* **2019**, *490*, 3196–3233. [CrossRef]
75. Feldmann, R.; Carollo, C.M.; Mayer, L.; Renzini, A.; Lake, G.; Quinn, T.; Stinson, G.S.; Yepes, G. The Evolution of Central Group Galaxies in Hydrodynamical Simulations. *Astrophys. J.* **2010**, *709*, 218–240. [CrossRef]
76. Oppenheimer, B.D.; Crain, R.A.; Schaye, J.; Rahmati, A.; Richings, A.J.; Trayford, J.W.; Tumlinson, J.; Bower, R.G.; Schaller, M.; Theuns, T. Bimodality of low-redshift circumgalactic O VI in non-equilibrium EAGLE zoom simulations. *Mon. Not. R. Astron. Soc.* **2016**, *460*, 2157–2179. [CrossRef]
77. Joshi, G.D.; Parker, L.C.; Wadsley, J.; Keller, B.W. The trajectories of galaxies in groups: Mass-loss and preprocessing. *Mon. Not. R. Astron. Soc.* **2019**, *483*, 235–248. [CrossRef]
78. Eckert, D.; Gaspari, M.; Gastaldello, F.; Le Brun, A.M.C.; O'Sullivan, E. Feedback from Active Galactic Nuclei in Galaxy Groups. *Universe* **2021**, *7*, 142. [CrossRef]
79. Dolag, K.; Stasyszyn, F. An MHD GADGET for cosmological simulations. *Mon. Not. R. Astron. Soc.* **2009**, *398*, 1678–1697. [CrossRef]
80. Pakmor, R.; Springel, V. Simulations of magnetic fields in isolated disc galaxies. *Mon. Not. R. Astron. Soc.* **2013**, *432*, 176–193. [CrossRef]
81. Frenk, C.S.; White, S.D.M.; Bode, P.; Bond, J.R.; Bryan, G.L.; Cen, R.; Couchman, H.M.P.; Evrard, A.E.; Gnedin, N.; Jenkins, A.; et al. The Santa Barbara Cluster Comparison Project: A Comparison of Cosmological Hydrodynamics Solutions. *Astrophys. J.* **1999**, *525*, 554–582. [CrossRef]
82. Wadsley, J.W.; Veeravalli, G.; Couchman, H.M.P. On the treatment of entropy mixing in numerical cosmology. *Mon. Not. R. Astron. Soc.* **2008**, *387*, 427–438. [CrossRef]
83. Rasia, E.; Borgani, S.; Murante, G.; Planelles, S.; Beck, A.M.; Biffi, V.; Ragone-Figueroa, C.; Granato, G.L.; Steinborn, L.K.; Dolag, K. Cool Core Clusters from Cosmological Simulations. *Astrophys. J. Lett.* **2015**, *813*, L17. [CrossRef]
84. Hahn, O.; Martizzi, D.; Wu, H.Y.; Evrard, A.E.; Teyssier, R.; Wechsler, R.H. rhapsody-g simulations—I. The cool cores, hot gas and stellar content of massive galaxy clusters. *Mon. Not. R. Astron. Soc.* **2017**, *470*, 166–186. [CrossRef]
85. Sembolini, F.; Yepes, G.; Pearce, F.R.; Knebe, A.; Kay, S.T.; Power, C.; Cui, W.; Beck, A.M.; Borgani, S.; Dalla Vecchia, C.; et al. nIFTy galaxy cluster simulations—I. Dark matter and non-radiative models. *Mon. Not. R. Astron. Soc.* **2016**, *457*, 4063–4080. [CrossRef]
86. Haardt, F.; Madau, P. Radiative Transfer in a Clumpy Universe. IV. New Synthesis Models of the Cosmic UV/X-Ray Background. *Astrophys. J.* **2012**, *746*, 125. [CrossRef]
87. Ferland, G.J.; Porter, R.L.; van Hoof, P.A.M.; Williams, R.J.R.; Abel, N.P.; Lykins, M.L.; Shaw, G.; Henney, W.J.; Stancil, P.C. The 2013 Release of Cloudy. *Rev. Mex. Astron. Astrofis.* **2013**, *49*, 137–163.

88. Wiersma, R.P.C.; Schaye, J.; Smith, B.D. The effect of photoionization on the cooling rates of enriched, astrophysical plasmas. *Mon. Not. R. Astron. Soc.* **2009**, *393*, 99–107. [CrossRef]
89. Smith, B.D.; Bryan, G.L.; Glover, S.C.O.; Goldbaum, N.J.; Turk, M.J.; Regan, J.; Wise, J.H.; Schive, H.Y.; Abel, T.; Emerick, A.; et al. GRACKLE: A chemistry and cooling library for astrophysics. *Mon. Not. R. Astron. Soc.* **2017**, *466*, 2217–2234. [CrossRef]
90. Vogelsberger, M.; Genel, S.; Sijacki, D.; Torrey, P.; Springel, V.; Hernquist, L. A model for cosmological simulations of galaxy formation physics. *Mon. Not. R. Astron. Soc.* **2013**, *436*, 3031–3067. [CrossRef]
91. Cen, R.; Fang, T. Where Are the Baryons? III. Nonequilibrium Effects and Observables. *Astrophys. J.* **2006**, *650*, 573–591. [CrossRef]
92. Gnat, O.; Sternberg, A. Time-dependent Ionization in Radiatively Cooling Gas. *Astrophys. J. Suppl.* **2007**, *168*, 213–230. [CrossRef]
93. Gnat, O.; Ferland, G.J. Ion-by-ion Cooling Efficiencies. *Astrophys. J. Suppl.* **2012**, *199*, 20. [CrossRef]
94. Oppenheimer, B.D.; Schaye, J. Non-equilibirum ionization and cooling of metal-enriched gas in the presence of a photoionization background. *Mon. Not. R. Astron. Soc.* **2013**, *434*, 1043–1062. [CrossRef]
95. Lewis, G.F.; Babul, A.; Katz, N.; Quinn, T.; Hernquist, L.; Weinberg, D.H. The effects of gasdynamics, cooling, star formation, and numerical resolution in simulations of cluster formation. *Astrophys. J.* **2000**, *536*, 623. [CrossRef]
96. Voit, G.M.; Kay, S.T.; Bryan, G.L. The baseline intracluster entropy profile from gravitational structure formation. *Mon. Not. R. Astron. Soc.* **2005**, *364*, 909–916. [CrossRef]
97. Taylor, J.E.; Babul, A. The evolution of substructure in galaxy, group and cluster haloes - I. Basic dynamics. *Mon. Not. R. Astron. Soc.* **2004**, *348*, 811–830. [CrossRef]
98. Barnes, E.I.; Williams, L.L.R.; Babul, A.; Dalcanton, J.J. Density Profiles of Collisionless Equilibria. II. Anisotropic Spherical Systems. *Astrophys. J.* **2007**, *654*, 814–824. [CrossRef]
99. Springel, V.; Hernquist, L. Cosmological smoothed particle hydrodynamics simulations: A hybrid multiphase model for star formation. *Mon. Not. R. Astron. Soc.* **2003**, *339*, 289–311. [CrossRef]
100. Schaye, J.; Dalla Vecchia, C. On the relation between the Schmidt and Kennicutt-Schmidt star formation laws and its implications for numerical simulations. *Mon. Not. R. Astron. Soc.* **2008**, *383*, 1210–1222. [CrossRef]
101. Davé, R.; Thompson, R.; Hopkins, P.F. MUFASA: Galaxy formation simulations with meshless hydrodynamics. *Mon. Not. R. Astron. Soc.* **2016**, *462*, 3265–3284. [CrossRef]
102. Kennicutt, J.; Robert, C. Star Formation in Galaxies Along the Hubble Sequence. *Annu. Rev. Astron. Astrophys.* **1998**, *36*, 189–232. [CrossRef]
103. Kroupa, P. On the variation of the initial mass function. *Mon. Not. R. Astron. Soc.* **2001**, *322*, 231–246. [CrossRef]
104. Chabrier, G. Galactic Stellar and Substellar Initial Mass Function. *Publ. Astron. Soc. Pac.* **2003**, *115*, 763–795. [CrossRef]
105. Gastaldello, F.; Simionescu, A.; Mernier, F.; Biffi, V.; Gaspari, M.; Sato, K.; Matsushita, K. The metal content of the hot atmospheres of galaxy groups. *Universe*, accepted.
106. Wiersma, R.P.C.; Schaye, J.; Theuns, T.; Dalla Vecchia, C.; Tornatore, L. Chemical enrichment in cosmological, smoothed particle hydrodynamics simulations. *Mon. Not. R. Astron. Soc.* **2009**, *399*, 574–600. [CrossRef]
107. Nelson, D.; Kauffmann, G.; Pillepich, A.; Genel, S.; Springel, V.; Pakmor, R.; Hernquist, L.; Weinberger, R.; Torrey, P.; Vogelsberger, M.; Marinacci, F. The abundance, distribution, and physical nature of highly ionized oxygen O VI, O VII, and O VIII in IllustrisTNG. *Mon. Not. R. Astron. Soc.* **2018**, *477*, 450–479. [CrossRef]
108. Tornatore, L.; Borgani, S.; Dolag, K.; Matteucci, F. Chemical enrichment of galaxy clusters from hydrodynamical simulations. *Mon. Not. R. Astron. Soc.* **2007**, *382*, 1050–1072. [CrossRef]
109. Davé, R.; Oppenheimer, B.D.; Sivanand am, S. Enrichment and pre-heating in intragroup gas from galactic outflows. *Mon. Not. R. Astron. Soc.* **2008**, *391*, 110–123. [CrossRef]
110. Peterson, J.R.; Kahn, S.M.; Paerels, F.B.S.; Kaastra, J.S.; Tamura, T.; Bleeker, J.A.M.; Ferrigno, C.; Jernigan, J.G. High-Resolution X-Ray Spectroscopic Constraints on Cooling-Flow Models for Clusters of Galaxies. *Astrophys. J.* **2003**, *590*, 207–224. [CrossRef]
111. Biffi, V.; Mernier, F.; Medvedev, P. Enrichment of the Hot Intracluster Medium: Numerical Simulations. *Space Sci. Rev.* **2018**, *214*, 123. [CrossRef]
112. Shen, S.; Wadsley, J.; Stinson, G. The enrichment of the intergalactic medium with adiabatic feedback - I. Metal cooling and metal diffusion. *Mon. Not. R. Astron. Soc.* **2010**, *407*, 1581–1596. [CrossRef]
113. Hopkins, P.F.; Wetzel, A.; Kereš, D.; Faucher-Giguère, C.A.; Quataert, E.; Boylan-Kolchin, M.; Murray, N.; Hayward, C.C.; Garrison-Kimmel, S.; Hummels, C.; et al. FIRE-2 simulations: Physics versus numerics in galaxy formation. *Mon. Not. R. Astron. Soc.* **2018**, *480*, 800–863. [CrossRef]
114. Rennehan, D. Mixing matters. *arXiv* **2021**, arXiv:2104.07673.
115. Oppenheimer, B.D.; Davé, R.; Kereš, D.; Fardal, M.; Katz, N.; Kollmeier, J.A.; Weinberg, D.H. Feedback and recycled wind accretion: Assembling the z = 0 galaxy mass function. *Mon. Not. R. Astron. Soc.* **2010**, *406*, 2325–2338. [CrossRef]
116. Schaye, J.; Dalla Vecchia, C.; Booth, C.M.; Wiersma, R.P.C.; Theuns, T.; Haas, M.R.; Bertone, S.; Duffy, A.R.; McCarthy, I.G.; van de Voort, F. The physics driving the cosmic star formation history. *Mon. Not. R. Astron. Soc.* **2010**, *402*, 1536–1560. [CrossRef]
117. Scannapieco, C.; Wadepuhl, M.; Parry, O.H.; Navarro, J.F.; Jenkins, A.; Springel, V.; Teyssier, R.; Carlson, E.; Couchman, H.M.P.; Crain, R.A.; et al. The Aquila comparison project: The effects of feedback and numerical methods on simulations of galaxy formation. *Mon. Not. R. Astron. Soc.* **2012**, *423*, 1726–1749. [CrossRef]

118. Haas, M.R.; Schaye, J.; Booth, C.M.; Dalla Vecchia, C.; Springel, V.; Theuns, T.; Wiersma, R.P.C. Physical properties of simulated galaxy populations at z = 2 − I. Effect of metal-line cooling and feedback from star formation and AGN. *Mon. Not. R. Astron. Soc.* **2013**, *435*, 2931–2954. [CrossRef]
119. Torrey, P.; Vogelsberger, M.; Genel, S.; Sijacki, D.; Springel, V.; Hernquist, L. A model for cosmological simulations of galaxy formation physics: multi-epoch validation. *Mon. Not. R. Astron. Soc.* **2014**, *438*, 1985–2004. [CrossRef]
120. White, S.D.M.; Frenk, C.S. Galaxy Formation through Hierarchical Clustering. *Astrophys. J.* **1991**, *379*, 52. [CrossRef]
121. Oppenheimer, B.D.; Davé, R.; Finlator, K. Tracing the re-ionization-epoch intergalactic medium with metal absorption lines. *Mon. Not. R. Astron. Soc.* **2009**, *396*, 729–758. [CrossRef]
122. Keating, L.C.; Puchwein, E.; Haehnelt, M.G.; Bird, S.; Bolton, J.S. Testing the effect of galactic feedback on the IGM at z ∼ 6 with metal-line absorbers. *Mon. Not. R. Astron. Soc.* **2016**, *461*, 606–626. [CrossRef]
123. Finlator, K.; Oppenheimer, B.D.; Davé, R.; Zackrisson, E.; Thompson, R.; Huang, S. The soft, fluctuating UVB at z ∼ 6 as traced by C IV, Si IV, and C II. *Mon. Not. R. Astron. Soc.* **2016**, *459*, 2299–2310. [CrossRef]
124. Rahmati, A.; Schaye, J.; Crain, R.A.; Oppenheimer, B.D.; Schaller, M.; Theuns, T. Cosmic distribution of highly ionized metals and their physical conditions in the EAGLE simulations. *Mon. Not. R. Astron. Soc.* **2016**, *459*, 310–332. [CrossRef]
125. Bird, S.; Rubin, K.H.R.; Suresh, J.; Hernquist, L. Simulating the carbon footprint of galactic haloes. *Mon. Not. R. Astron. Soc.* **2016**, *462*, 307–322. [CrossRef]
126. Cen, R.; Chisari, N.E. Star Formation Feedback and Metal-enrichment History of the Intergalactic Medium. *Astrophys. J.* **2011**, *731*, 11. [CrossRef]
127. Cooksey, K.L.; Thom, C.; Prochaska, J.X.; Chen, H.W. The Last Eight-Billion Years of Intergalactic C IV Evolution. [CrossRef]
128. Becker, G.D.; Rauch, M.; Sargent, W.L.W. High-Redshift Metals. I. The Decline of C IV at z > 5.3. *Astrophys. J.* **2009**, *698*, 1010–1019. [CrossRef]
129. Songaila, A. The Minimum Universal Metal Density between Redshifts of 1.5 and 5.5. *Astrophys. J. Lett.* **2001**, *561*, L153–L156. [CrossRef]
130. D'Odorico, V.; Cupani, G.; Cristiani, S.; Maiolino, R.; Molaro, P.; Nonino, M.; Centurión, M.; Cimatti, A.; di Serego Alighieri, S.; Fiore, F.; et al. Metals in the IGM approaching the re-ionization epoch: Results from X-shooter at the VLT. *Mon. Not. R. Astron. Soc.* **2013**, *435*, 1198–1232. [CrossRef]
131. Matejek, M.S.; Simcoe, R.A. A Survey of Mg II Absorption at 2 < z < 6 with Magellan/FIRE. I. Sample and Evolution of the Mg II Frequency. *Astrophys. J.* **2012**, *761*, 112. [CrossRef]
132. McCarthy, I.G.; Schaye, J.; Ponman, T.J.; Bower, R.G.; Booth, C.M.; Dalla Vecchia, C.; Crain, R.A.; Springel, V.; Theuns, T.; Wiersma, R.P.C. The case for AGN feedback in galaxy groups. *Mon. Not. R. Astron. Soc.* **2010**, *406*, 822–839. [CrossRef]
133. Planelles, S.; Borgani, S.; Fabjan, D.; Killedar, M.; Murante, G.; Granato, G.L.; Ragone-Figueroa, C.; Dolag, K. On the role of AGN feedback on the thermal and chemodynamical properties of the hot intracluster medium. *Mon. Not. R. Astron. Soc.* **2014**, *438*, 195–216. [CrossRef]
134. Oppenheimer, B.D.; Davé, R. Cosmological simulations of intergalactic medium enrichment from galactic outflows. *Mon. Not. R. Astron. Soc.* **2006**, *373*, 1265–1292. [CrossRef]
135. Murray, N.; Quataert, E.; Thompson, T.A. On the Maximum Luminosity of Galaxies and Their Central Black Holes: Feedback from Momentum-driven Winds. *Astrophys. J.* **2005**, *618*, 569–585. [CrossRef]
136. Dalla Vecchia, C.; Schaye, J. Simulating galactic outflows with thermal supernova feedback. *Mon. Not. R. Astron. Soc.* **2012**, *426*, 140–158. [CrossRef]
137. Oppenheimer, B.D.; Davé, R. Mass, metal, and energy feedback in cosmological simulations. *Mon. Not. R. Astron. Soc.* **2008**, *387*, 577–600. [CrossRef]
138. Muratov, A.L.; Kereš, D.; Faucher-Giguère, C.A.; Hopkins, P.F.; Quataert, E.; Murray, N. Gusty, gaseous flows of FIRE: Galactic winds in cosmological simulations with explicit stellar feedback. *Mon. Not. R. Astron. Soc.* **2015**, *454*, 2691–2713. [CrossRef]
139. Hopkins, P.F.; Kereš, D.; Oñorbe, J.; Faucher-Giguère, C.A.; Quataert, E.; Murray, N.; Bullock, J.S. Galaxies on FIRE (Feedback In Realistic Environments): Stellar feedback explains cosmologically inefficient star formation. *Mon. Not. R. Astron. Soc.* **2014**, *445*, 581–603. [CrossRef]
140. Davies, J.J.; Crain, R.A.; Oppenheimer, B.D.; Schaye, J. The quenching and morphological evolution of central galaxies is facilitated by the feedback-driven expulsion of circumgalactic gas. *Mon. Not. R. Astron. Soc.* **2020**, *491*, 4462–4480. [CrossRef]
141. Stinson, G.; Seth, A.; Katz, N.; Wadsley, J.; Governato, F.; Quinn, T. Star formation and feedback in smoothed particle hydrodynamic simulations—I. Isolated galaxies. *Mon. Not. R. Astron. Soc.* **2006**, *373*, 1074–1090. [CrossRef]
142. Crain, R.A.; Theuns, T.; Dalla Vecchia, C.; Eke, V.R.; Frenk, C.S.; Jenkins, A.; Kay, S.T.; Peacock, J.A.; Pearce, F.R.; Schaye, J.; et al. Galaxies-intergalactic medium interaction calculation—I. Galaxy formation as a function of large-scale environment. *Mon. Not. R. Astron. Soc.* **2009**, *399*, 1773–1794. [CrossRef]
143. Bower, R.G.; Schaye, J.; Frenk, C.S.; Theuns, T.; Schaller, M.; Crain, R.A.; McAlpine, S. The dark nemesis of galaxy formation: Why hot haloes trigger black hole growth and bring star formation to an end. *Mon. Not. R. Astron. Soc.* **2017**, *465*, 32–44. [CrossRef]
144. Booth, C.M.; Schaye, J. Cosmological simulations of the growth of supermassive black holes and feedback from active galactic nuclei: Method and tests. *Mon. Not. R. Astron. Soc.* **2009**, *398*, 53–74. [CrossRef]

145. Di Matteo, T.; Colberg, J.; Springel, V.; Hernquist, L.; Sijacki, D. Direct Cosmological Simulations of the Growth of Black Holes and Galaxies. *Astrophys. J.* **2008**, *676*, 33–53. [CrossRef]
146. Hirschmann, M.; Dolag, K.; Saro, A.; Bachmann, L.; Borgani, S.; Burkert, A. Cosmological simulations of black hole growth: AGN luminosities and downsizing. *Mon. Not. R. Astron. Soc.* **2014**, *442*, 2304–2324. [CrossRef]
147. Tremmel, M.; Governato, F.; Volonteri, M.; Quinn, T.R. Off the beaten path: A new approach to realistically model the orbital decay of supermassive black holes in galaxy formation simulations. *Mon. Not. R. Astron. Soc.* **2015**, *451*, 1868–1874. [CrossRef]
148. Springel, V.; Di Matteo, T.; Hernquist, L. Modelling feedback from stars and black holes in galaxy mergers. *Mon. Not. R. Astron. Soc.* **2005**, *361*, 776–794. [CrossRef]
149. Steinborn, L.K.; Dolag, K.; Hirschmann, M.; Prieto, M.A.; Remus, R.S. A refined sub-grid model for black hole accretion and AGN feedback in large cosmological simulations. *Mon. Not. R. Astron. Soc.* **2015**, *448*, 1504–1525. [CrossRef]
150. Gaspari, M.; Ruszkowski, M.; Oh, S.P. Chaotic cold accretion on to black holes. *Mon. Not. R. Astron. Soc.* **2013**, *432*, 3401–3422. [CrossRef]
151. Rosas-Guevara, Y.M.; Bower, R.G.; Schaye, J.; Furlong, M.; Frenk, C.S.; Booth, C.M.; Crain, R.A.; Dalla Vecchia, C.; Schaller, M.; Theuns, T. The impact of angular momentum on black hole accretion rates in simulations of galaxy formation. *Mon. Not. R. Astron. Soc.* **2015**, *454*, 1038–1057. [CrossRef]
152. Hopkins, P.F.; Quataert, E. An analytic model of angular momentum transport by gravitational torques: From galaxies to massive black holes. *Mon. Not. R. Astron. Soc.* **2011**, *415*, 1027–1050. [CrossRef]
153. Anglés-Alcázar, D.; Davé, R.; Faucher-Giguère, C.A.; Özel, F.; Hopkins, P.F. Gravitational torque-driven black hole growth and feedback in cosmological simulations. *Mon. Not. R. Astron. Soc.* **2017**, *464*, 2840–2853. [CrossRef]
154. Davé, R.; Crain, R.A.; Stevens, A.R.H.; Narayanan, D.; Saintonge, A.; Catinella, B.; Cortese, L. Galaxy cold gas contents in modern cosmological hydrodynamic simulations. *Mon. Not. R. Astron. Soc.* **2020**, *497*, 146–166. [CrossRef]
155. Lin, Y.T.; Mohr, J.J. K-band Properties of Galaxy Clusters and Groups: Brightest Cluster Galaxies and Intracluster Light. *Astrophys. J.* **2004**, *617*, 879–895. [CrossRef]
156. Shakura, N.I.; Sunyaev, R.A. Reprint of 1973A&A....24..337S. Black holes in binary systems. Observational appearance. *Astron. Astrophys.* **1973**, *500*, 33–51.
157. Weinberger, R.; Springel, V.; Hernquist, L.; Pillepich, A.; Marinacci, F.; Pakmor, R.; Nelson, D.; Genel, S.; Vogelsberger, M.; Naiman, J.; et al. Simulating galaxy formation with black hole driven thermal and kinetic feedback. *Mon. Not. R. Astron. Soc.* **2017**, *465*, 3291–3308. [CrossRef]
158. McCarthy, I.G.; Schaye, J.; Bower, R.G.; Ponman, T.J.; Booth, C.M.; Dalla Vecchia, C.; Springel, V. Gas expulsion by quasar-driven winds as a solution to the overcooling problem in galaxy groups and clusters. *Mon. Not. R. Astron. Soc.* **2011**, *412*, 1965–1984. [CrossRef]
159. Sijacki, D.; Springel, V.; Di Matteo, T.; Hernquist, L. A unified model for AGN feedback in cosmological simulations of structure formation. *Mon. Not. R. Astron. Soc.* **2007**, *380*, 877–900. [CrossRef]
160. Genel, S.; Vogelsberger, M.; Springel, V.; Sijacki, D.; Nelson, D.; Snyder, G.; Rodriguez-Gomez, V.; Torrey, P.; Hernquist, L. Introducing the Illustris project: The evolution of galaxy populations across cosmic time. *Mon. Not. R. Astron. Soc.* **2014**, *445*, 175–200. [CrossRef]
161. Dubois, Y.; Devriendt, J.; Slyz, A.; Teyssier, R. Self-regulated growth of supermassive black holes by a dual jet-heating active galactic nucleus feedback mechanism: Methods, tests and implications for cosmological simulations. *Mon. Not. R. Astron. Soc.* **2012**, *420*, 2662–2683. [CrossRef]
162. Davies, J.J.; Crain, R.A.; McCarthy, I.G.; Oppenheimer, B.D.; Schaye, J.; Schaller, M.; McAlpine, S. The gas fractions of dark matter haloes hosting simulated ~L* galaxies are governed by the feedback history of their black holes. *Mon. Not. R. Astron. Soc.* **2019**, *485*, 3783–3793. [CrossRef]
163. Terrazas, B.A.; Bell, E.F.; Pillepich, A.; Nelson, D.; Somerville, R.S.; Genel, S.; Weinberger, R.; Habouzit, M.; Li, Y.; Hernquist, L.; et al. The relationship between black hole mass and galaxy properties: Examining the black hole feedback model in IllustrisTNG. *Mon. Not. R. Astron. Soc.* **2020**, *493*, 1888–1906. [CrossRef]
164. Nulsen, P.E.J. Transport processes and the stripping of cluster galaxies. *Mon. Not. R. Astron. Soc.* **1982**, *198*, 1007–1016. [CrossRef]
165. Sijacki, D.; Springel, V. Physical viscosity in smoothed particle hydrodynamics simulations of galaxy clusters. *Mon. Not. R. Astron. Soc.* **2006**, *371*, 1025–1046. [CrossRef]
166. Hopkins, P.F. Anisotropic diffusion in mesh-free numerical magnetohydrodynamics. *Mon. Not. R. Astron. Soc.* **2017**, *466*, 3387–3405. [CrossRef]
167. Barnes, D.J.; Kannan, R.; Vogelsberger, M.; Pfrommer, C.; Puchwein, E.; Weinberger, R.; Springel, V.; Pakmor, R.; Nelson, D.; Marinacci, F.; et al. Enhancing AGN efficiency and cool-core formation with anisotropic thermal conduction. *Mon. Not. R. Astron. Soc.* **2019**, *488*, 3003–3013. [CrossRef]
168. Rennehan, D.; Babul, A.; Hopkins, P.F.; Davé, R.; Moa, B. Dynamic localized turbulent diffusion and its impact on the galactic ecosystem. *Mon. Not. R. Astron. Soc.* **2019**, *483*, 3810–3831. [CrossRef]
169. Fabian, A.C.; Sanders, J.S.; Ettori, S.; Taylor, G.B.; Allen, S.W.; Crawford, C.S.; Iwasawa, K.; Johnstone, R.M.; Ogle, P.M. Chandra imaging of the complex X-ray core of the Perseus cluster. *Mon. Not. R. Astron. Soc.* **2000**, *318*, L65–L68. [CrossRef]
170. Brüggen, M.; Ruszkowski, M.; Hallman, E. Active Galactic Nuclei Heating and Dissipative Processes in Galaxy Clusters. *Astrophys. J.* **2005**, *630*, 740–749. [CrossRef]

171. Dursi, L.J.; Pfrommer, C. Draping of Cluster Magnetic Fields over Bullets and Bubbles—Morphology and Dynamic Effects. *Astrophys. J.* **2008**, *677*, 993–1018. [CrossRef]
172. Fabian, A.C.; Sanders, J.S.; Allen, S.W.; Crawford, C.S.; Iwasawa, K.; Johnstone, R.M.; Schmidt, R.W.; Taylor, G.B. A deep Chandra observation of the Perseus cluster: Shocks and ripples. *Mon. Not. R. Astron. Soc.* **2003**, *344*, L43–L47. [CrossRef]
173. Ruszkowski, M.; Brüggen, M.; Begelman, M.C. Three-Dimensional Simulations of Viscous Dissipation in the Intracluster Medium. *Astrophys. J.* **2004**, *615*, 675–680. [CrossRef]
174. Jiang, Y.F.; Oh, S.P. A New Numerical Scheme for Cosmic-Ray Transport. *Astrophys. J.* **2018**, *854*, 5. [CrossRef]
175. Chan, T.K.; Kereš, D.; Hopkins, P.F.; Quataert, E.; Su, K.Y.; Hayward, C.C.; Faucher-Giguère, C.A. Cosmic ray feedback in the FIRE simulations: Constraining cosmic ray propagation with GeV γ-ray emission. *Mon. Not. R. Astron. Soc.* **2019**, *488*, 3716–3744. [CrossRef]
176. Thomas, T.; Pfrommer, C. Cosmic-ray hydrodynamics: Alfvén-wave regulated transport of cosmic rays. *Mon. Not. R. Astron. Soc.* **2019**, *485*, 2977–3008. [CrossRef]
177. Zweibel, E.G. The microphysics and macrophysics of cosmic rays. *Phys. Plasmas* **2013**, *20*, 055501. [CrossRef]
178. Zweibel, E.G. The basis for cosmic ray feedback: Written on the wind. *Phys. Plasmas* **2017**, *24*, 055402. [CrossRef]
179. Wiener, J.; Pfrommer, C.; Oh, S.P. Cosmic ray-driven galactic winds: Streaming or diffusion? *Mon. Not. R. Astron. Soc.* **2017**, *467*, 906–921. [CrossRef]
180. Butsky, I.S.; Quinn, T.R. The Role of Cosmic-ray Transport in Shaping the Simulated Circumgalactic Medium. *Astrophys. J.* **2018**, *868*, 108. [CrossRef]
181. Hopkins, P.F.; Chan, T.K.; Squire, J.; Quataert, E.; Ji, S.; Kereš, D.; Faucher-Giguère, C.A. Effects of different cosmic ray transport models on galaxy formation. *Mon. Not. R. Astron. Soc.* **2021**, *501*, 3663–3669. [CrossRef]
182. Ipavich, F.M. Galactic winds driven by cosmic rays. *Astrophys. J.* **1975**, *196*, 107–120. [CrossRef]
183. Uhlig, M.; Pfrommer, C.; Sharma, M.; Nath, B.B.; Enßlin, T.A.; Springel, V. Galactic winds driven by cosmic ray streaming. *Mon. Not. R. Astron. Soc.* **2012**, *423*, 2374–2396. [CrossRef]
184. Pakmor, R.; Pfrommer, C.; Simpson, C.M.; Springel, V. Galactic Winds Driven by Isotropic and Anisotropic Cosmic-Ray Diffusion in Disk Galaxies. *Astrophys. J. Lett.* **2016**, *824*, L30. [CrossRef]
185. Ruszkowski, M.; Yang, H.Y.K.; Zweibel, E. Global Simulations of Galactic Winds Including Cosmic-ray Streaming. *Astrophys. J.* **2017**, *834*, 208. [CrossRef]
186. Hopkins, P.F.; Chan, T.K.; Ji, S.; Hummels, C.B.; Kereš, D.; Quataert, E.; Faucher-Giguère, C.A. Cosmic ray driven outflows to Mpc scales from L_* galaxies. *Mon. Not. R. Astron. Soc.* **2021**, *501*, 3640–3662. [CrossRef]
187. Girichidis, P.; Naab, T.; Hanasz, M.; Walch, S. Cooler and smoother—The impact of cosmic rays on the phase structure of galactic outflows. *Mon. Not. R. Astron. Soc.* **2018**, *479*, 3042–3067. [CrossRef]
188. Jacob, S.; Pakmor, R.; Simpson, C.M.; Springel, V.; Pfrommer, C. The dependence of cosmic ray-driven galactic winds on halo mass. *Mon. Not. R. Astron. Soc.* **2018**, *475*, 570–584. [CrossRef]
189. Guo, F.; Oh, S.P. Feedback heating by cosmic rays in clusters of galaxies. *Mon. Not. R. Astron. Soc.* **2008**, *384*, 251–266. [CrossRef]
190. Enßlin, T.; Pfrommer, C.; Miniati, F.; Subramanian, K. Cosmic ray transport in galaxy clusters: Implications for radio halos, gamma-ray signatures, and cool core heating. *Astron. Astrophys.* **2011**, *527*, A99. [CrossRef]
191. Wiener, J.; Oh, S.P.; Guo, F. Cosmic ray streaming in clusters of galaxies. *Mon. Not. R. Astron. Soc.* **2013**, *434*, 2209–2228. [CrossRef]
192. Su, K.Y.; Hopkins, P.F.; Hayward, C.C.; Faucher-Giguère, C.A.; Kereš, D.; Ma, X.; Orr, M.E.; Chan, T.K.; Robles, V.H. Cosmic rays or turbulence can suppress cooling flows (where thermal heating or momentum injection fail). *Mon. Not. R. Astron. Soc.* **2020**, *491*, 1190–1212. [CrossRef]
193. Jacob, S.; Pfrommer, C. Cosmic ray heating in cool core clusters II: Self-regulation cycle and non-thermal emission. *Mon. Not. R. Astron. Soc.* **2017**. [CrossRef]
194. Salem, M.; Bryan, G.L.; Corlies, L. Role of cosmic rays in the circumgalactic medium. *Mon. Not. R. Astron. Soc.* **2016**, *456*, 582–601. [CrossRef]
195. Ji, S.; Chan, T.K.; Hummels, C.B.; Hopkins, P.F.; Stern, J.; Kereš, D.; Quataert, E.; Faucher-Giguère, C.A.; Murray, N. Properties of the circumgalactic medium in cosmic ray-dominated galaxy haloes. *Mon. Not. R. Astron. Soc.* **2020**, *496*, 4221–4238. [CrossRef]
196. Buck, T.; Pfrommer, C.; Pakmor, R.; Grand, R.J.J.; Springel, V. The effects of cosmic rays on the formation of Milky Way-mass galaxies in a cosmological context. *Mon. Not. R. Astron. Soc.* **2020**, *497*, 1712–1737. [CrossRef]
197. Sharma, P.; Parrish, I.J.; Quataert, E. Thermal Instability with Anisotropic Thermal Conduction and Adiabatic Cosmic Rays: Implications for Cold Filaments in Galaxy Clusters. *Astrophys. J.* **2010**, *720*, 652–665. [CrossRef]
198. Ruszkowski, M.; Yang, H.Y.K.; Reynolds, C.S. Powering of Hα Filaments by Cosmic Rays. *Astrophys. J.* **2018**, *858*, 64. [CrossRef]
199. Butsky, I.S.; Fielding, D.B.; Hayward, C.C.; Hummels, C.B.; Quinn, T.R.; Werk, J.K. The Impact of Cosmic Rays on Thermal Instability in the Circumgalactic Medium. *Astrophys. J.* **2020**, *903*, 77. [CrossRef]
200. Helsdon, S.F.; Ponman, T.J. The intragroup medium in loose groups of galaxies. *Mon. Not. R. Astron. Soc.* **2000**, *315*, 356–370. [CrossRef]
201. White, D.A.; Jones, C.; Forman, W. An investigation of cooling flows and general cluster properties from an X-ray image deprojection analysis of 207 clusters of galaxies. *Mon. Not. R. Astron. Soc.* **1997**, *292*, 419–467. [CrossRef]

202. Balogh, M.L.; Babul, A.; Patton, D.R. Pre-heated isentropic gas in groups of galaxies. *Mon. Not. R. Astron. Soc.* **1999**, *307*, 463–479. [CrossRef]
203. Bryan, G.L. Explaining the Entropy Excess in Clusters and Groups of Galaxies without Additional Heating. *Astrophys. J. Lett.* **2000**, *544*, L1–L5. [CrossRef]
204. Davé, R.; Katz, N.; Weinberg, D.H. X-Ray Scaling Relations of Galaxy Groups in a Hydrodynamic Cosmological Simulation. *Astrophys. J.* **2002**, *579*, 23–41. [CrossRef]
205. Gonzalez, A.H.; Zaritsky, D.; Zabludoff, A.I. A Census of Baryons in Galaxy Clusters and Groups. *Astrophys. J.* **2007**, *666*, 147–155. [CrossRef]
206. Balogh, M.L.; McCarthy, I.G.; Bower, R.G.; Eke, V.R. Testing cold dark matter with the hierarchical build-up of stellar light. *Mon. Not. R. Astron. Soc.* **2008**, *385*, 1003–1014. [CrossRef]
207. Kravtsov, A.V.; Vikhlinin, A.A.; Meshcheryakov, A.V. Stellar Mass—Halo Mass Relation and Star Formation Efficiency in High-Mass Halos. *Astron. Lett.* **2018**, *44*, 8–34. [CrossRef]
208. Ponman, T.J.; Sanderson, A.J.R.; Finoguenov, A. The Birmingham-CfA cluster scaling project—III. Entropy and similarity in galaxy systems. *Mon. Not. R. Astron. Soc.* **2003**, *343*, 331–342. [CrossRef]
209. Fabian, A.C.; Nulsen, P.E.J.; Canizares, C.R. Cooling flows in clusters of galaxies. *Nature* **1984**, *310*, 733–740. [CrossRef]
210. Giodini, S.; Pierini, D.; Finoguenov, A.; Pratt, G.W.; Boehringer, H.; Leauthaud, A.; Guzzo, L.; Aussel, H.; Bolzonella, M.; Capak, P.; et al. Stellar and Total Baryon Mass Fractions in Groups and Clusters Since Redshift 1. *Astrophys. J.* **2009**, *703*, 982–993. [CrossRef]
211. Schaller, M.; Frenk, C.S.; Bower, R.G.; Theuns, T.; Jenkins, A.; Schaye, J.; Crain, R.A.; Furlong, M.; Dalla Vecchia, C.; McCarthy, I.G. Baryon effects on the internal structure of ΛCDM haloes in the EAGLE simulations. *Mon. Not. R. Astron. Soc.* **2015**, *451*, 1247–1267. [CrossRef]
212. Sun, M.; Sehgal, N.; Voit, G.M.; Donahue, M.; Jones, C.; Forman, W.; Vikhlinin, A.; Sarazin, C. The Pressure Profiles of Hot Gas in Local Galaxy Groups. *Astrophys. J. Lett.* **2011**, *727*, L49. [CrossRef]
213. Babul, A.; Balogh, M.L.; Lewis, G.F.; Poole, G.B. Physical implications of the X-ray properties of galaxy groups and clusters. *Mon. Not. R. Astron. Soc.* **2002**, *330*, 329–343. [CrossRef]
214. Barnes, D.J.; Vogelsberger, M.; Kannan, R.; Marinacci, F.; Weinberger, R.; Springel, V.; Torrey, P.; Pillepich, A.; Nelson, D.; Pakmor, R.; et al. A census of cool-core galaxy clusters in IllustrisTNG. *Mon. Not. R. Astron. Soc.* **2018**, *481*, 1809–1831. [CrossRef]
215. Robson, D.; Davé, R. X-ray emission from hot gas in galaxy groups and clusters in SIMBA. *Mon. Not. R. Astron. Soc.* **2020**, *498*, 3061–3076. [CrossRef]
216. McCarthy, I.G.; Babul, A.; Bower, R.G.; Balogh, M.L. Towards a holistic view of the heating and cooling of the intracluster medium. *Mon. Not. R. Astron. Soc.* **2008**, *386*, 1309–1331. [CrossRef]
217. Gitti, M.; Nulsen, P.E.J.; David, L.P.; McNamara, B.R.; Wise, M.W. A Chandra Study of the Large-scale Shock and Cool Filaments in Hydra A: Evidence for Substantial Gas Dredge-up by the Central Outburst. *Astrophys. J.* **2011**, *732*, 13. [CrossRef]
218. Vantyghem, A.N.; McNamara, B.R.; Russell, H.R.; Main, R.A.; Nulsen, P.E.J.; Wise, M.W.; Hoekstra, H.; Gitti, M. Cycling of the powerful AGN in MS 0735.6+7421 and the duty cycle of radio AGN in clusters. *Mon. Not. R. Astron. Soc.* **2014**, *442*, 3192–3205. [CrossRef]
219. David, L.P.; Nulsen, P.E.J.; McNamara, B.R.; Forman, W.; Jones, C.; Ponman, T.; Robertson, B.; Wise, M. A High-Resolution Study of the Hydra A Cluster with Chandra: Comparison of the Core Mass Distribution with Theoretical Predictions and Evidence for Feedback in the Cooling Flow. *Astrophys. J.* **2001**, *557*, 546–559. [CrossRef]
220. Hlavacek-Larrondo, J.; McDonald, M.; Benson, B.A.; Forman, W.R.; Allen, S.W.; Bleem, L.E.; Ashby, M.L.N.; Bocquet, S.; Brodwin, M.; Dietrich, J.P.; et al. X-Ray Cavities in a Sample of 83 SPT-selected Clusters of Galaxies: Tracing the Evolution of AGN Feedback in Clusters of Galaxies out to z = 1.2. *Astrophys. J.* **2015**, *805*, 35. [CrossRef]
221. Cavaliere, A.; Menci, N.; Tozzi, P. Diffuse Baryons in Groups and Clusters of Galaxies. *Astrophys. J.* **1998**, *501*, 493–508. [CrossRef]
222. Voit, G.M.; Bryan, G.L. Regulation of the X-ray luminosity of clusters of galaxies by cooling and supernova feedback. *Nature* **2001**, *414*, 425–427. [CrossRef]
223. Voit, G.M.; Bryan, G.L.; Balogh, M.L.; Bower, R.G. Modified Entropy Models for the Intracluster Medium. *Astrophys. J.* **2002**, *576*, 601–624. [CrossRef]
224. Borgani, S.; Murante, G.; Springel, V.; Diaferio, A.; Dolag, K.; Moscardini, L.; Tormen, G.; Tornatore, L.; Tozzi, P. X-ray properties of galaxy clusters and groups from a cosmological hydrodynamical simulation. *Mon. Not. R. Astron. Soc.* **2004**, *348*, 1078–1096. [CrossRef]
225. Brough, S.; Forbes, D.A.; Kilborn, V.A.; Couch, W. Southern GEMS groups—I. Dynamical properties. *Mon. Not. R. Astron. Soc.* **2006**, *370*, 1223–1246. [CrossRef]
226. Brough, S.; Couch, W.J.; Collins, C.A.; Jarrett, T.; Burke, D.J.; Mann, R.G. The luminosity-halo mass relation for brightest cluster galaxies. *Mon. Not. R. Astron. Soc.* **2008**, *385*, L103–L107. [CrossRef]
227. Von Der Linden, A.; Best, P.N.; Kauffmann, G.; White, S.D.M. How special are brightest group and cluster galaxies? *Mon. Not. R. Astron. Soc.* **2007**, *379*, 867–893. [CrossRef]
228. Weinmann, S.M.; van den Bosch, F.C.; Yang, X.; Mo, H.J. Properties of galaxy groups in the Sloan Digital Sky Survey—I. The dependence of colour, star formation and morphology on halo mass. *Mon. Not. R. Astron. Soc.* **2006**, *366*, 2–28. [CrossRef]

229. Gozaliasl, G.; Finoguenov, A.; Khosroshahi, H.G.; Mirkazemi, M.; Erfanianfar, G.; Tanaka, M. Brightest group galaxies: Stellar mass and star formation rate (paper I). *Mon. Not. R. Astron. Soc.* **2016**, *458*, 2762–2775. [CrossRef]
230. Cougo, J.; Rembold, S.B.; Ferrari, F.; Kaipper, A.L.P. Morphometric analysis of brightest cluster galaxies. *Mon. Not. R. Astron. Soc.* **2020**, *498*, 4433–4449. [CrossRef]
231. Loubser, S.I.; Hoekstra, H.; Babul, A.; O'Sullivan, E. Diversity in the stellar velocity dispersion profiles of a large sample of brightest cluster galaxies $z \leq 0.3$. *Mon. Not. R. Astron. Soc.* **2018**, *477*, 335–358. [CrossRef]
232. Murante, G.; Giovalli, M.; Gerhard, O.; Arnaboldi, M.; Borgani, S.; Dolag, K. The importance of mergers for the origin of intracluster stars in cosmological simulations of galaxy clusters. *Mon. Not. R. Astron. Soc.* **2007**, *377*, 2–16. [CrossRef]
233. Dubinski, J. The Origin of the Brightest Cluster Galaxies. *Astrophys. J.* **1998**, *502*, 141–149. [CrossRef]
234. De Lucia, G.; Blaizot, J. The hierarchical formation of the brightest cluster galaxies. *Mon. Not. R. Astron. Soc.* **2007**, *375*, 2–14. [CrossRef]
235. Cooper, A.P.; Gao, L.; Guo, Q.; Frenk, C.S.; Jenkins, A.; Springel, V.; White, S.D.M. Surface photometry of brightest cluster galaxies and intracluster stars in ΛCDM. *Mon. Not. R. Astron. Soc.* **2015**, *451*, 2703–2722. [CrossRef]
236. Nipoti, C. The special growth history of central galaxies in groups and clusters. *Mon. Not. R. Astron. Soc.* **2017**, *467*, 661–673. [CrossRef]
237. Ragone-Figueroa, C.; Granato, G.L.; Ferraro, M.E.; Murante, G.; Biffi, V.; Borgani, S.; Planelles, S.; Rasia, E. BCG mass evolution in cosmological hydro-simulations. *Mon. Not. R. Astron. Soc.* **2018**, *479*, 1125–1136. [CrossRef]
238. Clauwens, B.; Schaye, J.; Franx, M.; Bower, R.G. The three phases of galaxy formation. *Mon. Not. R. Astron. Soc.* **2018**, *478*, 3994–4009. [CrossRef]
239. Davison, T.A.; Norris, M.A.; Pfeffer, J.L.; Davies, J.J.; Crain, R.A. An EAGLE's view of ex situ galaxy growth. *Mon. Not. R. Astron. Soc.* **2020**, *497*, 81–93. [CrossRef]
240. Henden, N.A.; Puchwein, E.; Sijacki, D. The baryon content of groups and clusters of galaxies in the FABLE simulations. *Mon. Not. R. Astron. Soc.* **2020**, *498*, 2114–2137. [CrossRef]
241. Jackson, T.M.; Pasquali, A.; Pacifici, C.; Engler, C.; Pillepich, A.; Grebel, E.K. The stellar mass assembly of low-redshift, massive, central galaxies in SDSS and the TNG300 simulation. *Mon. Not. R. Astron. Soc.* **2020**, *497*, 4262–4275. [CrossRef]
242. Katsianis, A.; Xu, H.; Yang, X.; Luo, Y.; Cui, W.; Davé, R.; Lagos, C.D.P.; Zheng, X.; Zhao, P. The specific star formation rate function at different mass scales and quenching: A comparison between cosmological models and SDSS. *Mon. Not. R. Astron. Soc.* **2021**, *500*, 2036–2048. [CrossRef]
243. Remus, R.S.; Dolag, K.; Hoffmann, T. The Outer Halos of Very Massive Galaxies: BCGs and their DSC in the Magneticum Simulations. *Galaxies* **2017**, *5*, 49. [CrossRef]
244. Tacchella, S.; Diemer, B.; Hernquist, L.; Genel, S.; Marinacci, F.; Nelson, D.; Pillepich, A.; Rodriguez-Gomez, V.; Sales, L.V.; Springel, V.; Vogelsberger, M. Morphology and star formation in IllustrisTNG: The build-up of spheroids and discs. *Mon. Not. R. Astron. Soc.* **2019**, *487*, 5416–5440. [CrossRef]
245. Gonzalez, A.H.; Sivanandam, S.; Zabludoff, A.I.; Zaritsky, D. Galaxy Cluster Baryon Fractions Revisited. *Astrophys. J.* **2013**, *778*, 14. [CrossRef]
246. Coupon, J.; Arnouts, S.; van Waerbeke, L.; Moutard, T.; Ilbert, O.; van Uitert, E.; Erben, T.; Garilli, B.; Guzzo, L.; Heymans, C.; et al. The galaxy–halo connection from a joint lensing, clustering and abundance analysis in the CFHTLenS/VIPERS field. *Mon. Not. R. Astron. Soc.* **2015**, *449*, 1352–1379. [CrossRef]
247. Girelli, G.; Pozzetti, L.; Bolzonella, M.; Giocoli, C.; Marulli, F.; Baldi, M. The stellar-to-halo mass relation over the past 12 Gyr. I. Standard ΛCDM model. *Astron. Astrophys.* **2020**, *634*, A135. [CrossRef]
248. Yang, X.; Mo, H.J.; van den Bosch, F.C. Galaxy Groups in the SDSS DR4. II. Halo Occupation Statistics. *Astrophys. J.* **2008**, *676*, 248–261. [CrossRef]
249. Yang, X.; Mo, H.J.; van den Bosch, F.C.; Zhang, Y.; Han, J. Evolution of the Galaxy-Dark Matter Connection and the Assembly of Galaxies in Dark Matter Halos. *Astrophys. J.* **2012**, *752*, 41. [CrossRef]
250. Moster, B.P.; Naab, T.; White, S.D.M. Galactic star formation and accretion histories from matching galaxies to dark matter haloes. *Mon. Not. R. Astron. Soc.* **2013**, *428*, 3121–3138. [CrossRef]
251. Van Uitert, E.; Cacciato, M.; Hoekstra, H.; Brouwer, M.; Sifón, C.; Viola, M.; Baldry, I.; Bland-Hawthorn, J.; Brough, S.; Brown, M.J.I.; et al. The stellar-to-halo mass relation of GAMA galaxies from 100 deg^2 of KiDS weak lensing data. *Mon. Not. R. Astron. Soc.* **2016**, *459*, 3251–3270. [CrossRef]
252. Erfanianfar, G.; Finoguenov, A.; Furnell, K.; Popesso, P.; Biviano, A.; Wuyts, S.; Collins, C.A.; Mirkazemi, M.; Comparat, J.; Khosroshahi, H.; et al. Stellar mass-halo mass relation for the brightest central galaxies of X-ray clusters since $z \sim 0.65$. *Astron. Astrophys.* **2019**, *631*, A175. [CrossRef]
253. Kolokythas, K.; Vaddi, S. Star-formation & galaxy growth in the dominant galaxies of CLoGS sample: Feedback implications in Local Universe galaxy groups. In preparation.
254. Sand, D.J.; Graham, M.L.; Bildfell, C.; Zaritsky, D.; Pritchet, C.; Hoekstra, H.; Just, D.W.; Herbert-Fort, S.; Sivanandam, S.; Foley, R.J.; et al. The Multi-Epoch nearby Cluster Survey: Type Ia Supernova Rate Measurement in $z \sim 0.1$ Clusters and the Late-time Delay Time Distribution. *Astrophys. J.* **2012**, *746*, 163. [CrossRef]
255. Bildfell, C.; Hoekstra, H.; Babul, A.; Mahdavi, A. Resurrecting the red from the dead: Optical properties of BCGs in X-ray luminous clusters. *Mon. Not. R. Astron. Soc.* **2008**, *389*, 1637–1654. [CrossRef]

256. Mahdavi, A.; Hoekstra, H.; Babul, A.; Bildfell, C.; Jeltema, T.; Henry, J.P. Joint Analysis of Cluster Observations. II. Chandra/XMM-Newton X-Ray and Weak Lensing Scaling Relations for a Sample of 50 Rich Clusters of Galaxies. *Astrophys. J.* **2013**, *767*, 116. [CrossRef]
257. Hoekstra, H.; Herbonnet, R.; Muzzin, A.; Babul, A.; Mahdavi, A.; Viola, M.; Cacciato, M. The Canadian Cluster Comparison Project: Detailed study of systematics and updated weak lensing masses. *Mon. Not. R. Astron. Soc.* **2015**, *449*, 685–714. [CrossRef]
258. Loubser, S.I.; Babul, A.; Hoekstra, H.; Mahdavi, A.; Donahue, M.; Bildfell, C.; Voit, G.M. The regulation of star formation in cool-core clusters: Imprints on the stellar populations of brightest cluster galaxies. *Mon. Not. R. Astron. Soc.* **2016**, *456*, 1565–1578. [CrossRef]
259. Herbonnet, R.; Sifón, C.; Hoekstra, H.; Bahé, Y.; van der Burg, R.F.J.; Melin, J.B.; von der Linden, A.; Sand, D.; Kay, S.; Barnes, D. CCCP and MENeaCS: (updated) weak-lensing masses for 100 galaxy clusters. *Mon. Not. R. Astron. Soc.* **2020**, *497*, 4684–4703. [CrossRef]
260. Mittal, R.; Whelan, J.T.; Combes, F. Constraining star formation rates in cool-core brightest cluster galaxies. *Mon. Not. R. Astron. Soc.* **2015**, *450*, 2564–2592. [CrossRef]
261. Cooke, K.C.; Fogarty, K.; Kartaltepe, J.S.; Moustakas, J.; O'Dea, C.P.; Postman, M. Stellar Mass and 3.4 μm M/L Ratio Evolution of Brightest Cluster Galaxies in COSMOS since $z \sim 1.0$. *Astrophys. J.* **2018**, *857*, 122. [CrossRef]
262. Whitaker, K.E.; van Dokkum, P.G.; Brammer, G.; Franx, M. The Star Formation Mass Sequence Out to z = 2.5. *Astrophys. J. Lett.* **2012**, *754*, L29. [CrossRef]
263. McDonald, M.; Veilleux, S.; Rupke, D.S.N.; Mushotzky, R.; Reynolds, C. Star formation efficiency in the cool cores of galaxy clusters. *Astrophys. J.* **2011**, *734*, 95. [CrossRef]
264. McCarthy, I.G.; Balogh, M.L.; Babul, A.; Poole, G.B.; Horner, D.J. Models of the Intracluster Medium with Heating and Cooling: Explaining the Global and Structural X-Ray Properties of Clusters. *Astrophys. J.* **2004**, *613*, 811–830. [CrossRef]
265. Andrade-Santos, F.; Jones, C.; Forman, W.R.; Lovisari, L.; Vikhlinin, A.; Weeren, R.J.V.; Murray, S.S.; Kraft, R.; Mazzotta, P.; David, L.; et al. The Fraction of Cool-core Clusters in X-ray versus SZ Samples Using Chandra Observations. *Astrophys. J.* **2017**, *843*. [CrossRef]
266. Bonjean, V.; Aghanim, N.; Salomé, P.; Beelen, A.; Douspis, M.; Soubrié, E. Star formation rates and stellar masses from machine learning. *Astron. Astrophys.* **2019**, *622*, A137. [CrossRef]
267. Kennicutt, R.C.; Evans, N.J. Star Formation in the Milky Way and Nearby Galaxies. *Annu. Rev. Astron. Astrophys.* **2012**, *50*, 531–608. [CrossRef]
268. Pipino, A.; Kaviraj, S.; Bildfell, C.; Babul, A.; Hoekstra, H.; Silk, J. Evidence for recent star formation in BCGs: A correspondence between blue cores and UV excess. *Mon. Not. R. Astron. Soc.* **2009**, *395*, 462–471. [CrossRef]
269. Fogarty, K.; Postman, M.; Larson, R.; Donahue, M.; Moustakas, J. The Relationship Between Brightest Cluster Galaxy Star Formation and the Intracluster Medium in CLASH. *Astrophys. J.* **2017**, *846*, 103. [CrossRef]
270. Mittal, R.; McDonald, M.; Whelan, J.T.; Bruzual, G. The challenging task of determining star formation rates: The case of a massive stellar burst in the brightest cluster galaxy of Phoenix galaxy cluster. *Mon. Not. R. Astron. Soc.* **2017**, *465*, 3143–3153. [CrossRef]
271. Loubser, S.I.; Sánchez-Blázquez, P. The ultraviolet upturn in brightest cluster galaxies. *Mon. Not. R. Astron. Soc.* **2011**, *410*, 2679–2689. [CrossRef]
272. Li, Y.T.; Chen, L.W. First ranked galaxies of non-elliptical morphology. *Mon. Not. R. Astron. Soc.* **2019**, *482*, 4084–4095. [CrossRef]
273. Scannapieco, C.; Gadotti, D.A.; Jonsson, P.; White, S.D.M. An observer's view of simulated galaxies: Disc-to-total ratios, bars and (pseudo-)bulges. *Mon. Not. R. Astron. Soc.* **2010**, *407*, L41–L45. [CrossRef]
274. Bottrell, C.; Torrey, P.; Simard, L.; Ellison, S.L. Galaxies in the Illustris simulation as seen by the Sloan Digital Sky Survey—II. Size-luminosity relations and the deficit of bulge-dominated galaxies in Illustris at low mass. *Mon. Not. R. Astron. Soc.* **2017**, *467*, 2879–2895. [CrossRef]
275. Zhu, L.; van de Ven, G.; Méndez-Abreu, J.; Obreja, A. Morphology and kinematics of orbital components in CALIFA galaxies across the Hubble sequence. *Mon. Not. R. Astron. Soc.* **2018**, *479*, 945–960. [CrossRef]
276. Valageas, P.; Silk, J. The entropy history of the universe. *Astron. Astrophys.* **1999**, *350*, 725–742.
277. Nath, B.B.; Roychowdhury, S. Heating of the intracluster medium by quasar outflows. *Mon. Not. R. Astron. Soc.* **2002**, *333*, 145–155. [CrossRef]
278. Pointon, S.K.; Nielsen, N.M.; Kacprzak, G.G.; Muzahid, S.; Churchill, C.W.; Charlton, J.C. The Impact of the Group Environment on the O VI Circumgalactic Medium. *Astrophys. J.* **2017**, *844*, 23. [CrossRef]
279. Nielsen, N.M.; Kacprzak, G.G.; Pointon, S.K.; Churchill, C.W.; Murphy, M.T. MAGIICAT VI. The Mg II Intragroup Medium Is Kinematically Complex. *Astrophys. J.* **2018**, *869*, 153. [CrossRef]
280. Yoon, J.H.; Putman, M.E.; Thom, C.; Chen, H.W.; Bryan, G.L. Warm Gas in the Virgo Cluster. I. Distribution of Lyα Absorbers. *Astrophys. J.* **2012**, *754*, 84. [CrossRef]
281. Manuwal, A.; Narayanan, A.; Muzahid, S.; Charlton, J.C.; Khaire, V.; Chand, H. C IV absorbers tracing cool gas in dense galaxy group/cluster environments. *Mon. Not. R. Astron. Soc.* **2019**, *485*, 30–46. [CrossRef]
282. Yoon, J.H.; Putman, M.E. Lyα Absorbers and the Coma Cluster. *Astrophys. J.* **2017**, *839*, 117. [CrossRef]
283. Connor, T.; Zahedy, F.S.; Chen, H.W.; Cooper, T.J.; Mulchaey, J.S.; Vikhlinin, A. COS Observations of the Cosmic Web: A Search for the Cooler Components of a Hot, X-Ray Identified Filament. *Astrophys. J. Lett.* **2019**, *884*, L20. [CrossRef]

284. Muzahid, S.; Charlton, J.; Nagai, D.; Schaye, J.; Srianand, R. Discovery of an H i-rich Gas Reservoir in the Outskirts of SZ-effect-selected Clusters. *Astrophys. J. Lett.* **2017**, *846*, L8. [CrossRef]
285. Pradeep, J.; Narayanan, A.; Muzahid, S.; Nagai, D.; Charlton, J.C.; Srianand, R. Detection of metal-rich, cool-warm gas in the outskirts of galaxy clusters. *Mon. Not. R. Astron. Soc.* **2019**, *488*, 5327–5339. [CrossRef]
286. Zahedy, F.S.; Chen, H.W.; Johnson, S.D.; Pierce, R.M.; Rauch, M.; Huang, Y.H.; Weiner, B.J.; Gauthier, J.R. Characterizing circumgalactic gas around massive ellipticals at z $\sim$ 0.4—II. Physical properties and elemental abundances. *Mon. Not. R. Astron. Soc.* **2019**, *484*, 2257–2280. [CrossRef]
287. Biffi, V.; Planelles, S.; Borgani, S.; Rasia, E.; Murante, G.; Fabjan, D.; Gaspari, M. The origin of ICM enrichment in the outskirts of present-day galaxy clusters from cosmological hydrodynamical simulations. *Mon. Not. R. Astron. Soc.* **2018**, *476*, 2689–2703. [CrossRef]
288. McCourt, M.; Sharma, P.; Quataert, E.; Parrish, I.J. Thermal instability in gravitationally stratified plasmas: implications for multiphase structure in clusters and galaxy haloes. *Mon. Not. R. Astron. Soc.* **2012**, *419*, 3319–3337. [CrossRef]
289. Esmerian, C.J.; Kravtsov, A.V.; Hafen, Z.; Faucher-Giguère, C.A.; Quataert, E.; Stern, J.; Kereš, D.; Wetzel, A. Thermal instability in the CGM of $L_\star$ galaxies: Testing 'precipitation' models with the FIRE simulations. *Mon. Not. R. Astron. Soc.* **2021**. [CrossRef]
290. Sharma, P.; McCourt, M.; Quataert, E.; Parrish, I.J. Thermal instability and the feedback regulation of hot haloes in clusters, groups and galaxies. *Mon. Not. R. Astron. Soc.* **2012**, *420*, 3174–3194. [CrossRef]
291. Nelson, D.; Sharma, P.; Pillepich, A.; Springel, V.; Pakmor, R.; Weinberger, R.; Vogelsberger, M.; Marinacci, F.; Hernquist, L. Resolving small-scale cold circumgalactic gas in TNG50. *Mon. Not. R. Astron. Soc.* **2020**, *498*, 2391–2414. [CrossRef]
292. Tumlinson, J.; Thom, C.; Werk, J.K.; Prochaska, J.X.; Tripp, T.M.; Katz, N.; Davé, R.; Oppenheimer, B.D.; Meiring, J.D.; Ford, A.B.; et al. The COS-Halos Survey: Rationale, Design, and a Census of Circumgalactic Neutral Hydrogen. *Astrophys. J.* **2013**, *777*, 59. [CrossRef]
293. Viola, M.; Cacciato, M.; Brouwer, M.; Kuijken, K.; Hoekstra, H.; Norberg, P.; Robotham, A.S.G.; van Uitert, E.; Alpaslan, M.; Baldry, I.K.; et al. Dark matter halo properties of GAMA galaxy groups from 100 square degrees of KiDS weak lensing data. *Mon. Not. R. Astron. Soc.* **2015**, *452*, 3529–3550. [CrossRef]
294. Jeltema, T.E.; Binder, B.; Mulchaey, J.S. The Hot Gas Halos of Galaxies in Groups. *Astrophys. J.* **2008**, *679*, 1162–1172. [CrossRef]
295. Bamford, S.P.; Nichol, R.C.; Baldry, I.K.; Land, K.; Lintott, C.J.; Schawinski, K.; Slosar, A.; Szalay, A.S.; Thomas, D.; Torki, M.; et al. Galaxy Zoo: The dependence of morphology and colour on environment. *Mon. Not. R. Astron. Soc.* **2009**, *393*, 1324–1352. [CrossRef]
296. Bahé, Y.M.; McCarthy, I.G.; Crain, R.A.; Theuns, T. The competition between confinement and ram pressure and its implications for galaxies in groups and clusters. *Mon. Not. R. Astron. Soc.* **2012**, *424*, 1179–1186. [CrossRef]
297. Zinger, E.; Dekel, A.; Kravtsov, A.V.; Nagai, D. Quenching of satellite galaxies at the outskirts of galaxy clusters. *Mon. Not. R. Astron. Soc.* **2018**, *475*, 3654–3681. [CrossRef]
298. Bahé, Y.M.; McCarthy, I.G. Star formation quenching in simulated group and cluster galaxies: when, how, and why? *Mon. Not. R. Astron. Soc.* **2015**, *447*, 969–992. [CrossRef]
299. Applebaum, E.; Brooks, A.M.; Christensen, C.R.; Munshi, F.; Quinn, T.R.; Shen, S.; Tremmel, M. Ultrafaint Dwarfs in a Milky Way Context: Introducing the Mint Condition DC Justice League Simulations. *Astrophys. J.* **2021**, *906*, 96. [CrossRef]
300. Blitz, L.; Rosolowsky, E. The Role of Pressure in GMC Formation II: The H_2-Pressure Relation. *Astrophys. J.* **2006**, *650*, 933–944. [CrossRef]
301. Rahmati, A.; Pawlik, A.H.; Raičević, M.; Schaye, J. On the evolution of the H I column density distribution in cosmological simulations. *Mon. Not. R. Astron. Soc.* **2013**, *430*, 2427–2445. [CrossRef]
302. Gnedin, N.Y.; Draine, B.T. Line Overlap and Self-Shielding of Molecular Hydrogen in Galaxies. *Astrophys. J.* **2014**, *795*, 37. [CrossRef]
303. Marasco, A.; Crain, R.A.; Schaye, J.; Bahé, Y.M.; van der Hulst, T.; Theuns, T.; Bower, R.G. The environmental dependence of H I in galaxies in the EAGLE simulations. *Mon. Not. R. Astron. Soc.* **2016**, *461*, 2630–2649. [CrossRef]
304. Stevens, A.R.H.; Diemer, B.; Lagos, C.d.P.; Nelson, D.; Pillepich, A.; Brown, T.; Catinella, B.; Hernquist, L.; Weinberger, R.; Vogelsberger, M.; et al. Atomic hydrogen in IllustrisTNG galaxies: The impact of environment parallelled with local 21-cm surveys. *Mon. Not. R. Astron. Soc.* **2019**, *483*, 5334–5354. [CrossRef]
305. Diemer, B.; Stevens, A.R.H.; Forbes, J.C.; Marinacci, F.; Hernquist, L.; Lagos, C.d.P.; Sternberg, A.; Pillepich, A.; Nelson, D.; Popping, G.; et al. Modeling the Atomic-to-molecular Transition in Cosmological Simulations of Galaxy Formation. *Astrophys. J. Suppl.* **2018**, *238*, 33. [CrossRef]
306. Bahé, Y.M.; Crain, R.A.; Kauffmann, G.; Bower, R.G.; Schaye, J.; Furlong, M.; Lagos, C.; Schaller, M.; Trayford, J.W.; Dalla Vecchia, C.; et al. The distribution of atomic hydrogen in EAGLE galaxies: Morphologies, profiles, and H I holes. *Mon. Not. R. Astron. Soc.* **2016**, *456*, 1115–1136. [CrossRef]
307. Stevens, A.R.H.; Lagos, C.d.P.; Cortese, L.; Catinella, B.; Diemer, B.; Nelson, D.; Pillepich, A.; Hernquist, L.; Marinacci, F.; Vogelsberger, M. Molecular hydrogen in IllustrisTNG galaxies: Carefully comparing signatures of environment with local CO and SFR data. *Mon. Not. R. Astron. Soc.* **2021**, *502*, 3158–3178. [CrossRef]
308. Saintonge, A.; Catinella, B.; Tacconi, L.J.; Kauffmann, G.; Genzel, R.; Cortese, L.; Davé, R.; Fletcher, T.J.; Graciá-Carpio, J.; Kramer, C.; et al. xCOLD GASS: The Complete IRAM 30 m Legacy Survey of Molecular Gas for Galaxy Evolution Studies. *Astrophys. J. Suppl.* **2017**, *233*, 22. [CrossRef]

309. Genel, S. How Environment Affects Galaxy Metallicity through Stripping and Formation History: Lessons from the Illustris Simulation. *Astrophys. J.* **2016**, *822*, 107. [CrossRef]
310. Bahé, Y.M.; Schaye, J.; Crain, R.A.; McCarthy, I.G.; Bower, R.G.; Theuns, T.; McGee, S.L.; Trayford, J.W. The origin of the enhanced metallicity of satellite galaxies. *Mon. Not. R. Astron. Soc.* **2017**, *464*, 508–529. [CrossRef]
311. Gupta, A.; Yuan, T.; Torrey, P.; Vogelsberger, M.; Martizzi, D.; Tran, K.V.H.; Kewley, L.J.; Marinacci, F.; Nelson, D.; Pillepich, A.; et al. Chemical pre-processing of cluster galaxies over the past 10 billion years in the IllustrisTNG simulations. *Mon. Not. R. Astron. Soc.* **2018**, *477*, L35–L39. [CrossRef]
312. Pasquali, A.; Gallazzi, A.; van den Bosch, F.C. The gas-phase metallicity of central and satellite galaxies in the Sloan Digital Sky Survey. *Mon. Not. R. Astron. Soc.* **2012**, *425*, 273–286. [CrossRef]
313. van de Voort, F.; Bahé, Y.M.; Bower, R.G.; Correa, C.A.; Crain, R.A.; Schaye, J.; Theuns, T. The environmental dependence of gas accretion on to galaxies: quenching satellites through starvation. *Mon. Not. R. Astron. Soc.* **2017**, *466*, 3460–3471. [CrossRef]
314. Watts, A.B.; Power, C.; Catinella, B.; Cortese, L.; Stevens, A.R.H. Global H I asymmetries in IllustrisTNG: A diversity of physical processes disturb the cold gas in galaxies. *Mon. Not. R. Astron. Soc.* **2020**, *499*, 5205–5219. [CrossRef]
315. Chung, A.; van Gorkom, J.H.; Kenney, J.D.P.; Vollmer, B. Virgo Galaxies with Long One-sided H I Tails. *Astrophys. J. Lett.* **2007**, *659*, L115–L119. [CrossRef]
316. Ricarte, A.; Tremmel, M.; Natarajan, P.; Quinn, T. A Link between Ram Pressure Stripping and Active Galactic Nuclei. *Astrophys. J. Lett.* **2020**, *895*, L8. [CrossRef]
317. Poggianti, B.M.; Jaffé, Y.L.; Moretti, A.; Gullieuszik, M.; Radovich, M.; Tonnesen, S.; Fritz, J.; Bettoni, D.; Vulcani, B.; Fasano, G.; et al. Ram-pressure feeding of supermassive black holes. *Nature* **2017**, *548*, 304–309. [CrossRef]
318. Troncoso-Iribarren, P.; Padilla, N.; Santander, C.; Lagos, C.D.P.; García-Lambas, D.; Rodríguez, S.; Contreras, S. The better half—Asymmetric star formation due to ram pressure in the EAGLE simulations. *Mon. Not. R. Astron. Soc.* **2020**, *497*, 4145–4161. [CrossRef]
319. Feldmann, R.; Carollo, C.M.; Mayer, L. The Hubble Sequence in Groups: The Birth of the Early-type Galaxies. *Astrophys. J.* **2011**, *736*, 88. [CrossRef]
320. Joshi, G.D.; Pillepich, A.; Nelson, D.; Marinacci, F.; Springel, V.; Rodriguez-Gomez, V.; Vogelsberger, M.; Hernquist, L. The fate of disc galaxies in IllustrisTNG clusters. *Mon. Not. R. Astron. Soc.* **2020**, *496*, 2673–2703. [CrossRef]
321. van den Bosch, F.C.; Ogiya, G. Dark matter substructure in numerical simulations: A tale of discreteness noise, runaway instabilities, and artificial disruption. *Mon. Not. R. Astron. Soc.* **2018**, *475*, 4066–4087. [CrossRef]
322. Engler, C.; Pillepich, A.; Joshi, G.D.; Nelson, D.; Pasquali, A.; Grebel, E.K.; Lisker, T.; Zinger, E.; Donnari, M.; Marinacci, F.; et al. The distinct stellar-to-halo mass relations of satellite and central galaxies: Insights from the IllustrisTNG simulations. *Mon. Not. R. Astron. Soc.* **2021**, *500*, 3957–3975. [CrossRef]
323. Pallero, D.; Gómez, F.A.; Padilla, N.D.; Bahé, Y.M.; Vega-Martínez, C.A.; Torres-Flores, S. Too dense to go through: The importance of low-mass clusters for satellite quenching. *arXiv* **2020**, arXiv:2012.08593.
324. Jung, S.L.; Choi, H.; Wong, O.I.; Kimm, T.; Chung, A.; Yi, S.K. On the Origin of Gas-poor Galaxies in Galaxy Clusters Using Cosmological Hydrodynamic Simulations. *Astrophys. J.* **2018**, *865*, 156. [CrossRef]
325. Bahé, Y.M.; McCarthy, I.G.; Balogh, M.L.; Font, A.S. Why does the environmental influence on group and cluster galaxies extend beyond the virial radius? *Mon. Not. R. Astron. Soc.* **2013**, *430*, 3017–3031. [CrossRef]
326. Peebles, P.J.E. *The Large-Scale Structure of the Universe*; Princeton University Press: Princeton, NJ, USA, 1980.
327. Bond, J.R.; Efstathiou, G.; Silk, J. Massive neutrinos and the large-scale structure of the universe. *Phys. Rev. Lett.* **1980**, *45*, 1980–1984. [CrossRef]
328. Davis, M.; Efstathiou, G.; Frenk, C.S.; White, S.D.M. The evolution of large-scale structure in a universe dominated by cold dark matter. *Astrophys. J.* **1985**, *292*, 371–394. [CrossRef]
329. Kaiser, N. Clustering in real space and in redshift space. *Mon. Not. R. Astron. Soc.* **1987**, *227*, 1–21. [CrossRef]
330. Peacock, J.A.; Dodds, S.J. Reconstructing the Linear Power Spectrum of Cosmological Mass Fluctuations. *Mon. Not. R. Astron. Soc.* **1994**, *267*, 1020. [CrossRef]
331. Mead, A.J.; Heymans, C.; Lombriser, L.; Peacock, J.A.; Steele, O.I.; Winther, H.A. Accurate halo-model matter power spectra with dark energy, massive neutrinos and modified gravitational forces. *Mon. Not. R. Astron. Soc.* **2016**, *459*, 1468–1488. [CrossRef]
332. Takahashi, R.; Sato, M.; Nishimichi, T.; Taruya, A.; Oguri, M. Revising the Halofit Model for the Nonlinear Matter Power Spectrum. *Astrophys. J.* **2012**, *761*, 152. [CrossRef]
333. van Daalen, M.P.; Schaye, J.; Booth, C.M.; Dalla Vecchia, C. The effects of galaxy formation on the matter power spectrum: A challenge for precision cosmology. *Mon. Not. R. Astron. Soc.* **2011**, *415*, 3649–3665. [CrossRef]
334. Schneider, A.; Teyssier, R. A new method to quantify the effects of baryons on the matter power spectrum. *J. Cosmol. Astropart. Phys.* **2015**, *12*, 049. [CrossRef]
335. Mummery, B.O.; McCarthy, I.G.; Bird, S.; Schaye, J. The separate and combined effects of baryon physics and neutrino free streaming on large-scale structure. *Mon. Not. R. Astron. Soc.* **2017**, *471*, 227–242. [CrossRef]
336. Semboloni, E.; Hoekstra, H.; Schaye, J.; van Daalen, M.P.; McCarthy, I.G. Quantifying the effect of baryon physics on weak lensing tomography. *Mon. Not. R. Astron. Soc.* **2011**, *417*, 2020–2035. [CrossRef]
337. Debackere, S.N.B.; Schaye, J.; Hoekstra, H. How baryons can significantly bias cluster count cosmology. *Mon. Not. R. Astron. Soc.* **2021**. [CrossRef]

338. Mead, A.J.; Tröster, T.; Heymans, C.; Van Waerbeke, L.; McCarthy, I.G. A hydrodynamical halo model for weak-lensing cross correlations. *Astron. Astrophys.* **2020**, *641*, A130. [CrossRef]
339. van Daalen, M.P.; Schaye, J. The contributions of matter inside and outside of haloes to the matter power spectrum. *Mon. Not. R. Astron. Soc.* **2015**, *452*, 2247–2257. [CrossRef]
340. Huterer, D.; Takada, M. Calibrating the nonlinear matter power spectrum: Requirements for future weak lensing surveys. *Astropart. Phys.* **2005**, *23*, 369–376. [CrossRef]
341. Hearin, A.P.; Zentner, A.R.; Ma, Z. General requirements on matter power spectrum predictions for cosmology with weak lensing tomography. *J. Cosmol. Astropart. Phys.* **2012**, *2012*, 034. [CrossRef]
342. Puchwein, E.; Sijacki, D.; Springel, V. Simulations of AGN Feedback in Galaxy Clusters and Groups: Impact on Gas Fractions and the L_X-T Scaling Relation. *Astrophys. J. Lett.* **2008**, *687*, L53. [CrossRef]
343. Chisari, N.E.; Richardson, M.L.A.; Devriendt, J.; Dubois, Y.; Schneider, A.; Le Brun, A.M.C.; Beckmann, R.S.; Peirani, S.; Slyz, A.; Pichon, C. The impact of baryons on the matter power spectrum from the Horizon-AGN cosmological hydrodynamical simulation. *Mon. Not. R. Astron. Soc.* **2018**, *480*, 3962–3977. [CrossRef]
344. van Daalen, M.P.; McCarthy, I.G.; Schaye, J. Exploring the effects of galaxy formation on matter clustering through a library of simulation power spectra. *Mon. Not. R. Astron. Soc.* **2020**, *491*, 2424–2446. [CrossRef]
345. Schneider, A.; Stoira, N.; Refregier, A.; Weiss, A.J.; Knabenhans, M.; Stadel, J.; Teyssier, R. Baryonic effects for weak lensing. Part I. Power spectrum and covariance matrix. *J. Cosmol. Astropart. Phys.* **2020**, *2020*, 019. [CrossRef]
346. Aricò, G.; Angulo, R.E.; Contreras, S.; Ondaro-Mallea, L.; Pellejero-Ibañez, M.; Zennaro, M. The BACCO Simulation Project: A baryonification emulator with Neural Networks. *arXiv* **2020**, arXiv:2011.15018.
347. McCarthy, I.G.; Bird, S.; Schaye, J.; Harnois-Deraps, J.; Font, A.S.; van Waerbeke, L. The BAHAMAS project: The CMB-large-scale structure tension and the roles of massive neutrinos and galaxy formation. *Mon. Not. R. Astron. Soc.* **2018**, *476*, 2999–3030. [CrossRef]
348. White, S.D.M.; Navarro, J.F.; Evrard, A.E.; Frenk, C.S. The baryon content of galaxy clusters: A challenge to cosmological orthodoxy. *Nature* **1993**, *366*, 429–433. [CrossRef]
349. Pillepich, A.; Porciani, C.; Reiprich, T.H. The X-ray cluster survey with eRosita: Forecasts for cosmology, cluster physics and primordial non-Gaussianity. *Mon. Not. R. Astron. Soc.* **2012**, *422*, 44–69. [CrossRef]
350. Pierre, M.; Pacaud, F.; Adami, C.; Alis, S.; Altieri, B.; Baran, N.; Benoist, C.; Birkinshaw, M.; Bongiorno, A.; Bremer, M.N.; et al. The XXL Survey. I. Scientific motivations—XMM-Newton observing plan—Follow-up observations and simulation programme. *Astron. Astrophys.* **2016**, *592*, A1. [CrossRef]
351. Bocquet, S.; Saro, A.; Dolag, K.; Mohr, J.J. Halo mass function: Baryon impact, fitting formulae, and implications for cluster cosmology. *Mon. Not. R. Astron. Soc.* **2016**, *456*, 2361–2373. [CrossRef]
352. Pillepich, A.; Reiprich, T.H.; Porciani, C.; Borm, K.; Merloni, A. Forecasts on dark energy from the X-ray cluster survey with eROSITA: constraints from counts and clustering. *Mon. Not. R. Astron. Soc.* **2018**, *481*, 613–626. [CrossRef]
353. Wijers, N.A.; Schaye, J.; Oppenheimer, B.D. The warm-hot circumgalactic medium around EAGLE-simulation galaxies and its detection prospects with X-ray and UV line absorption. *Mon. Not. R. Astron. Soc.* **2020**, *498*, 574–598. [CrossRef]
354. Yoshida, N.; Furlanetto, S.R.; Hernquist, L. The Temperature Structure of the Warm-Hot Intergalactic Medium. *Astrophys. J. Lett.* **2005**, *618*, L91–L94. [CrossRef]
355. Biffi, V.; Dolag, K.; Böhringer, H.; Lemson, G. Observing simulated galaxy clusters with PHOX: A novel X-ray photon simulator. *Mon. Not. R. Astron. Soc.* **2012**, *420*, 3545–3556. [CrossRef]
356. Biffi, V.; Dolag, K.; Böhringer, H. Investigating the velocity structure and X-ray observable properties of simulated galaxy clusters with PHOX. *Mon. Not. R. Astron. Soc.* **2013**, *428*, 1395–1409. [CrossRef]
357. Zuhone, J.A.; Hallman, E.J. pyXSIM: Synthetic X-ray observations generator. *arXiv* **2016**, arXiv:1608.002.
358. Dauser, T.; Falkner, S.; Lorenz, M.; Kirsch, C.; Peille, P.; Cucchetti, E.; Schmid, C.; Brand, T.; Oertel, M.; Smith, R.; et al. SIXTE: A generic X-ray instrument simulation toolkit. *Astron. Astrophys.* **2019**, *630*, A66. [CrossRef]
359. Merloni, A.; Predehl, P.; Becker, W.; Böhringer, H.; Boller, T.; Brunner, H.; Brusa, M.; Dennerl, K.; Freyberg, M.; Friedrich, P.; et al. eROSITA Science Book: Mapping the Structure of the Energetic Universe. *arXiv* **2012**, arXiv:1209.3114.
360. Biffi, V.; Dolag, K.; Merloni, A. AGN contamination of galaxy-cluster thermal X-ray emission: Predictions for eRosita from cosmological simulations. *Mon. Not. R. Astron. Soc.* **2018**, *481*, 2213–2227. [CrossRef]
361. Oppenheimer, B.D.; Bogdán, Á.; Crain, R.A.; ZuHone, J.A.; Forman, W.R.; Schaye, J.; Wijers, N.A.; Davies, J.J.; Jones, C.; Kraft, R.P.; et al. EAGLE and Illustris-TNG Predictions for Resolved eROSITA X-Ray Observations of the Circumgalactic Medium around Normal Galaxies. *Astrophys. J. Lett.* **2020**, *893*, L24. [CrossRef]
362. Mazzotta, P.; Rasia, E.; Moscardini, L.; Tormen, G. Comparing the temperatures of galaxy clusters from hydrodynamical N-body simulations to Chandra and XMM-Newton observations. *Mon. Not. R. Astron. Soc.* **2004**, *354*, 10–24. [CrossRef]
363. Rasia, E.; Ettori, S.; Moscardini, L.; Mazzotta, P.; Borgani, S.; Dolag, K.; Tormen, G.; Cheng, L.M.; Diaferio, A. Systematics in the X-ray cluster mass estimators. *Mon. Not. R. Astron. Soc.* **2006**, *369*, 2013–2024. [CrossRef]
364. Barnes, D.J.; Vogelsberger, M.; Pearce, F.A.; Pop, A.R.; Kannan, R.; Cao, K.; Kay, S.T.; Hernquist, L. Characterizing hydrostatic mass bias with MOCK-X. *Mon. Not. R. Astron. Soc.* **2021**. [CrossRef]

365. Avila, S.; Crocce, M.; Ross, A.J.; García-Bellido, J.; Percival, W.J.; Banik, N.; Camacho, H.; Kokron, N.; Chan, K.C.; Andrade-Oliveira, F.; et al. Dark Energy Survey Year-1 results: Galaxy mock catalogues for BAO. *Mon. Not. R. Astron. Soc.* **2018**, *479*, 94–110. [CrossRef]
366. Korytov, D.; Hearin, A.; Kovacs, E.; Larsen, P.; Rangel, E.; Hollowed, J.; Benson, A.J.; Heitmann, K.; Mao, Y.Y.; Bahmanyar, A.; et al. CosmoDC2: A Synthetic Sky Catalog for Dark Energy Science with LSST. *Astrophys. J. Suppl.* **2019**, *245*, 26. [CrossRef]
367. To, C.H.; Krause, E.; Rozo, E.; Wu, H.Y.; Gruen, D.; DeRose, J.; Rykoff, E.; Wechsler, R.H.; Becker, M.; Costanzi, M.; et al. Combination of cluster number counts and two-point correlations: validation on mock Dark Energy Survey. *Mon. Not. R. Astron. Soc.* **2021**, *502*, 4093–4111. [CrossRef]
368. Arnaud, M.; Pratt, G.W.; Piffaretti, R.; Böhringer, H.; Croston, J.H.; Pointecouteau, E. The universal galaxy cluster pressure profile from a representative sample of nearby systems (REXCESS) and the Y_{SZ} - M_{500} relation. *Astron. Astrophys.* **2010**, *517*, A92. [CrossRef]
369. Le Brun, A.M.C.; McCarthy, I.G.; Melin, J.B. Testing Sunyaev-Zel'dovich measurements of the hot gas content of dark matter haloes using synthetic skies. *Mon. Not. R. Astron. Soc.* **2015**, *451*, 3868–3881. [CrossRef]
370. McQuinn, M. Locating the "Missing" Baryons with Extragalactic Dispersion Measure Estimates. *Astrophys. J. Lett.* **2014**, *780*, L33. [CrossRef]
371. Prochaska, J.X.; Zheng, Y. Probing Galactic haloes with fast radio bursts. *Mon. Not. R. Astron. Soc.* **2019**, *485*, 648–665. [CrossRef]
372. Gaspari, M.; Eckert, D.; Ettori, S.; Tozzi, P.; Bassini, L.; Rasia, E.; Brighenti, F.; Sun, M.; Borgani, S.; Johnson, S.D.; et al. The X-Ray Halo Scaling Relations of Supermassive Black Holes. *Astrophys. J.* **2019**, *884*, 169. [CrossRef]
373. Bogdán, Á.; Lovisari, L.; Volonteri, M.; Dubois, Y. Correlation between the Total Gravitating Mass of Groups and Clusters and the Supermassive Black Hole Mass of Brightest Galaxies. *Astrophys. J.* **2018**, *852*, 131. [CrossRef]
374. Booth, C.M.; Schaye, J. Dark matter haloes determine the masses of supermassive black holes. *Mon. Not. R. Astron. Soc.* **2010**, *405*, L1–L5. [CrossRef]
375. Truong, N.; Pillepich, A.; Werner, N. Correlations between supermassive black holes and hot gas atmospheres in IllustrisTNG and X-ray observations. *Mon. Not. R. Astron. Soc.* **2021**, *501*, 2210–2230. [CrossRef]
376. Gaspari, M.; Tombesi, F.; Cappi, M. Linking macro-, meso- and microscales in multiphase AGN feeding and feedback. *Nature Astronomy* **2020**, *4*, 10–13. [CrossRef]
377. Bassini, L.; Rasia, E.; Borgani, S.; Ragone-Figueroa, C.; Biffi, V.; Dolag, K.; Gaspari, M.; Granato, G.L.; Murante, G.; Taffoni, G.; Tornatore, L. Black hole mass of central galaxies and cluster mass correlation in cosmological hydro-dynamical simulations. *Astron. Astrophys.* **2019**, *630*, A144. [CrossRef]
378. Cowie, L.L.; Binney, J. Radiative regulation of gas flow within clusters of galaxies: A model for cluster X-ray sources. *Astrophys. J.* **1977**, *215*, 723–732. [CrossRef]
379. Fabian, A.C. Cooling Flows in Clusters of Galaxies. *Annu. Rev. Astron. Astrophys.* **1994**, *32*, 277–318. [CrossRef]
380. Peterson, J.; Fabian, A. X-ray spectroscopy of cooling clusters. *Phys. Rep.* **2006**, *427*, 1–39. [CrossRef]
381. Gaspari, M.; Ruszkowski, M.; Sharma, P. Cause and Effect of Feedback: Multiphase Gas in Cluster Cores Heated by AGN Jets. *Astrophys. J.* **2012**, *746*, 94. [CrossRef]
382. Li, Y.; Bryan, G.L. Modeling Active Galactic Nucleus Feedback in Cool-core Clusters: The Balance between Heating and Cooling. *Astrophys. J.* **2014**, *789*, 54. [CrossRef]
383. Li, Y.; Bryan, G.L.; Ruszkowski, M.; Voit, G.M.; O'Shea, B.W.; Donahue, M. Cooling, AGN Feedback, and Star Formation in Simulated Cool-core Galaxy Clusters. *Astrophys. J.* **2015**, *811*, 73. [CrossRef]
384. Prasad, D.; Sharma, P.; Babul, A. Cool Core Cycles: Cold Gas and AGN Jet Feedback in Cluster Cores. *Astrophys. J.* **2015**, *811*, 108. [CrossRef]
385. Prasad, D.; Sharma, P.; Babul, A. AGN jet-driven stochastic cold accretion in cluster cores. *Mon. Not. R. Astron. Soc.* **2017**, *471*, 1531–1542. [CrossRef]
386. Prasad, D.; Sharma, P.; Babul, A. Cool-core Clusters: The Role of BCG, Star Formation, and AGN-driven Turbulence. *Astrophys. J.* **2018**, *863*, 62. [CrossRef]
387. Nemmen, R.S.; Tchekhovskoy, A. On the efficiency of jet production in radio galaxies. *Mon. Not. R. Astron. Soc.* **2015**, *449*, 316–327. [CrossRef]
388. McNamara, B.R.; Rohanizadegan, M.; Nulsen, P.E.J. Are Radio Active Galactic Nuclei Powered by Accretion or Black Hole Spin? *Astrophys. J.* **2011**, *727*, 39. [CrossRef]
389. Hummels, C.B.; Smith, B.D.; Hopkins, P.F.; O'Shea, B.W.; Silvia, D.W.; Werk, J.K.; Lehner, N.; Wise, J.H.; Collins, D.C.; Butsky, I.S. The Impact of Enhanced Halo Resolution on the Simulated Circumgalactic Medium. *Astrophys. J.* **2019**, *882*, 156. [CrossRef]
390. Hopkins, P.F.; Quataert, E. How do massive black holes get their gas? *Mon. Not. R. Astron. Soc.* **2010**, *407*, 1529–1564. [CrossRef]
391. Gaspari, M.; Brighenti, F.; Temi, P.; Ettori, S. Can AGN Feedback Break the Self-similarity of Galaxies, Groups, and Clusters? *Astrophys. J. Lett.* **2014**, *783*, L10. [CrossRef]
392. Meece, G.R.; Voit, G.M.; O'Shea, B.W. Triggering and Delivery Algorithms for AGN Feedback. *Astrophys. J.* **2017**, *841*, 133. [CrossRef]
393. Vernaleo, J.C.; Reynolds, C.S. AGN Feedback and Cooling Flows: Problems with Simple Hydrodynamic Models. *Astrophys. J.* **2006**, *645*, 83–94. [CrossRef]

394. Cielo, S.; Babul, A.; Antonuccio-Delogu, V.; Silk, J.; Volonteri, M. Feedback from reorienting AGN jets. I. Jet-ICM coupling, cavity properties and global energetics. *Astron. Astrophys.* **2018**, *617*, A58. [CrossRef]
395. Babul, A.; Sharma, P.; Reynolds, C.S. Isotropic Heating of Galaxy Cluster Cores via Rapidly Reorienting Active Galactic Nucleus Jets. *Astrophys. J.* **2013**, *768*, 11. [CrossRef]
396. Su, K.Y.; Hopkins, P.F.; Bryan, G.L.; Somerville, R.S.; Hayward, C.C.; Anglés-Alcázar, D.; Faucher-Giguère, C.A.; Wellons, S.; Stern, J.; Terrazas, B.A.; et al. Which AGN Jets Quench Star Formation in Massive Galaxies? *arXiv* **2021**, arXiv:2102.02206.
397. Käfer, F.; Finoguenov, A.; Eckert, D.; Clerc, N.; Ramos-Ceja, M.E.; Sanders, J.S.; Ghirardini, V. Toward the low-scatter selection of X-ray clusters. Galaxy cluster detection with eROSITA through cluster outskirts. *Astron. Astrophys.* **2020**, *634*, A8. [CrossRef]
398. Pearson, R.J.; Ponman, T.J.; Norberg, P.; Robotham, A.S.G.; Babul, A.; Bower, R.G.; McCarthy, I.G.; Brough, S.; Driver, S.P.; Pimbblet, K. Galaxy And Mass Assembly: Search for a population of high-entropy galaxy groups. *Mon. Not. R. Astron. Soc.* **2017**, *469*, 3489–3504. [CrossRef]
399. Moretti, A.; Paladino, R.; Poggianti, B.M.; Serra, P.; Ramatsoku, M.; Franchetto, A.; Deb, T.; Gullieuszik, M.; Tomičić, N.; Mingozzi, M.; et al. The High Molecular Gas Content, and the Efficient Conversion of Neutral into Molecular Gas, in Jellyfish Galaxies. *Astrophys. J. Lett.* **2020**, *897*, L30. [CrossRef]
400. Tonnesen, S.; Bryan, G.L. Gas Stripping in Simulated Galaxies with a Multiphase Interstellar Medium. *Astrophys. J.* **2009**, *694*, 789–804. [CrossRef]
401. Ploeckinger, S.; Schaye, J. Radiative cooling rates, ion fractions, molecule abundances, and line emissivities including self-shielding and both local and metagalactic radiation fields. *Mon. Not. R. Astron. Soc.* **2020**, *497*, 4857–4883. [CrossRef]
402. Ahn, C.P.; Alexandroff, R.; Allende Prieto, C.; Anders, F.; Anderson, S.F.; Anderton, T.; Andrews, B.H.; Aubourg, É.; Bailey, S.; Bastien, F.A.; et al. The Tenth Data Release of the Sloan Digital Sky Survey: First Spectroscopic Data from the SDSS-III Apache Point Observatory Galactic Evolution Experiment. *Astrophys. J. Suppl.* **2014**, *211*, 17. [CrossRef]
403. Amodeo, S.; Battaglia, N.; Schaan, E.; Ferraro, S.; Moser, E.; Aiola, S.; Austermann, J.E.; Beall, J.A.; Bean, R.; Becker, D.T.; et al. Atacama Cosmology Telescope: Modeling the gas thermodynamics in BOSS CMASS galaxies from kinematic and thermal Sunyaev-Zel'dovich measurements. *Phys. Rev. D* **2021**, *103*, 063514. [CrossRef]
404. Schaan, E.; Ferraro, S.; Amodeo, S.; Battaglia, N.; Aiola, S.; Austermann, J.E.; Beall, J.A.; Bean, R.; Becker, D.T.; Bond, R.J.; et al. Atacama Cosmology Telescope: Combined kinematic and thermal Sunyaev-Zel'dovich measurements from BOSS CMASS and LOWZ halos. *Phys. Rev. D* **2021**, *103*, 063513. [CrossRef]
405. Navarro, J.F.; Frenk, C.S.; White, S.D.M. A Universal Density Profile from Hierarchical Clustering. *Astrophys. J.* **1997**, *490*, 493–508. [CrossRef]
406. Lim, S.H.; Barnes, D.; Vogelsberger, M.; Mo, H.J.; Nelson, D.; Pillepich, A.; Dolag, K.; Marinacci, F. Properties of the ionized CGM and IGM: Tests for galaxy formation models from the Sunyaev-Zel'dovich effect. *Mon. Not. R. Astron. Soc.* **2021**, *504*, 5131–5143. [CrossRef]
407. Planck Collaboration; Ade, P.A.R.; Aghanim, N.; Arnaud, M.; Ashdown, M.; Atrio-Barandela, F.; Aumont, J.; Baccigalupi, C.; Balbi, A.; Banday, A.J.; et al. Planck intermediate results. XI. The gas content of dark matter halos: The Sunyaev-Zeldovich-stellar mass relation for locally brightest galaxies. *Astron. Astrophys.* **2013**, *557*, A52. [CrossRef]
408. The EAGLE team. The EAGLE simulations of galaxy formation: Public release of particle data. *arXiv* **2017**, arXiv:1706.09899.
409. Nelson, D.; Springel, V.; Pillepich, A.; Rodriguez-Gomez, V.; Torrey, P.; Genel, S.; Vogelsberger, M.; Pakmor, R.; Marinacci, F.; Weinberger, R.; et al. The IllustrisTNG simulations: Public data release. *Comput. Astrophys. Cosmol.* **2019**, *6*, 2. [CrossRef]

MDPI

Review

The Metal Content of the Hot Atmospheres of Galaxy Groups

Fabio Gastaldello [1,*], Aurora Simionescu [2,3,4], Francois Mernier [2,5], Veronica Biffi [6,7], Massimo Gaspari [8,9], Kosuke Sato [10] and Kyoko Matsushita [11]

1 Istituto di Astrofisica Spaziale e Fisica Cosmica (IASF)—Milano, National Institute of Astrophysics (INAF), via A. Corti 12, I-20133 Milano, Italy
2 SRON Netherlands Institute for Space Research, Sorbonnelaan 2, 3584 CA Utrecht, The Netherlands; a.simionescu@sron.nl (A.S.); francois.mernier@esa.int (F.M.)
3 Leiden Observatory, Leiden University, P.O. Box 9513, 2300 RA Leiden, The Netherlands
4 Kavli Institute for the Physics and Mathematics of the Universe (WPI), The University of Tokyo, Kashiwa, Chiba 277-8583, Japan
5 European Space Agency (ESA), European Space Research and Technology Center (ESTEC), Keplerlaan 1, 2201 AZ Noordwijk, The Netherlands
6 INAF—Osservatorio Astronomico di Trieste, via Tiepolo 11, I-34143 Trieste, Italy; veronica.biffi@inaf.it
7 IFPU—Institute for Fundamental Physics of the Universe, via Beirut 2, I-34014 Trieste, Italy
8 INAF—Osservatorio di Astrofisica e Scienza dello Spazio di Bologna, via Piero Gobetti 93/3, I-40129 Bologna, Italy; massimo.gaspari@inaf.it
9 Deparment of Astrophysical Sciences, Princeton University, 4 Ivy Lane, Princeton, NJ 08544, USA
10 Department of Physics, Saitama University, 255 Shimo-okubo, Sakura-ku, Saitama-shi , Saitama 338-8520, Japan; ksksato@phy.saitama-u.ac.jp
11 Department of Physics, Tokyo University of Science, 1-3 Kagurazaka, Shinjuku-ku, Tokyo 162-8601, Japan; matusita@rs.tus.ac.jp
* Correspondence: fabio.gastaldello@inaf.it

Citation: Gastaldello, F.; Simionescu, A.; Mernier, F.; Biffi, V.; Gaspari, M.; Sato, K.; Matsushita, K. The Metal Content of the Hot Atmospheres of Galaxy Groups. *Universe* **2021**, *7*, 208. https://doi.org/10.3390/universe7070208

Academic Editor: Francesco Shankar

Received: 31 March 2021
Accepted: 16 June 2021
Published: 24 June 2021

Abstract: Galaxy groups host the majority of matter and more than half of all the galaxies in the Universe. Their hot (10^7 K), X-ray emitting intra-group medium (IGrM) reveals emission lines typical of many elements synthesized by stars and supernovae. Because their gravitational potentials are shallower than those of rich galaxy clusters, groups are ideal targets for studying, through X-ray observations , feedback effects, which leave important marks on their gas and metal contents. Here, we review the history and present status of the chemical abundances in the IGrM probed by X-ray spectroscopy. We discuss the limitations of our current knowledge, in particular due to uncertainties in the modeling of the Fe-L shell by plasma codes, and coverage of the volume beyond the central region. We further summarize the constraints on the abundance pattern at the group mass scale and the insight it provides to the history of chemical enrichment. Parallel to the observational efforts, we review the progress made by both cosmological hydrodynamical simulations and controlled high-resolution 3D simulations to reproduce the radial distribution of metals in the IGrM, the dependence on system mass from group to cluster scales, and the role of AGN and SN feedback in producing the observed phenomenology. Finally, we highlight future prospects in this field, where progress will be driven both by a much richer sample of X-ray emitting groups identified with eROSITA, and by a revolution in the study of X-ray spectra expected from micro-calorimeters onboard XRISM and ATHENA.

Keywords: galaxies:abundances; galaxies:clusters:intracluster medium; X-rays:galaxies

1. Introduction

Two major astrophysical discoveries have provided key answers to the fundamental question of the origin of the chemical elements in the past century: the discovery that stellar nucleosynthesis is responsible for the production of all the heavy elements from lithium to uranium [1–3] and the detection of line emission due to highly ionized iron in the X-ray spectra of the intra-cluster medium (ICM) [4,5]. The impact of these two discoveries was extraordinary. The first one demonstrated that all the elements (with the exception of hydrogen, helium, and traces of lithium and berillium produced by the Big Bang nucleosynthesis) are forged in the cores of stars and in supernovae (SNe) and that when a star explodes as a supernova it enriches the surrounding interstellar medium with freshly created elements. The second one showed that galaxies lost part of their synthesized elements and that there has been a considerable exchange of chemical elements between stars, galaxies and the hot plasma surrounding them. This also means that the chemical elements trace the formation and evolution of structure which is shaped by the physical processes occurring on a very wide range of spatial scales, from the size of single supernova remnants to cosmological volumes.

The improvements in the stellar and supernova nucleosynthesis theory and modelization (e.g., References [6,7] and references therein) established that the major astrophysical sources of the chemical elements are: (i) Core-collapse supernovae (SNcc) and their massive progenitors ($\gtrsim$8–10 $M_{\odot}$) synthesize most of the O, Ne, and Mg of the Universe and a considerable fraction of Si and S (collectively called α elements as they are the result of fusion process involving the capture of α particles); (ii) Type Ia supernovae (SNIa), whose progenitors are generally believed to be exploding white dwarfs in binary systems, synthesize Ar, Fe, and the other Fe-peak elements, such as Cr and Ni, and the remaining fraction of Si and S; (iii) Asymptotic giant branch (AGB) stars produce mainly C, N which are ejected through stellar winds.

In astrophysics, the term "metals" refers to all the elements heavier than helium, in contrast to the terminology adopted in other scientific disciplines, in part because all these elements make up a small contribution in number and mass with respect to H and He. A considerable fraction of the metals (and most of the iron) do not reside in the galaxies of groups and clusters and they have been expelled in their surrounding X-ray emitting hot atmospheres (for the purpose of this review, we will define the hot atmosphere of groups as intra-group medium, IGrM). Indeed, the iron share, i.e., the ratio of iron in the hot atmosphere and the iron locked up in stars in the galaxies, ranges from 1 up to 10 (see, for recent measurements, References [8,9]). Therefore, the key question is to understand the main transport mechanisms responsible for that unbalance (see, for a more detailed discussion, References [10,11]). There are two broad categories of mechanisms: (i) extraction by ram pressure stripping and galaxy-galaxy interactions; (ii) ejection by galactic winds powered from inside the galaxies themselves either by SNe (stellar feedback) or by the supermassive black hole (SMBH) at their center (in the so-called active galactic nucleus, AGN, feedback). Other important processes redistributing the metals within the hot atmospheres are the central AGN uplift of metals (see the reviews by References [12–15] and the companion review by Eckert et al.) and sloshing, i.e., the offset of the bulk of central part of the hot atmosphere from its hydrostatic equilibrium in its gravitational potential and the subsequent oscillations that may broaden the original distribution at larger scales (see the reviews by References [16,17]). Another source of metals could be the diffuse stellar component not associated to any single galaxy but to the global halo of the cluster or group (also know as intra-cluster light, ICL) polluting the ICM and the IGrM in situ [18].

The purpose of this paper is to review the status of the metal abundance measurements in the IGrM and the progress made by simulations to reproduce and interpret those measurements. It is a companion of the other reviews in this series addressing the scaling relations of these systems (Lovisari et al.), the impact of AGN feedback (Eckert et al.), the overall insight provided by simulations (Oppenheimter et al.), and the properties of the particular class of fossil groups (Aguerri et al.).

Groups of galaxies (which can be defined as objects with total masses M_{500} in the range 10^{13}–10^{14} $M_{\odot}$, though see Lovisari et al. for the unavoidable ambiguity of the definition of a galaxy group) bridge the mass spectrum between L* galaxies and galaxy clusters. They are known to host a significant fraction of the number of galaxies in the Universe (e.g., References [19,20]), they form in the filaments of the cosmic web and not only in the nodes (e.g., Reference [21]), and they are bright enough to be relatively easily observable in

X-rays, while having low enough masses such that complex baryonic physics (e.g., cooling, galactic winds, AGN feedback) begins to dominate above gravity, making these objects more than simple scaled down versions of galaxy clusters (e.g., References [22,23] and the aforementioned reviews in this series). Groups of galaxies appear as critical systems to understand the process of structure formation, the dynamical assembly of baryons in the dark matter halos, and the complex physical processes affecting both the gas and the stellar components. For all the above reasons, it is important to study metals in groups: if, for example, it is more controversial with respect to clusters that they can be considered as "closed box", they can provide key information on the processes, resulting in the redistribution and loss of metals, and it is clearly instructive to investigate the metal budget in different types of systems (see the interesting discussion in the review by Reference [24]).

Many excellent reviews exist already focusing either on the more general topic of X-ray spectroscopy, mainly of the ICM, or directly on metal abundances both observationally and theoretically [10,11,25–32] that ease our work which will then focus on the topics more directly related to the recent updates about metal abundances in galaxy groups. In this respect the only reviews of the topic are about the early history of the metal abundance measurements in the IGrM described in Reference [19] and a more recent update including the *Chandra*and XMM-*Newton* data can be found in Reference [33]. A comparison of the metal budget in groups compared to the one in clusters and elliptical galaxies has been discussed in Reference [31].

The review is organized as follows. In Section 2, we review the observational measurements, from the early *ROSAT* and *ASCA* results to the more recent *Chandra*, XMM-*Newton* and *Suzaku* CCD and high spectral resolution (with the RGS instrument on board XMM-*Newton*) results. In Section 3, we discuss the theoretical framework and the insight from numerical simulations. In Section 4, we discuss the most relevant upcoming missions which will provide a key contribution to the field, and ,in Section 5, we present our final remarks.

2. X-ray Observations

2.1. The Observational Signatures of Metals in the IGrM

The X-ray spectra of the hot, diffuse gas that fills the dark matter halos of galaxies, galaxy groups, and galaxy clusters, is typically described as an optically thin plasma in collisional ionization equilibrium (CIE), composed mainly of primordial hydrogen and helium gas but containing trace amounts of heavier elements from C up to Ni. These approximations usually provide a sufficient description of the bulk of the emission, although subtle effects due to various deviations from a simple thermal model can sometimes become relevant—we refer the reader to Gu et al. [34] for a review.

There are two noteworthy differences between the X-ray emission from the ICM and IGrM. Firstly, in the hot ICM of clusters of galaxies, the free-free bremsstrahlung continuum is the dominant radiation process (see, e.g., Reference [35]). Conversely, for plasma temperatures around and below 1 keV, which are typical of galaxy groups, increasing contributions to the continuum level come also from (i) recombination radiation, caused by the capture of an electron by an ion, leading to a spectral shape characterized by sharp ionization edges, and (ii) the slow transition from the 2 s to the 1 s state, which is forbidden by angular momentum conservation but can happen as a very slow two-photon process giving rise to continuum emission (see, e.g., Figure 6 of Reference [28]). This makes it more difficult to determine the equivalent width of a given spectral line (EW[1]), a quantity that serves as a main diagnostic of the abundance of the chemical element from which that line originates.

Secondly, at higher plasma temperatures ($\sim$4 keV and above), abundance measurements are typically driven by the signal obtained from the Fe XXV He-α line at 6.7 keV (rest frame). By comparison, below temperatures of around 2 keV, i.e., in the low-mass cluster and group regime, the most prominent diagnostic of the metallicity comes instead from the Fe-L line complex at 0.7–1.2 keV. This blend of emission lines originating from the L-shell transitions of Fe XVII–Fe XXIV is completely unresolved at the spectral resolution of CCD cameras; the lines are so closely spaced together that, in many cases, they remain blended even for the XMM-*Newton* RGS and the upcoming high-spectral resolution microcalorimeter onboard *XRISM*. Next to the dominant Fe-L lines, emission from Ne X and the L-shell of Ni at similar energies is also blended within the same spectral structure.

In practice, elemental abundances are usually estimated by fitting the X-ray spectra with models of CIE emitting plasma, that account for the presence of various emission lines within the Fe-L blend (as well as the recombination and two-photon continua). The two plasma radiation codes most commonly used today in X-ray astronomy are AtomDB [36,37] and SPEXACT [38,39]. These models have evolved considerably over the last 4 decades, since the Fe-L emission was first discovered with the Solid-State Spectrometer onboard the *Einstein* satellite [40,41].

In Figure 1, we illustrate the historic development of these two model 'flavors' over time, using the specific example of a 1 keV plasma with Solar elemental composition, folded through the spectral response of a CCD camera. Simply and perhaps simplistically put, transitions belonging to lower ionization states of Fe (whose emission lines lie at lower energies within the Fe-L 'bump') require more complex computations and were, therefore, not fully accounted for in earlier models. This is why, both for AtomDB and SPEX, the shape of the lower energy wing of the Fe-L blend particularly appears to evolve significantly, and great care must be taken when discussing results obtained with older precursors of these plasma emission codes. It is encouraging that the models seem to be converging in recent years and, at least at CCD-level spectral resolution, the latest versions of AtomDB and SPEX only differ at about the $\sim$5–10% level for a 1 keV plasma. Nevertheless, larger differences still remain for lower temperatures (up to 20% at kT < 0.4 keV when folded through a CCD response), and when viewing the models at higher spectral resolution, where the brightest lines in the Fe-L complex can be distinguished from the blend (for a general and up-to-date discussion of the discrepancies of the two plasma codes, see the discussion in Reference [42]).

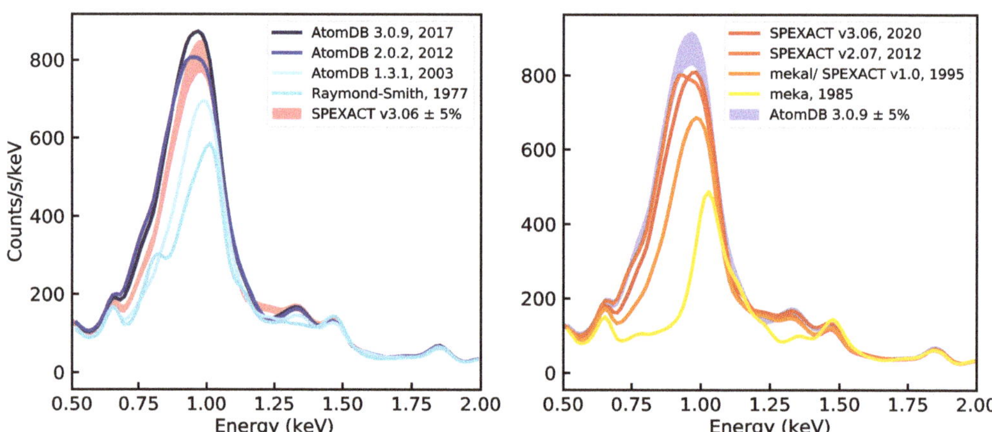

Figure 1. Historic evolution of the two most commonly used plasma emission models. For illustration, we focus on the specific case of a 1 keV plasma with Solar composition (following the Solar abundance units of Asplund et al. [43]). All models have the same emission measure $\int n_e n_H dV$ and have been folded through the XMM-Newton MOS detector response. (**Left**) AtomDB [37] can be viewed as the replacement for/development of the Raymond-Smith code [44]. (**Right**) SPEXACT originated from the 'meka' model developed by Rolf Mewe and Jelle Kaastra, which later became the 'mekal' model; the addition of the final 'l' comes from Duane Liedahl who calculated the atomic parameters for a large number of Fe L-shell ions [35,45,46].

Observationally speaking, the Fe-L complex is both a blessing and a curse. On one hand, as mentioned above, it involves modeling the emission of Fe ions with many remaining electrons, which is significantly more difficult than H- or He-like transitions. On the other hand, because these lines are very bright, the total 0.5–2.0 keV flux of a kT = 0.6–0.8 keV IGrM plasma with a Solar composition is *more than three times brighter* than a 4 keV ICM plasma with the same emission measure (defined as $\int n_e n_H dV$). Quite literally, metals make the IGrM shine, and emission from Fe, in particular, is crucial for detecting what would otherwise be extremely faint and diffuse plasma in galaxy groups. However, this also means that, unlike the case of the ICM, a knowledge of the metallicity in the IGrM is indispensable for converting the observed X-ray flux into a particle number density. Furthermore, since the total flux and shape of the Fe-L bump are extremely sensitive to the plasma temperature (e.g., see the difference between the Fe-L model for a 0.6 versus a 1.2 keV plasma in Figure 2), significant biases arise when a spectrum containing multiple temperature components (for instance due to projection effects or radiative cooling in the core of a group) is approximated by a single-temperature model. This effect, dubbed 'the Fe bias', is discussed in detail already in References [47,48] using ASCA data of elliptical galaxies and galaxy groups. In that work, and many subsequent references thereto, it is consistently shown that, if the abundance measurements are driven by the Fe-L signal, a two-temperature model can yield best-fit metallicities more than twice higher than a single temperature approximation. Although this conclusion was originally reached with old versions of the Fe-L plasma emission, it remains true today, as we illustrate in Figure 2. It is also worth mentioning in passing that, at least at CCD spectral resolution, multi-temperature models can only be constrained if the abundances of the two components are coupled to each other, which need not be true in nature. In addition, beside the IGrM being intrinsically multi-phase, the unresolved emission from low-mass X-ray binaries (LMXB) may be an important spectral component, at least near the center of the brightest group galaxy (BGG) (e.g., Reference [49]); if unaccounted for, this may lead to biases in the measured abundances as well.

As a last cautionary note in terms of interpreting various metal abundance measurements quoted in the literature, it is important to remember that these are customarily reported with respect to the Solar number ratio of that element compared to H; however, this reference point, too, has evolved over the past few decades. While the Solar photospheric units of Reference [50] are still the default in the Xspec fitting package and, therefore, widely used, this reference value for Fe/H is between 43 and 48% higher than reported by the more recent work of References [43,51], respectively. Hence, the Fe abundances reported by various groups can be considerably different depending on the Solar units assumed, and care must be taken when comparing the results. Here, we choose to normalize all quoted abundances to the units of Reference [43].

Moreover, the absolute values of, e.g., O/H (often assumed to be Solar, with the implied variations/uncertainties just stated above), also affect the way that absorption edges from our own Milky Way influence the spectrum; since most of the emission from the IGrM is in the soft band, using the correct Galactic $n_{\rm H}$ (see, for example, the discussion in Reference [52]), as well as the correct chemical composition of the absorbing gas, is important for obtaining a robust characterization of the spectral properties of the IGrM.

Armed with this overview of the spectral characteristics and potential pitfalls of modeling the IGrM, in the following sub-sections, we discuss how our observational picture of metals in galaxy groups has evolved over the past several decades.

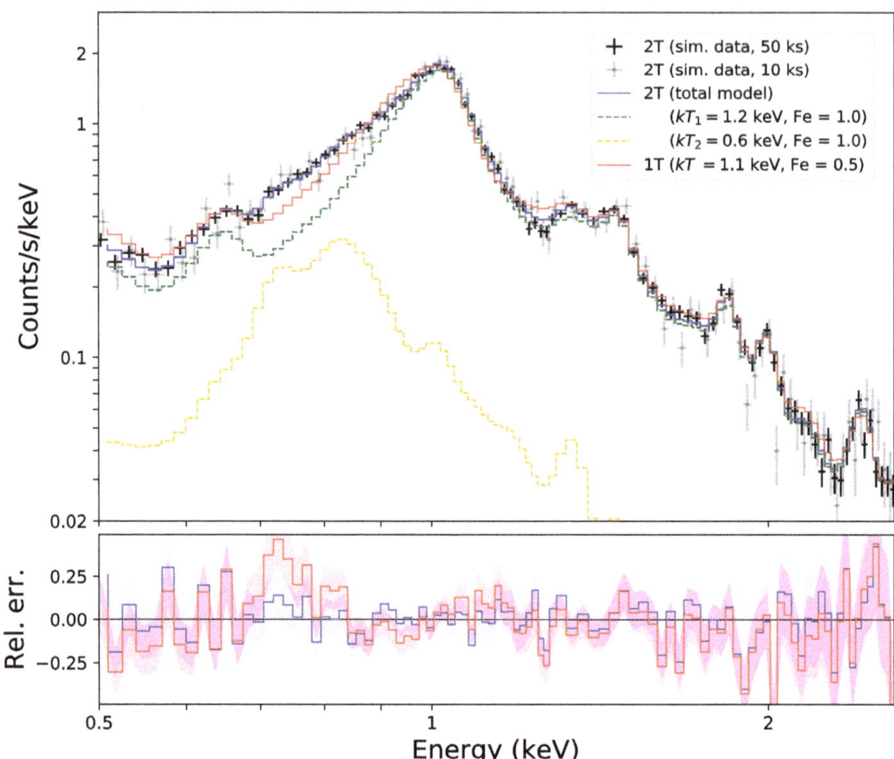

Figure 2. Simulated XMM-*Newton* MOS observations of a multi-temperature plasma, illustrating the effect of the Fe bias. A mix of 0.6 and 1.2 keV plasmas with emission measures in a proportion of 1:10, Solar abundances in the units of Asplund et al. [43], and a total flux similar to that of NGC5846 (integrated within $0.05r_{500}$), was simulated using SPEXACT v3.0.6. Despite the fact that the low temperature component has only 1/10th of the emission measure of the hotter gas, a single temperature fit results in a best fit metallicity that is half of the value input to the simulation. The bottom panel shows the fit residuals for a 1T and 2T model for the 10 ks observation; while there is still a hint that the 1T model does not perfectly describe the data (given the positive residuals around 0.7 keV), this could easily be missed for fainter/more distant targets or when using smaller extraction regions for creating radial profiles or maps.

2.2. The Early Measurements of Global Metallicity

The first milestone discoveries of a true detection of the IGrM in the two galaxy groups NGC 2300 [53] (see Figure 3) and HCG 62 [54] done with *ROSAT* pointed to a surprisingly low abundance of the plasma of 0.09 Solar for NGC 2300 and 0.22 Solar for HCG 62, assuming the RS thermal plasma model [44]. The entire detectable extent of the emission was fitted in a single aperture (25′ for NGC 2300 and 18′ for HCG 62, in the case of the analysis of NGC 2300 excluding the emission around the central galaxy itself). The *ROSAT* observation of NGC 5044 [55] made it possible to measure spatially resolved temperatures and abundances, with super-Solar abundances in the inner 6′ and beyond that radius consistent with a uniform distribution of 1.2 Solar. Very low abundance, <0.12, was reported in the NGC 4261 group from a fit with a RS model to a spectrum extracted from 40′ [56]. In the first sample of HCGs [57], a solid detection of extended emission in 3 of them (the already mentioned HCG 62, HCG 92, and HCG 97) and a possible detection in another 3, were all well fitted by a RS model with very low metal abundances. Additional *ROSAT* observations allowed spatially resolved measurements for NGC 2300 confirming the low abundance, less than 0.16 Solar [58]. In a more complete survey of 85 HCGs with either deeper pointed (in 32 cases) or survey *ROSAT* PSPC observations, extended emission from an IGrM was detected in 22 of them, including in the group emission also the emission spatially located on the dominant central elliptical [59]. The metallicity derived for the 12 spectra of enough quality within an aperture of 200 kpc (in their cosmology) all pointed to a low abundance with a weighted mean of 0.27 Solar. Interestingly enough, Ponman et al. [59] commented to treat these results with caution because the inferred low metal abundances rely heavily on the isothermal assumption: when temperature variations in the gas are taken into account, metallicities several times higher can be inferred [60], extending the search for *ROSAT* observation of galaxy groups beyond HCGs with other optical catalogues in a sample of an additional 14 groups, finding emission from 4 of them, and extending the census of the IGrM to 25 of the 48 groups analyzed at that time. Some conclusions were starting to be made, with the general lower abundance of the IGrM with respect to the ICM, despite the more equal share between gas and stellar mass, possibly suggesting that the IGrM may be largely primordial.

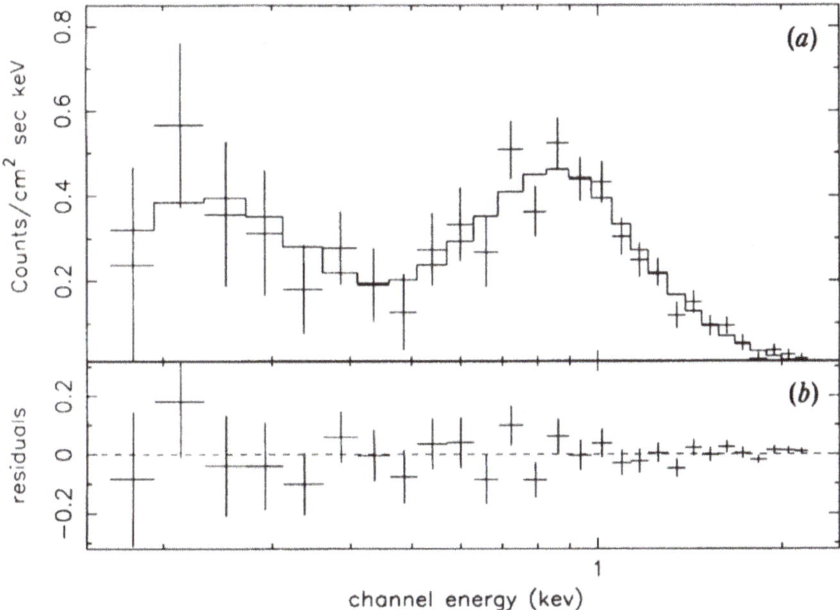

Figure 3. The *ROSAT* X-ray spectrum of the group NGC 2300 plotted together with the beat-fit Raymond-Smith model (**a**) and residuals from the best fit model (**b**). Figure reproduced with permission from Reference [53].

However, concerns were starting to increase about the ability to model the dominant Fe-L emission in the IGrM by the available plasma codes. The first *ASCA* CCD spectra (with the SIS instrument) of the cores of cool core clusters, Perseus, Abell 1795, and the Centaurus cluster exposed the limitations of both the RS and MEKA [35,45] models [61], causing a major revision of the modeling of the Fe-L shell emission [46] which later was incorporated in the MEKAL code. These concerns were reinforced by the discrepancy between the low metallicity found also in the inter-stellar medium of elliptical galaxies and the super-Solar abundances expected just by stellar mass loss [62,63].

ASCA measurements also reported a great scatter in the metallicity of the IGrM. The *ASCA* study of NGC 5044 and HCG 51 reported metal abundances significantly higher than those of NGC 2300 and HCG 62 (also performed with *ASCA*) and more similar to clusters [64]. Ref. [65] analyzed *ASCA* data for 17 groups with single apertures ranging from 4′ to 30′, finding, in general, low abundances in the range 0.15–0.6 Solar. The higher temperature and mass objects with *ASCA* measurements reported by Reference [66] showed an average abundance of 0.44 Solar consistent with that observed in rich clusters and, therefore, clearly highlighting the 1 keV regime as the one showing the spread to lower values of the abundance measurement.

The overall summary as done by Mulchaey [19] is that of a surprising scatter in the measured metallicities in groups, from low (0.15–0.3 Solar) to higher than the values determined in clusters in those days (0.7–0.9) with *ROSAT* and *ASCA*.

A key insight was provided by a series of papers showing the biases introduced by fitting with a single isothermal model complex spectra with multi-temperature components. In the spectra extracted from large apertures in the bright cores of galaxy groups and ellipticals, temperature and abundance gradients are present [47,48,67,68]. The very sub-Solar abundances obtained from previous studies were an artefact of fitting isothermal models and two-temperature models provided better fits to the data and higher metallicities. This is the Fe-bias also described in the previous section and demonstrated by means of simulations of *ASCA* spectra [48]. The discovery of the Fe bias highlighted the importance of the ability of performing spatially resolved spectroscopy and the difficulties in the modelization of the thermal and abundance structure in the cores of galaxy, groups and clusters, as it was shown at those times by the early results of M87 with XMM-*Newton* [69].

The last influential paper dealing with single measurements of metal abundances is Baumgartner et al. [70], presenting an analysis of the *ASCA* spectra of 273 groups and clusters with the largest possible aperture collecting all the detectable flux and stacked in bins of temperature. That work found a constant Fe abundance value of 0.3 Solar for hot clusters and for groups with an increase up to a factor of 3 with respect to the average value in the range 2–4 keV. This is a manifestation of the "inverse" Fe-bias [71–73] which overestimates the abundances in multi-temperature plasma (ranging from about 1–2 keV to about 5 keV) resulting in a mean global temperature in the range 2–4 keV as found by Baumgartner et al. [70]. Although, in practice, when this occurs, the spectra show the presence of both Fe-L and Fe-K lines, the inverse Fe-bias is essentially weighted by the higher statistics of the Fe-L complex. In this regime, the fitting procedure increases the estimated Fe abundance to overcome the weaker Fe-lines expected in the single temperature plasma (for more details, see Reference [73]).

2.3. *The Spatial Distribution of the Metals in the IGrM*

Succeeding the *ROSAT* and *ASCA* eras, which allowed to set a first light on global metallicities of the IGrM, the advent of CCD instruments offered by the generation of early 2000s X-ray observatories (*Chandra*, XMM-*Newton*, *Suzaku*) brought a significant progress. They allowed not only to reveal the spatial distribution of metals across galaxy groups, but also to focus on elements other than Fe—hence exploring groups' chemical history with respect to its SNIa *and* SNcc components. In the following sub-sections, we tackle these two aspects in more detail.

2.3.1. Radial Profiles of Iron Abundance

The essence of this sub-section is summarized in Figure 4, where we show a few recent radial metallicity (i.e., Fe) profiles of galaxy groups (from both individual and sample measurements), with comparison with typical cluster profiles. These profiles, along with a number of other ones reported in the literature, are further discussed below.

Radial metallicity profiles of individual sources have been in fact investigated by many authors using either XMM-*Newton*, *Chandra*, or *Suzaku* (or even earlier with *ROSAT* [48]). This is the case for systems, such as NGC 5044 [74,75], RX J1159+5531 [76,77], AMW 4 [78], HCG 62 [75,79–81], MKW 4 [75,82], NGC 1399 [83–85], and UGC 03957 [86].

The vast majority of these studies show a gradual metallicity increase towards the core of the systems, with the maximum value spanning from half a Solar to slightly super-Solar values. This picture is qualitatively in line with the centrally peaked Fe abundance profiles that are typically found in relaxed clusters (e.g., References [9,52,87–89]). Quantitatively, samples including groups *and* clusters are valuable to provide comprehensive comparisons. For instance, Johnson et al. [90,91] studied 28 galaxy groups and concluded that (i) systems with lower level of feedback impact are on average more metal rich within $\sim 0.03 R_{500}$, and (ii) systems classified as "cool-cores"[2] are, on average, more enriched in their cores ($\sim 0.1 R_{500}$) than clusters. Similar conclusions were reached for 43 *Chandra* groups (re-) analyzed by Sun [33], although a significant increase of average metallicity with group temperature (hence, mass) was also reported. More recently, results from the CHEERS sample—consisting of 21 "groups/ellipticals" and 23 "clusters"[3]—suggest for instance a similar decreasing profile for both types, with the former being on average slightly less enriched than the latter [89]. On the other hand, Lovisari and Reiprich [52] analyzed a sample of 207 systems and concluded that, despite their scatter, on average "groups/ellipticals" (defined in the same way) have a slightly higher metallicity than clusters within $0.1 R_{r500}$. A recent re-analysis of the CHEERS sample within $0.1 R_{500}$ using an updated SPEX version (v3.0.4, also more consistent with the apec v3.0.9 version used in Lovisari and Reiprich [52]) find more consistent results, with groups being at least as enriched as clusters within that limit [92]. More detailed discussions and interpretations on the absolute metallicities in groups versus more massive systems is addressed in Section 2.4 (observations) and Section 3.2.1 (cosmological simulations).

In addition, quite remarkably, Figure 4 suggests that the sample-averaged metallicity gradient measured from these different authors all have a similar slope. One notable exception (not shown here) is perhaps the sample of 15 nearby groups observed with *Chandra* by Rasmussen and Ponman [93,94], whose average profile exhibits a significantly sharper central peak (as already pointed outby Reference [33]). This difference might originate from spectral modeling (including outdated atomic data and/or multi-temperature biases), instrumental calibration, or subtle background effects, which were all less understood at that time.

It is worth noting that the metallicity does not always increase with decreasing radius. In fact, for a number of systems, the Fe abundance was found to peak a few kpc outside of the core while *decreasing* towards its very center. Although historically discovered and investigated in the Centaurus cluster [95], these drops are more commonly found in lower-mass systems (e.g., References [79,80,89,93,96]). Whether the presence of these drops is truly related to mass of the system (and/or the strength of their cool core) is not clear yet. Indeed, abundance drop detections might be affected by selection biases—originating from either the usually larger distance of clusters (resulting in a poorer spatial resolution, hence no detected drop), or the selection itself of the currently most studied systems.

Such low abundances are in fact surprising and intriguing, as they cannot be easily explained by classical models of IGrM formation and enrichment. Although, in some cases, drops were found to be the result of spectroscopic biases (e.g., multi-temperature bias [97]), no evidence points toward the latter as being the sole explanation. Similarly, resonant scattering seems to be excluded from the culprit list in at least a few specific cases [98], and possible helium sedimentation—leading to an incorrect estimate of the continuum—should provide limited effects only, if not largely inhibited by thermal diffusion [99,100] and references therein). Interestingly, however, probing the *chemical composition* in the central low-Fe regions of these systems may provide an interesting hint toward the physical nature of these drops. This is further discussed in Section 2.3.2.

Another important open debate concerns the comparison of clusters' and groups' metallicities in their outskirts ($\sim R_{500}$ and beyond). Whereas there is now striking evidence for clusters having their metallicity flattening with radius and converging toward an universal value of $\sim$0.3 Solar [9,101,102], groups and elliptical galaxies have sometimes been measured with an uninterrupted decrease of metallicity down to at most 0.1–0.2 Solar (e.g., References [77,84,103,104]). The trend seems to be followed by the sample results of Rasmussen and Ponman [93,94] and of Sun [33]. These results, however, should be interpreted with caution, given how recent atomic codes improvements changed our view on the Fe-L complex, its modeling at moderate resolution, and its associated abundance

(Section 2.1). Moreover, some past measurements may have been affected by the Fe-bias discussed above, as moderate exposures available per source did not necessarily allow to model outer regions with more than one temperature. We note, for instance, that the more recent sample measurements of Mernier et al. [89] and Lovisari and Reiprich [52] show hints of a flattening beyond $\sim 0.3R_{500}$ that remains formally consistent with the 0.3 Solar value reported in clusters (References [9,102]; see Figure 4). These results are in agreement with Thölken et al. [86], who reported that the metallicity profile of the group UGC 03957 does not decrease further below 0.3 Solar, even at distances *beyond* R_{200}. Other measurements, on the other hand, show in-between results, with evidence of a flattening though around 0.2 Solar, i.e., *below* the universal value. This is the case for the galaxy group RX J1159+5531 [77], as well as (perhaps even more intriguingly) for the Virgo cluster [105].

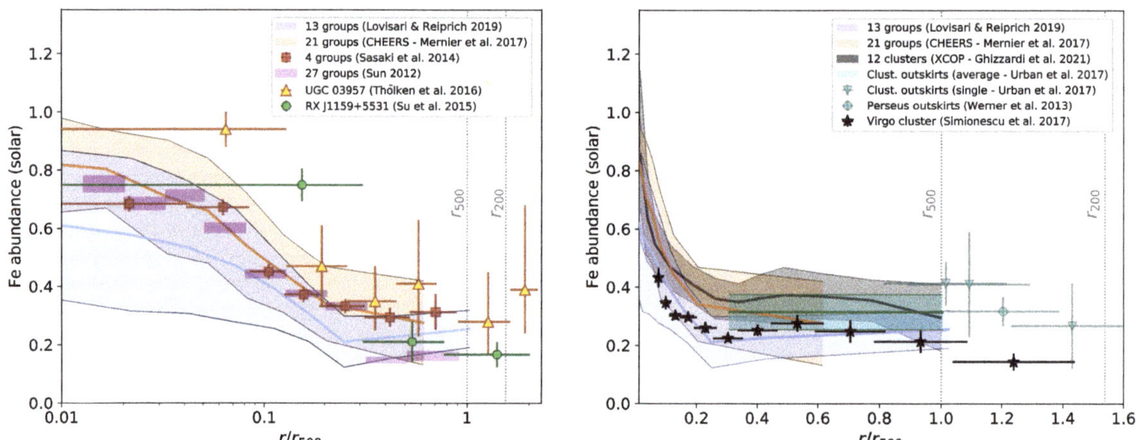

Figure 4. Fe abundance radial profiles in various galaxy groups (and clusters) from the literature. (**Left**) The recent average profiles of (Reference Mernier et al. [89], the 21 CHEERS groups) and (Reference Lovisari and Reiprich [52], 13 groups—excluding systems overlapping with the CHEERS observations) are compared with those of (Sasaki et al. [75], 4 groups) and (Sun [33] 27 groups—below $kT_{500} = 1.9$ keV), as well as with independent measurements of UGC 03957 [86] and RX J1159+5531 (Reference [77], azimuthally averaged). (**Right**) The same two average group profiles are compared with measurements of more massive systems—i.e., the XCOP sample (Reference [9], 12 clusters), the Fe abundance in cluster outskirts (Urban et al. [102], with averaged and single measurements shown, respectively, below and beyond r_{500}—also including the outermost Perseus value from Werner et al. [101]), as well as the (azimuthally averaged) profiles of the Virgo cluster [105]. For consistency, the scatter envelope of the samples of Urban et al. [102] and Lovisari and Reiprich [52] have been computed following that of the CHEERS sample Mernier et al. [89]. All measurements have been re-scaled into radial units of r_{500} (following the values given in the corresponding papers and/or the conversion proposed by Reference [106]) and into Solar units of Asplund et al. [43].

The question of whether groups and clusters have their outskirts enriched at similar levels is of crucial importance. Besides the fact that outskirts represent by far the largest volume of these systems (hence, the bulk of their metal masses), they are direct witnesses of freshly accreted gas through the gravitational potential of these systems and, thus, constitute a fossil record of the enrichment of these systems at their formation epoch. In fact, the (radially *and* azimuthally) uniform metallicity distribution measured in clusters outskirts constitutes by far our best evidence in favor of an "early-enrichment" scenario, in which supermassive black hole feedback played a fundamental role in ejecting and mixing freshly produced metals out of their galaxy hosts during or before their assembly into larger scale structures and the formation of their hot ICM, i.e., at $z \gtrsim 2$–3 (for recent reviews, see, e.g., References [31,107]). Quite remarkably, this redshift range also corresponds to the peak of star formation activity (for a review, see, e.g., Reference [108]), as well as to an epoch of enhanced AGN accretion and activity—not only at cosmic scale (for a review, see, e.g., Reference [109]) but also (and especially) in clusters and groups (e.g., References [110–112]), naturally leading to the picture of their higher feedback to efficiently stir the freshly produced metals. Robust measurements revealing a uniform metal distribution in the IGrM as in the ICM will constitute decisive evidence towards this scenario and its "universal" 0.3 Solar value. On the contrary, significantly lower abundances measured in the IGrM outside $\sim R_{500}$ would challenge this scenario and would require to rethink our global picture of chemical enrichment at galactic scales and beyond. High resolution spectroscopy coupled to high throughput will be essential to bring our current measurements up to the required accuracy (Section 4).

2.3.2. Chemical Composition and Its Radial Dependence

Since Fe has the strongest emission lines in the IGrM, it typically dominates the abundance measurements reported in the literature. For low-statistics spectra, it is common to assume that the abundances of other elements with respect to Fe follow the Solar ratio. However, important information about the metal enrichment history of the IGrM is encoded in its chemical composition, in particular since the O/Fe, Mg/Fe, and/or Si/Fe ratios are good tracers of the relative contribution of SNcc and SNIa. This relative contribution is expressed in various ways throughout the relevant literature, for example as the ratio between the numbers of different supernova explosions (either N_{cc}/N_{Ia} or $f_{Ia} = N_{Ia}/N_{Ia}+N_{cc}$); or as the fraction of Fe supplied by SNIa, $f_{\rm Fe,Ia} = N_{Ia} * y_{Ia,Fe}/(N_{cc} * y_{cc,Fe} + N_{Ia} * y_{Ia,Fe})$, where $y_{SN,i}$ represents the mass of element i produced by a supernova of type SN. The details of this decomposition depend on the exact model yields $y_{SN,i}$, which are subject to remaining uncertainties in stellar astrophysics, and furthermore rely on various assumptions about the initial metallicity and mass function of the supernova progenitors. Nevertheless, the general trend wherein light-α elements are almost exclusively produced by SNcc, while Fe-group elements are mainly supplied by SNIa is robust among the current chemical evolution models, lending credibility to this type of analysis.

Back to the *ROSAT* and *ASCA* era, Finoguenov and Ponman [113] reported from a sample of four galaxy groups that SNcc products (i.e., Si and Mg) were found to be more uniformly spread, while SNIa products (namely, Fe) showed a more peaked distribution. These results were naturally interpreted as the bulk of SNcc having exploded, gotten mixed with, and enriched their surroundings earlier than the bulk of SNIa (the latter being more likely to originate from long-lived low-mass star populations in the red-and-dead central dominant galaxy). That interpretation was later supported by Rasmussen and Ponman [93,94] who measured a radial increase of the Si/Fe ratio with *Chandra* observations of 15 groups.

However, these initial conclusions do not appear to have stood the test of time. Some early XMM-*Newton* data already provided results that conflicted with the initial paradigm of a relatively uniform α-element and peaked Fe distribution in the IGrM: no gradient in Si/Fe was seen in NGC5044, comparing the regions within and beyond 48 kpc from the BGG [74]; a constant and close to Solar α/Fe out to at least 100 kpc was reported in NGC507 [114]; and Xue et al. [115] found that *all* measured abundances (O, Mg, Si, S, and Fe) in the group RGH80 showed a monotonic decrease with radius. In all these three cases, $f_{\rm Fe,Ia}$ was inferred to be in the range of 70–85%, assuming a model consisting of simple linear combinations of SNIa and SNcc. Therefore, although most Fe is being supplied indeed by SNIa, there did not appear to be a significant change in $f_{\rm Fe,Ia}$ with radius, or from system to system. Similar conclusions were starting to be reached in galaxy clusters, as well (e.g., Reference [72,97,116,117]).

The low instrumental background of *Suzaku*, and the superior low-energy response of the XIS CCDs particularly in the first few years after launch, shed additional light on this topic: the radial profiles of Mg/Fe, Si/Fe, and S/Fe were consistently shown to remain uniform (i.e., all four elements showed a radially decreasing profile) in HCG 62 [118], NGC 5044 [119], NGC 507 [120], and NGC 1550 [121] over the entire area probed by the *Suzaku* observations. A sample of 4 groups consisting of MKW4, HCG62, NGC1550, and NGC5044, was covered by *Suzaku* out to as far as 0.5 r_{180}, confirming that the Mg/Fe and Si/Fe ratios remain nearly constant and close to the Solar ratio (assuming the units of Reference [122]) out to a significant fraction of the virial radius [75]. All these measurements are consistent with an N_{cc}/N_{Ia} of 3–4 [119,121,123]. For the supernova yields assumed in these works, $N_{cc}/N_{Ia} = 3$ corresponds to a $f_{\rm Fe,Ia}$ of 80%, in line with the XMM-*Newton* results discussed in the previous paragraph. A point of contention in the *Suzaku* results remained the O abundance, that seemed to have much shallower radial gradients than all other α-elements (a conclusion shared by all references mentioned earlier in this paragraph). This would imply increasing O/Mg and O/Si ratios as a function of radius; since all these three elements are predominantly produced by SNcc, it is impossible to reconcile these measurements with a simple model where the relative contribution from SNIa and SNcc varies with distance from the BGG. It is likely that residuals in modeling the Galactic absorption and/or OVIII foreground emission, or issues related to Solar Wind Charge Exchange (whose strongest emission also comes from O), may have affected the measurements.

More recently, results from the XMM-*Newton* CHEERS sample reported similar radial distributions of the O, Mg, Si, S, Ar, Ca, and Fe abundances in clusters *and* groups separately. While the covered radial range does not extend as far as that from *Suzaku* studies, the agreement between the radial trends of all measured elemental abundances, together with the larger sample size, provides solid and cohesive evidence for a lack of significant spatial variation of the SNcc versus SNIa contributions to the enrichment across the mass scale [89].

With the latest advancements in our knowledge of spectral modeling, multi-temperature biases, and/or instrumental calibration, current measurements, thus, favor a uniform chemical composition over the entire volume of clusters and groups, as early suggested for the cores of both systems (and ellipticals) by de Grandi and Molendi [124]. Interestingly, in both regimes the chemical composition is also remarkably close to that of our own Solar System. Indeed, *Hitomi* confirmed that all the investigated X/Fe ratios of the Perseus Cluster are consistent with Solar at very high precision [125,126] and detailed investigations of the CHEERS sample found the same trend for groups and ellipticals, as well [127,128]. This is further illustrated in Figure 5, where we compiled the average chemical composition of the 21 CHEERS low-mass systems.

It is worth noting that the chemical composition of the ICM/IGrM must, therefore, be markedly different than that of the stars in the BCG/BGG: as shown by References [129,130], massive early-type galaxies (ETGs) with a velocity dispersion above 200 km/s typically have high α/Fe ratios up to twice the Solar value, which is inconsistent with the abundance pattern of the hot diffuse gas in their immediate vicinity (see Figure 5). High values of α/Fe are usually associated with a very short starburst: BCG/BGGs may have made most of their stars before SNIa had time to explode, so that very few Fe-group elements are incorporated into the stars themselves. Nearly all SNIa later polluted the central ICM/IGrM instead, gradually lowering its α/Fe ratios.

But although the SNIa contribution cannot have come too quickly (else the stars in the central galaxy would not have such a high α/Fe), it also cannot have happened too slowly, or else the observed radial distribution of Fe should follow the present-day stellar light, which is not observed (Section 2.4). A significant late-time input of SNIa products would also modify the radial trends of α/Fe in the ICM/IGrM, contradicting the constant near-Solar α/Fe ratios measured throughout the volumes of clusters and groups.

This suggests that most (if not all) SNIa contributing to the enrichment exploded not much later than the peak of cosmic star formation ($z \simeq$ 2–3 [108]). Several studies of the SNIa delay-time distribution in fact support this picture, finding that a significant number of such explosions occur as early as 100 Myr after a star formation event (Totani et al. [131], Maoz et al. [132]; for a review on SNIa delay-time distribution and its interpretations, see Maoz et al. [133]).

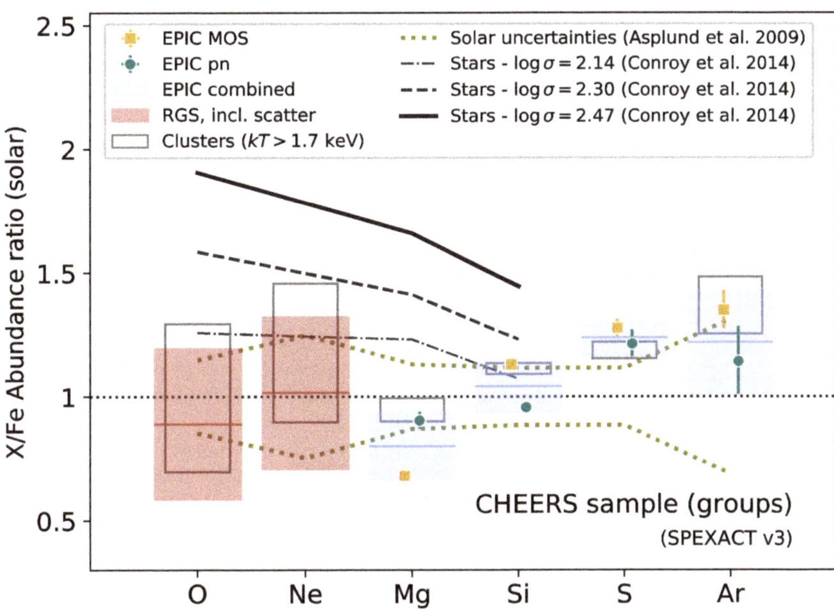

Figure 5. Average chemical composition (expressed as X/Fe abundance ratios in units of Reference [43]) within the central $0.05r_{500}$ of the 21 galaxy groups and massive ellipticals from the CHEERS sample (defined as $kT_{\mathrm{mean}} < 1.7$ keV). The O/Fe and Ne/Fe ratios (including their intrinsic scatter) were measured using the XMM-*Newton* RGS instruments, while the other ratios are measured using the XMM-*Newton* EPIC MOS and pn instruments. To be conservative, the EPIC "combined" measurements cover the entire MOS-pn discrepancies, which fully accounts for the intrinsic scatter, as well. Comparison with the average chemical composition of clusters shows values that are similar and near-Solar in both regimes. For comparison, Solar uncertainties are also shown, as well as stellar abundances in ETGs measured for different bins of stellar velocity dispersion σ [129]. Adapted from Mernier et al. [128].

Nevertheless, the confirmation (or rejection) of this new paradigm will be crucial to achieve with future missions. Metal abundances determined with CCD spectrometers in clusters of galaxies are still subject to systematic uncertainties in the range of ~20% [124,126]; given the still ongoing challenge to derive accurate abundances from unresolved line complexes, these uncertainties may be even more important for the IGrM. Ultimately, the stellar population histories of typical central dominant galaxies are fundamentally different from that of the Milky Way. Future measurements using high-resolution spectroscopy will reach percent-level accuracy in determining the abundance of numerous chemical elements in gaseous halos of varying mass; it would be nothing short of a stunning cosmic conspiracy if, as smaller and smaller spatial scales start to be probed at such a level of precision, the central abundances in groups and clusters remain in agreement with the Solar composition.

Besides quantifying the relative contribution of SNIa and SNcc to the enrichment of the IGrM, elemental abundance ratios may also reveal the nature of the so far unexplained abundance drops that are sometimes observed (see Section 2.3.1) in the very inner centers of groups and clusters. Under the assumption that these abundance drops have an astrophysical origin, an interesting scenario proposed by Panagoulia et al. [80,134] considers that IGrM-phase metals may deplete into dust and then become invisible to the X-ray window. As a second step, AGN jets and buoyantly rising bubbles may contribute to move this dust mass away, before eventually re-heating it to the X-ray phase outside of the very core. If true, an interesting corollary of this scenario concerns the Ne and Ar abundance. As these two elements are noble gases, they cannot be incorporated into dust; hence, they should *not* exhibit any central decrease. Although a few authors have investigated this issue [89,135,136], no real consensus is established yet given the sensitivity of the measurements to systematic effects. Ar and Ne lines should be easily measurable with future micro-calorimeters (Section 4). Provided that atomic codes continue to converge (Section 2.1) in the years to come, *Athena* (and possibly *XRISM* for very nearby systems) will provide a definitive answer to this question.

Clarifying these outstanding issues will allow us to identify which combinations of theoretical supernova yield models, $y_{SN,i}$, provide the best fit to the observations of the ICM and IGrM (avoiding regions affected by dust depletion if applicable), offering important clues about open aspects of stellar astrophysics. For instance, it was realized very early on that abundance ratios measured from X-ray spectra of the ICM could be used to distinguish between various SNIa explosion mechanisms, preferring a deflagration over a delayed detonation model [137]. Similar conclusions were tentatively reached for the IGrM [74,123]. It is becoming clear, however, that both an improvement in the data quality and increased accuracy in the yield models are necessary before robust conclusions can be drawn (de Grandi and Molendi [124], Mernier et al. [138], Hitomi Collaboration [125], Simionescu et al. [126]; for a review, see Mernier et al. [31]). Significant progress is expected to be driven in this sense by upcoming high-resolution X-ray spectroscopy studies.

2.3.3. 2-Dimensional Metallicity Maps

In addition to the radial dependence of the metal abundances, important information can be inferred also from the azimuthal substructure revealed by 2D maps; typically, these are of course only available for very deep observations of the brightest systems. Early work by Finoguenov et al. [139] presented a systematic analysis of the metallicity distribution in NGC5846, NGC4636, and NGC5044 using XMM-*Newton*. It was shown that, while the profiles are consistent with a linear decrease with radius, the scatter of the data points was as high as 30–50%. This pointed towards a patchiness of the 2D metal abundance using typical spatial resolution elements of 2–10 kpc, which cannot be explained solely by the satellite subhalos. Later studies revealed that, very generally speaking, the main physical mechanisms responsible for such a 2D metallicity substructure are related either to AGN feedback or to ongoing mergers.

In terms of AGN feedback, in the case of clusters of galaxies, it is now well established that the buoyantly rising bubbles produced by the activity of the supermassive black hole in the BCG are able to uplift metals in their wake, leading to an abundance enhancement along the axis corresponding to the radio jets compared to the perpendicular direction [72,140–142]. Given the shallower gravitational potential wells of galaxy groups, one might expect this effect to be even more pronounced, and even more important for the physical evolution of the IGrM, as metals produced in the BGG may even escape the group halo through the action of the AGN. However, to our knowledge, a systematic study of the metal asymmetry in groups (i.e., an equivalent to the sample study of References [141,142] which focused primarily on galaxy clusters) is still lacking. The main impediment is likely related to the fact that the region of uplift is also generally expected to be multi-phase (see, especially, Reference [140]) which, as discussed in Section 2.1, significantly complicates the determination of an exact Fe budget in a given spatial region.

Nonetheless, hints that the relativistic radio lobes of the central AGN do have an impact on the metallicity distribution in groups have been obtained in a few objects. Perhaps the clearest example so far is that of AWM 4, where O'Sullivan et al. [143] found a metal enhancement along the inner jet of the central dominant galaxy, NGC6051, corresponding to an excess mass of iron in the entrained gas of $\sim 1.4 \times 10^6$ $M_\odot$. Another case is that of NGC4636, where O'Sullivan et al. [144] report a plume of cool, metal-rich gas extending beyond a known AGN lobe to the southwest of the galaxy center, and interpret this to be the product of metal entrainment by past AGN activity. Less clear is the scenario in NGC4325;

here Laganá et al. [145] reported an elongated Fe-rich filament to the south/southeast of the central galaxy which could be due to metal entrainment by the AGN; however, no X-ray cavity is found in this system that would confirm this interpretation. Finally, there are the interesting examples of rather clearly detected *anti*correlations, i.e., a low metallicity corresponding to the radio lobes in NGC5813 [146] and inner radio lobes of M49 [96]—although, in both cases, only a single-temperature fit was used to create the 2D maps; hence, the Fe bias may be the reason for these results.

Mergers on the other hand typically result in tails and arcs of enhanced metallicity in the IGrM, depending on the merger stage and geometry. In a simple case where a subgroup is falling towards a larger cluster of galaxies, a ram-pressure stripped tail exhibiting an orderly head-tail morphology is often seen; as metals are stripped from the central group galaxy, the elemental abundances in this tail are expected to be higher than those of the surrounding diffuse medium. This has been confirmed by X-ray spectral mapping of the metallicity in a handful of cases. One of the clearest examples is that of the M86 group falling into the Virgo Cluster; this is a very rare case where a metal abundance map from a *two-temperature* model is available for the IGrM [147], showing a long, 100–150 kpc tail of near-Solar abundance (in units of Grevesse and Sauval [148]) trailing M86. The abundance in the ram-pressure stripped tail is about twice higher than the off-tail regions, demonstrating how infalling groups contribute to the metal budgets of the ICM. Another remarkable system is the northeastern group falling into Abell 2142, which exhibits a long, straight, narrow tail that flares out after about 300 kpc from the BGG. The metallicity map published in Eckert et al. [149] shows a significant enrichment along most of the narrow tail, with the transition between the straight tail and the irregular diffuse tail corresponding to a marked abundance drop. Recent spectral maps by O'Sullivan et al. [150] also show tails of cooler, lower entropy, metal-enriched gas behind both cores in a group-group (as opposed to group-cluster) merger in NGC 6338.

In later merger stages, after the first pericenter passage of the sub- and main halo, internal gas sloshing or tidal (also known as 'slingshot') tails [151,152] can instead be recognized as arc-shaped high metallicity 'fronts'. Internal gas sloshing is likely responsible for the high abundance arc in HCG 62 [79,81,153,154] and for the abundance map asymmetry in NGC5044 [155]; although no 2D metal abundance map is available, radial profiles of an azimuthally resolved wedge in the NGC 7618–UGC 12491 pair [156] suggest a metal enhancement that was originally attributed to ram-pressure stripping but later recognized as rather due to a slingshot tail [152].

Of course, mergers and AGN feedback can work in unison. For instance, in M49, the 2D Fe abundance map derived by Su et al. [157] using XMM-Newton (covering a significantly larger field than that in Reference [96] discussed above) suggests both the presence of a metal enriched tail to the southwest, and a metal enhancement aligned with two outer ghost X-ray cavities along the NE-SW axis on smaller spatial scales (see Figure 6). The authors conclude that the tail gas can be traced back to the cooler and enriched gas uplifted from the BGG center by buoyant bubbles, implying that active galactic nucleus outbursts may have intensified the stripping process. On the other hand, Sheardown et al. [152] argue instead that M49 may host a slingshot rather than a ram-pressure tail. A similar case may be that of NGC507; again, no 2D metal abundance map is available, but Kraft et al. [158] report a gradient in the elemental abundance across a sharp arc-like X-ray surface brightness discontinuity with opening angle of 125 deg. Because that discontinuity is aligned with a low surface brightness radio lobe, the authors conclude that this 'abundance front' can be explained by the transport of high-abundance material from the center of the galaxy due to the transonic inflation of the radio lobe; however, it was subsequently realized (e.g. see previous paragraph) that classical cold fronts are likely to produce such abundance arcs as well. The abundance feature in NGC507 is therefore not unusual and could be simply due to classical sloshing, or to an interaction between AGN feedback and past merging activity.

Figure 6. XMM-*Newton* spectroscopic map of the Fe abundance in M49 in units of the Solar abundance of Reference [43], derived assuming a single temperature model. X-ray contours in the 0.7–1.3 keV energy band are overlaid in black. The Fe distribution is elongated in the direction of the AGN ghost cavities (denoted by white dashed circles), with an additional extension towards the west/southwest on larger scales, likely related to a ram-pressure or slingshot tail as the galaxy is falling into the Virgo Cluster. Figure reproduced with permission from Reference [157].

2.4. Metal Budgets

In the previous sections, we reviewed the measurements of the abundances probing a fraction of the IGrM volume. As pointed out early by Arnaud et al. [159], the physically meaningful quantity for the study of the IGrM (and of the ICM) are the metal (iron or other chemical elements) mass and stellar mass present in the groups (and clusters). The ratio of the iron mass and stellar mass is directly linked to a fundamental quantity in chemical evolution models, the iron (or other chemical elements when measured) yield which is the ratio of the total iron mass released by stars to the total stellar mass formed for a given stellar population (see References [8,9,160] and references therein):

$$\mathcal{Y}_{\rm Fe} = \frac{M^{\rm star}_{\rm Fe,500} + M_{\rm Fe,500}}{M_{\rm star,500}(0)}, \tag{1}$$

where $M_{\rm Fe,500}$ is the iron mass enclosed within r_{500} in the ICM/IGrM, $M^{\rm star}_{\rm Fe,500}$ is the iron mass locked into stars, $M_{\rm star,500}(0)$ is the mass of gas that went into stars whose present mass is reduced to $M_{\rm star,500}$ by the mass return from stellar mass loss, i.e., $M_{\rm star,500}(0) = r_o M_{\rm star,500}$, where r_o is the return factor. We take $r_o = 1/0.58$ following Renzini and Andreon [8] and Maraston [161]. A caveat should be made that the iron yield can be matched to a theoretical prediction only if we are able to make a full inventory under the assumption of a closed system. If iron can leave the system or just does not reside within the radius used to make the estimate, we cannot draw a conclusive inference. This is particularly the case at the scale of groups as we discuss further in this section.

One can then measure $M_{\rm Fe,500}$ either by multiplying a representative deprojected gas-mass weighted iron abundance times the total gas mass of the system within r_{500} [8] or by taking fully into account the radial dependence of the deprojected iron abundance and gas mass [9,162]:

$$M_{\rm Fe}(<R) = 4\pi A_{\rm Fe} m_{\rm H} Z_{\odot} \int_0^R Z_{\rm depro}(r)\, n_{\rm H}(r)\, r^2 dr, \tag{2}$$

where $Z_{\rm depro}$ is the deprojected abundance profile, $A_{\rm Fe}$ is the atomic weight of iron, and $m_{\rm H}$ is the atomic unit mass. The hydrogen density $n_{\rm H}$ is derived from the gas density $n_{\rm gas}$ through the usual relation $n_{\rm gas} = (1 + n_e/n_{\rm H})n_{\rm H} = 2.21 n_{\rm H}$, where n_e is the electron density; $n_{\rm gas}$ is obtained through deprojection. In the latter case $M_{\rm Fe,500} = M_{\rm Fe}(< r_{500})$. For the measurement of $M^{\rm star}_{\rm Fe,500}$ it is usually assumed that the average iron abundance in clusters and groups stars is solar (for the validity and limitations of this assumption, see, for example, Reference [163] and references therein). This iron abundance of the stars is then multiplied by the total stellar mass enclosed within r_{500}. The latter value can again be estimated through two approaches. The first one performs a flux measurement for each galaxy in a given optical band and calculates the mass of the galaxy through the Spectral Energy Distribution (SED) fitting, the mass of the galaxies are then summed together [9,164]. The second approach calculates the total luminosity in a given optical band by integrating the luminosity function of the red cluster galaxies, summing the contribution of the BCG and possibly of ICL and then multiply for an assumed stellar-mass-to-light ratio in the same optical band (see, for example, Reference [8] and references therein). Both quantities are deprojected assuming a generalized or a simple Navarro-Frenk-White (NFW) distribution for the galaxy and optical light distribution. These observational estimates can be compared with the expected theoretical estimate based on the current understanding of stellar nucleosynthesis. We take the values reported in Ghizzardi et al. [9] for the $\mathcal{Y}_{\rm Fe}$ based on the derivation by Renzini and Andreon [8] and Maoz and Graur [165]. $\mathcal{Y}_{\rm Fe}$ is computed as the product of the Fe mass produced by a SN explosion, y, and the number of SN events produced per unit mass of gas turned into stars, k. Both contributions from Ia and CC SN are considered. Thus, $\mathcal{Y}_{\rm Fe}$ can be written as:

$$\mathcal{Y}_{\rm Fe} = y_{\rm Ia} \cdot k_{\rm Ia} + y_{\rm CC} \cdot k_{\rm CC}, \tag{3}$$

where Ia and CC subscripts refer to the two different SN types. For Ia, we assume $y_{\rm Ia} = 0.7\, M_{\odot}$ and $k_{\rm Ia} = 1.3 \times 10^{-3}\, M_{\odot}^{-1}$. Renzini and Andreon [8], as well as Greggio and Renzini [166], suggest a possible higher $k_{\rm Ia}$ value of $2.5 \times 10^{-3}\, M_{\odot}^{-1}$. For CC SN we assume $y_{\rm CC} = 0.074\, M_{\odot}$ and $k_{\rm CC} = 1.0 \times 10^{-2}\, M_{\odot}^{-1}$. Substituting the above values in Equation (3) and dividing by the solar abundance we get $\mathcal{Y}_{\rm Fe,\odot} = 0.93\, Z_{\odot}$. An higher estimate can be obtained by assuming that SNIa rate is higher in clusters with respect to the field [165,167,168]. If, following Freundlich and Maoz [168], we assume a SNIa rate per unit mass of $k_{\rm Ia} = 3.1 \pm 1.1 \times 10^{-3}\, M_{\odot}^{-1}$, we derive $\mathcal{Y}_{\rm Fe,\odot} = 2.34 \pm 0.62\, Z_{\odot}$.

We report in Figure 7 the effective iron yield for the sample of clusters studied in Ghizzardi et al. [9], for the sample of groups studied in Renzini and Andreon [8] with iron abundance measurements in the IGrM derived by Sun [33] and for the objects in Sasaki et al. [75]. This plot seems to show apparently that for groups there is no discrepancy with the theoretical expectations as it is dramatically the case at the cluster scale. The large amount of intra-cluster iron is difficult to reconcile with the metal production of stars seen in clusters today and it is posing a long standing challenge, with several unconventional solutions proposed, such as a very different IMF in clusters or a significant contribution by pop III stars or pair-instability supernovae (see the review by Reference [31] and references therein). However, these results should be treated with caution as not only the measurement of $\mathcal{Y}_{\mathrm{Fe},\odot}$ is prone to systematic errors (see the exhaustive discussion in Reference [9]) but also the different trends of stellar and gas mass as a function of total mass play a fundamental role. In particular, the low baryonic fraction of groups within r_{500} with respect to the cosmic baryon fraction does not allow to draw strong conclusions. Clearly, this plot should be more populated, particularly at the mass scale of groups with robust measurements of the iron abundance consistently out to r_{500}.

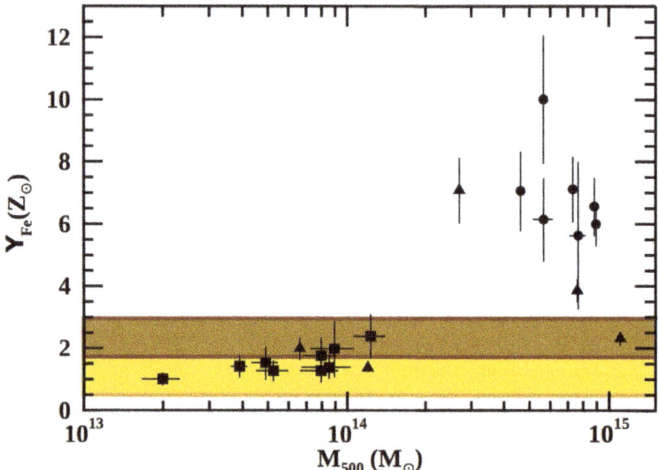

Figure 7. Effective iron yield $\mathcal{Y}_{\mathrm{Fe},\odot}$ for the clusters in the sample of Ghizzardi et al. [9] (circles), for the groups in the sample of Renzini and Andreon [8] (squares) and for the additional objects (NGC 1550, MKW 4, Hydra A, Perseus, and Coma in order of increasing mass) in the sample of Sasaki et al. [75] (triangles) as a function of the total mass of the system. We estimated stellar masses for the objects of Sasaki et al. [75] converting their L_K optical luminosities to stellar masses, assuming a stellar mass-to-light ratio in the K-band of 1 consistent with stellar population models for a Kroupa IMF [161] and with observational results [169–171]. The yellow band shows the expected value computed through the SN yields derived from Maoz and Graur [165] and Renzini and Andreon [8]; the brown band represents the expected value derived assuming a higher SNIa rate in galaxy clusters than in the field, following Freundlich and Maoz [168].

Indeed, both the iron abundance and to a lesser extent the total iron mass in the IGrM do not depend only on the total amount of iron produced, but also on its dilution with pristine gas, its ejection due to non-gravitational feedback by AGN and SN and the different phases in the gas. All these effects should be taken into account (e.g., Reference [172]).

Another related quantity exploits directly the luminosity of the system either in the optical (B) or infrared (K) bands: it is the ratio of the iron mass to the total light of the cluster/group (IMLR [62,173]). If different metals in addition to iron are measured then the specific element MLR can be estimated, like, for example, O and Si MLRs. These quantities are even more important to consider given the trends of stellar and gas fractions (and their sum, the baryon fraction) as a function of total mass which clearly mark the scale of groups as a crucial one in comparison with clusters. The derived baryon fraction for rich clusters is consistent with the cosmic baryon fraction, $\Omega_b/\Omega_M \sim 0.15$ [174] as obtained with X-ray, optical and infra-red observations (e.g., References [175–181]; also see the companion reviews by Lovisari et al. and Eckert et al.) On the other hand, groups are characterized by higher stellar mass and lower IGrM mass fractions than rich clusters, and the number of baryon fraction tends to be lower with smaller groups (e.g., References [8,182–184]). Explanations of this discrepancy are suggested in the above references as follows: 1. different physical processes depending on the system mass, like, for example, a different efficiency of baryon-to-stars conversion; 2. observational data missing for fainter sources (as, for example, the intra-cluster light component) and for the IGRM at large radii; 3. systematic errors for the mass estimations; 4. non-gravitational heating and metal mixing by AGN feed back (e.g., Reference [172,185]; see Eckert et al.), but a definitive solution has not been reached yet. As for the non-gravitational effect, such as AGN feedback, the entropy profiles are a good probe to estimate for each group and cluster, and we describe it in the following paragraph.

Historically, Arnaud et al. [159] found that the total iron mass in the ICM is proportional to the total luminosity of the early type galaxies in the clusters. And the IMLR in the B-band had larger values in clusters than in groups, mainly caused by the biased low early global abundance measurements (see Section 2.2) at the group scale (e.g., References [186,187]). The current state of the art of measurements of the IMLRs is achieved by combining near-infrared (K-band) luminosities (more directly related to the bulk of the stellar mass in early type galaxies [171,188]) obtained with the two micron all sky survey (2MASS) and the measurements performed by Suzaku extending to the outer regions of nearby clusters and groups (e.g., References [120,121,189]). Figure 8 reports the IMLRs thus obtained: there is a general increase with radius and poorer clusters and groups also have lower IMLRs within the 0.2 r_{180} region. On the other hand, the IMLRs of groups and clusters in the outer region at $r > 0.5\ r_{180}$ seem to be closer to each other. Figure 8 suggests that poorer systems (groups and clusters with fewer member galaxies) could not hold the gas including metals due to the relatively shallower potential in their assembly process. In a following work, Sasaki et al. [75] showed that the same systems have lower IMLRs and larger entropy excess, which is a signature of the non-gravitational energy input in the central regions of groups during the assembly stage.

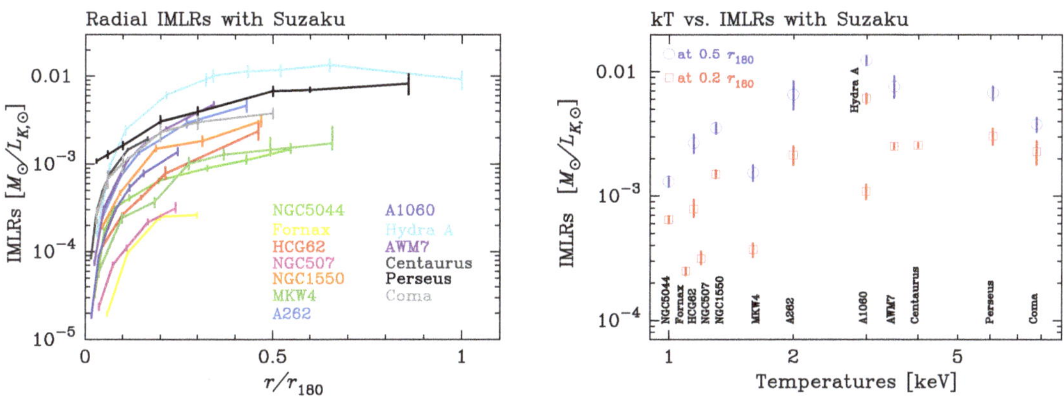

Figure 8. (**Left**) Radial IMLR profiles with Suzaku X-ray observations and K-band luminosity of the member galaxies from 2MASS data [75,120,121,189–191]. (**Right**) Temperature dependence of the IMLRs for groups and clusters in 0.2 or 0.5 r_{180} with Suzaku.

Abundance ratios of silicon to iron in clusters and groups look similar to each other in $r < r_{500}$ (see Section 2.3.2), and the silicon mass in poorer systems has a lower value than those in larger systems (the same trend as for the iron mass, caused by the lower gas mass). Because significant fractions of oxygen and silicon are mainly synthesized by SNcc, they are good indicators to estimate the amount of massive stars in the past. Renzini [192] calculated the oxygen and silicon MLRs under the assumption of the Salpeter initial mass function with the slope of the power-law shape to be -2.35. If we assumed the silicon to iron abundance ratios to be $\sim$1 up to the virial radius (e.g., References [89,189]), the expected silicon mass to light ratios (SiMLR) in rich clusters agree with the estimate derived by Renzini [192], as shown in Matsushita et al. [189] and Sasaki et al. (2021).

However, for groups, neither iron nor silicon have been observed yet out to and beyond r_{500} with the exception of a handful of systems. To study the metal enrichment history of the ICM and IGrM, we need to measure oxygen profile, mainly produced by SNcc, to the virial radius and beyond since the whole clusters and group include the metals synthesized in the cluster and group formation phase. To summarize, in order to progress in the understanding of the chemical enrichment of both groups and clusters, we need to measure the gas and metal MLRs to the virial radius for both clusters and groups to high redshift of $z \sim 2$ with the next generation of X-ray instruments (see Section 4).

2.5. High Resolution Spectroscopy: Current Observational Results with RGS

Undoubtedly, the need for high spectral resolution (particularly the capability to resolve the Fe-L complex) is absolutely crucial to provide accurate constraints on chemical abundances (and the physics of the enrichment) in groups and elliptical galaxies. While waiting for the exploitation of micro-calorimeters onboard *XRISM* and *Athena*, one should keep in mind the valuable potential of the Reflection Grating Spectrometer (RGS) instrument onboard XMM-*Newton* to deliver high resolution spectra. Formally, the RGS has a spectral resolution of ~3 eV. In the case of extended sources, however, the slit-less characteristic of this instrument induces a convolution of a given line profile with the spatial distribution of its surface brightness along the dispersion direction of the detector. Although this makes the RGS abundance measurements of extended clusters rather challenging (yet still feasible, e.g., Reference [126]), groups are by definition more compact and are less affected by this instrumental broadening. The spectral window of the instrument (typically ~6–30 Å, corresponding to ~0.4–2 keV) is both an advantage—as it covers the O VIII (and O VII), N VII, and even C VI lines which are difficult or impossible to detect with CCD instrumental responses—and a drawback—as the continuum is challenging to constrain within this band. Consequently, the power of RGS resides less in the measurements of the IGrM absolute abundances than in the measurements of their (relative) N/Fe, O/Fe, Ne/Fe, and Mg/Fe ratios. Although discrepancies of absolute Fe measurements between RGS and CCD-like instruments may in principle lead to discrepancies in their respective X/Fe ratios, we note a generally good agreement in the latter case (e.g., Reference [127]).

Abundances measured using RGS have been, for instance, reported on individual poor systems [193–195], as well as in larger samples [196–198]. The CHEERS sample, which was constructed specifically to ensure a $>5\sigma$ detection of the O VIII line in each system, provided interesting measurements in this respect. Mao et al. [199] obtained reliable constraints ($>3\sigma$) on the N/Fe ratio in six galaxies (M 49, NGC 4636, NGC 4649) and groups (NGC 5044, NGC 5813, and NGC 5846), as well as in M 87 and one cluster (A 3526). Unlike all the other ratios known in the IGrM which are typically close to Solar (Section 2.3.2 and Figure 5), the average N/Fe is clearly super-Solar. This strongly suggests that the bulk of nitrogen originates from an enrichment channel that is separate from the usual SNcc and SNIa contributions—very likely AGB stars. The O/Fe ratio was investigated (in groups and ellipticals, but also in more massive systems) by de Plaa et al. [200], and was found to be consistent with Solar, thus being in line with the picture of the hot gas chemical composition remaining uniform with mass (Section 2.3.2), despite a significant scatter from system to system.

Despite the valuable ability of RGS to measure accurately important ratios (particularly N/Fe and O/Fe, see Figure 9), its limited spectral window—coupled to its sensitivity to the spatial extent of the source and the difficulty to perform spatial spectroscopy (see, however, Reference [201,202])—makes this instrument taken at its best advantage when combined with CCD measurements. Nevertheless, RGS offers the unique advantage to reveal a glimpse of the main transitions populating the Fe-L complex at groups (and clusters) temperature regime(s). This is particularly essential, not only to refine our science prediction expected with micro-calorimeter instruments, like Resolve onboard *XRISM* (Section 4.2), but also to pursue the improvement of our spectral models in this crucial spectral band even before the release of *XRISM*, particularly by comparing updated atomic calculations with (i) laboratory measurements and (ii) state-of-the-art observational data [203,204].

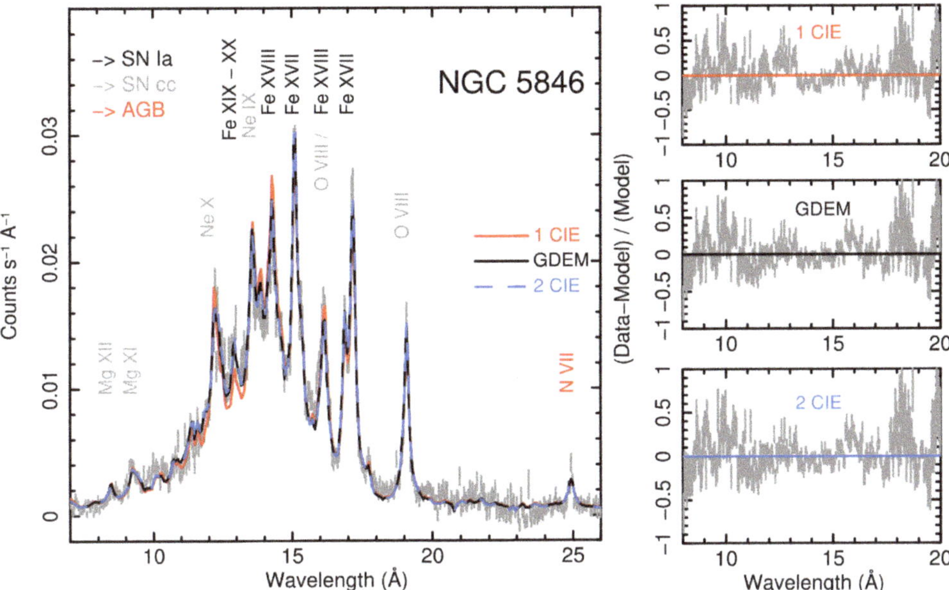

Figure 9. An example of RGS spectrum from a deep observation of NGC 5846, with indication of its main emission lines (and the stellar origin of their corresponding element). Spectra from the RGS 1 and RGS 2 instruments were combined before successive fits assuming models with one temperature (1 CIE), two temperatures (2 CIE), and a Gaussian-like multi-temperature structure (GDEM). Residuals are shown in the right panels. Figure reproduced with permission from de Plaa et al. [200].

3. Theoretical Framework and Simulations

3.1. High-Resolution Hydrodynamical Simulations and Small-Scale Astrophysics

While Section 3.2 focuses on the large-scale metal origins and evolution from the high-z universe via cosmological particle simulations, here we discuss the metal abundance evolution in terms of small-scale (astro)physics and tracer advection by means of high-resolution (HR) hydrodynamical (HD) grid simulations in single halos. Resolutions in such simulations can reach down to the pc scale and often employ adaptive/static mesh refinement, together with finite-volume Godunov methods with third-order (or higher) accuracy. Given the high resolution (and small timesteps), such simulations are mainly focusing on the cores of galaxy groups ($r < 0.1\,R_{500}$) and are well suited to track the turbulence mixing, shocks, entrainment, and detailed feedback/feeding imprints, such as AGN cavities and thermal instabilities. We note that, in this section, Section 3.1, we take a more physically-oriented approach, rather than following an historical sequence.

Before tackling typical HD simulation results, it is worth to review a few common physical and numerical properties of metals leveraged by investigations of different groups. A key small-scale feature of metals is that they are passive tracers of the hydrodynamical evolution; thus, they can be used akin to dyes/pollutants in fluid dynamics studies (e.g., Reference [205]). In hydrodynamics (e.g., Reference [206]), the equation describing the temporal rate of change of the metal tracer density is given by (in the Eulerian/grid framework):

$$\frac{\partial \rho_Z}{\partial t} + \nabla \cdot (\rho_Z \boldsymbol{v}) = S_Z, \tag{4}$$

where S_Z is a general metal abundance source term; in Equation (6) and the related paragraph below, we will unpack such a source term, which, in the IGrM, is mainly shaped by stellar feedback processes, such as supernovae and stellar winds. In fluid dynamics, Equation (4) is also known as a conservation equation. In the weak compressibility case, the metals are purely advected along the Lagrangian stream, reducing Equation (4) to

$$\frac{d\rho_Z}{dt} \equiv \frac{\partial \rho_Z}{\partial t} + \boldsymbol{v} \cdot \boldsymbol{\nabla} \rho_Z = S_Z. \tag{5}$$

We note that, in localized IGrM regions with $\boldsymbol{\nabla} \cdot \boldsymbol{v} \neq 0$, the pollutants may also experience compressions or rarefactions, hence tracing not only smooth bulk processes (subsonic turbulent eddies) but also nonlinear in-situ features (shocks and cold fronts). In the HD grid simulations, such iron density is implemented via a normalized scalar $Z \in [0,1]$ tied to each cell gas density, such as $\rho_Z = Z\,\rho$ (see Reference [207]). Ought to high-order Godunov schemes, such as the Piecewise-Parabolic Method (PPM; Colella and Woodward [208]), numerical diffusion is kept at low levels compared with physical diffusion, e.g., due to turbulence. Further, despite the large diversity of elements, IGrM HR numerical studies often use the approximation $Z \approx Z_{\rm Fe}$, since iron has one of the strongest line emissions among heavy elements—especially in hot plasma halos—which can be robustly constrained via X-ray spectra (Section 2). We note that, while metals are a dynamical passive tracer, they contribute significantly to the line cooling of the gas below $T < 1$ keV (e.g., Reference [209]), hence accelerating the IGrM condensation cascade.

Typical HR HD simulations focus on the group core, where the central galaxy contributes to a substantial amount of metals later dispersed in the diffuse IGrM, which are mainly produced via supernovae (SN) explosions and winds from red giant stars (SWs). An exemplary and well modeled system is the nearby galaxy group NGC 5044, with the homonymous central galaxy dominating over the many smaller satellites. In several ETGs/BGGs analytic studies and non-cosmological HD simulations (e.g., Loewenstein and Mathews [210], Ciotti et al. [173], Renzini and Ciotti [62], Brighenti and Mathews [211], Brighenti and Mathews [212], Mathews and Brighenti [213], Gaspari et al. [214], Gaspari et al. [215], Pellegrini et al. [216]), a common implementation of Equation (5) is to recast it in terms of the astrophysical IGrM abundance of the ith-element (by mass in Solar unit):

$$\frac{dZ_i}{dt} = (\underbrace{N_*\,\alpha_*}_{\text{stellar winds}} + \underbrace{N_{\rm SN}\,\alpha_{\rm SN}}_{\text{supernovae}})\,\frac{\rho_*}{\rho}, \tag{6}$$

where the two normalization factors related to stellar winds and supernovae are, respectively, $N_* = Z_{*,i} - Z_i$ (with $Z_{*,i}$ the stellar abundance) and $N_{\rm SN} = y_{{\rm SN},i}/(Z_{i,\odot} M_{\rm SN})$ (with $y_{{\rm SN},i}$ the SN yield in $M_\odot$ and $M_{\rm SN}$ the ejected supernova mass). The BGG stellar density ρ_* is usually modeled via a de Vaucoulers profile [217]. The specific injection rates due to SWs and SN are $\alpha_* \approx 4.7 \times 10^{-20} (t/t_{\rm now})^{-1.3}\,{\rm s}^{-1}$ and $\alpha_{\rm SN} \simeq 3.2 \times 10^{-20}\, r_{\rm SN}(t) (M_{\rm SN}/M_\odot)\, \Upsilon_B^{-1}\,{\rm s}^{-1}$, where Υ_B^{-1} is the optical stellar mass-to-light ratio in the B band (e.g., References [213,218–220]), respectively. As introduced above, iron is one of the best metals to leverage as dynamical tracer in HR simulations: in BGGs/ETGs (which have star-formation histories peaked at early times) SNII are mostly consumed at high redshifts, while SNIa—exploding in binary systems with a white dwarf—drive the BGG long-term iron enrichment. The local SNIa rate is $r_{\rm SNIa} \sim 0.1 (t/t_{\rm now})^{-1.1}$ SNU (per 100 year and stellar luminosity $10^{10}\,L_{{\rm B},\odot}$, e.g., Greggio [218], Mannucci et al. [219], Humphrey and Buote [220]). The iron normalization values for the SNIa ($M_{\rm SNIa} = 1.4\,M_\odot$) are $y_{\rm SNIa,Fe} = 0.7\,M_\odot \sim 10\,y_{\rm SNII,Fe}$ and $Z_{{\rm Fe},\odot} = 1.83 \times 10^{-3}$.

We now review the results of HD simulations leveraging the metal tracing framework. In HR HD simulations, metals are a crucial tool to unveil and understand the kinematics of major physical processes, such as AGN feedback, SMBH feeding, and IGrM turbulence. While the detailed AGN self-regulation thermodynamics (cooling and heating cycle) is tackled in the companion Eckert et al. review (also see the unification diagram in Gaspari et al. [221]), here, we focus on the main IGrM cores kinematics features. The SMBH/AGN at the center of each BGG or ETG is fed recurrently via chaotic cold accretion (CCA [222]), i.e., the filamentary/cloudy condensation rain that is generated via *nonlinear* turbulent thermal instability in the IGrM hot halo (Gaspari et al. [223], Voit [224]; also see McCourt et al. [225], Sharma et al. [226] for linear thermal instability simulations). Such frequent CCA clouds trigger the AGN feedback response by re-ejecting substantial amount of mass and energy via ultrafast outflows and relativistic jets (e.g., References [227,228]). At the macro scale of tens kpc, such entrained outflows/jets use their mechanical ram pressure to generate a diverse range of astrophysical phenomena (buoyant X-ray cavities, weak transonic shocks, turbulent eddies), which recurrently reheat the IGrM halo and quench cooling flows/star formation throughout the several Gyr evolution (e.g., Churazov et al. [229], Brüggen [230], Brighenti and Mathews [231], McNamara and Nulsen [12], McNamara and Nulsen [13], Gitti et al. [15], Gaspari et al. [232], Barai et al. [233], Yang et al. [234]).

The macro-scale AGN feedback deposition channels are difficult to disentangle through simple temperature or surface brightness maps. The metal tracers are instead able to unveil in a clear manner such feedback features. Indeed, while the BGG produces a continuous reservoir of metals/iron in the core of the galaxy group, the self-regulated AGN outflows uplift them (Equation (4)) outwards, on top of the low-Z background, thus creating key contrast patterns and imprints. Figure 10 exemplifies this during a typical HR HD simulation [215] of AGN feedback—self-regulated via CCA—in a galaxy group akin to NGC 5044 (with a central BGG $M_* \simeq 3.4 \times 10^{11}\, M_\odot$). In the right panel, the meso AGN outflows have just uplifted the iron generated in the core of the BGG (magenta) up to several tens kpc. The pattern is highly anisotropic and the enhancement can reach up to $\sim$2$\times$ values compared with the pristine background (<0.3 $Z_\odot$, e.g., Ghizzardi et al. [9]). Inhomogeneities are also visible, particularly the thin metal-rich rim that envelopes the inflated buoyant bubble. At variance, in the left panel, as the AGN outflows subside and CCA feeding is quenched via the previous AGN outburst, the cascading subsonic turbulence drives mixing, eventually washing out the inhomogeneities (cavity, cocoon, jet channel) and restoring the azimuthally symmetric IGrM halo 'weather'. Subsequently, this enables another phase of gradual IGrM precipitation and, hence, boosted accretion, with the triggering of another AGN feedback cycle via the condensed material. Both the above-shown anisotropic metal outflows and turbulent distributions have been found by a wide range of hot halo observations involving mechanical AGN feedback (Section 2.3.3) and a diverse range of HD numerical studies and groups [207,235–240]. We note that galactic SN-driven outflows, albeit weaker and difficult to spatially detect, can enhance the anisotropic enrichment, especially in low-mass halos (e.g., References [241,242]).

Metal maps not only give us constraints on the effects of AGN feedback, but can also constrain different AGN feeding modes. Figure 11 shows two HR HD simulations testing two different models of AGN self-regulation and feeding in a massive galaxy group with extended IGrM ($L_X \sim 10^{43}\,\mathrm{erg\,s^{-1}}$). In the left panel, CCA feeding/cold mode self-regulation drives a very intermittent duty cycle tightly related to the recent cooling rate in the group core (typically with pink-noise time power spectrum; Gaspari et al. [223]). The rapid flickering of the AGN enables the formation of characteristic AGN feedback imprints, such as buoyant bubbles and cocoon shocks that are encased by metal-rich rims (cyan) protruding on top of the low abundance background. In addition, the buoyant bubble can dredge out trailing filaments of metals via the hydrodynamical Darwin [243] effect ($V_{\rm trail} \sim 0.5\, V_{\rm b}$, where $V_{\rm b}$ is the bubble volume). Conversely, hot-mode feeding (e.g., Bondi or ADAF; Bondi [244], Narayan and Fabian [245]) drives a continuous AGN feedback evolution with a perennial, monolithic, wide and uniform cylinder of metals, without any signs of cavities or shocks. As for the thermodynamics (cf. the companion Eckert et al. review), observational constraints favor the former intermittent bubble duty cycle (via cold mode/CCA), rather than the latter quasi-continuous triggering mode (Bondi or hot-mode feeding)—Simionescu et al. [140], Hlavacek-Larrondo et al. [246], McNamara and Nulsen [13], Gitti et al. [15], Liu et al. [247], Gaspari et al. [248].

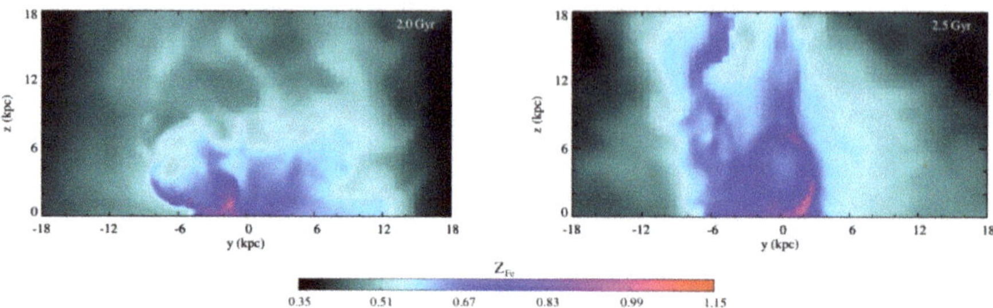

Figure 10. Emission-weighted iron abundance during a typical HR HD 3D simulation of self-regulated AGN feedback in a median 1 keV galaxy group akin to NGC 5044 (adapted from Gaspari et al. [215]), showing two typical stages. The black regions denote the diffuse, primordial iron background ($Z_{Fe} < 0.3\,Z_{\odot}$). (**Left**) turbulence-driven period, during which mixing dominates, gradually washing out the anisotropic features and restoring azimuthal symmetry. (**Right**) the meso-scale (sub-kpc) AGN outflows have inflated a common X-ray cavity in the IGrM, generating a thin metal-rich rim (coincident with the compressive cocoon shock) and anisotropic iron uplift from the BGG outwards in the extended IGrM.

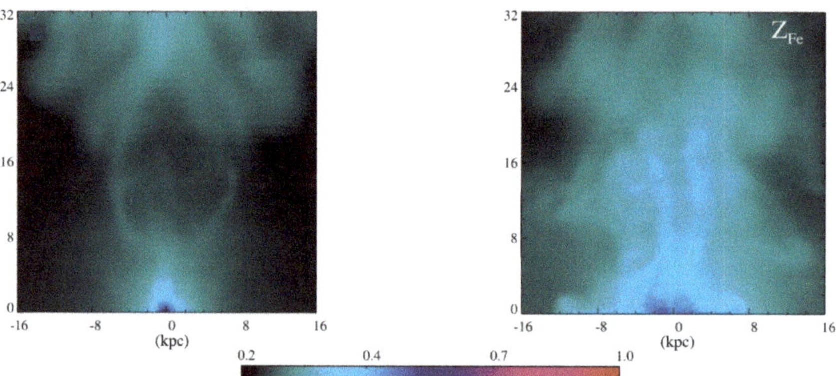

Figure 11. The iron abundance projected maps can differentiate between different models of AGN feeding triggering and self-regulation, here in a common HR HD simulation of a massive galaxy group (adapted from Gaspari et al. [214]). (**Left**) CCA feeding mode, driving intermittent and frequent AGN features, such as cavities with metal-rich rims and trailing filaments. (**Right**) Bondi feeding mode, driving a perennial, wide monolithic cylinder of metals into the group core (with no bubbles or cocoons).

As introduced above, a key component of the metal circulation is turbulence, either generated by the AGN feedback or by the large-scale cosmological evolution (Section 3.2), which is worth to further dissect. Remarkably, turbulent motions generate two (seemingly) contrasting effects, but on different scales. On the one hand, turbulent motions promote the diffusion of metals, tending to equalize the radial abundance profile from a negative to null gradient (e.g., References [249,250]). On the other hand, turbulence induces local chaotic relative density fluctuations $\delta\rho/\rho$. As per the above Equation (5), metals can be considered on average akin to passive tracers of the HD density field; thus, $\delta\rho_Z/\rho_Z \sim \delta\rho/\rho$. As shown by numerical and analytic studies (e.g., References [251,252]), such stratified hot-halo fluctuations are linearly tied to the turbulent Mach number $\mathrm{Ma_t}$; hence, the relative metal abundance can help us to constrain the level of turbulence in the IGrM too, $\delta\rho_Z/\rho_Z \propto \mathrm{Ma_t}$, with the slope of the Fourier spectrum constraining plasma processes, such as thermal conduction [253]. In the IGrM, the inferred 3D sonic Mach number of turbulence is $Ma_t \sim 0.3$–0.5 [254,255], i.e., σ_v of a few $100\,\mathrm{km\,s^{-1}}$. This can complement upcoming spectral X-ray IFU studies carried out via *XRISM* and *Athena* (see Section 4), with detailed synthetic observations already highlighting unprecedented features of metals and turbulence in hot halos [256–259]. Moreover, constraining the turbulent metal evolution in the IGrM plasma phase enables to assess the kinematics of the top-down multiphase rain, since the condensed warm (Hα+[NII]) filaments and cold (CO, HI) clouds share analogous ensemble velocity dispersion [260–263].

While the large-scale cosmological evolution is discussed in the upcoming Section 3.2, it is worth to note here that at $r \gtrsim 100\,\mathrm{kpc}$ (and with Gyr frequency), the infalling substructures and interacting galaxies (particularly dry dark matter halos; see the HD simulation review by Zuhone and Roediger [17]) can induce significant amount of sloshing in the IGrM, hence creating large-scale metal anisotropies and tails that are often correlated with cold fronts/contact discontinuities or ram-pressure stripping features [149,155,264–269]. For the observational insights on related metallicity maps, we refer the interested reader to Section 2.3.3.

3.2. Cosmological Simulations and Large-Scale Evolution

The metal content of cosmic structures has been addressed via complex large-scale cosmological hydrodynamical simulations, as well [32]. Cosmological simulations [270] allow us to study and predict the formation and evolution of galaxies and galaxy systems, such as groups and clusters, within the large-scale cosmological framework [271,272], while consistently accounting for a large variety of physical processes shaping the baryonic matter component—from gas cooling, to star-formation and BH evolution, to energy feedback. In particular, chemical evolution models are needed to consistently follow the production and evolution of the metal content in the stellar and gaseous components, which has important consequences on the cooling properties of the gas, on the conversion of gas into stars, and, therefore, is linked to the thermo-dynamical structure of galaxy systems. Given the large dynamical ranges spanned in simulations of cosmological volumes, chemical evolution is typically treated via a sub-resolution model, similarly to other small-scale physical processes (like star formation or energy feedback).

Chemical evolution models have been introduced in cosmological simulations starting from the 1990s [273,274]. While early studies including chemical enrichment mainly focused on galaxies [273,275], soon Smoothed-Particle Hydrodynamics (SPH) simulations of large-scale structure and galaxy clusters started to include chemical evolution models as well [276–278]. Despite different level of complexity, already in the early implementations, the metal production associated to both SNIa and SNcc was included, and the metal content of the gas was taken into account in the cooling process (e.g., References [279,280]). In Reference [278], the contribution to chemical enrichment due to low- and intermediate-mass stars undergoing the AGB phase, as well as the treatment of metallicity-dependent stellar yields and mass-dependent stellar life-times were also included [281]. Starting from the initial models that focused on total metallicity or iron abundance, an increasing level of detail has been reached over the years, with modern simulations typically following the production and evolution of several metal species separately (e.g., oxygen, silicon, etc.). Chemical enrichment models are typically based on three fundamental pillars: the initial mass function (IMF) (e.g., References [282–284]), and mass-dependent stellar life-times (e.g., References [285–288]) and metal yields (e.g., References [6,289–292]). In SPH simulations [293], in particular, chemical evolution models are coupled directly with the star formation model, where every stellar particle is representative of a simple stellar population (SSP), that is a population of stars all characterized by the same age and metallicity. The assumptions on the IMF and on the stellar yields are required to predict the amount of metals generated by each SSP and the time-scale on which different enrichment channels (primarily SNIa, SNcc and AGB stars) release the metal mass into the surrounding gas elements (e.g., References [294,295], for more details on the principal equations that describe the stellar evolution and metal production).

Results from cosmological hydrodynamical simulations on the chemical enrichment of cosmic structures, from galaxies to groups and clusters, can be very sensitive to the specific assumptions on the IMF or of stellar yields. In particular, these can affect the normalization of metallicity profiles and the value of global abundances. Changes in the underlying IMF functional form, affect directly the final ICM metallicity and abundance ratio profiles, for instance, due to different relative amounts of low- and high-mass stars [281,288,296]. The yield tables are also an important source of uncertainty [289] in the predicted integrated level of enrichment, as well as the supernova rates (see a recent investigation based on the Illustris Simulations by Reference [297]). More importantly, the complex interplay with other gas thermal and dynamical processes treated in the simulations, such as energetic feedback or merging processes, can substantially impact the spatial distribution of metals and, therefore, the gradients of the radial profiles, as further discussed in Section 3.2.1 (also see Section 3.1).

In the last decade, more and more cosmological simulation codes have combined chemical enrichment models with many other important physical processes describing the evolution of gas and stars, with the principal aim of reaching an increasingly detailed and realistic picture of cosmic structures, from galaxies to galaxy clusters and cosmic filaments, to be compared against observational findings [172,297–303].

Nonetheless, most of the numerical studies based on cosmological simulations so far have concentrated on the case of (massive) galaxy clusters, for which the impact of resolution, feedback processes and interplay with the member galaxy population can be better constrained. Given their special position at the crossroad between smaller-scale physics and cosmic evolution, galaxy groups represent in fact a rather challenging, albeit crucial, target: capturing correctly the effects of feedback from central galaxies and BHs, given the shallower potential wells of groups compared to clusters, is of great importance. This has been in fact the main source of discrepancy in the comparison with observational findings, and consequently one of the crucial testbeds for cosmological simulations and the physical models therein included (for a thorough discussion, see the companion review by Oppenheimer et al.).

3.2.1. Results from Cosmological Simulations

Cosmological simulations can be extremely powerful resources to study and predict the detailed enrichment history of the gas in cosmic structures, as well as to investigate the expected spatial distribution of metals. It is, therefore, crucial to assess their reliability by comparing simulated results to observational findings.

As a consequence of the interplay of different physical and dynamical processes, especially energetic feedback, simulations allow us to explore the expected observable signatures on the resulting distribution of metals in the IGrM gas. Numerical investigations showed, for instance, that feedback from AGNs is crucial to reproduce the large-scale homogeneous enrichment observed in the outer periphery of galaxy systems, such as groups and clusters [107]. In such simulations, metallicity profiles show a relatively flat trend out to large distances from the center in clusters, and a very similar enrichment level in smaller structures, as well. Consistent findings emerge from recent observational studies, as discussed in detail in Section 2.3.1 [31].

This effect of early AGN feedback promoting a more homogeneous enrichment and shallower radial profiles, was already observed in the simulations by Reference [172], for both massive clusters and lower temperature systems. At the group regime ($T \lesssim 3\,\mathrm{keV}$), they found that the flat profile of iron abundance in the outskirts of the simulated groups was in contrast with the observational results by Reference [93], employed for comparison (also see Reference [300]). The simulation results are instead more in line with recent observational data, e.g., by Reference [52,89]. Already in Reference [172], it was shown that also the the silicon-to-iron abundance ratio in simulations was found to be flat out to large distances from the center in groups, as well as in clusters, indicating also a similar contribution of SNIa and SNcc to the gas metal enrichment. A relatively flat silicon-to-iron ratio was also confirmed by independent results obtained by Reference [23] on a set of simulated galaxy groups extracted from the OverWhelmingly Large Simulations (OWLS) project [304], despite finding typically more pronounced radial gradients of the IGrM metallicity. Several independent simulation studies converge on the idea that energy feedback solely due to supernova winds typically produces clumpier distributions of α-elements, like oxygen or silicon, and steeper, decreasing radial metallicity profiles. The reason for the different metal distribution can be related to the origin of the α-elements, mostly produced by SNcc and, therefore, confined in the vicinity of star-formation sites, where they can be efficiently locked back into newly formed stars unless an efficient mechanism intervenes to quickly distribute them far enough. The steeper profiles in absence of AGN feedback was also mildly noted in Reference [305], although the galactic outflow model used in their cosmological simulations was able to reproduce the global iron content in group-size halos, together with various observations of cosmic chemical enrichment. Those authors concluded, as well, that an efficient outflow mechanism, able to displace pre-enriched gas out of galaxies, must be in place already at early times, in order to explain the observed chemical enrichment of the inter-galactic medium at $z \sim 6$ (also see Reference [280]).

These general trends have been also confirmed by following simulation campaigns [297,298,306]. In addition to consistent results on the relation between AGN feedback effects and the enrichment level at large distances from the center (i.e., beyond $\sim$0.3 R_{500}), these recent simulations also finally reproduced the diversity of thermal and chemical properties found in the core ($\lesssim$0.1 $\times$ R_{500}) of observed systems [303,306]. Some differences, e.g., on the metallicity profile normalization and, thus, on the global enrichment level, is nonetheless still present depending on the specific set of simulations analyzed [301], and consequently on the set of stellar yields and supernova rates adopted. In addition, the modeling of dust and metal spreading within cosmological simulations can further impact the details of the spatial distribution of metals that remain in the gaseous phase, for which further dedicated studies are needed.

When comparing simulation results with observations, it is important to pay attention to the way quantities are evaluated in simulations. In particular, estimates of metallicity (as well as other thermal properties) can be derived in different ways depending on the weight w used to compute the average Z value, that is $Z_w = \int wZ\mathrm{d}V / \int w\mathrm{d}V$. Typical weights are the mass of the gas or its emissivity in the X-ray band. In general, flatter metallicity profiles in simulations are better reproduced when a projected emission-weighted estimate is pursued, whereas the mass-weighted three-dimensional metallicity typically shows a somewhat steeper decrease with radius [107]. In this perspective, the issue of a fair comparison between gas properties in simulations and observations has also been tackled via the generation of detailed X-ray synthetic observations out of numerical simulations, properly taking into account the specific characteristics of X-ray telescopes. With such techniques, Ref. [71] showed that an observational-like reconstruction of the iron and oxygen abundances with mock XMM-*Newton* observations of simulated clusters and groups is in good agreement with the intrinsic simulation value. More recently, Ref. [257] employed synthetic observations of the X-ray Integral Field Unit (X-IFU) on board the next-generation European X-ray observatory Athena to reconstruct chemical properties of the ICM from simulated galaxy clusters. The authors showed that the metallicity values obtained from X-IFU spectra match well the emission-measure-estimate computed directly from the simulations.

The homogeneous enrichment of the intra-cluster and intra-group gas on large scales, as indicated by the little scatter around very shallow radial profiles in the outskirts, is also strongly connected to the history of the chemical enrichment. The so-called pre-enrichment scenario implies that the gas enriched within proto-groups and proto-cluster galaxies has been displaced beyond their shallower potential wells by some efficient mechanism at early times ($z \gtrsim 3$)—while stellar feedback alone is not sufficient, many recent state-of-the-art cosmological simulations identify the responsible mechanism with early AGN feedback. This allows the bulk of the diffuse inter-galactic gas to be pre-enriched and then re-accreted into the assembling galaxy group or cluster. This further supports the idea that a significant fraction of the gas chemical enrichment happens at those early times, as confirmed by the little evolution of the gas metallicity below $z \sim 2$, especially on the large scales, found in observations (e.g., References [307–309]) and also in simulation studies [107,297,298]. At the group scale, cosmological simulations predict a gas metallicity evolution below $z \lesssim 1$–2 that is very similar to the one found, and observed, in more massive clusters. Consistently with the results discussed above, simulations including AGN feedback find a shallow dependence on redshift, especially when the global metallicity within R_{500} or the enrichment level in the outer regions ($\gtrsim 0.3 \times R_{500}$) is concerned. Ref. [299] show that the evolution of the metallicity in different radial range is similarly mild at groups scales, as well, unless only stellar feedback is included in the simulations. In that case, they note again an effective reduction of the gas iron and oxygen content which is more severe particularly in the group regime. This is interpreted as a more substantial, unsuppressed, star formation activity which efficiently, and preferentially, consumes metal-rich gas. In the galactic-outflow model by Reference [305] a higher growth rate, with respect to observations, was in fact observed in the simulated groups.

The similarities between chemical enrichment of the IGrM in lower-temperature groups and the ICM in massive objects is further supported by the shallow dependence of gas global metallicity on the system mass (or temperature). In contrast to observational findings, where lower iron abundances were typically observed in group-size systems compared to clusters, Ref. [310] report a shallow, mildly anti-correlating, metallicity-temperature relation, employing semi-analytic models of galaxy evolution. Simulation results, like those presented in References [297,302], also predict a shallow anti-relation between metallicity and temperature, that extends without breaks from clusters down to groups. More recently, Reference [299] compared the relation between temperature and iron abundance in the core (i.e., $<0.1\,R_{500}$) of simulated groups and clusters with recent results from the CHEERS sample [92], also finding a shallow anti-correlation overall, with a mass-invariance of the IGrM and ICM iron abundance in cool-core clusters, as shown in Figure 12. In the figure, in particular, the simulated data (star symbols) are reported together with the best-fit relation for the whole sample, and for the CC and NCC subsamples, with the former in better agreement with the CHEERS results (also see Reference [31]).

Ref. [302] show that a flat relation with temperature applies, as well, when abundance ratios relative to iron are investigated (e.g., O/Fe, Si/Fe, S/Fe, etc.). Interestingly, this is also in line with recent observational findings (e.g., Reference [128]; also see discussion in Section 2.3.2).

So far, the main limitation to study simulated groups of galaxies in more detail has been the lack of simulations within cosmological context able to consistently resolve and match the stellar and gas properties of massive galaxy clusters and galaxies, simultaneously. Groups, in this respect, have often rather served as crucial testbed for assessing the prediction power of physical models embedded into cosmological simulations, especially related to feedback from central galaxies (see the companion reviews by Oppenheimer et al. and Eckert et al.). Nonetheless, given the recent encouraging results obtained on the chemical and thermal properties of the diffuse gas in various cosmic structures, especially with improvements in the description of chemo-energetic feedback from galaxies in group and cluster cores, more dedicated studies specifically focusing on the group regime in cosmological simulations are definitely needed to explore the details of their formation and evolution.

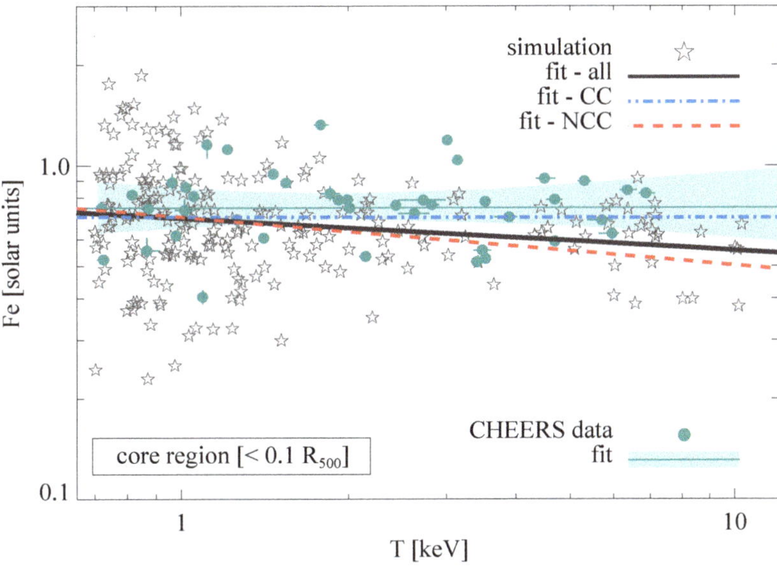

Figure 12. Relation between gas iron abundance and temperature in the core ($r < 0.1R_{500}$) of groups and clusters. Comparison between cosmological simulations (empty stars) and X-ray observational results from the CHEERS sample (filled circles). We also report best-fit relations for the whole simulated sample (solid grey line), and for the CC and NCC subsamples (blue dot-dashed and red dashed lines, respectively), as well as the relation determined from the CHEERS data (turquoise line and shaded area, for the associated 68.3% confidence region). Consistently with previous sections, iron abundances, relative to hydrogen, are reported with respect to the Solar reference value by Reference [43]. Adapted from Reference [299].

4. Future X-ray Missions

Beyond the impressive observational efforts provided by the community to best characterize metals in the IGrM using the past and current fleet of X-ray observatories (Section 2.2 and 2.3), it is clear that the next generation of X-ray missions is essential to overcome the current limitations and to advance our knowledge. As we further discuss in this section, higher spectral resolution, higher throughput, and larger sky coverage will be key factors.

4.1. *eROSITA*

The present knowledge of the physical properties of the IGrM, both thermodynamical and chemical, is limited to archival studies with known systems mainly at the high end of the mass and luminosity ranges. X-ray selection provides a more reliable way than optical selection of identifying virialized groups with a bona fide IGrM, but groups are typically at the limit sensitivity of the *ROSAT* All Sky Survey (RASS) and their detection is biased towards peaked surface brightness objects (the X-ray cool-core bias [311,312]). Even though the advent of XMM-*Newton* and *Chandra* made available deep surveys of limited areas providing less biased samples of groups, these systems are typically at moderate redshifts and they lack even global abundance measurements (e.g., References [313,314]). Optically selected systems with a dedicated X-ray follow-up can circumvent some of the biases of the RASS X-ray selection and provide some partial answer to the characteristic of the general population of groups (see, for example, Reference [315] and the interesting discussion therein).

Spectrum Roentgen Gamma (SRG, launched in 2019) hosts the eROSITA instrument, a set of seven co-aligned soft X-ray telescopes covering the 0.2–10 keV band, with a field of view of 1° and ∼15″ spatial resolution equipped with CCDs with spectral resolution of 60–80 eV in the 0.5–2 keV band [316]. At the end of 2023, after four years surveying the whole sky once every 6 months, eROSITA will build up an all-sky survey 25× deeper than RASS in the 0.5–2 keV band [316,317].

eROSITA will, therefore, provide for the first time a large, homogeneously selected X-ray sample of groups for detailed studies with future generations of X-ray instruments (Reference [315,317,318], and also see the companion reviews, particularly Eckert et al.) detailed in the next sections and for the pointed phase of eROSITA itself, in a similar fashion as *ROSAT*. In particular, given the large field of view, pointed observations of eROSITA will provide valuable information for the outer regions of groups and their metallicity. These upcoming data sets are all the more interesting in the context of the expected spectral model improvements driven by microcalorimeter data, which will further ensure that the results derived from lower-spectral resolution CCD observations are robust.

4.2. XRISM

Non-dispersive, high-spectral resolution X-ray micro-calorimeters will enable a giant leap in our understanding of the metal content of the diffuse IGrM. The next such detector slated for launch is the *Resolve* instrument onboard the X-ray Imaging and Spectroscopy Mission (*XRISM*), a JAXA-led satellite with contributions from NASA and ESA [319] with a launch expected around 2023. The *XRISM*/*Resolve* instrument, covering a 3 × 3 arcmin field of view with 35 micro-calorimeter pixels, will carry forward the seminal observations begun by the *Hitomi*/*SXS* on the ICM of the Perseus Cluster [125]. With each pixel delivering a spectral resolution of <7 eV, more than 10 times better than conventional CCDs, and given its non-dispersive nature meaning that the spectra of extended sources are not blurred by the size of the target, *XRISM*/*Resolve* will reveal emission lines from various chemical elements in the IGrM with unprecedented sharpness (see, for example, Figure 13). Due to the relatively modest spatial resolution (with a HPD of 1.7 arcmin) and effective area, *XRISM* observations are ideally suited for studying the centers of nearby clusters and groups. Using these targets, together with dedicated laboratory measurements driven by these new observations (e.g., Reference [204,320]), it is expected that remaining uncertainties and differences between AtomDB and SPEXACT in modeling the Fe-L line emission will be ironed out early during the lifetime of the mission, whose expected launch is currently set for Japanese fiscal year 2022.

For the bright, line-rich cores of galaxy groups, 100 ks observations with *XRISM*/*Resolve* can determine the abundances of Fe, O, Ne, Mg, Si, and S with statistical precisions of 5% and systematic uncertainties that are far reduced compared to those from fitting CCD spectra. Weaker lines from other elements like N, Al, Ar, Ca, and Ni (based on the Ni L-shell emission) may also be detected in the IGrM. This will provide a fantastic and stringent test of the current picture that the chemical composition of the ICM/IGrM is consistent with the Solar values (see Section 2.3.2).

Furthermore, precise measurements of the metal abundance patterns of the ICM and IGrM are expected to offer new tests of stellar astrophysics. Elemental abundances in the Sun and in the stars in the Local Group have so far been the most commonly used points of comparison in order to check whether or not current theoretical nucleosynthesis yields provide a self-consistent picture that can appropriately describe Galactic chemical evolution (see, e.g., Reference [7,321]). High-resolution X-ray spectroscopy with *XRISM*, followed by the *Athena* X-IFU (see Section 4.3) will provide complementary measurements of the enrichment history of the hot-diffuse gas, with a precision rivaling optical and infrared stellar spectroscopy, ushering in a new era of extra-galactic archaeology.

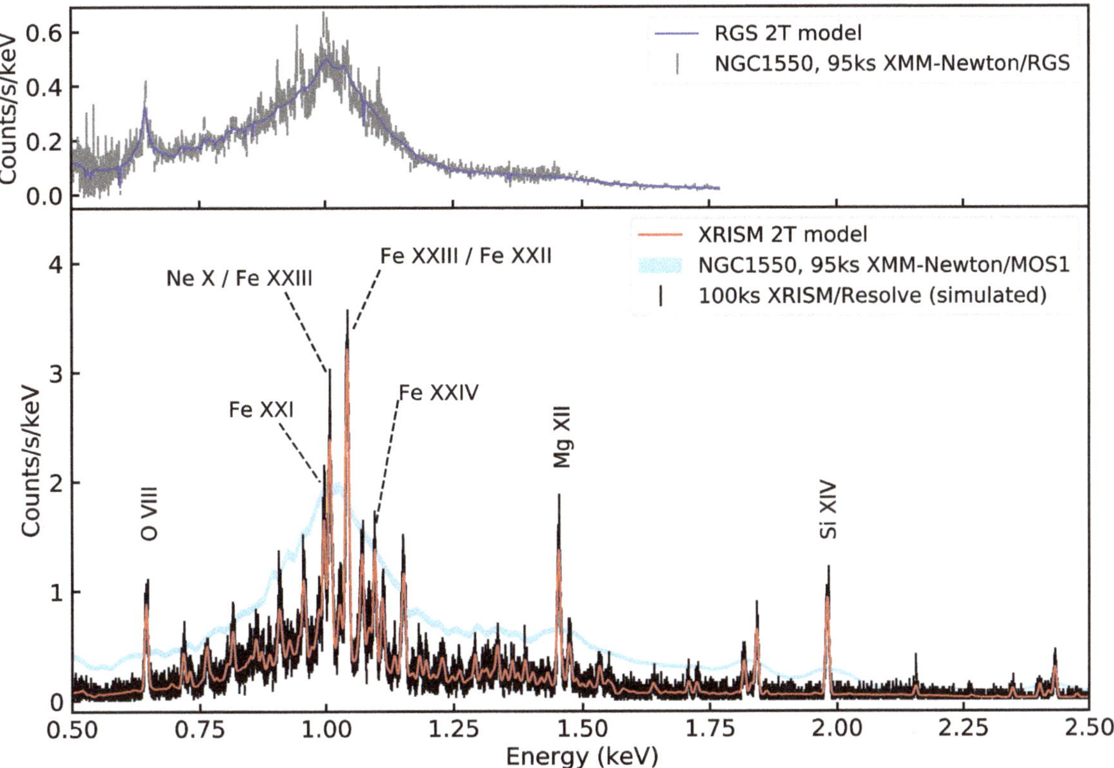

Figure 13. A comparison between existing archival XMM-*Newton* data of the central regions of the galaxy group NGC1550, and predictions for *XRISM/Resolve* using a similar exposure time and spectral model. (**Top panel**) The RGS spectrum extracted from a 3.4 arcmin-wide stripe in the cross-dispersion direction; spectra from RGS1 and RGS2 and from the three different existing observations have been stacked for display purposes (for details of the data reduction, see Reference [322]). The blue curve shows the best-fit 2T model using SPEXACT v3.0.6, including the effect of line broadening due to the extent of the source. (**Bottom panel**) The XMM-*Newton*-MOS1 spectrum from the central 0.05 r_{500} region of NGC1550 (adapted from Reference [128]) is shown in blue. This region roughly corresponds to the FoV of *XRISM/Resolve*. In red, we show the predicted model obtained by re-scaling the 2T RGS model from the top panel to match the XMM-*Newton*-MOS1 flux, and folding this through the *XRISM/Resolve* response. This re-scaling is necessary because the absolute flux determined by XMM-*Newton*-RGS for an extended source is uncertain, since technically a region up to 10 arcmin along the dispersion direction can contribute to the observed count rate. A simulated 100 ks *XRISM/Resolve* spectrum using this model is shown in black. The 5 brightest Fe lines, and lines from all elements other than Fe with a line flux exceeding 5×10^{-17} photons/s/cm^3, are labeled.

4.3. Athena

In the continuation of the high resolution spectroscopy era to be settled by *XRISM*, the future European mission *Athena* will be a game changer for our understanding of the chemical enrichment of hot halos pervading large-scale structures, from individual galaxies to rich clusters. Currently planned to be launched around 2033, *Athena* will embark two revolutionary instruments: the Wide Field Imager (WFI [323,324]) and the X-ray Integral Field Unit (X-IFU [325,326]). The former consists of a DEPFET (depleted p-channel field-effect transistor) camera able to perform imaging and moderate-resolution spectroscopy over an impressive field of view of $40' \times 40'$, whereas the latter is a cryogenic spectrometer made of a $\sim$5$'$ diameter array of more than 3000 TES (transition edge sensors)—each of which offering an exquisite spectral resolution of 2.5 eV over a required spatial resolution of 5 arcsec half energy width (thus allowing to probe the spatial substructure if the IGrM at levels comparable to XMM-*Newton*). Concretely, both instruments will be nicely complementary. While the WFI is expected to discover a large number of high-redshift clusters and groups, the X-IFU will be able to investigate them spectroscopically with unprecedented resolving power. The promises of the latter have been thoroughly demonstrated in the case of clusters, in terms of spatial distribution of metals (Reference [257]; also see Section 3.2.1) but also of chemical composition and underlying stellar sources [259].

The outstanding scientific potential offered by *Athena* for exploring the metal content of the IGrM is illustrated in Figure 14, where we simulated 100 ks of WFI and X-IFU exposures (in comparison to that of XMM-*Newton*/pn) for the core (<0.05 r_{500}) of a typical NGC 1550-like group assumed at various redshifts. Whereas, for this specific case, pn cannot provide significant metal constraints at $z = 0.5$ and beyond, we calculate that the WFI and the X-IFU should be able to track the overall metallicity up to $z = 1$ within $\sim$40% or less. An interesting feature immediately visible from this figure is the possibility of performing spectroscopy on high-redshift groups with the WFI. As the effective spectral resolution of any instrument naturally tends to deteriorate with decreasing flux (in order to keep enough meaningful statistics per spectral bin), at $z = 1$ typical WFI and X-IFU spectra of groups are expected to deliver very similar spectral information. Combined with its ability to detect and image several groups simultaneously over a large region of the sky (more than 10,000 systems at $z > 0.5$ with $M \geq 10^{13} M_{\odot}$ over the nominal four years of the mission [327]), this will make the WFI a highly valuable instrument to trace the metal content of *many* (local and distant) groups at once. This actually provides a unique synergy with the X-IFU, as the latter will be invaluable to resolve a plethora of metal lines at low and moderate redshifts (in order to derive absolute and relative abundances with exquisite accuracy, but also to further refine atomic calculations and make the available spectral codes further converge).

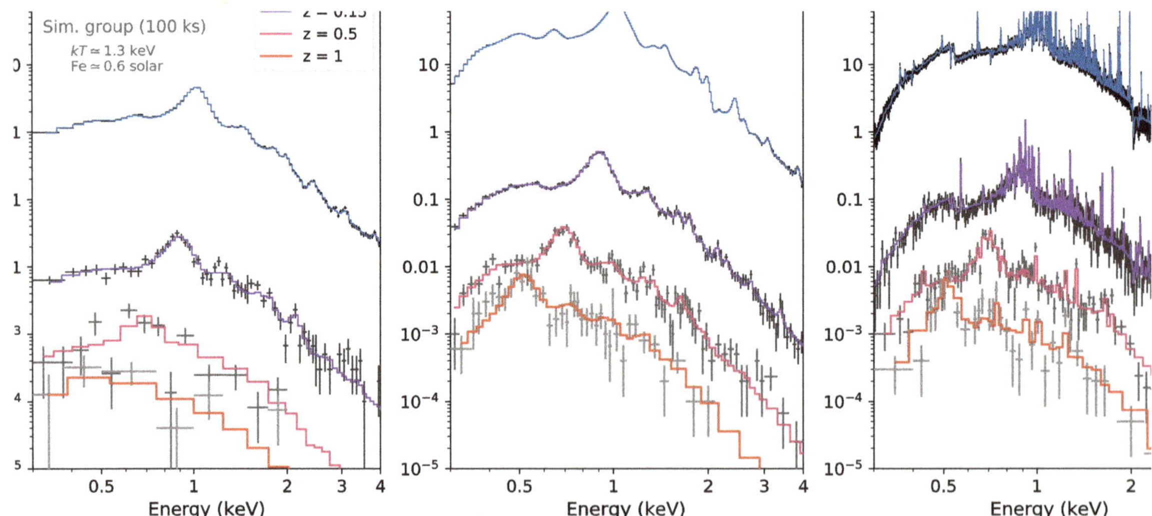

Figure 14. Simulated (100 ks) spectra of the core of a typical NGC 1550-like group ($L_{0.3-2\text{keV}} \simeq 2.1 \times 10^{42}$ erg/s, $kT \simeq 1.3$ keV, Fe $\simeq$ 0.6 Solar) set at various redshifts, as seen by the XMM-*Newton*/pn, *Athena*/WFI, and *Athena*/X-IFU instruments. Each spectrum has been appropriately re-binned for clarity.

Another particularly interesting possibility offered by the X-IFU instrument resides in the determination of the abundances ratios. By giving us access to even more metals with even fainter lines than *XRISM*, the X-IFU will offer the best diagnostics of the SN explosion mechanisms, initial metallicity of the progenitor stars contributing to the enrichment of the cosmos, and especially the slope of the IMF, with the aim to contribute in a significant way to the debate about its universality. Higher line emissivities of a few key elements (e.g., O, Mg) at the groups regime are critical for constraining the IMF and will nicely complement the case of clusters, which will be thoroughly studied, as well (see, e.g., Reference [259]).

Although more tailored WFI and X-IFU predictions (including, e.g., cosmological evolution of groups, the effects of the background and its reproducibility and/or other instrumental effects) are left for future dedicated work, it is clear that *Athena* will push our understanding of the chemistry of large-scale structures to the next level, even at and below groups scales.

4.4. HUBS and Super DIOS

While *XRISM* and *Athena* will lead to significant advancements in our understanding of the precise chemical make-up of the centers of galaxy groups and its redshift evolution, the high spectral resolution IFUs onboard both of these future missions have a limited field of view. This means that studies of the metal abundance ratios in the outskirts of groups and clusters would be very expensive in terms of observing time: nearby objects would be too large on the sky, and require mosaics composed of an unwieldy number of pointings, while high-redshift objects whose outskirts do fit within the FoV would be significantly dimmer. To give a concrete example, covering the entire area between 1–1.5 R_{500} of NGC 5846[4] would require no less than 423 observations with the *Athena*/X-IFU.

Two future missions currently under study promise to address this issue by offering capabilities that are complementary to those of *Athena*: the Hot Universe Baryon Surveyor (*HUBS*) [329], which is a project of the Chinese National Space Administration (CNSA), and the JAXA-led *Super DIOS* ("Diffuse Intergalactic Oxygen Surveyor", Sato et al. [330]) mission. Both have expected launch dates in the 2030s. By prioritizing a shorter mirror focal length (which implies a smaller on-axis effective area) but using larger pixels and covering a larger field of view (of order $\sim$1 deg^2), these future satellites will be able to survey a wider area of the sky more efficiently while maintaining a high spectral resolution ($\Delta E \lesssim 2$ eV). Due to the shorter focal length and their planned deployment in low-Earth orbit which both help to minimize the detector background, these missions are optimized to enable detailed high-resolution spectroscopy of the faint outskirts of nearby groups and clusters of galaxies, among several other science topics. In Figure 15, we summarize the capabilities and advantages of planned X-ray IFUs over the next $\sim$decade. Remarkable to note is that the *Super DIOS* concept plans to employ nearly an order of magnitude more TES pixels than *Athena* with developments of a new TES readout system, enabling a 10–15 arcsec resolution over 0.5–1 deg^2; while HUBS has a similar number of pixels as *Athena* (and, therefore, a poorer spatial resolution of $\sim$1 arcmin over a 1 deg^2 FoV), it carries a central 12 by 12 arcmin sub-array with an energy resolution of 0.6 eV, that will be the first detector to reach a resolving power of $E/\Delta E > 1000$ around the Fe-L complex.

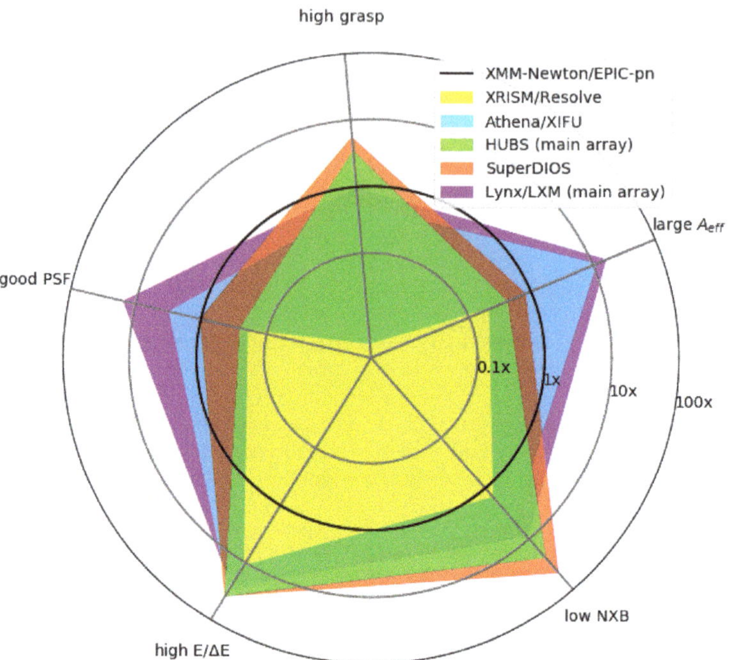

Figure 15. Summary of the capabilities of future missions and mission concepts carrying high-spectral resolution X-ray integral field unit detectors, using the *XMM-Newton* EPIC/pn as a reference value of 1 along each axis. The grasp is defined as the effective area A_{eff} integrated over the field of view; plotted along the 'low NXB' axis are the non-dimensional figures of merit F_{NXB} for detector-background-limited observations (also known as non-X-ray background), expressed as $F_{NXB} = A_{eff}/(F^2B)$ with F the mirror focal length and B=1 in low-Earth orbit and 4 for high orbits, as defined by, e.g., Reference [331].

4.5. *Arcus*

Arcus is a proposed NASA Medium Explorer mission that aims to significantly improve our spectroscopic capabilities using soft X-ray gratings. With spectral resolution $R > 2500$ between 12–50 Å, the Arcus proposal possesses the ability to resolve the absorption of a diverse range of metal species in the extended halos of galaxies, groups, and clusters along sight lines toward distant quasars [332]. X-ray grating spectroscopy provides a critical complementary probe to planned micro-calorimeters by having $\sim 10\times$ greater resolution and probing the diffuse hot gas comprising the majority of missing metals and baryons [333,334]. Arcus is designed to achieve a 10-fold increase in sensitivity (i.e., square root of effective area times resolution) over existing grating spectrometers on *Chandra* and XMM-*Newton*.

According to cosmological simulations [335], the IGrM has the greatest potential to show a rich variety of ion species. O VIII should be the most detected species within group virial radii, while O VII should be detected in lower mass groups at $M_{500} \lesssim 10^{13.5} M_{\odot}$. In its main mission, Arcus should detect the $T = 10^{5.4-6.8}$ K IGrM in at least 10 group halos at a 3 mÅ equivalent width threshold (at 5σ) for O VII and O VIII. Interestingly, the Fe XVII should be detectable within R_{500} for groups, probing gas up to $\approx 10^7$ K for at least 5 group halos. Integrating longer on the brightest X-ray quasars opens up the possibility to detect more species using a 1 or 2 mÅ detection threshold, including Ne IX and Ne X. C V and C VI (and maybe N VII) may also be detectable in the group outskirts for deeper observations probing gas down to 10^5 K. Sensitive UV absorption line spectroscopy brought about by the Cosmic Origins Spectrograph on *Hubble* discovered that the objects with the richest and most diverse set of species probing gaseous halos are star-forming, L^* galaxies [336]. It is expected that groups will be the analogous richly detected, multi-species objects to be probed via X-ray absorption line spectroscopy.

4.6. *Lynx*

The *Lynx* mission concept [337] proposed to the US 2020 Decadal Survey is designed to carry a new generation of X-ray telescopes enabling a sub-arcsecond resolution over a $22' \times 22'$ field of view and an effective area of 2 m^2 at 1 keV. It is designed to be equipped with three complementary instruments. An active pixel array (the high-definition X-ray imager, HDXI) that would provide wide-field CCD-like spectral imaging, but with $0.3''$ pixels to take advantage of the exquisite spatial resolution. The *Lynx* X-ray Microcalorimeter (LXM) which would bring 3 eV spectral resolution on $1''$ spatial scales over a $5'$ field of view (0.3 eV in a ultra-high resolution sub-array). Finally, an X-ray grating spectrometer (XRG) with an effective area of 4000 cm^2 and resolving power greater than 5000 to exploit the better spectral resolution of gratings for point sources in the soft energy band. The key improvements *Lynx* will bring if approved concern the ability to resolve the metal abundance structure of the gas near and outside the virial radius of groups at low redshift by combining a study of emission and absorption against background AGNs. *Lynx* will also push the realm of enrichment studies in the IGrM to the z = 2–3 range where strong trends are expected. *Lynx* is specifically targeting observations of high-redshift groups down to a mass scale of $M_{500} = 2 \times 10^{13}$ $M_{\odot}$ at z > 3 (https://www.lynxobservatory.com/report, accessed on 21 June 2021. [338]).

5. Concluding Remarks

Throughout this review, we have seen how crucial the case of galaxy groups is in order to complete our understanding of the journey of metals—from stars and supernovae to the largest scales of our Universe. Groups are in fact a unique piece of the puzzle to relate chemical enrichment at (sub-) galactic scales and at cluster scales (Section 2.4), making IGrM abundance studies particularly valuable in this respect. As we have also discussed, however, such studies are still in their pathfinder steps, and enrichment in groups remains much less explored (hence, understood) than in clusters. The reasons are diverse, and include notably on the observational side: (i) the intrinsic faintness of the IGrM with respect to the ICM, making observational studies more demanding in terms of exposure time; (ii) the lack of well defined samples of groups in the X-ray band, marking again a stark contrast with respect to clusters; (iii) our lack of spectral knowledge of the Fe-L complex (ruling almost all the X-ray emission of the IGrM) and of the likely multi-temperature structure of the gas (Section 2.1). As we have seen in Section 4, in the next few years, each of the above limitations will be tackled by higher quality data in terms of sizes of the sample of groups to be targeted and available high-throughput and high-resolution spectra. Those data will be coupled with an improved theoretical understanding of the spectral modeling. It is also essential to continue the efforts to improve the accuracy of the theoretical SN yields and the measurements of SN rates as a function of the cosmic time. On the simulation side, the main challenge will be to reproduce key stellar and gaseous properties on both galactic and Mpc scales, simultaneously including micro-scale physics and the large-scale cosmological context (Sections 3.1 and 3.2). With the ongoing exponential high-performance computing advancements, we are getting closer to a quantum leap in terms of a single high-resolution hydrodynamical cosmological simulation reaching this goal, thus providing more detailed predictions for the regime of galaxy groups. It is a reasonable bet to forecast the coming era as a golden age of maturity for the study of metal abundances in galaxy groups.

Author Contributions: F.G.: lead author, Sect. 1, 2.2; A.S.: Sect. 2.1, 2.3.2, 2.3.3; F.M.: Sect. 2.3.1, 2.5; V.B.: Sect. 3.2; M.G.: Sect. 3.1; K.S. and K.M.: Sect. 2.4 with contribution from F.G. F.G., A.S., F.M. and K.S contributed to Sect. 4. All authors contributed to Sect.5. All authors have read and agreed to the published version of the manuscript.

Funding: This research received no external funding.

Institutional Review Board Statement: Not applicable.

Informed Consent Statement: Not applicable.

Data Availability Statement: The review is based on public data and/or published papers. The results collected from literature and used to generate some of the figures in this review can be found at this link 10.5281/zenodo.5011831.

Acknowledgments: We thank the two anonymous referees for helpful reports which improved the quality of this review. We would like to thank Ben Oppenheimer for providing the section of the future mission Arcus and for useful discussions. We would like to thank David Buote, Silvano Molendi, Simona Ghizzardi, Fabrizio Brighenti, William Mathews for comments and suggestions. FG acknowledges financial contribution from INAF "Call per interventi aggiuntivi a sostegno della ricerca di main stream di INAF". AS is supported by the Women In Science Excel (WISE) programme of the NWO, and acknowledges the World Premier Research Center Initiative (WPI) and the Kavli IPMU for the continued hospitality. SRON Netherlands Institute for Space Research is supported financially by NWO. MG acknowledges partial support by NASA Chandra GO8-19104X/GO9-20114X and HST GO-15890.020-A grants.

Conflicts of Interest: The authors declare no conflict of interest.

Notes

1 $EW = \int {}^{(I_\nu - I_\nu^0)}/I_\nu^0 \, d(h\nu)$, where I_ν^0 is the continuum intensity without the line
2 In Johnson et al. [90], a "cool-core" group is defined as such when its temperature profile shows a clear central decrease out to at least $\sim 0.1 R_{500}$. This definition is of course arbitrary and may differ from other proposed ones.
3 In Mernier et al. [89], a "group/elliptical" is defined as such when the mean temperature of the system within $0.05 R_{500}$ does not exceed 1.7 keV. Here, again, this definition is arbitrary and a given rich group could be defined as a poor cluster in other studies.
4 For the R_{500} value in Reference [328].

References

1. Eddington, A.S. The internal constitution of the stars. *Observatory* **1920**, *43*, 341–358.
2. Hoyle, F. The synthesis of the elements from hydrogen. *Mon. Not. R. Astron. Soc.* **1946**, *106*, 343. [CrossRef]
3. Burbidge, E.M.; Burbidge, G.R.; Fowler, W.A.; Hoyle, F. Synthesis of the Elements in Stars. *Rev. Mod. Phys.* **1957**, *29*, 547–650. [CrossRef]
4. Mitchell, R.J.; Culhane, J.L.; Davison, P.J.N.; Ives, J.C. Ariel 5 observations of the X-ray spectrum of the Perseus cluster. *Mon. Not. R. Astron. Soc.* **1976**, *175*, 29P–34P. [CrossRef]
5. Serlemitsos, P.J.; Smith, B.W.; Boldt, E.A.; Holt, S.S.; Swank, J.H. X-radiation from clusters of galaxies: Spectral evidence for a hot evolved gas. *Astrophys. J.* **1977**, *211*, L63–L66. [CrossRef]
6. Nomoto, K.; Kobayashi, C.; Tominaga, N. Nucleosynthesis in Stars and the Chemical Enrichment of Galaxies. *Annu. Rev. Astron. Astrophys.* **2013**, *51*, 457–509. [CrossRef]
7. Kobayashi, C.; Karakas, A.I.; Lugaro, M. The Origin of Elements from Carbon to Uranium. *Astrophys. J.* **2020**, *900*, 179. [CrossRef]
8. Renzini, A.; Andreon, S. Chemical evolution on the scale of clusters of galaxies: A conundrum? *Mon. Not. R. Astron. Soc.* **2014**, *444*, 3581–3591. [CrossRef]
9. Ghizzardi, S.; Molendi, S.; van der Burg, R.; De Grandi, S.; Bartalucci, I.; Gastaldello, F.; Rossetti, M.; Biffi, V.; Borgani, S.; Eckert, D.; et al. Iron in X-COP: Tracing enrichment in cluster outskirts with high accuracy abundance profiles. *Astron. Astrophys.* **2021**, *646*, A92. [CrossRef]
10. Renzini, A., Metal Content and Production in Clusters of Galaxies. In *A Pan-Chromatic View of Clusters of Galaxies and the Large-Scale Structure*; Plionis, M., López-Cruz, O., Hughes, D., Eds.; Springer: Berlin, Germany, 2008; Volume 740, p. 24._6. [CrossRef]
11. Schindler, S.; Diaferio, A. Metal Enrichment Processes. *Space Sci. Rev.* **2008**, *134*, 363–377. [CrossRef]
12. McNamara, B.R.; Nulsen, P.E.J. Heating Hot Atmospheres with Active Galactic Nuclei. *Annu. Rev. Astron. Astrophys.* **2007**, *45*, 117–175. [CrossRef]
13. McNamara, B.R.; Nulsen, P.E.J. Mechanical feedback from active galactic nuclei in galaxies, groups and clusters. *New J. Phys.* **2012**, *14*, 055023. [CrossRef]

14. Fabian, A.C. Observational Evidence of Active Galactic Nuclei Feedback. *Annu. Rev. Astron. Astrophys.* **2012**, *50*, 455–489. [CrossRef]
15. Gitti, M.; Brighenti, F.; McNamara, B.R. Evidence for AGN Feedback in Galaxy Clusters and Groups. *Adv. Astron.* **2012**, *2012*, 950641. [CrossRef]
16. Markevitch, M.; Vikhlinin, A. Shocks and cold fronts in galaxy clusters. *Phys. Rep.* **2007**, *443*, 1–53. doi:10.1016/j.physrep.2007.01.001. [CrossRef]
17. Zuhone, J.A.; Roediger, E. Cold fronts: Probes of plasma astrophysics in galaxy clusters. *J. Plasma Phys.* **2016**, *82*, 535820301. [CrossRef]
18. Sivanandam, S.; Zabludoff, A.I.; Zaritsky, D.; Gonzalez, A.H.; Kelson, D.D. The Enrichment of the Intracluster Medium. *Astrophys. J.* **2009**, *691*, 1787–1806. [CrossRef]
19. Mulchaey, J.S. X-ray Properties of Groups of Galaxies. *Annu. Rev. Astron. Astrophys.* **2000**, *38*, 289–335. [CrossRef]
20. Eke, V.R.; Baugh, C.M.; Cole, S.; Frenk, C.S.; Navarro, J.F. Galaxy groups in the 2dF Galaxy Redshift Survey: The number density of groups. *Mon. Not. R. Astron. Soc.* **2006**, *370*, 1147–1158. [CrossRef]
21. Tempel, E.; Kipper, R.; Saar, E.; Bussov, M.; Hektor, A.; Pelt, J. Galaxy filaments as pearl necklaces. *Astron. Astrophys.* **2014**, *572*, A8. [CrossRef]
22. Ponman, T.J.; Sanderson, A.J.R.; Finoguenov, A. The Birmingham-CfA cluster scaling project—III. Entropy and similarity in galaxy systems. *Mon. Not. R. Astron. Soc.* **2003**, *343*, 331–342. [CrossRef]
23. McCarthy, I.G.; Schaye, J.; Ponman, T.J.; Bower, R.G.; Booth, C.M.; Dalla Vecchia, C.; Crain, R.A.; Springel, V.; Theuns, T.; Wiersma, R.P.C. The case for AGN feedback in galaxy groups. *Mon. Not. R. Astron. Soc.* **2010**, *406*, 822–839. [CrossRef]
24. Maiolino, R.; Mannucci, F. De re metallica: The cosmic chemical evolution of galaxies. *Astron. Astrophys. Rev.* **2019**, *27*, 3. [CrossRef]
25. Renzini, A. *The Chemistry of Galaxy Clusters. Clusters of Galaxies: Probes of Cosmological Structure and Galaxy Evolution*; Mulchaey, J.S., Dressler, A., Oemler, A., Eds.; Cambridge University Press: Cambridge, UK, 2004; p. 260.
26. Werner, N.; Durret, F.; Ohashi, T.; Schindler, S.; Wiersma, R.P.C. Observations of Metals in the Intra-Cluster Medium. *Space Sci. Rev.* **2008**, *134*, 337–362. [CrossRef]
27. Borgani, S.; Diaferio, A.; Dolag, K.; Schindler, S. Thermodynamical Properties of the ICM from Hydrodynamical Simulations. *Clust. Galaxies* **2008**, *134*, 269–293. [CrossRef]
28. Böhringer, H.; Werner, N. X-ray spectroscopy of galaxy clusters: Studying astrophysical processes in the largest celestial laboratories. *Astron. Astrophys. Rev.* **2010**, *18*, 127–196. [CrossRef]
29. Kim, D.W. Metal Abundances in the Hot ISM of Elliptical Galaxies. *Astrophys. Space Sci. Libr.* **2012**, *378*, 121._5. [CrossRef]
30. de Plaa, J. The origin of the chemical elements in cluster cores. *Astron. Nachrichten* **2013**, *334*, 416. [CrossRef]
31. Mernier, F.; Biffi, V.; Yamaguchi, H.; Medvedev, P.; Simionescu, A.; Ettori, S.; Werner, N.; Kaastra, J.S.; de Plaa, J.; Gu, L. Enrichment of the Hot Intracluster Medium: Observations. *Space Sci. Rev.* **2018**, *214*, 129. [CrossRef]
32. Biffi, V.; Mernier, F.; Medvedev, P. Enrichment of the Hot Intracluster Medium: Numerical Simulations. *Space Sci. Rev.* **2018**, *214*, 123. [CrossRef]
33. Sun, M. Hot gas in galaxy groups: Recent observations. *New J. Phys.* **2012**, *14*, 045004. [CrossRef]
34. Gu, L.; Zhuravleva, I.; Churazov, E.; Paerels, F.; Kaastra, J.; Yamaguchi, H. X-Ray Spectroscopy of Galaxy Clusters: Beyond the CIE Modeling. *Space Sci. Rev.* **2018**, *214*, 108. [CrossRef]
35. Mewe, R.; Lemen, J.R.; van den Oord, G.H.J. Calculated X-radiation from optically thin plasmas. VI—Improved calculations for continuum emission and approximation formulae for nonrelativistic average Gaunt actors. *Astron. Astrophys. Suppl. Ser.* **1986**, *65*, 511–536.
36. Smith, R.K.; Brickhouse, N.S.; Liedahl, D.A.; Raymond, J.C. Collisional Plasma Models with APEC/APED: Emission-Line Diagnostics of Hydrogen-like and Helium-like Ions. *Astrophys. J.* **2001**, *556*, L91–L95. [CrossRef]
37. Foster, A.R.; Ji, L.; Smith, R.K.; Brickhouse, N.S. Updated Atomic Data and Calculations for X-Ray Spectroscopy. *Astrophys. J.* **2012**, *756*, 128. [CrossRef]
38. Kaastra, J.S.; Mewe, R.; Nieuwenhuijzen, H. Spex: A New Code for Spectral Analysis of X and UV Spectra. In *UV and X-ray Spectroscopy of Astrophysical and Laboratory Plasmas*; Yamashita, K., Watanabe, T., Eds.; Universal Academy Press: Tokyo, Japan, 1996; p. 411.
39. Kaastra, J.S.; Raassen, A.J.J.; de Plaa, J.; Gu, L. SPEX X-ray Spectral Fitting Package. *Zenodo* **2020**. [CrossRef]
40. Mushotzky, R.F.; Holt, S.S.; Boldt, E.A.; Serlemitsos, P.J.; Smith, B.W. Observation of the core of the Perseus cluster with the Einstein solid state spectrometer—Cooling gas and elemental abundances. *Astrophys. J.* **1981**, *244*, L47–L51. [CrossRef]
41. Lea, S.M.; Mushotzky, R.; Holt, S.S. Einstein Observatory solid state spectrometer observations of M87 and the Virgo cluster. *Astrophys. J.* **1982**, *262*, 24–32. [CrossRef]
42. Mernier, F.; Werner, N.; Lakhchaura, K.; de Plaa, J.; Gu, L.; Kaastra, J.S.; Mao, J.; Simionescu, A.; Urdampilleta, I. How do atomic code uncertainties affect abundance measurements in the intracluster medium? *Astron. Nachr.* **2020**, *341*, 203–209. [CrossRef]
43. Asplund, M.; Grevesse, N.; Sauval, A.J.; Scott, P. The Chemical Composition of the Sun. *Annu. Rev. Astron. Astrophys.* **2009**, *47*, 481–522. [CrossRef]
44. Raymond, J.C.; Smith, B.W. Soft X-ray spectrum of a hot plasma. *Astrophys. J. Suppl. Ser.* **1977**, *35*, 419–439. [CrossRef]

45. Mewe, R.; Gronenschild, E.H.B.M.; van den Oord, G.H.J. Calculated X-radiation from optically thin plasmas. V. *Astron. Astrophys. Suppl. Ser.* **1985**, *62*, 197–254.
46. Liedahl, D.A.; Osterheld, A.L.; Goldstein, W.H. New Calculations of Fe L-Shell X-Ray Spectra in High-Temperature Plasmas. *Astrophys. J.* **1995**, *438*, L115. [CrossRef]
47. Buote, D.A.; Fabian, A.C. X-ray spectral analysis of elliptical galaxies from ASCA: The Fe abundance in a multiphase medium. *Mon. Not. R. Astron. Soc.* **1998**, *296*, 977–994. [CrossRef]
48. Buote, D.A. Iron Gradients in Cooling Flow Galaxies and Groups. *Astrophys. J.* **2000**, *539*, 172–186. [CrossRef]
49. Irwin, J.A.; Athey, A.E.; Bregman, J.N. X-Ray Spectral Properties of Low-Mass X-Ray Binaries in Nearby Galaxies. *Astrophys. J.* **2003**, *587*, 356–366. [CrossRef]
50. Anders, E.; Grevesse, N. Abundances of the elements: Meteoritic and solar. *Geochim. Cosmochim. Acta* **1989**, *53*, 197–214. [CrossRef]
51. Lodders, K.; Palme, H.; Gail, H.P. Abundances of the Elements in the Solar System. *Landolt Börnstein* **2009**. [CrossRef]
52. Lovisari, L.; Reiprich, T.H. The non-uniformity of galaxy cluster metallicity profiles. *Mon. Not. R. Astron. Soc.* **2019**, *483*, 540–557. [CrossRef]
53. Mulchaey, J.S.; Davis, D.S.; Mushotzky, R.F.; Burstein, D. Diffuse X-ray emission from the NGC 2300 group of galaxies - Implications for dark matter and galaxy evolution in small groups. *Astrophys. J.* **1993**, *404*, L9–L12. [CrossRef]
54. Ponman, T.J.; Bertram, D. Hot gas and dark matter in a compact galaxy group. *Nature* **1993**, *363*, 51–54. [CrossRef]
55. David, L.P.; Jones, C.; Forman, W.; Daines, S. Mapping the dark matter in the NGC 5044 group with ROSAT: Evidence for a nearly homogeneous cooling flow with a cooling wake. *Astrophys. J.* **1994**, *428*, 544–554. [CrossRef]
56. Davis, D.S.; Mushotzky, R.F.; Mulchaey, J.S.; Worrall, D.M.; Birkinshaw, M.; Burstein, D. Diffuse Hot Gas in the NGC 4261 Group of Galaxies. *Astrophys. J.* **1995**, *444*, 582. [CrossRef]
57. Saracco, P.; Ciliegi, P. ROSAT observations of compact groups of galaxies. *Astron. Astrophys.* **1995**, *301*, 348.
58. Davis, D.S.; Mulchaey, J.S.; Mushotzky, R.F.; Burstein, D. Spectral and Spatial Analysis of the Intragroup Medium in the NGC 2300 Group. *Astrophys. J.* **1996**, *460*, 601. [CrossRef]
59. Ponman, T.J.; Bourner, P.D.J.; Ebeling, H.; Böhringer, H. A ROSAT survey of Hickson's compact galaxy groups. *Mon. Not. R. Astron. Soc.* **1996**, *283*, 690–708. [CrossRef]
60. Mulchaey, J.S.; Davis, D.S.; Mushotzky, R.F.; Burstein, D. The Intragroup Medium in Poor Groups of Galaxies. *Astrophys. J.* **1996**, *456*, 80. [CrossRef]
61. Fabian, A.C. Cooling Flows in Clusters of Galaxies. *Annu. Rev. Astron. Astrophys.* **1994**, *32*, 277–318. 194.001425. [CrossRef]
62. Renzini, A.; Ciotti, L. Transverse Dissections of the Fundamental Planes of Elliptical Galaxies and Clusters of Galaxies. *Astrophys. J.* **1993**, *416*, L49. [CrossRef]
63. Arimoto, N.; Matsushita, K.; Ishimaru, Y.; Ohashi, T.; Renzini, A. The Iron Discrepancy in Elliptical Galaxies after ASCA. *Astrophys. J.* **1997**, *477*, 128–143. [CrossRef]
64. Fukazawa, Y.; Makishima, K.; Matsushita, K.; Yamasaki, N.; Ohashi, T.; Mushotzky, R.F.; Sakima, Y.; Tsusaka, Y.; Yamashita, K. Metal Abundance of an X-Ray Emitting Gas in Two Groups of Galaxies: The NGC 5044 Group and HCG 51. *Publ. Astron. Soc. Jpn.* **1996**, *48*, 395–407. [CrossRef]
65. Davis, D.S.; Mulchaey, J.S.; Mushotzky, R.F. The Enrichment History of Hot Gas in Poor Galaxy Groups. *Astrophys. J.* **1999**, *511*, 34–40. [CrossRef]
66. Hwang, U.; Mushotzky, R.F.; Burns, J.O.; Fukazawa, Y.; White, R.A. Mass and Metallicity of Five X-Ray-bright Galaxy Groups. *Astrophys. J.* **1999**, *516*, 604–618. [CrossRef]
67. Buote, D.A. X-ray evidence for multiphase hot gas with solar abundances in the brightest elliptical galaxies. *Mon. Not. R. Astron. Soc.* **1999**, *309*, 685–714. [CrossRef]
68. Buote, D.A. X-ray evidence for multiphase hot gas with nearly solar Fe abundances in the brightest groups of galaxies. *Mon. Not. R. Astron. Soc.* **2000**, *311*, 176–200. [CrossRef]
69. Molendi, S.; Gastaldello, F. On the metal abundance in the core of M 87. *Astron. Astrophys.* **2001**, *375*, L14–L17. [CrossRef]
70. Baumgartner, W.H.; Loewenstein, M.; Horner, D.J.; Mushotzky, R.F. Intermediate-Element Abundances in Galaxy Clusters. *Astrophys. J.* **2005**, *620*, 680–696. [CrossRef]
71. Rasia, E.; Mazzotta, P.; Bourdin, H.; Borgani, S.; Tornatore, L.; Ettori, S.; Dolag, K.; Moscardini, L. X-MAS2: Study Systematics on the ICM Metallicity Measurements. *Astrophys. J.* **2008**, *674*, 728–741. [CrossRef]
72. Simionescu, A.; Werner, N.; Böhringer, H.; Kaastra, J.S.; Finoguenov, A.; Brüggen, M.; Nulsen, P.E.J. Chemical enrichment in the cluster of galaxies Hydra A. *Astron. Astrophys.* **2009**, *493*, 409–424. [CrossRef]
73. Gastaldello, F.; Ettori, S.; Balestra, I.; Brighenti, F.; Buote, D.A.; de Grandi, S.; Ghizzardi, S.; Gitti, M.; Tozzi, P. Apparent high metallicity in 3-4 keV galaxy clusters: The inverse iron-bias in action in the case of the merging cluster Abell 2028. *Astron. Astrophys.* **2010**, *522*, A34. [CrossRef]
74. Buote, D.A.; Lewis, A.D.; Brighenti, F.; Mathews, W.G. XMM-Newton and Chandra Observations of the Galaxy Group NGC 5044. II. Metal Abundances and Supernova Fraction. *Astrophys. J.* **2003**, *595*, 151–166. [CrossRef]
75. Sasaki, T.; Matsushita, K.; Sato, K. Metal Distributions out to 0.5 r_{180} in the Intracluster Medium of Four Galaxy Groups Observed with Suzaku. *Astrophys. J.* **2014**, *781*, 36. [CrossRef]

76. Humphrey, P.J.; Buote, D.A.; Brighenti, F.; Flohic, H.M.L.G.; Gastaldello, F.; Mathews, W.G. Tracing the Gas to the Virial Radius (R_{100}) in a Fossil Group. *Astrophys. J.* **2012**, *748*, 11. [CrossRef]
77. Su, Y.; Buote, D.; Gastaldello, F.; Brighenti, F. The Entire Virial Radius of the Fossil Cluster RX J1159+5531: I. Gas Properties. *Astrophys. J.* **2015**, *805*, 104. [CrossRef]
78. O'Sullivan, E.; Vrtilek, J.M.; Kempner, J.C.; David, L.P.; Houck, J.C. AWM 4—An isothermal cluster observed with XMM-Newton. *Mon. Not. R. Astron. Soc.* **2005**, *357*, 1134–1150. [CrossRef]
79. Rafferty, D.A.; Bîrzan, L.; Nulsen, P.E.J.; McNamara, B.R.; Brandt, W.N.; Wise, M.W.; Röttgering, H.J.A. A deep Chandra observation of the active galactic nucleus outburst and merger in Hickson compact group 62. *Mon. Not. R. Astron. Soc.* **2013**, *428*, 58–70. [CrossRef]
80. Panagoulia, E.K.; Sanders, J.S.; Fabian, A.C. A volume-limited sample of X-ray galaxy groups and clusters—III. Central abundance drops. *Mon. Not. R. Astron. Soc.* **2015**, *447*, 417–436. [CrossRef]
81. Hu, D.; Xu, H.; Kang, X.; Li, W.; Zhu, Z.; Ma, Z.; Shan, C.; Zhang, Z.; Gu, L.; Liu, C.; et al. A Study of the Merger History of the Galaxy Group HCG 62 Based on X-Ray Observations and Smoothed Particle Hydrodynamic Simulations. *Astrophys. J.* **2019**, *870*, 61. [CrossRef]
82. O'Sullivan, E.; Vrtilek, J.M.; Read, A.M.; David, L.P.; Ponman, T.J. An XMM-Newton observation of the galaxy group MKW 4. *Mon. Not. R. Astron. Soc.* **2003**, *346*, 525–539. [CrossRef]
83. Buote, D.A. An XMM-Newton Observation of NGC 1399 Reveals Two Phases of Hot Gas and Supersolar Abundances in the Central Regions. *Astrophys. J.* **2002**, *574*, L135–L138. [CrossRef]
84. O'Sullivan, E.; Ponman, T.J.; Collins, R.S. X-ray scaling properties of early-type galaxies. *Mon. Not. R. Astron. Soc.* **2003**, *340*, 1375–1399. [CrossRef]
85. Su, Y.; Nulsen, P.E.J.; Kraft, R.P.; Forman, W.R.; Jones, C.; Irwin, J.A.; Randall, S.W.; Churazov, E. Buoyant AGN Bubbles in the Quasi-isothermal Potential of NGC 1399. *Astrophys. J.* **2017**, *847*, 94. [CrossRef]
86. Thölken, S.; Lovisari, L.; Reiprich, T.H.; Hasenbusch, J. X-ray analysis of the galaxy group UGC 03957 beyond R_{200} with Suzaku. *Astron. Astrophys.* **2016**, *592*, A37. [CrossRef]
87. De Grandi, S.; Molendi, S. Metallicity Gradients in X-Ray Clusters of Galaxies. *Astrophys. J.* **2001**, *551*, 153–159. [CrossRef]
88. Leccardi, A.; Molendi, S. Radial metallicity profiles for a large sample of galaxy clusters observed with XMM-Newton. *Astron. Astrophys.* **2008**, *487*, 461–466. [CrossRef]
89. Mernier, F.; de Plaa, J.; Kaastra, J.S.; Zhang, Y.Y.; Akamatsu, H.; Gu, L.; Kosec, P.; Mao, J.; Pinto, C.; Reiprich, T.H.; et al. Radial metal abundance profiles in the intra-cluster medium of cool-core galaxy clusters, groups, and ellipticals. *Astron. Astrophys.* **2017**, *603*, A80. [CrossRef]
90. Johnson, R.; Ponman, T.J.; Finoguenov, A. A statistical analysis of the Two-Dimensional XMM-Newton Group Survey: The impact of feedback on group properties. *Mon. Not. R. Astron. Soc.* **2009**, *395*, 1287–1308. [CrossRef]
91. Johnson, R.; Finoguenov, A.; Ponman, T.J.; Rasmussen, J.; Sanderson, A.J.R. Abundance profiles and cool cores in galaxy groups. *Mon. Not. R. Astron. Soc.* **2011**, *413*, 2467–2480. [CrossRef]
92. Mernier, F.; de Plaa, J.; Werner, N.; Kaastra, J.S.; Raassen, A.J.J.; Gu, L.; Mao, J.; Urdampilleta, I.; Truong, N.; Simionescu, A. Mass-invariance of the iron enrichment in the hot haloes of massive ellipticals, groups, and clusters of galaxies. *Mon. Not. R. Astron. Soc.* **2018**, *478*, L116–L121. [CrossRef]
93. Rasmussen, J.; Ponman, T.J. Temperature and abundance profiles of hot gas in galaxy groups—I. Results and statistical analysis. *Mon. Not. R. Astron. Soc.* **2007**, *380*, 1554–1572. [CrossRef]
94. Rasmussen, J.; Ponman, T.J. Temperature and abundance profiles of hot gas in galaxy groups—II. Implications for feedback and ICM enrichment. *Mon. Not. R. Astron. Soc.* **2009**, *399*, 239–263. [CrossRef]
95. Sanders, J.S.; Fabian, A.C. Spatially resolved X-ray spectroscopy of the core of the Centaurus cluster. *Mon. Not. R. Astron. Soc.* **2002**, *331*, 273–283. [CrossRef]
96. Gendron-Marsolais, M.; Kraft, R.P.; Bogdan, A.; Hlavacek-Larrondo, J.; Forman, W.R.; Jones, C.; Su, Y.; Nulsen, P.; Randall, S.W.; Roediger, E. Uplift, Feedback, and Buoyancy: Radio Lobe Dynamics in NGC 4472. *Astrophys. J.* **2017**, *848*, 26. [CrossRef]
97. Werner, N.; de Plaa, J.; Kaastra, J.S.; Vink, J.; Bleeker, J.A.M.; Tamura, T.; Peterson, J.R.; Verbunt, F. XMM-Newton spectroscopy of the cluster of galaxies 2A 0335+096. *Astron. Astrophys.* **2006**, *449*, 475–491. [CrossRef]
98. Sanders, J.S.; Fabian, A.C. Resonance scattering, absorption and off-centre abundance peaks in clusters of galaxies. *Mon. Not. R. Astron. Soc.* **2006**, *370*, 63–73. [CrossRef]
99. Ettori, S.; Fabian, A.C. Effects of sedimented helium on the X-ray properties of galaxy clusters. *Mon. Not. R. Astron. Soc.* **2006**, *369*, L42–L46. [CrossRef]
100. Medvedev, P.; Gilfanov, M.; Sazonov, S.; Shtykovskiy, P. Impact of thermal diffusion and other abundance anomalies on cosmological uses of galaxy clusters. *Mon. Not. R. Astron. Soc.* **2014**, *440*, 2464–2473. [CrossRef]
101. Werner, N.; Urban, O.; Simionescu, A.; Allen, S.W. A uniform metal distribution in the intergalactic medium of the Perseus cluster of galaxies. *Nature* **2013**, *502*, 656–658. [CrossRef]
102. Urban, O.; Werner, N.; Allen, S.W.; Simionescu, A.; Mantz, A. A uniform metallicity in the outskirts of massive, nearby galaxy clusters. *Mon. Not. R. Astron. Soc.* **2017**, *470*, 4583–4599. [CrossRef]
103. Buote, D.A.; Brighenti, F.; Mathews, W.G. Ultralow Iron Abundances in the Distant Hot Gas in Galaxy Groups. *Astrophys. J.* **2004**, *607*, L91–L94. [CrossRef]

104. O'Sullivan, E.; Vrtilek, J.M.; Harris, D.E.; Ponman, T.J. On the Anomalous Temperature Distribution of the Intergalactic Medium in the NGC 3411 Group of Galaxies. *Astrophys. J.* **2007**, *658*, 299–313. [CrossRef]
105. Simionescu, A.; Werner, N.; Mantz, A.; Allen, S.W.; Urban, O. Witnessing the growth of the nearest galaxy cluster: Thermodynamics of the Virgo Cluster outskirts. *Mon. Not. R. Astron. Soc.* **2017**, *469*, 1476–1495. [CrossRef]
106. Reiprich, T.H.; Basu, K.; Ettori, S.; Israel, H.; Lovisari, L.; Molendi, S.; Pointecouteau, E.; Roncarelli, M. Outskirts of Galaxy Clusters. *Space Sci. Rev.* **2013**, *177*, 195–245. [CrossRef]
107. Biffi, V.; Planelles, S.; Borgani, S.; Rasia, E.; Murante, G.; Fabjan, D.; Gaspari, M. The origin of ICM enrichment in the outskirts of present-day galaxy clusters from cosmological hydrodynamical simulations. *Mon. Not. R. Astron. Soc.* **2018**, *476*, 2689–2703. [CrossRef]
108. Madau, P.; Dickinson, M. Cosmic Star-Formation History. *Annu. Rev. Astron. Astrophys.* **2014**, *52*, 415–486. [CrossRef]
109. Hickox, R.C.; Alexander, D.M. Obscured Active Galactic Nuclei. *Annu. Rev. Astron. Astrophys.* **2018**, *56*, 625–671. [CrossRef]
110. Martini, P.; Miller, E.D.; Brodwin, M.; Stanford, S.A.; Gonzalez, A.H.; Bautz, M.; Hickox, R.C.; Stern, D.; Eisenhardt, P.R.; Galametz, A.; et al. The Cluster and Field Galaxy Active Galactic Nucleus Fraction at z = 1-1.5: Evidence for a Reversal of the Local Anticorrelation between Environment and AGN Fraction. *Astrophys. J.* **2013**, *768*, 1. [CrossRef]
111. Bufanda, E.; Hollowood, D.; Jeltema, T.E.; Rykoff, E.S.; Rozo, E.; Martini, P.; Abbott, T.M.C.; Abdalla, F.B.; Allam, S.; Banerji, M.; et al. The evolution of active galactic nuclei in clusters of galaxies from the Dark Energy Survey. *Mon. Not. R. Astron. Soc.* **2017**, *465*, 2531–2539. [CrossRef]
112. Krishnan, C.; Hatch, N.A.; Almaini, O.; Kocevski, D.; Cooke, E.A.; Hartley, W.G.; Hasinger, G.; Maltby, D.T.; Muldrew, S.I.; Simpson, C. Enhancement of AGN in a protocluster at z = 1.6. *Mon. Not. R. Astron. Soc.* **2017**, *470*, 2170–2178. [CrossRef]
113. Finoguenov, A.; Ponman, T.J. Constraining the role of Type IA and Type II supernovae in galaxy groups by spatially resolved analysis of ROSAT and ASCA observations. *Mon. Not. R. Astron. Soc.* **1999**, *305*, 325–337. [CrossRef]
114. Kim, D.W.; Fabbiano, G. XMM-Newton Observations of NGC 507: Supersolar Metal Abundances in the Hot Interstellar Medium. *Astrophys. J.* **2004**, *613*, 933–947. [CrossRef]
115. Xue, Y.J.; Böhringer, H.; Matsushita, K. An XMM-Newton study of the RGH 80 galaxy group. *Astron. Astrophys.* **2004**, *420*, 833–845. [CrossRef]
116. Gastaldello, F.; Molendi, S. Abundance Gradients and the Role of Supernovae in M87. *Astrophys. J.* **2002**, *572*, 160–168. [CrossRef]
117. de Plaa, J.; Werner, N.; Bykov, A.M.; Kaastra, J.S.; Méndez, M.; Vink, J.; Bleeker, J.A.M.; Bonamente, M.; Peterson, J.R. Chemical evolution in Sérsic 159-03 observed with XMM-Newton. *Astron. Astrophys.* **2006**, *452*, 397–412. [CrossRef]
118. Tokoi, K.; Sato, K.; Ishisaki, Y.; Ohashi, T.; Yamasaki, N.Y.; Nakazawa, K.; Matsushita, K.; Fukazawa, Y.; Hoshino, A.; Tamura, T.; et al. Suzaku Observation of HCG 62: Temperature, Abundance, and Extended Hard X-Ray Emission Profiles. *Publ. Astron. Soc. Jpn.* **2008**, *60*, S317. [CrossRef]
119. Komiyama, M.; Sato, K.; Nagino, R.; Ohashi, T.; Matsushita, K. Suzaku Observations of Metallicity Distribution in the Intracluster Medium of the NGC 5044 Group. *Publ. Astron. Soc. Jpn.* **2009**, *61*, S337. [CrossRef]
120. Sato, K.; Matsushita, K.; Ishisaki, Y.; Yamasaki, N.Y.; Ishida, M.; Ohashi, T. Suzaku Observation of Group of Galaxies NGC 507: Temperature and Metal Distributions in the Intra-Cluster Medium. *Publ. Astron. Soc. Jpn.* **2009**, *61*, S353. [CrossRef]
121. Sato, K.; Kawaharada, M.; Nakazawa, K.; Matsushita, K.; Ishisaki, Y.; Yamasaki, N.Y.; Ohashi, T. Metallicity of the Fossil Group NGC 1550 Observed with Suzaku. *Publ. Astron. Soc. Jpn.* **2010**, *62*, 1445. [CrossRef]
122. Lodders, K. Solar System Abundances and Condensation Temperatures of the Elements. *Astrophys. J.* **2003**, *591*, 1220–1247. [CrossRef]
123. Sato, K.; Tokoi, K.; Matsushita, K.; Ishisaki, Y.; Yamasaki, N.Y.; Ishida, M.; Ohashi, T. Type Ia and II Supernovae Contributions to Metal Enrichment in the Intracluster Medium Observed with Suzaku. *Astrophys. J.* **2007**, *667*, L41–L44. [CrossRef]
124. de Grandi, S.; Molendi, S. Metal abundances in the cool cores of galaxy clusters. *Astron. Astrophys.* **2009**, *508*, 565–574. [CrossRef]
125. Hitomi Collaboration. Solar abundance ratios of the iron-peak elements in the Perseus cluster. *Nature* **2017**, *551*, 478–480. [CrossRef]
126. Simionescu, A.; Nakashima, S.; Yamaguchi, H.; Matsushita, K.; Mernier, F.; Werner, N.; Tamura, T.; Nomoto, K.; de Plaa, J.; Leung, S.C.; et al. Constraints on the chemical enrichment history of the Perseus Cluster of galaxies from high-resolution X-ray spectroscopy. *Mon. Not. R. Astron. Soc.* **2019**, *483*, 1701–1721. [CrossRef]
127. Mernier, F.; de Plaa, J.; Pinto, C.; Kaastra, J.S.; Kosec, P.; Zhang, Y.Y.; Mao, J.; Werner, N. Origin of central abundances in the hot intra-cluster medium. I. Individual and average abundance ratios from XMM-Newton EPIC. *Astron. Astrophys.* **2016**, *592*, A157. [CrossRef]
128. Mernier, F.; Werner, N.; de Plaa, J.; Kaastra, J.S.; Raassen, A.J.J.; Gu, L.; Mao, J.; Urdampilleta, I.; Simionescu, A. Solar chemical composition in the hot gas of cool-core ellipticals, groups, and clusters of galaxies. *Mon. Not. R. Astron. Soc.* **2018**, *480*, L95–L100. [CrossRef]
129. Conroy, C.; Graves, G.J.; van Dokkum, P.G. Early-type Galaxy Archeology: Ages, Abundance Ratios, and Effective Temperatures from Full-spectrum Fitting. *Astrophys. J.* **2014**, *780*, 33. [CrossRef]
130. Johansson, J.; Thomas, D.; Maraston, C. Chemical element ratios of Sloan Digital Sky Survey early-type galaxies. *Mon. Not. R. Astron. Soc.* **2012**, *421*, 1908–1926. [CrossRef]
131. Totani, T.; Morokuma, T.; Oda, T.; Doi, M.; Yasuda, N. Delay Time Distribution Measurement of Type Ia Supernovae by the Subaru/XMM-Newton Deep Survey and Implications for the Progenitor. *Publ. Astron. Soc. Jpn.* **2008**, *60*, 1327. [CrossRef]

132. Maoz, D.; Mannucci, F.; Brandt, T.D. The delay-time distribution of Type Ia supernovae from Sloan II. *Mon. Not. R. Astron. Soc.* **2012**, *426*, 3282–3294. [CrossRef]
133. Maoz, D.; Mannucci, F.; Nelemans, G. Observational Clues to the Progenitors of Type Ia Supernovae. *Annu. Rev. Astron. Astrophys.* **2014**, *52*, 107–170. [CrossRef]
134. Panagoulia, E.K.; Fabian, A.C.; Sanders, J.S. Searching for the missing iron mass in the core of the Centaurus cluster. *Mon. Not. R. Astron. Soc.* **2013**, *433*, 3290–3296. [CrossRef]
135. Lakhchaura, K.; Mernier, F.; Werner, N. Possible depletion of metals into dust grains in the core of the Centaurus cluster of galaxies. *Astron. Astrophys.* **2019**, *623*, A17. [CrossRef]
136. Liu, A.; Zhai, M.; Tozzi, P. On the origin of central abundance drops in the intracluster medium of galaxy groups and clusters. *Mon. Not. R. Astron. Soc.* **2019**, *485*, 1651–1664. [CrossRef]
137. Dupke, R.A.; White, R.E.I. Constraints on Type IA Supernova Models from X-Ray Spectra of Galaxy Clusters. *Astrophys. J.* **2000**, *528*, 139–144. [CrossRef]
138. Mernier, F.; de Plaa, J.; Pinto, C.; Kaastra, J.S.; Kosec, P.; Zhang, Y.Y.; Mao, J.; Werner, N.; Pols, O.R.; Vink, J. Origin of central abundances in the hot intra-cluster medium. II. Chemical enrichment and supernova yield models. *Astron. Astrophys.* **2016**, *595*, A126. [CrossRef]
139. Finoguenov, A.; Davis, D.S.; Zimer, M.; Mulchaey, J.S. The Two-dimensional XMM-Newton Group Survey: Z < 0.012 Groups. *Astrophys. J.* **2006**, *646*, 143–160. [CrossRef]
140. Simionescu, A.; Werner, N.; Finoguenov, A.; Böhringer, H.; Brüggen, M. Metal-rich multi-phase gas in M 87. AGN-driven metal transport, magnetic-field supported multi-temperature gas, and constraints on non-thermal emission observed with XMM-Newton. *Astron. Astrophys.* **2008**, *482*, 97–112. [CrossRef]
141. Kirkpatrick, C.C.; McNamara, B.R.; Cavagnolo, K.W. Anisotropic Metal-enriched Outflows Driven by Active Galactic Nuclei in Clusters of Galaxies. *Astrophys. J.* **2011**, *731*, L23. [CrossRef]
142. Kirkpatrick, C.C.; McNamara, B.R. Hot outflows in galaxy clusters. *Mon. Not. R. Astron. Soc.* **2015**, *452*, 4361–4376. [CrossRef]
143. O'Sullivan, E.; Giacintucci, S.; David, L.P.; Vrtilek, J.M.; Raychaudhury, S. A deep Chandra observation of the poor cluster AWM 4—II. The role of the radio jets in enriching the intracluster medium. *Mon. Not. R. Astron. Soc.* **2011**, *411*, 1833–1842. [CrossRef]
144. O'Sullivan, E.; Vrtilek, J.M.; Kempner, J.C. Active Galactic Nucleus Feedback and Gas Mixing in the Core of NGC 4636. *Astrophys. J.* **2005**, *624*, L77–L80. [CrossRef]
145. Laganá, T.F.; Lovisari, L.; Martins, L.; Lanfranchi, G.A.; Capelato, H.V.; Schellenberger, G. A metal-rich elongated structure in the core of the group NGC 4325. *Astron. Astrophys.* **2015**, *573*, A66. [CrossRef]
146. Randall, S.W.; Nulsen, P.E.J.; Jones, C.; Forman, W.R.; Bulbul, E.; Clarke, T.E.; Kraft, R.; Blanton, E.L.; David, L.; Werner, N.; et al. A Very Deep Chandra Observation of the Galaxy Group NGC 5813: AGN Shocks, Feedback, and Outburst History. *Astrophys. J.* **2015**, *805*, 112. [CrossRef]
147. Ehlert, S.; Werner, N.; Simionescu, A.; Allen, S.W.; Kenney, J.D.P.; Million, E.T.; Finoguenov, A. Ripping apart at the seams: The network of stripped gas surrounding M86. *Mon. Not. R. Astron. Soc.* **2013**, *430*, 2401–2410. [CrossRef]
148. Grevesse, N.; Sauval, A.J. Standard Solar Composition. *Space Sci. Rev.* **1998**, *85*, 161–174.:1005161325181. [CrossRef]
149. Eckert, D.; Gaspari, M.; Owers, M.S.; Roediger, E.; Molendi, S.; Gastaldello, F.; Paltani, S.; Ettori, S.; Venturi, T.; Rossetti, M.; et al. Deep Chandra observations of the stripped galaxy group falling into Abell 2142. *Astron. Astrophys.* **2017**, *605*, A25. [CrossRef]
150. O'Sullivan, E.; Schellenberger, G.; Burke, D.J.; Sun, M.; Vrtilek, J.M.; David, L.P.; Sarazin, C. Building a cluster: Shocks, cavities, and cooling filaments in the group-group merger NGC 6338. *Mon. Not. R. Astron. Soc.* **2019**, *488*, 2925–2946. [CrossRef]
151. Hallman, E.J.; Markevitch, M. Chandra Observation of the Merging Cluster A168: A Late Stage in the Evolution of a Cold Front. *Astrophys. J.* **2004**, *610*, L81–L84. [CrossRef]
152. Sheardown, A.; Fish, T.M.; Roediger, E.; Hunt, M.; ZuHone, J.; Su, Y.; Kraft, R.P.; Nulsen, P.; Churazov, E.; Forman, W.; et al. A New Class of X-Ray Tails of Early-type Galaxies and Subclusters in Galaxy Clusters: Slingshot Tails versus Ram Pressure Stripped Tails. *Astrophys. J.* **2019**, *874*, 112. [CrossRef]
153. Gu, J.; Xu, H.; Gu, L.; An, T.; Wang, Y.; Zhang, Z.; Wu, X.P. A High-Abundance Arc in the Compact Group of Galaxies HCG 62: An AGN- or Merger-Induced Metal Outflow? *Astrophys. J.* **2007**, *659*, 275–282. [CrossRef]
154. Gitti, M.; O'Sullivan, E.; Giacintucci, S.; David, L.P.; Vrtilek, J.; Raychaudhury, S.; Nulsen, P.E.J. Cavities and Shocks in the Galaxy Group HCG 62 as Revealed by Chandra, XMM-Newton, and Giant Metrewave Radio Telescope Data. *Astrophys. J.* **2010**, *714*, 758–771. [CrossRef]
155. O'Sullivan, E.; David, L.P.; Vrtilek, J.M. The impact of sloshing on the intragroup medium and old radio lobe of NGC 5044. *Mon. Not. R. Astron. Soc.* **2014**, *437*, 730–739. [CrossRef]
156. Kraft, R.P.; Jones, C.; Nulsen, P.E.J.; Hardcastle, M.J. The Complex X-Ray Morphology of NGC 7618: A Major Group-Group Merger in the Local Universe? *Astrophys. J.* **2006**, *640*, 762–767. [CrossRef]
157. Su, Y.; Kraft, R.P.; Nulsen, P.E.J.; Jones, C.; Maccarone, T.J.; Mernier, F.; Lovisari, L.; Sheardown, A.; Randall, S.W.; Roediger, E.; et al. Extended X-Ray Study of M49: The Frontier of the Virgo Cluster. *Astron. J.* **2019**, *158*, 6. [CrossRef]
158. Kraft, R.P.; Forman, W.R.; Churazov, E.; Laslo, N.; Jones, C.; Markevitch, M.; Murray, S.S.; Vikhlinin, A. An Unusual Discontinuity in the X-Ray Surface Brightness Profile of NGC 507: Evidence of an Abundance Gradient? *Astrophys. J.* **2004**, *601*, 221–227. [CrossRef]

159. Arnaud, M.; Rothenflug, R.; Boulade, O.; Vigroux, L.; Vangioni-Flam, E. Some constraints on the origin of the iron enriched intra-cluster medium. *Astron. Astrophys.* **1992**, *254*, 49–64.
160. Portinari, L.; Moretti, A.; Chiosi, C.; Sommer-Larsen, J. Can a "Standard" Initial Mass Function Explain the Metal Enrichment in Clusters of Galaxies? *Astrophys. J.* **2004**, *604*, 579–595. [CrossRef]
161. Maraston, C. Evolutionary population synthesis: Models, analysis of the ingredients and application to high-z galaxies. *Mon. Not. R. Astron. Soc.* **2005**, *362*, 799–825. [CrossRef]
162. De Grandi, S.; Ettori, S.; Longhetti, M.; Molendi, S. On the iron content in rich nearby clusters of galaxies. *Astron. Astrophys.* **2004**, *419*, 7–18. [CrossRef]
163. Maoz, D.; Sharon, K.; Gal-Yam, A. The Supernova Delay Time Distribution in Galaxy Clusters and Implications for Type-Ia Progenitors and Metal Enrichment. *Astrophys. J.* **2010**, *722*, 1879–1894. [CrossRef]
164. van der Burg, R.F.J.; Hoekstra, H.; Muzzin, A.; Sifón, C.; Balogh, M.L.; McGee, S.L. Evidence for the inside-out growth of the stellar mass distribution in galaxy clusters since z ~1. *Astron. Astrophys.* **2015**, *577*, A19. [CrossRef]
165. Maoz, D.; Graur, O. Star Formation, Supernovae, Iron, and α: Consistent Cosmic and Galactic Histories. *Astrophys. J.* **2017**, *848*, 25. [CrossRef]
166. Greggio, L.; Renzini, A. *Stellar Populations. A User Guide from Low to High Redshift*; John Wiley & Sons: Hoboken, NJ, USA, 2011.
167. Friedmann, M.; Maoz, D. The rate of Type-Ia supernovae in galaxy clusters and the delay-time distribution out to redshift 1.75. *Mon. Not. R. Astron. Soc.* **2018**, *479*, 3563–3581. [CrossRef]
168. Freundlich, J.; Maoz, D. The delay time distribution of Type-Ia supernovae in galaxy clusters: The impact of extended star-formation histories. *Mon. Not. R. Astron. Soc.* **2021**, *502*, 5882–5895. [CrossRef]
169. Humphrey, P.J.; Buote, D.A.; Gastaldello, F.; Zappacosta, L.; Bullock, J.S.; Brighenti, F.; Mathews, W.G. A Chandra View of Dark Matter in Early-Type Galaxies. *Astrophys. J.* **2006**, *646*, 899–918. [CrossRef]
170. Gastaldello, F.; Buote, D.A.; Humphrey, P.J.; Zappacosta, L.; Bullock, J.S.; Brighenti, F.; Mathews, W.G. Probing the Dark Matter and Gas Fraction in Relaxed Galaxy Groups with X-Ray Observations from Chandra and XMM-Newton. *Astrophys. J.* **2007**, *669*, 158–183. [CrossRef]
171. Nagino, R.; Matsushita, K. Gravitational potential and X-ray luminosities of early-type galaxies observed with XMM-Newton and Chandra. *Astron. Astrophys.* **2009**, *501*, 157–169. [CrossRef]
172. Fabjan, D.; Borgani, S.; Tornatore, L.; Saro, A.; Murante, G.; Dolag, K. Simulating the effect of active galactic nuclei feedback on the metal enrichment of galaxy clusters. *Mon. Not. R. Astron. Soc.* **2010**, *401*, 1670–1690. [CrossRef]
173. Ciotti, L.; D'Ercole, A.; Pellegrini, S.; Renzini, A. Winds, outflows, and inflows in X-ray elliptical galaxies. *Astrophys. J.* **1991**, *376*, 380–403. [CrossRef]
174. Planck Collaboration; Aghanim, N.; Akrami, Y.; Ashdown, M.; Aumont, J.; Baccigalupi, C.; Ballardini, M.; Banday, A.J.; Barreiro, R.B.; Bartolo, N.; et al. Planck 2018 results. VI. Cosmological parameters. *Astron. Astrophys.* **2020**, *641*, A6. [CrossRef]
175. Vikhlinin, A.; Kravtsov, A.; Forman, W.; Jones, C.; Markevitch, M.; Murray, S.S.; Van Speybroeck, L. Chandra Sample of Nearby Relaxed Galaxy Clusters: Mass, Gas Fraction, and Mass-Temperature Relation. *Astrophys. J.* **2006**, *640*, 691–709. [CrossRef]
176. Gonzalez, A.H.; Zaritsky, D.; Zabludoff, A.I. A Census of Baryons in Galaxy Clusters and Groups. *Astrophys. J.* **2007**, *666*, 147–155. [CrossRef]
177. Pratt, G.W.; Croston, J.H.; Arnaud, M.; Böhringer, H. Galaxy cluster X-ray luminosity scaling relations from a representative local sample (REXCESS). *Astron. Astrophys.* **2009**, *498*, 361–378. [CrossRef]
178. Ichikawa, K.; Matsushita, K.; Okabe, N.; Sato, K.; Zhang, Y.Y.; Finoguenov, A.; Fujita, Y.; Fukazawa, Y.; Kawaharada, M.; Nakazawa, K.; et al. Suzaku Observations of the Outskirts of A1835: Deviation from Hydrostatic Equilibrium. *Astrophys. J.* **2013**, *766*, 90. [CrossRef]
179. Okabe, N.; Umetsu, K.; Tamura, T.; Fujita, Y.; Takizawa, M.; Zhang, Y.Y.; Matsushita, K.; Hamana, T.; Fukazawa, Y.; Futamase, T.; et al. Universal profiles of the intracluster medium from Suzaku X-ray and Subaru weak-lensing observations*. *Publ. Astron. Soc. Jpn.* **2014**, *66*, 99. [CrossRef]
180. Ghirardini, V.; Eckert, D.; Ettori, S.; Pointecouteau, E.; Molendi, S.; Gaspari, M.; Rossetti, M.; De Grandi, S.; Roncarelli, M.; Bourdin, H.; et al. Universal thermodynamic properties of the intracluster medium over two decades in radius in the X-COP sample. *Astron. Astrophys.* **2019**, *621*, A41. [CrossRef]
181. Eckert, D.; Ghirardini, V.; Ettori, S.; Rasia, E.; Biffi, V.; Pointecouteau, E.; Rossetti, M.; Molendi, S.; Vazza, F.; Gastaldello, F.; et al. Non-thermal pressure support in X-COP galaxy clusters. *Astron. Astrophys.* **2019**, *621*, A40. [CrossRef]
182. Lin, Y.T.; Mohr, J.J.; Stanford, S.A. Near-Infrared Properties of Galaxy Clusters: Luminosity as a Binding Mass Predictor and the State of Cluster Baryons. *Astrophys. J.* **2003**, *591*, 749–763. [CrossRef]
183. Giodini, S.; Pierini, D.; Finoguenov, A.; Pratt, G.W.; Boehringer, H.; Leauthaud, A.; Guzzo, L.; Aussel, H.; Bolzonella, M.; Capak, P.; et al. Stellar and Total Baryon Mass Fractions in Groups and Clusters Since Redshift 1. *Astrophys. J.* **2009**, *703*, 982–993. [CrossRef]
184. Dai, X.; Bregman, J.N.; Kochanek, C.S.; Rasia, E. On the Baryon Fractions in Clusters and Groups of Galaxies. *Astrophys. J.* **2010**, *719*, 119–125. [CrossRef]
185. McCarthy, I.G.; Bower, R.G.; Balogh, M.L. Revisiting the baryon fractions of galaxy clusters: A comparison with WMAP 3-yr results. *Mon. Not. R. Astron. Soc.* **2007**, *377*, 1457–1463. [CrossRef]
186. Renzini, A. Iron as a Tracer in Galaxy Clusters and Groups. *Astrophys. J.* **1997**, *488*, 35–43. [CrossRef]

187. Makishima, K.; Ezawa, H.; Fukuzawa, Y.; Honda, H.; Ikebe, Y.; Kamae, T.; Kikuchi, K.; Matsushita, K.; Nakazawa, K.; Ohashi, T.; et al. X-Ray Probing of the Central Regions of Clusters of Galaxies. *Publ. Astron. Soc. Jpn.* **2001**, *53*, 401–420. [CrossRef]
188. Lin, Y.T.; Mohr, J.J. K-band Properties of Galaxy Clusters and Groups: Brightest Cluster Galaxies and Intracluster Light. *Astrophys. J.* **2004**, *617*, 879–895. [CrossRef]
189. Matsushita, K.; Sakuma, E.; Sasaki, T.; Sato, K.; Simionescu, A. Metal-mass-to-light Ratios of the Perseus Cluster Out to the Virial Radius. *Astrophys. J.* **2013**, *764*, 147. [CrossRef]
190. Murakami, H.; Komiyama, M.; Matsushita, K.; Nagino, R.; Sato, T.; Sato, K.; Kawaharada, M.; Nakazawa, K.; Ohashi, T.; Takei, Y. Suzaku and XMM-Newton Observations of the Fornax Cluster: Temperature and Metallicity Distribution. *Publ. Astron. Soc. Jpn.* **2011**, *63*, S963–S977. [CrossRef]
191. Sakuma, E.; Ota, N.; Sato, K.; Sato, T.; Matsushita, K. Suzaku Observations of Metal Distributions in the Intracluster Medium of the Centaurus Cluster. *Publ. Astron. Soc. Jpn.* **2011**, *63*, S979–S990. [CrossRef]
192. Renzini, A., Steeper, Flatter, OR Just Salpeter? Evidence From Galaxy Evolution and Galaxy Clusters. In *The Initial Mass Function 50 Years Later*; Corbelli, E., Palla, F., Zinnecker, H., Eds.; Springer: Berlin, Germany. 2005; Volume 327, p. 221._43. [CrossRef]
193. Xu, H.; Kahn, S.M.; Peterson, J.R.; Behar, E.; Paerels, F.B.S.; Mushotzky, R.F.; Jernigan, J.G.; Brinkman, A.C.; Makishima, K. High-Resolution Observations of the Elliptical Galaxy NGC 4636 with the Reflection Grating Spectrometer on Board XMM-Newton. *Astrophys. J.* **2002**, *579*, 600–606. [CrossRef]
194. Werner, N.; Zhuravleva, I.; Churazov, E.; Simionescu, A.; Allen, S.W.; Forman, W.; Jones, C.; Kaastra, J.S. Constraints on turbulent pressure in the X-ray haloes of giant elliptical galaxies from resonant scattering. *Mon. Not. R. Astron. Soc.* **2009**, *398*, 23–32. [CrossRef]
195. Grange, Y.G.; de Plaa, J.; Kaastra, J.S.; Werner, N.; Verbunt, F.; Paerels, F.; de Vries, C.P. The metal contents of two groups of galaxies. *Astron. Astrophys.* **2011**, *531*, A15. [CrossRef]
196. Tamura, T.; Kaastra, J.S.; Makishima, K.; Takahashi, I. High resolution soft X-ray spectroscopy of the elliptical galaxy NGC 5044. Results from the reflection grating spectrometer on-board XMM-Newton. *Astron. Astrophys.* **2003**, *399*, 497–504. [CrossRef]
197. Sanders, J.S.; Fabian, A.C.; Frank, K.A.; Peterson, J.R.; Russell, H.R. Deep high-resolution X-ray spectra from cool-core clusters. *Mon. Not. R. Astron. Soc.* **2010**, *402*, 127–144. [CrossRef]
198. Sanders, J.S.; Fabian, A.C.; Smith, R.K. Constraints on turbulent velocity broadening for a sample of clusters, groups and elliptical galaxies using XMM-Newton. *Mon. Not. R. Astron. Soc.* **2011**, *410*, 1797–1812. [CrossRef]
199. Mao, J.; de Plaa, J.; Kaastra, J.S.; Pinto, C.; Gu, L.; Mernier, F.; Yan, H.L.; Zhang, Y.Y.; Akamatsu, H. Nitrogen abundance in the X-ray halos of clusters and groups of galaxies. *Astron. Astrophys.* **2019**, *621*, A9. [CrossRef]
200. de Plaa, J.; Kaastra, J.S.; Werner, N.; Pinto, C.; Kosec, P.; Zhang, Y.Y.; Mernier, F.; Lovisari, L.; Akamatsu, H.; Schellenberger, G.; et al. CHEERS: The chemical evolution RGS sample. *Astron. Astrophys.* **2017**, *607*, A98. [CrossRef]
201. Werner, N.; Böhringer, H.; Kaastra, J.S.; de Plaa, J.; Simionescu, A.; Vink, J. XMM-Newton high-resolution spectroscopy reveals the chemical evolution of M 87. *Astron. Astrophys.* **2006**, *459*, 353–360. [CrossRef]
202. Ahoranta, J.; Finoguenov, A.; Pinto, C.; Sanders, J.; Kaastra, J.; de Plaa, J.; Fabian, A. Observations of asymmetric velocity fields and gas cooling in the NGC 4636 galaxy group X-ray halo. *Astron. Astrophys.* **2016**, *592*, A145. [CrossRef]
203. Gu, L.; Raassen, A.J.J.; Mao, J.; de Plaa, J.; Shah, C.; Pinto, C.; Werner, N.; Simionescu, A.; Mernier, F.; Kaastra, J.S. X-ray spectra of the Fe-L complex. *Astron. Astrophys.* **2019**, *627*, A51. [CrossRef]
204. Gu, L.; Shah, C.; Mao, J.; Raassen, T.; de Plaa, J.; Pinto, C.; Akamatsu, H.; Werner, N.; Simionescu, A.; Mernier, F.; et al. X-ray spectra of the Fe-L complex. II. Atomic data constraints from the EBIT experiment and X-ray grating observations of Capella. *Astron. Astrophys.* **2020**, *641*, A93. [CrossRef]
205. Scheeler, M.W.; van Rees, W.M.; Kedia, H.; Kleckner, D.; Irvine, W.T.M. Complete measurement of helicity and its dynamics in vortex tubes. *Science* **2017**, *357*, 487–491. [CrossRef]
206. Warhaft, Z. Passive Scalars in Turbulent Flows. *Annu. Rev. Fluid Mech.* **2000**, *32*, 203–240. [CrossRef]
207. Gaspari, M.; Melioli, C.; Brighenti, F.; D'Ercole, A. The dance of heating and cooling in galaxy clusters: Three-dimensional simulations of self-regulated active galactic nuclei outflows. *Mon. Not. R. Astron. Soc.* **2011**, *411*, 349–372. [CrossRef]
208. Colella, P.; Woodward, P.R. The Piecewise Parabolic Method (PPM) for Gas-Dynamical Simulations. *J. Comput. Phys.* **1984**, *54*, 174–201. [CrossRef]
209. Sutherland, R.S.; Dopita, M.A. Cooling functions for low-density astrophysical plasmas. *Astrophys. J. Suppl. Ser.* **1993**, *88*, 253–327. [CrossRef]
210. Loewenstein, M.; Mathews, W.G. Hot gas metallicity and the history of supernova activity in elliptical galaxies. *Astrophys. J.* **1991**, *373*, 445–451. [CrossRef]
211. Brighenti, F.; Mathews, W.G. Evolution of Hot Gas and Dark Halos in Group-dominant Elliptical Galaxies: Influence of Cosmic Inflow. *Astrophys. J.* **1999**, *512*, 65–78. [CrossRef]
212. Brighenti, F.; Mathews, W.G. Heated Cooling Flows. *Astrophys. J.* **2002**, *573*, 542–561. [CrossRef]
213. Mathews, W.G.; Brighenti, F. Hot Gas in and around Elliptical Galaxies. *Annu. Rev. Astron. Astrophys.* **2003**, *41*, 191–239. [CrossRef]
214. Gaspari, M.; Brighenti, F.; D'Ercole, A.; Melioli, C. AGN feedback in galaxy groups: The delicate touch of self-regulated outflows. *Mon. Not. R. Astron. Soc.* **2011**, *415*, 1549–1568. [CrossRef]

215. Gaspari, M.; Brighenti, F.; Temi, P. Mechanical AGN feedback: Controlling the thermodynamical evolution of elliptical galaxies. *Mon. Not. R. Astron. Soc.* **2012**, *424*, 190–209. [CrossRef]
216. Pellegrini, S.; Gan, Z.; Ostriker, J.P.; Ciotti, L. Metal abundances in the MACER simulations of the hot interstellar medium. *Astron. Nachrichten* **2020**, *341*, 184–190. [CrossRef]
217. Mellier, Y.; Mathez, G. Deprojection of the de Vaucouleurs R exp 1/4 brightness profile. *Astron. Astrophys.* **1987**, *175*, 1–3.
218. Greggio, L. The rates of type Ia supernovae. I. Analytical formulations. *Astron. Astrophys.* **2005**, *441*, 1055–1078. [CrossRef]
219. Mannucci, F.; Della Valle, M.; Panagia, N.; Cappellaro, E.; Cresci, G.; Maiolino, R.; Petrosian, A.; Turatto, M. The supernova rate per unit mass. *Astron. Astrophys.* **2005**, *433*, 807–814. [CrossRef]
220. Humphrey, P.J.; Buote, D.A. A Chandra Survey of Early-Type Galaxies. I. Metal Enrichment in the Interstellar Medium. *Astrophys. J.* **2006**, *639*, 136–156. [CrossRef]
221. Gaspari, M.; Tombesi, F.; Cappi, M. Linking macro-, meso- and microscales in multiphase AGN feeding and feedback. *Nat. Astron.* **2020**, *4*, 10–13. [CrossRef]
222. Gaspari, M.; Ruszkowski, M.; Oh, S.P. Chaotic cold accretion on to black holes. *Mon. Not. R. Astron. Soc.* **2013**, *432*, 3401–3422. [CrossRef]
223. Gaspari, M.; Temi, P.; Brighenti, F. Raining on black holes and massive galaxies: The top-down multiphase condensation model. *Mon. Not. R. Astron. Soc.* **2017**, *466*, 677–704. [CrossRef]
224. Voit, G.M. A Role for Turbulence in Circumgalactic Precipitation. *Astrophys. J.* **2018**, *868*, 102. [CrossRef]
225. McCourt, M.; Sharma, P.; Quataert, E.; Parrish, I.J. Thermal instability in gravitationally stratified plasmas: Implications for multiphase structure in clusters and galaxy haloes. *Mon. Not. R. Astron. Soc.* **2012**, *419*, 3319–3337. [CrossRef]
226. Sharma, P.; McCourt, M.; Quataert, E.; Parrish, I.J. Thermal instability and the feedback regulation of hot haloes in clusters, groups and galaxies. *Mon. Not. R. Astron. Soc.* **2012**, *420*, 3174–3194. [CrossRef]
227. Tombesi, F.; Cappi, M.; Reeves, J.N.; Nemmen, R.S.; Braito, V.; Gaspari, M.; Reynolds, C.S. Unification of X-ray winds in Seyfert galaxies: From ultra-fast outflows to warm absorbers. *Mon. Not. R. Astron. Soc.* **2013**, *430*, 1102–1117. [CrossRef]
228. Sądowski, A.; Gaspari, M. Kinetic and radiative power from optically thin accretion flows. *Mon. Not. R. Astron. Soc.* **2017**, *468*, 1398–1404. [CrossRef]
229. Churazov, E.; Brüggen, M.; Kaiser, C.R.; Böhringer, H.; Forman, W. Evolution of Buoyant Bubbles in M87. *Astrophys. J.* **2001**, *554*, 261–273. [CrossRef]
230. Brüggen, M. Simulations of Buoyant Bubbles in Galaxy Clusters. *Astrophys. J.* **2003**, *592*, 839–845. [CrossRef]
231. Brighenti, F.; Mathews, W.G. Stopping Cooling Flows with Jets. *Astrophys. J.* **2006**, *643*, 120–127. [CrossRef]
232. Gaspari, M.; Ruszkowski, M.; Sharma, P. Cause and Effect of Feedback: Multiphase Gas in Cluster Cores Heated by AGN Jets. *Astrophys. J.* **2012**, *746*, 94. [CrossRef]
233. Barai, P.; Murante, G.; Borgani, S.; Gaspari, M.; Granato, G.L.; Monaco, P.; Ragone-Figueroa, C. Kinetic AGN feedback effects on cluster cool cores simulated using SPH. *Mon. Not. R. Astron. Soc.* **2016**, *461*, 1548–1567. [CrossRef]
234. Yang, H.Y.K.; Gaspari, M.; Marlow, C. The Impact of Radio AGN Bubble Composition on the Dynamics and Thermal Balance of the Intracluster Medium. *Astrophys. J.* **2019**, *871*, 6. [CrossRef]
235. Barai, P.; Viel, M.; Murante, G.; Gaspari, M.; Borgani, S. Kinetic or thermal AGN feedback in simulations of isolated and merging disc galaxies calibrated by the M-σ relation. *Mon. Not. R. Astron. Soc.* **2014**, *437*, 1456–1475. [CrossRef]
236. Taylor, P.; Kobayashi, C. Quantifying AGN-driven metal-enhanced outflows in chemodynamical simulations. *Mon. Not. R. Astron. Soc.* **2015**, *452*, L59–L63. [CrossRef]
237. Valentini, M.; Brighenti, F. AGN-stimulated cooling of hot gas in elliptical galaxies. *Mon. Not. R. Astron. Soc.* **2015**, *448*, 1979–1998. [CrossRef]
238. Duan, X.; Guo, F. Metal-rich Trailing Outflows Uplifted by AGN Bubbles in Galaxy Clusters. *Astrophys. J.* **2018**, *861*, 106. [CrossRef]
239. Wittor, D.; Gaspari, M. Dissecting the turbulent weather driven by mechanical AGN feedback. *Mon. Not. R. Astron. Soc.* **2020**, *498*, 4983–5002. [CrossRef]
240. Choi, E.; Brennan, R.; Somerville, R.S.; Ostriker, J.P.; Hirschmann, M.; Naab, T. The Impact of Outflows Driven by Active Galactic Nuclei on Metals in and around Galaxies. *Astrophys. J.* **2020**, *904*, 8. [CrossRef]
241. Melioli, C.; Brighenti, F.; D'Ercole, A. Galactic fountains and outflows in star-forming dwarf galaxies: Interstellar medium expulsion and chemical enrichment. *Mon. Not. R. Astron. Soc.* **2015**, *446*, 299–316. [CrossRef]
242. Husemann, B.; Scharwächter, J.; Davis, T.A.; Pérez-Torres, M.; Smirnova-Pinchukova, I.; Tremblay, G.R.; Krumpe, M.; Combes, F.; Baum, S.A.; Busch, G.; et al. The Close AGN Reference Survey (CARS). A massive multi-phase outflow impacting the edge-on galaxy HE 1353-1917. *Astron. Astrophys.* **2019**, *627*, A53. [CrossRef]
243. Darwin, C. Note on hydrodynamics. *Proc. Camb. Philos. Soc.* **1953**, *49*, 342. [CrossRef]
244. Bondi, H. On spherically symmetrical accretion. *Mon. Not. R. Astron. Soc.* **1952**, *112*, 195. [CrossRef]
245. Narayan, R.; Fabian, A.C. Bondi flow from a slowly rotating hot atmosphere. *Mon. Not. R. Astron. Soc.* **2011**, *415*, 3721–3730. [CrossRef]
246. Hlavacek-Larrondo, J.; Fabian, A.C.; Sanders, J.S.; Taylor, G.B. AGN feedback and iron enrichment in the powerful radio galaxy, 4C+55.16. *Mon. Not. R. Astron. Soc.* **2011**, *415*, 3520–3530. [CrossRef]

247. Liu, W.; Sun, M.; Nulsen, P.; Clarke, T.; Sarazin, C.; Forman, W.; Gaspari, M.; Giacintucci, S.; Lal, D.V.; Edge, T. AGN feedback in galaxy group 3C 88: Cavities, shock, and jet reorientation. *Mon. Not. R. Astron. Soc.* **2019**, *484*, 3376–3392. [CrossRef]
248. Gaspari, M.; Eckert, D.; Ettori, S.; Tozzi, P.; Bassini, L.; Rasia, E.; Brighenti, F.; Sun, M.; Borgani, S.; Johnson, S.D.; et al. The X-Ray Halo Scaling Relations of Supermassive Black Holes. *Astrophys. J.* **2019**, *884*, 169. [CrossRef]
249. Rebusco, P.; Churazov, E.; Böhringer, H.; Forman, W. Impact of stochastic gas motions on galaxy cluster abundance profiles. *Mon. Not. R. Astron. Soc.* **2005**, *359*, 1041–1048. [CrossRef]
250. Rebusco, P.; Churazov, E.; Böhringer, H.; Forman, W. Effect of turbulent diffusion on iron abundance profiles. *Mon. Not. R. Astron. Soc.* **2006**, *372*, 1840–1850. [CrossRef]
251. Gaspari, M.; Churazov, E. Constraining turbulence and conduction in the hot ICM through density perturbations. *Astron. Astrophys.* **2013**, *559*, A78. [CrossRef]
252. Zhuravleva, I.; Churazov, E.M.; Schekochihin, A.A.; Lau, E.T.; Nagai, D.; Gaspari, M.; Allen, S.W.; Nelson, K.; Parrish, I.J. The Relation between Gas Density and Velocity Power Spectra in Galaxy Clusters: Qualitative Treatment and Cosmological Simulations. *Astrophys. J.* **2014**, *788*, L13. [CrossRef]
253. Gaspari, M.; Churazov, E.; Nagai, D.; Lau, E.T.; Zhuravleva, I. The relation between gas density and velocity power spectra in galaxy clusters: High-resolution hydrodynamic simulations and the role of conduction. *Astron. Astrophys.* **2014**, *569*, A67. [CrossRef]
254. Hofmann, F.; Sanders, J.S.; Nandra, K.; Clerc, N.; Gaspari, M. Thermodynamic perturbations in the X-ray halo of 33 clusters of galaxies observed with Chandra ACIS. *Astron. Astrophys.* **2016**, *585*, A130. [CrossRef]
255. Ogorzalek, A.; Zhuravleva, I.; Allen, S.W.; Pinto, C.; Werner, N.; Mantz, A.B.; Canning, R.E.A.; Fabian, A.C.; Kaastra, J.S.; de Plaa, J. Improved measurements of turbulence in the hot gaseous atmospheres of nearby giant elliptical galaxies. *Mon. Not. R. Astron. Soc.* **2017**, *472*, 1659–1676. [CrossRef]
256. Lau, E.T.; Gaspari, M.; Nagai, D.; Coppi, P. Physical Origins of Gas Motions in Galaxy Cluster Cores: Interpreting Hitomi Observations of the Perseus Cluster. *Astrophys. J.* **2017**, *849*, 54. [CrossRef]
257. Cucchetti, E.; Pointecouteau, E.; Peille, P.; Clerc, N.; Rasia, E.; Biffi, V.; Borgani, S.; Tornatore, L.; Dolag, K.; Roncarelli, M.; et al. Athena X-IFU synthetic observations of galaxy clusters to probe the chemical enrichment of the Universe. *Astron. Astrophys.* **2018**, *620*, A173. [CrossRef]
258. Roncarelli, M.; Gaspari, M.; Ettori, S.; Biffi, V.; Brighenti, F.; Bulbul, E.; Clerc, N.; Cucchetti, E.; Pointecouteau, E.; Rasia, E. Measuring turbulence and gas motions in galaxy clusters via synthetic Athena X-IFU observations. *Astron. Astrophys.* **2018**, *618*, A39. [CrossRef]
259. Mernier, F.; Cucchetti, E.; Tornatore, L.; Biffi, V.; Pointecouteau, E.; Clerc, N.; Peille, P.; Rasia, E.; Barret, D.; Borgani, S.; et al. Constraining the origin and models of chemical enrichment in galaxy clusters using the Athena X-IFU. *Astron. Astrophys.* **2020**, *642*, A90. [CrossRef]
260. Gaspari, M.; McDonald, M.; Hamer, S.L.; Brighenti, F.; Temi, P.; Gendron-Marsolais, M.; Hlavacek-Larrondo, J.; Edge, A.C.; Werner, N.; Tozzi, P.; et al. Shaken Snow Globes: Kinematic Tracers of the Multiphase Condensation Cascade in Massive Galaxies, Groups, and Clusters. *Astrophys. J.* **2018**, *854*, 167. [CrossRef]
261. Tremblay, G.R.; Combes, F.; Oonk, J.B.R.; Russell, H.R.; McDonald, M.A.; Gaspari, M.; Husemann, B.; Nulsen, P.E.J.; McNamara, B.R.; Hamer, S.L.; et al. A Galaxy-scale Fountain of Cold Molecular Gas Pumped by a Black Hole. *Astrophys. J.* **2018**, *865*, 13. [CrossRef]
262. Rose, T.; Edge, A.C.; Combes, F.; Gaspari, M.; Hamer, S.; Nesvadba, N.; Peck, A.B.; Sarazin, C.; Tremblay, G.R.; Baum, S.A.; et al. Constraining cold accretion on to supermassive black holes: Molecular gas in the cores of eight brightest cluster galaxies revealed by joint CO and CN absorption. *Mon. Not. R. Astron. Soc.* **2019**, *489*, 349–365. [CrossRef]
263. Simionescu, A.; ZuHone, J.; Zhuravleva, I.; Churazov, E.; Gaspari, M.; Nagai, D.; Werner, N.; Roediger, E.; Canning, R.; Eckert, D.; et al. Constraining Gas Motions in the Intra-Cluster Medium. *Space Sci. Rev.* **2019**, *215*, 24. [CrossRef]
264. Ettori, S.; Gastaldello, F.; Gitti, M.; O'Sullivan, E.; Gaspari, M.; Brighenti, F.; David, L.; Edge, A.C. Cold fronts and metal anisotropies in the X-ray cool core of the galaxy cluster Zw 1742+3306. *Astron. Astrophys.* **2013**, *555*, A93. [CrossRef]
265. Gastaldello, F.; Di Gesu, L.; Ghizzardi, S.; Giacintucci, S.; Girardi, M.; Roediger, E.; Rossetti, M.; Brighenti, F.; Buote, D.A.; Eckert, D.; et al. Sloshing Cold Fronts in Galaxy Groups and their Perturbing Disk Galaxies: An X-Ray, Optical, and Radio Case Study. *Astrophys. J.* **2013**, *770*, 56. [CrossRef]
266. Ghizzardi, S.; De Grandi, S.; Molendi, S. Metal distribution in sloshing galaxy clusters: The case of A496. *Astron. Astrophys.* **2014**, *570*, A117. [CrossRef]
267. De Grandi, S.; Eckert, D.; Molendi, S.; Girardi, M.; Roediger, E.; Gaspari, M.; Gastaldello, F.; Ghizzardi, S.; Nonino, M.; Rossetti, M. A textbook example of ram-pressure stripping in the Hydra A/A780 cluster. *Astron. Astrophys.* **2016**, *592*, A154. [CrossRef]
268. Clavico, S.; De Grandi, S.; Ghizzardi, S.; Rossetti, M.; Molendi, S.; Gastaldello, F.; Girardi, M.; Boschin, W.; Botteon, A.; Cassano, R.; et al. Growth and disruption in the Lyra complex. *Astron. Astrophys.* **2019**, *632*, A27. [CrossRef]
269. Tümer, A.; Tombesi, F.; Bourdin, H.; Ercan, E.N.; Gaspari, M.; Serafinelli, R. Exploring the multiphase medium in MKW 08: From the central active galaxy up to cluster scales. *Astron. Astrophys.* **2019**, *629*, A82. [CrossRef]
270. Dolag, K.; Borgani, S.; Schindler, S.; Diaferio, A.; Bykov, A.M. Simulation Techniques for Cosmological Simulations. *Space Sci. Rev.* **2008**, *134*, 229–268. [CrossRef]
271. Borgani, S.; Kravtsov, A. Cosmological Simulations of Galaxy Clusters. *Adv. Sci. Lett.* **2011**, *4*, 204–227. [CrossRef]

272. Vogelsberger, M.; Marinacci, F.; Torrey, P.; Puchwein, E. Cosmological simulations of galaxy formation. *Nat. Rev. Phys.* **2020**, *2*, 42–66. [CrossRef]
273. Steinmetz, M.; Mueller, E. The formation of disk galaxies in a cosmological context: Populations, metallicities and metallicity gradients. *Astron. Astrophys.* **1994**, *281*, L97–L100. [CrossRef]
274. Mosconi, M.B.; Tissera, P.B.; Lambas, D.G.; Cora, S.A. Chemical evolution using smooth particle hydrodynamical cosmological simulations—I. Implementation, tests and first results. *Mon. Not. R. Astron. Soc.* **2001**, *325*, 34–48. [CrossRef]
275. Kawata, D.; Gibson, B.K. GCD+: A new chemodynamical approach to modelling supernovae and chemical enrichment in elliptical galaxies. *Mon. Not. R. Astron. Soc.* **2003**, *340*, 908–922. [CrossRef]
276. Lia, C.; Portinari, L.; Carraro, G. Star formation and chemical evolution in smoothed particle hydrodynamics simulations: A statistical approach. *Mon. Not. R. Astron. Soc.* **2002**, *330*, 821–836. [CrossRef]
277. Valdarnini, R. Iron abundances and heating of the intracluster medium in hydrodynamical simulations of galaxy clusters. *Mon. Not. R. Astron. Soc.* **2003**, *339*, 1117–1134. [CrossRef]
278. Tornatore, L.; Borgani, S.; Matteucci, F.; Recchi, S.; Tozzi, P. Simulating the metal enrichment of the intracluster medium. *Mon. Not. R. Astron. Soc.* **2004**, *349*, L19–L24. [CrossRef]
279. Scannapieco, C.; Tissera, P.B.; White, S.D.M.; Springel, V. Feedback and metal enrichment in cosmological smoothed particle hydrodynamics simulations—I. A model for chemical enrichment. *Mon. Not. R. Astron. Soc.* **2005**, *364*, 552–564. [CrossRef]
280. Oppenheimer, B.D.; Davé, R. Cosmological simulations of intergalactic medium enrichment from galactic outflows. *Mon. Not. R. Astron. Soc.* **2006**, *373*, 1265–1292. [CrossRef]
281. Tornatore, L.; Borgani, S.; Dolag, K.; Matteucci, F. Chemical enrichment of galaxy clusters from hydrodynamical simulations. *Mon. Not. R. Astron. Soc.* **2007**, *382*, 1050–1072. [CrossRef]
282. Salpeter, E.E. The Luminosity Function and Stellar Evolution. *Astrophys. J.* **1955**, *121*, 161. [CrossRef]
283. Kroupa, P. On the variation of the initial mass function. *Mon. Not. R. Astron. Soc.* **2001**, *322*, 231–246. [CrossRef]
284. Chabrier, G. Galactic Stellar and Substellar Initial Mass Function. *Publ. Astron. Soc. Pac.* **2003**, *115*, 763–795. [CrossRef]
285. Tinsley, B.M. Stellar lifetimes and abundance ratios in chemical evolution. *Astrophys. J.* **1979**, *229*, 1046–1056. [CrossRef]
286. Padovani, P.; Matteucci, F. Stellar Mass Loss in Elliptical Galaxies and the Fueling of Active Galactic Nuclei. *Astrophys. J.* **1993**, *416*, 26. [CrossRef]
287. Portinari, L.; Chiosi, C.; Bressan, A. Galactic chemical enrichment with new metallicity dependent stellar yields. *Astron. Astrophys.* **1998**, *334*, 505–539. [CrossRef]
288. Romano, D.; Chiappini, C.; Matteucci, F.; Tosi, M. Quantifying the uncertainties of chemical evolution studies. I. Stellar lifetimes and initial mass function. *Astron. Astrophys.* **2005**, *430*, 491–505. [CrossRef]
289. Romano, D.; Karakas, A.I.; Tosi, M.; Matteucci, F. Quantifying the uncertainties of chemical evolution studies. II. Stellar yields. *Astron. Astrophys.* **2010**, *522*, A32. [CrossRef]
290. Karakas, A.I.; Lattanzio, J.C. The Dawes Review 2: Nucleosynthesis and Stellar Yields of Low- and Intermediate-Mass Single Stars. *Publ. Astron. Soc. Aust.* **2014**, *31*, e030. [CrossRef]
291. Nomoto, K.; Wanajo, S.; Kamiya, Y.; Tominaga, N.; Umeda, H. Chemical Yields from Supernovae and Hypernovae. In *The Galaxy Disk in Cosmological Context*; Andersen, J., Nordströara, M.B.; Bland-Hawthorn, J., Eds.; Cambridge University Press: Cambridge, UK, 2009; Volume 254, pp. 355–368. [CrossRef]
292. Nomoto, K.; Leung, S.C. Single Degenerate Models for Type Ia Supernovae: Progenitor's Evolution and Nucleosynthesis Yields. *Space Sci. Rev.* **2018**, *214*, 67. [CrossRef]
293. Springel, V. Smoothed Particle Hydrodynamics in Astrophysics. *Annu. Rev. Astron. Astrophys.* **2010**, *48*, 391–430. [CrossRef]
294. Tinsley, B.M. Evolution of the Stars and Gas in Galaxies. *Fundam. Cosmic Phys.* **1980**, *5*, 287–388. [CrossRef]
295. Matteucci, F. *The Chemical Evolution of the Galaxy*; Springer: Berlin, Germany. 2003.
296. Romeo, A.D.; Sommer-Larsen, J.; Portinari, L.; Antonuccio-Delogu, V. Simulating galaxy clusters—I. Thermal and chemical properties of the intracluster medium. *Mon. Not. R. Astron. Soc.* **2006**, *371*, 548–568. [CrossRef]
297. Vogelsberger, M.; Marinacci, F.; Torrey, P.; Genel, S.; Springel, V.; Weinberger, R.; Pakmor, R.; Hernquist, L.; Naiman, J.; Pillepich, A.; et al. The uniformity and time-invariance of the intra-cluster metal distribution in galaxy clusters from the IllustrisTNG simulations. *Mon. Not. R. Astron. Soc.* **2018**, *474*, 2073–2093. [CrossRef]
298. Biffi, V.; Planelles, S.; Borgani, S.; Fabjan, D.; Rasia, E.; Murante, G.; Tornatore, L.; Dolag, K.; Granato, G.L.; Gaspari, M.; et al. The history of chemical enrichment in the intracluster medium from cosmological simulations. *Mon. Not. R. Astron. Soc.* **2017**, *468*, 531–548. [CrossRef]
299. Truong, N.; Rasia, E.; Biffi, V.; Mernier, F.; Werner, N.; Gaspari, M.; Borgani, S.; Planelles, S.; Fabjan, D.; Murante, G. Mass-metallicity relation from cosmological hydrodynamical simulations and X-ray observations of galaxy groups and clusters. *Mon. Not. R. Astron. Soc.* **2019**, *484*, 2896–2913. [CrossRef]
300. Planelles, S.; Borgani, S.; Fabjan, D.; Killedar, M.; Murante, G.; Granato, G.L.; Ragone-Figueroa, C.; Dolag, K. On the role of AGN feedback on the thermal and chemodynamical properties of the hot intracluster medium. *Mon. Not. R. Astron. Soc.* **2014**, *438*, 195–216. [CrossRef]
301. Martizzi, D.; Hahn, O.; Wu, H.Y.; Evrard, A.E.; Teyssier, R.; Wechsler, R.H. RHAPSODY-G simulations—II. Baryonic growth and metal enrichment in massive galaxy clusters. *Mon. Not. R. Astron. Soc.* **2016**, *459*, 4408–4427. [CrossRef]

302. Dolag, K.; Mevius, E.; Remus, R.S. Distribution and Evolution of Metals in the Magneticum Simulations. *Galaxies* **2017**, *5*, 35. [CrossRef]
303. Barnes, D.J.; Kay, S.T.; Bahé, Y.M.; Dalla Vecchia, C.; McCarthy, I.G.; Schaye, J.; Bower, R.G.; Jenkins, A.; Thomas, P.A.; Schaller, M.; et al. The Cluster-EAGLE project: Global properties of simulated clusters with resolved galaxies. *Mon. Not. R. Astron. Soc.* **2017**, *471*, 1088–1106. [CrossRef]
304. Schaye, J.; Dalla Vecchia, C.; Booth, C.M.; Wiersma, R.P.C.; Theuns, T.; Haas, M.R.; Bertone, S.; Duffy, A.R.; McCarthy, I.G.; van de Voort, F. The physics driving the cosmic star formation history. *Mon. Not. R. Astron. Soc.* **2010**, *402*, 1536–1560. [CrossRef]
305. Davé, R.; Oppenheimer, B.D.; Sivanandam, S. Enrichment and pre-heating in intragroup gas from galactic outflows. *Mon. Not. R. Astron. Soc.* **2008**, *391*, 110–123. [CrossRef]
306. Rasia, E.; Borgani, S.; Murante, G.; Planelles, S.; Beck, A.M.; Biffi, V.; Ragone-Figueroa, C.; Granato, G.L.; Steinborn, L.K.; Dolag, K. Cool Core Clusters from Cosmological Simulations. *Astrophys. J.* **2015**, *813*, L17. [CrossRef]
307. Anderson, M.E.; Bregman, J.N.; Butler, S.C.; Mullis, C.R. Redshift Evolution in the Iron Abundance of the Intracluster Medium. *Astrophys. J.* **2009**, *698*, 317–323. [CrossRef]
308. McDonald, M.; Bulbul, E.; de Haan, T.; Miller, E.D.; Benson, B.A.; Bleem, L.E.; Brodwin, M.; Carlstrom, J.E.; Chiu, I.; Forman, W.R.; et al. The Evolution of the Intracluster Medium Metallicity in Sunyaev Zel'dovich-selected Galaxy Clusters at $0 < z < 1.5$. *Astrophys. J.* **2016**, *826*, 124. [CrossRef]
309. Mantz, A.B.; Allen, S.W.; Morris, R.G.; Simionescu, A.; Urban, O.; Werner, N.; Zhuravleva, I. The metallicity of the intracluster medium over cosmic time: Further evidence for early enrichment. *Mon. Not. R. Astron. Soc.* **2017**, *472*, 2877–2888. [CrossRef]
310. Yates, R.M.; Thomas, P.A.; Henriques, B.M.B. Iron in galaxy groups and clusters: Confronting galaxy evolution models with a newly homogenized data set. *Mon. Not. R. Astron. Soc.* **2017**, *464*, 3169–3193. [CrossRef]
311. Eckert, D.; Molendi, S.; Paltani, S. The cool-core bias in X-ray galaxy cluster samples. I. Method and application to HIFLUGCS. *Astron. Astrophys.* **2011**, *526*, A79. [CrossRef]
312. Rossetti, M.; Gastaldello, F.; Eckert, D.; Della Torre, M.; Pantiri, G.; Cazzoletti, P.; Molendi, S. The cool-core state of Planck SZ-selected clusters versus X-ray-selected samples: Evidence for cool-core bias. *Mon. Not. R. Astron. Soc.* **2017**, *468*, 1917–1930. [CrossRef]
313. Adami, C.; Giles, P.; Koulouridis, E.; Pacaud, F.; Caretta, C.A.; Pierre, M.; Eckert, D.; Ramos-Ceja, M.E.; Gastaldello, F.; Fotopoulou, S.; et al. The XXL Survey. XX. The 365 cluster catalogue. *Astron. Astrophys.* **2018**, *620*, A5. [CrossRef]
314. Gozaliasl, G.; Finoguenov, A.; Tanaka, M.; Dolag, K.; Montanari, F.; Kirkpatrick, C.C.; Vardoulaki, E.; Khosroshahi, H.G.; Salvato, M.; Laigle, C.; et al. Chandra centres for COSMOS X-ray galaxy groups: Differences in stellar properties between central dominant and offset brightest group galaxies. *Mon. Not. R. Astron. Soc.* **2019**, *483*, 3545–3565. [CrossRef]
315. O'Sullivan, E.; Ponman, T.J.; Kolokythas, K.; Raychaudhury, S.; Babul, A.; Vrtilek, J.M.; David, L.P.; Giacintucci, S.; Gitti, M.; Haines, C.P. The Complete Local Volume Groups Sample—I. Sample selection and X-ray properties of the high-richness subsample. *Mon. Not. R. Astron. Soc.* **2017**, *472*, 1482–1505. [CrossRef]
316. Predehl, P.; Andritschke, R.; Arefiev, V.; Babyshkin, V.; Batanov, O.; Becker, W.; Böhringer, H.; Bogomolov, A.; Boller, T.; Borm, K.; et al. The eROSITA X-ray telescope on SRG. *Astron. Astrophys.* **2021**, *647*, A1. [CrossRef]
317. Merloni, A.; Predehl, P.; Becker, W.; Böhringer, H.; Boller, T.; Brunner, H.; Brusa, M.; Dennerl, K.; Freyberg, M.; Friedrich, P.; et al. eROSITA Science Book: Mapping the Structure of the Energetic Universe. *arXiv* **2012**, arXiv:1209.3114. doi:10.1051/0004-6361/201936131.
318. Käfer, F.; Finoguenov, A.; Eckert, D.; Clerc, N.; Ramos-Ceja, M.E.; Sanders, J.S.; Ghirardini, V. Toward the low-scatter selection of X-ray clusters. Galaxy cluster detection with eROSITA through cluster outskirts. *Astron. Astrophys.* **2020**, *634*, A8. [CrossRef]
319. Tashiro, M.; Maejima, H.; Toda, K.; Kelley, R.; Reichenthal, L.; Hartz, L.; Petre, R.; Williams, B.; Guainazzi, M.; Costantini, E.; et al. Status of x-ray imaging and spectroscopy mission (XRISM). In Proceedings of the Space Telescopes and Instrumentation 2020: Ultraviolet to Gamma Ray, Online, 14-18 December 2020.
Volume 11444, p. 1144422. [CrossRef]
320. Betancourt-Martinez, G.; Akamatsu, H.; Barret, D.; Bautista, M.; Bernitt, S.; Bianchi, S.; Bodewits, D.; Brickhouse, N.; Brown, G.V.; Costantini, E.; et al. Unlocking the Capabilities of Future High-Resolution X-ray Spectroscopy Missions Through Laboratory Astrophysics. *BAAS* **2019**, *51*, 337. [CrossRef]
321. Griffith, E.; Johnson, J.A.; Weinberg, D.H. Abundance Ratios in GALAH DR2 and Their Implications for Nucleosynthesis. *Astrophys. J.* **2019**, *886*, 84. [CrossRef]
322. Pinto, C.; Sanders, J.S.; Werner, N.; de Plaa, J.; Fabian, A.C.; Zhang, Y.Y.; Kaastra, J.S.; Finoguenov, A.; Ahoranta, J. Chemical Enrichment RGS cluster Sample (CHEERS): Constraints on turbulence. *Astron. Astrophys.* **2015**, *575*, A38. [CrossRef]
323. Rau, A.; Meidinger, N.; Nandra, K.; Porro, M.; Barret, D.; Santangelo, A.; Schmid, C.; Struder, L.; Tenzer, C.; Wilms, J.; Amoros, C.; et al. The Hot and Energetic Universe: The Wide Field Imager (WFI) for Athena+. *arXiv* **2013**, arXiv:1308.6785.
324. Rau, A.; Nandra, K.; Aird, J.; Comastri, A.; Dauser, T.; Merloni, A.; Pratt, G.W.; Reiprich, T.H.; Fabian, A.C.; Georgakakis, A.; et al. Athena Wide Field Imager key science drivers. In Proceedings of the Space Telescopes and Instrumentation 2016: Ultraviolet to Gamma Ray, Edinburgh, UK, 26 June–1 July 2016; den Herder, J.W.A., Takahashi, T., Bautz, M., Eds.; Volume 9905, p. 99052B. [CrossRef]
325. Barret, D.; den Herder, J.W.; Piro, L.; Ravera, L.; Den Hartog, R.; Macculi, C.; Barcons, X.; Page, M.; Paltani, S.; Rauw, G.; et al. The Hot and Energetic Universe: The X-ray Integral Field Unit (X-IFU) for Athena+. *arXiv* **2013**, arXiv:1308.6784.

326. Barret, D.; Lam Trong, T.; den Herder, J.W.; Piro, L.; Cappi, M.; Houvelin, J.; Kelley, R.; Mas-Hesse, J.M.; Mitsuda, K.; Paltani, S.; et al. The ATHENA X-ray Integral Field Unit (X-IFU). In Proceedings of the Space Telescopes and Instrumentation 2018: Ultraviolet to Gamma Ray, Austin, TX, USA, 10–15 June 2018 ; den Herder, J.W.A., Nikzad, S., Nakazawa, K., Eds.; Volume 10699, p. 106991G. [CrossRef]
327. Zhang, C.; Ramos-Ceja, M.E.; Pacaud, F.; Reiprich, T.H. High-redshift galaxy groups as seen by ATHENA/WFI. *Astron. Astrophys.* **2020**, *642*, A17. [CrossRef]
328. Panagoulia, E.K.; Fabian, A.C.; Sanders, J.S. A volume-limited sample of X-ray galaxy groups and clusters—I. Radial entropy and cooling time profiles. *Mon. Not. R. Astron. Soc.* **2014**, *438*, 2341–2354. [CrossRef]
329. Cui, W.; Bregman, J.N.; Bruijn, M.P.; Chen, L.B.; Chen, Y.; Cui, C.; Fang, T.T.; Gao, B.; Gao, H.; Gao, J.R.; et al. HUBS: A dedicated hot circumgalactic medium explorer. In Proceedings of the Space Telescopes and Instrumentation 2020: Ultraviolet to Gamma Ray, Online, 14–18 December 2020; Volume 11444, p. 114442S. [CrossRef]
330. Sato, K.; Ohashi, T.; Ishisaki, Y.; Ezoe, Y.; Yamada, S.; Yamasaki, N.Y.; Mitsuda, K.; Ishida, M.; Maeda, Y.; Nakashima, Y.; et al. Super DIOS mission for exploring "dark baryon". In Proceedings of the Space Telescopes and Instrumentation 2020: Ultraviolet to Gamma Ray, Online, 14–18 December 2020; Volume 11444, p. 114445O. [CrossRef]
331. Mushotzky, R.F.; Aird, J.; Barger, A.J.; Cappelluti, N.; Chartas, G.; Corrales, L.; Eufrasio, R.; Fabian, A.C.; Falcone, A.D.; Gallo, E.; et al. The Advanced X-ray Imaging Satellite. *arXiv* **2019**, arXiv:1903.04083.
332. Smith, R.K.; Abraham, M.H.; Allured, R.; Bautz, M.; Bookbinder, J.; Bregman, J.N.; Brenneman, L.; Brickhouse, N.S.; Burrows, D.N.; Burwitz, V.; et al. Arcus: The X-ray grating spectrometer explorer. In Proceedings of the Space Telescopes and Instrumentation 2016: Ultraviolet to Gamma Ray, Edinburgh, UK, 26 June–1 July 2016; den Herder, J.W.A., Takahashi, T., Bautz, M., Eds.; Volume 9905, p. 99054M. [CrossRef]
333. Cen, R.; Fang, T. Where Are the Baryons? III. Nonequilibrium Effects and Observables. *Astrophys. J.* **2006**, *650*, 573–591. [CrossRef]
334. Wijers, N.A.; Schaye, J.; Oppenheimer, B.D.; Crain, R.A.; Nicastro, F. The abundance and physical properties of O VII and O VIII X-ray absorption systems in the EAGLE simulations. *Mon. Not. R. Astron. Soc.* **2019**, *488*, 2947–2969. [CrossRef]
335. Wijers, N.A.; Schaye, J.; Oppenheimer, B.D. The warm-hot circumgalactic medium around EAGLE-simulation galaxies and its detection prospects with X-ray and UV line absorption. *Mon. Not. R. Astron. Soc.* **2020**, *498*, 574–598. [CrossRef]
336. Werk, J.K.; Prochaska, J.X.; Tumlinson, J.; Peeples, M.S.; Tripp, T.M.; Fox, A.J.; Lehner, N.; Thom, C.; O'Meara, J.M.; Ford, A.B.; et al. The COS-Halos Survey: Physical Conditions and Baryonic Mass in the Low-redshift Circumgalactic Medium. *Astrophys. J.* **2014**, *792*, 8. [CrossRef]
337. Özel, F. The Lynx X-ray Surveyor. *Nat. Astron.* **2018**, *2*, 608–609. [CrossRef]
338. Gaskin, J.A.; Swartz, D.A.; Vikhlinin, A.; Özel, F.; Gelmis, K.E.; Arenberg, J.W.; Bandler, S.R.; Bautz, M.W.; Civitani, M.M.; Dominguez, A.; et al. Lynx X-Ray Observatory: An overview. *J. Astron. Telesc. Instrum. Syst.* **2019**, *5*, 021001. [CrossRef]

MDPI

Review

Properties of Fossil Groups of Galaxies

J. Alfonso L. Aguerri [1,2,†] and Stefano Zarattini [3,4,*,†]

1 Instituto de Astrofísica de Canarias, C/Vía Láctea s/n, E-38200 La Laguna, Spain; jalfonso@iac.es
2 Departamento de Astrofísica, Universidad de La Laguna, E-38206 La Laguna, Spain
3 Dipartimento di Fisica e Astronomia "G. Galilei", Università di Padova, Vicolo dell'Osservatorio 3, I-35122 Padova, Italy
4 INAF—Osservatorio Astronomico di Padova, Vicolo dell'Osservatorio 2, I-35122 Padova, Italy
* Correspondence: stefano.zarattini@unipd.it
† These authors contributed equally to this work.

Abstract: We review the formation and evolution of fossil groups and clusters from both the theoretical and the observational points of view. In the optical band, these systems are dominated by the light of the central galaxy. They were interpreted as old systems that had enough time to merge all the M* galaxies within the central one. During the last two decades, many observational studies were performed to prove the *old and relaxed* state of fossil systems. The majority of these studies that spans a wide range of topics including halos global scaling relations, dynamical substructures, stellar populations, and galaxy luminosity functions seem to challenge this scenario. The general picture that can be obtained by reviewing all the observational works is that the fossil state could be transitional. Indeed, the formation of the large magnitude gap observed in fossil systems could be related to internal processes rather than an old formation.

Keywords: fossil galaxy groups; galaxy clusters; galaxy groups; X-ray and optical observations; hydrodynamical simulations

Citation: Aguerri, J.A.L.; Zarattini, S. Properties of Fossil Groups of Galaxies. *Universe* **2021**, *7*, 132. https://doi.org/10.3390/universe7050132

Academic Editor: Francesco Shankar

Received: 26 March 2021
Accepted: 28 April 2021
Published: 4 May 2021

1. Introduction

The Lambda cold dark matter scenario (ΛCDM) predicts that structures in the Universe form following a hierarchical evolution: small objects collapsed first under their self-gravity and are then merged continuously to build larger structures. In this scenario, galaxy formed first, then merged in small groups and the process continues until the creation of massive galaxy clusters [1,2].

Ponman and Bertram [3] firstly suggested, while studying compact groups, that this building scenario could be taken to the extreme consequences. They predicted that, in some cases, all the main galaxies of a group could merge with one another, creating a giant galaxy embedded in an X-ray halo typical of a group. This prediction was supported, one year later, by the discovery of RX 11340.6 + 4018, an apparently isolated elliptical galaxy, at redshift $z = 0.171$, found in an extended X-ray halo. The estimated X-ray mass was 2.8×10^{13} $M_\odot$, making it a typical group-sized object [4]. These systems were named as "*fossil groups*" (FGs). They were thought to be the latest stage in the evolution of galaxy groups. For this reason, they were supposed to be old and dynamically relaxed systems [3].

The spatial density and fraction of FGs were estimated in different studies, both theoretically and observationally. The agreement on the spatial density is quite good, with estimations in the range $\sim$1–3 $\times 10^{-6}\, h_{50}^3$ Mpc^{-3}, see, e.g., in [5–7]. On the other hand, the fraction of FGs with respect to the total amount of clusters and groups is more debated. A comparison between different works can be found in Table 1 of Dariush et al. [8]: the estimated fractions varies between 1% and 40%, with a large scatter, and a mean value can be found at about $\sim$10–15%.

In this review, we show the main observational and theoretical works related to FGs written during the last three decades. Our aim is to show that these systems are fixed well in the current theory of structure formation in the Universe. They are just extreme systems produced following this theory. This review is structured in the following way: in Section 2, we will describe the search for FGs in the last ∼20 years and their observational definitions. In Section 3, we will present the theoretical scenarios proposed to describe these systems. In Section 4, we will discuss the properties of the intra-cluster medium and of the galaxy populations. Then, we will describe the possible progenitors of FGs in Section 5. Finally, in Section 6, we will propose a sample of "*genuine*" FGs and draw our conclusions.

2. The Search for Fossil Systems

Since their discovery in 1994, a strong observational effort was done to find FGs. Despite their supposed frequency, more than 10 years were needed before producing reasonably large catalogues of some tens of candidates. In this sense, one of the main obstacles for the community was to agree on a practical definition. In fact, at the beginning, the search was limited to isolated galaxies surrounded by an X-ray halo, but this definition was too loose [5]. More rigorous definitions, involving photometry, spectrocopy and X-ray data, were proposed and are actually used. We will discuss them in Section 2.1. However, these rigorous definitions had important observational limitations. The result of these constraints was that fossil samples were not selected in a homogeneous way and, in Section 2.2, we will discuss the strength and weaknesses of the different approaches adopted in the literature.

2.1. Operational Definitions of Fossil Groups

The most common operational definitions of FGs are based on the magnitude gap between the brightest member galaxy of the group/cluster and other galaxy sorted by their magnitude ($\Delta m_{1,j}$, where j represents the j-th ranked galaxy). In particular, Jones et al. [9] suggested that a group or a cluster of galaxies should be classified as fossil if the magnitude gap between its two brightest members (Δm_{12}) is larger than two magnitudes in the $r-$band. In addition, they also imposed that the two brightest galaxies should be located within half the (projected) virial radius (defined as r_{200}, the radius of a sphere whose mean density is 200 times the critical density of the Universe). Another common definition is the one presented in Dariush et al. [10]: a group/cluster of galaxies is classified as fossil if the magnitude gap between its first and fourth brightest member galaxies (Δm_{14}) is larger than 2.5 magnitudes in the $r-$band and within half the (projected) virial radius. Both definitions also require the presence of a diffuse X-ray halo, with $L_X \geq 10^{42}\, h_{50}^{-2}$ erg s^{-1}. This last criterium ensures that the system is located in a potential well similar in mass to groups or clusters of galaxies. An example of a typical FG and the color-magnitude diagram of its galaxy population is shown in Figure 1.

The original name and definition were thought to describe only galaxy groups. However, many studies, e.g., [11–13], found the existence of fossil clusters. This is due to the lack of an upper limit for the X-ray luminosity in both of the most common operational definitions. As a consequence, when describing the general properties of the population, *fossil groups* and *fossil clusters* are interchangeable in the literature. In addition, a more general *fossil systems* is also widely used. We invite the reader to consider these three definitions as equivalent along this review.

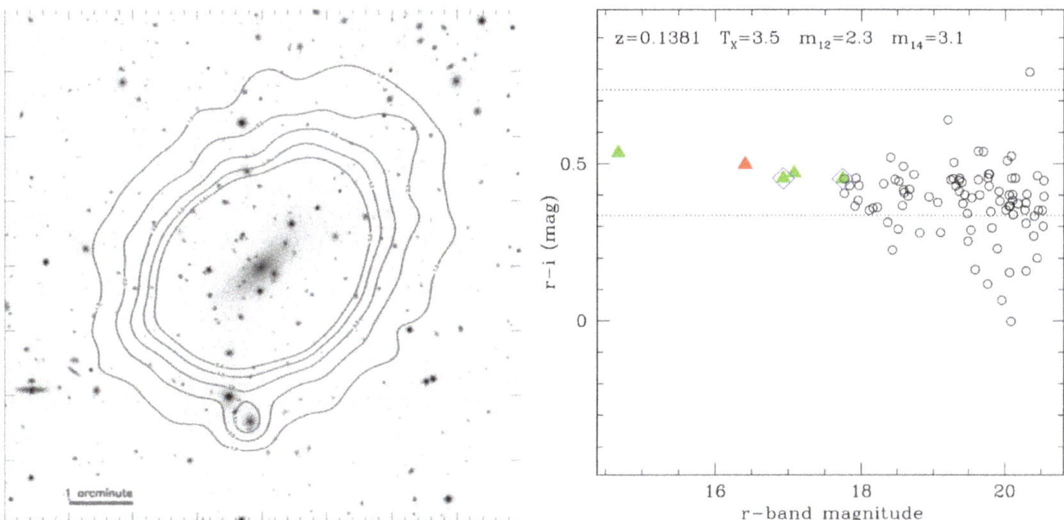

Figure 1. Left panel: SDSS image with X-ray contours of Abell 1068. **Right panel**: color-magnitude diagram of the same object. Green triangles are system members, whereas red triangles are those that are not. The redshift, $\Delta m12$, and $\Delta m14$ are reported in the right panel. The original image can be found in Harrison et al. [14]; in particular, it corresponds with their Figure B7.

2.2. Catalogues

We already mentioned that the prototype of FGs is RX 11340.6 + 4018, presented in Ponman et al. [4]. A few years later, Vikhlinin et al. [5] studied a sample of four *X-Ray Overluminous Elliptical Galaxies*: the authors claimed that these objects could have been part of the fossil category, but it was just a suggestion, since no operational definition was available until Jones et al. [9] proposed the use of the magnitude gap as the discriminating factor between fossils and non-fossils, proposing a sample of 5 FGs in their work.

Other studies presented small numbers of FGs candidates [11,12,15–18], amongst others, whereas the first large sample of FG candidates was presented in Santos et al. [6]: in this work, the authors selected 34 galaxy aggregations obtained from the Sloan Digital Sky Survey Data Release 5, SDSS DR5, [19] by cross matching the sample of SDSS *Luminous Red Galaxies* [20] with sources in the ROSAT all-sky *bright source catalogue* [21]. The result of the cross match was a list of elliptical galaxies surrounded by an X-ray halo. Then, the magnitude gap was computed within a fixed radius (500 kpc) and in a fixed redshift range ($\Delta z = 0.002$ when spectroscopic redshift is available, $\Delta z = 0.035$ when only photometric redshift is known). These 34 FG candidates were then studied in detail by the Fossil Group Origins (FOGO) project [13]. This observational project produced a set of results analyzing the properties of these systems. In the framework of this project, Zarattini et al. [22] confirmed that 15^{+8}_{-5} of the candidates are FGs according to the Jones et al. [9] or Dariush et al. [10] definitions. The uncertainties in the number of confirmed FGs reflect those on the definition of r_{200}. The large difference between the proposed candidates and the confirmed fossils can be explained with a stricter implementation of the definition criteria (e.g., differences in the search radius and membership definition). From the comparison between Santos et al. [6] and Zarattini et al. [22], it seems clear that three types of observations are needed in order to strictly define an FG: (i) X-ray data, required to estimate the mass of the system and define the virial radius, (ii) multi-object spectroscopy, in order to identify the real members of the FG, removing fore- and background objects, and (iii) an optical image, needed to measure the magnitude of each galaxy and to compute the magnitude gap between the brightest galaxy and the other members. The combination of the strong observational effort together with the demanding observational definition is probably the main limit for the building of a large and homogeneous dataset of fossil systems. However, the number of known FGs kept growing in the last decade. Without intending to be exhaustive, La Barbera et al. [7] used SDSS-DR4 and ROSAT to build a sample of 25 FG candidates, Tavasoli et al. [23] found 109 FGs obtained from SDSS-DR7 using a friend-of-friend algorithm, Makarov and Karachentsev [24] presented a catalogue of 395 nearby groups and claimed that $\sim$25% of those are FGs, Harrison et al. [14] presented a sample of 17 FG candidates, and Gozaliasl et al. [25] presented a catalogue of 129 groups, of which $22 \pm 6\%$ are fossils. While the number of candidates is rising, it become more and more complicated to have dedicated multi-object spectroscopic observations to constraint membership. The main drawback of this strategy is that the purity of the sample is difficult to control and non-fossil systems can dilute the statistical relevance of the results. It is thus complicated to estimate the number of "*genuine*" FGs, namely those that strictly accomplish the operational definitions, known up to date. However, in the last section of this review, we will try to define a sample of such genuine systems.

3. Theoretical Framework

Numerical simulations of the number of satellites in galaxy groups predict the presence of a huge number of dwarf galaxies, e.g., [26]. However, these predictions are not confirmed in observations. In fact, if the number of bright galaxies is in agreement with these predictions, the number of observed dwarfs is up to two orders of magnitude lower than expected, e.g., [27]. This is the so-called *missing satellite problem*. D'Onghia and Lake [28] pointed out that FGs could scale up the missing satellite problem at the mass scales of more massive galaxies. In particular, they compared the number of satellites predicted by ΛCDM at different halo mass scales and concluded that FGs shown smaller number of galaxies such as the Milky Way and the Large Magallanic Cloud than those predicted by the structure formation theory. They claimed that the reason of this lack of bright galaxies was due to over merging processes occurred in FGs. These early findings made FG extreme objects in the structure formation of the Universe. The over merging processes in FGs could be explained in terms of differences in the orbital structure between fossil and non-fossil systems. Sommer-Larsen [29] used cosmological TreeSPH simulations to study the orbital structure of the intra-group (IG) stars for a set of clusters with masses $\sim 10^{14}$ $M_{\odot}$. He concluded that the velocity distribution of the IG stars was significantly more radially anisotropic for fossil than for non-fossil systems. This pointed out that the initial velocity distribution of the group galaxies could play an important role in defining the fossil status of the system.

Several works have focused on the theoretical study of the mass assembly history of fossil and non-fossil systems. One of the first papers on this topic was done by D'Onghia et al. [30] where they analysed the mass assembly of systems with $M_{vir} \sim 10^{14}$ $M_{\odot}$. They found a correlation between Δm_{12} and the formation time of the group defined as the redshift at which 50% of the total mass of the system at $z = 0$ is already in place (z_{50}). In particular, they found that FGs have assembled more than half of their present mass at $z > 1$, with a subsequent growth by minor mergers alone. This early assembly gave enough time to FGs to merge their M^* galaxies (where M^* is the characteristic magnitude of the luminosity function, see Section 4.2.3 for details) and produce the large magnitude gaps observed at $z = 0$. The mass assembly of the low-mass group regime ($M_{vir} \sim 10^{13} - 10^{13.5} M_{\odot}$) was analysed by using the Millenium Simulation by Dariush et al. [8,10]. In particular, these authors found that the selection of the systems using their magnitude gap alone does not guarantee the selection of early formed systems. They observed that the majority of the objects that have assembled more than 50% of their halo mass at $z = 1$ are not fossil systems today. A similar result was found in Deason et al. [31]: 20% of the groups selected from the Millenium Simulation with large mass gap (similar to fossil systems) turned to be young objects. Raouf et al. [32] proposed that a combination of three observational parameters (magnitude gap, luminosity of the brightest cluster galaxy and its offset from the group luminosity centroid) considerably improve the selected rate of dynamically old systems. In Figure 2, it can be seen that the probability for a system with $\Delta m_{12} > 2.0$ and $\Delta m_{14} > 2.5$ to be old grows when the absolute magnitude of the BCG is smaller. In this case, Raouf et al. [32] defined a group as old if its halo has over 50% of its final mass at $z = 1$ and young if this fraction is less than 30%. Indeed, these plots demonstrate that Δm_{14} works better than Δm_{12}, when combined with the absolute magnitude of the central galaxies, in finding old systems. This result was recently confirmed in Zhoolideh Haghighi et al. [33], since the authors found a clear correlation between Δm_{12}, the offset of the luminosity centroid, and the dynamical age of a group/cluster. However, this criterium is less used in the literature, probably because it is newer than the Jones et al. [9] and Dariush et al. [10] ones.

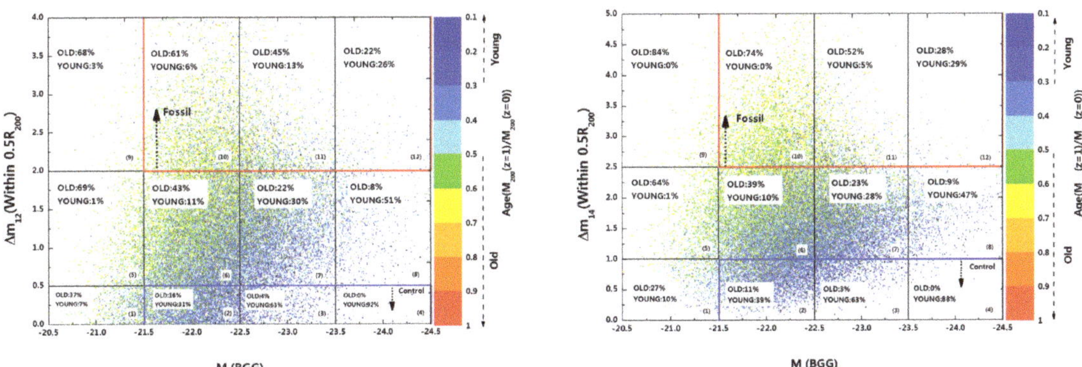

Figure 2. Distribution of the galaxy groups in the plane of luminosity gap Δm_{12} (**left panel**) and Δm_{14} (**right panel**) within 0.5 r_{200} and the $r-$band magnitude of the brightest group galaxy, in the Millennium simulations with Guo et al. (2011) semi-analytic model. Data points are colour-coded according to the ratio of the group halo mass at redshift z$\sim$1 to its mass at $z = 0$. The plane has been sub-divided into blocks within which the probability that the halo is old or young is given. In this diagram, panels (5), (9), and (10) contain mostly old systems while the panels (3), (4), and (8) are mostly occupied by young systems. The image is taken from Raouf et al. [32]; in particular, it corresponds to their Figure 1.

In addition, the Illustris Simulation was used to analyse the properties of the FGs in the mass regime $10^{13} - 10^{13.5} M_{\odot}$ [34]. The authors found that the magnitude gap of FGs identified at $z = 0$ were on average created about 3 Gyr ago. In addition, the fossil central galaxies became more massive than non-fossil ones. This difference was explained as due to differences in the mass acquired through mergers between z = 0.1–1, as can be seen in the left panel of Figure 3. Fossil BCGs also have a larger number of major mergers than non-fossil ones (right panel of Figure 3). Indeed, the last major merger of fossil BCGs was later (e.g., at lower redshift). No differences were found in the distribution of the time formation (z_{50}) of fossil and non-fossil halos. The group mass assembly of fossils and non-fossils differs only in the recent group accretion history, in particular in the formation time of the 80% of the mass of the halo. However, semi-analytical models studied the dynamical evolution of galaxies in groups with different formation epochs [35]. They found that BCGs of dynamically-young groups suffered the last major galaxy merger $\sim$2 Gyr more recently than their counterparts in dynamically old groups and that FGs are somewhere in the middle between the other two populations. However, these authors found a lack of recent major mergers in FGs that is in agreement with the evolution of their old systems.

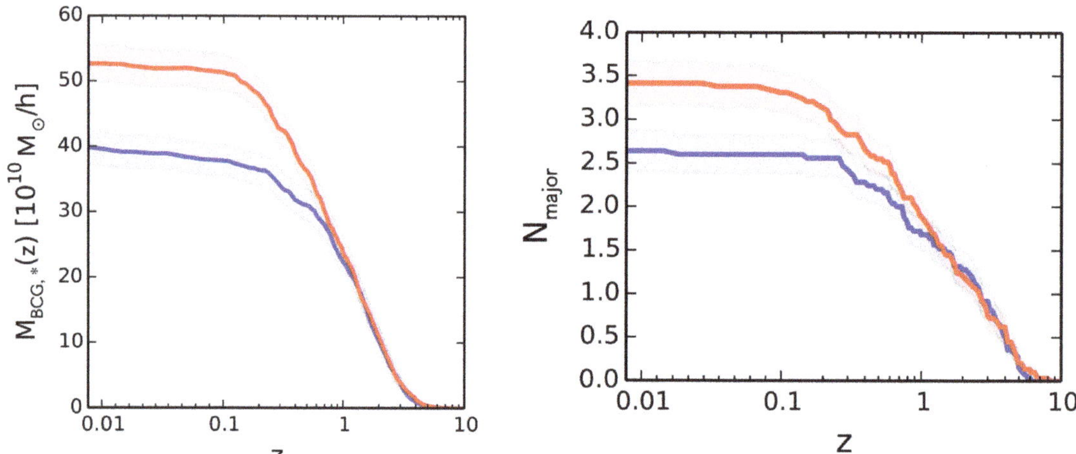

Figure 3. Left panel: average stellar mass assembly history for central galaxies within r_{200}. Shaded areas are 1σ errors calculated from 1000 bootstrap resamplings. **Right panel**: average number of cumulative major mergers occurred at the central galaxy across z. In both panels, fossil and non-fossil systems are represented in red and blue, respectively. The image is taken from Kundert et al. [34], in particular from their Figures 6 and 7.

Raouf et al. [36] also used the Illustris Simulation to study some properties of fossil systems. They found that this simulation overproduce FGs in comparison with observations and semi-analytical predictions. They also obtained that the intra-group medium (IGM) in dynamically evolved groups is hotter, for a given halo mass, than that in still evolving ones.

Several studies found that the fossil phase of a system could be transitional. Galaxy clusters and groups pass through fossil and non-fossil phases along their evolution. von Benda-Beckmann et al. [37] found a population of groups that presented a fossil phase at high redshift which is terminated later by the accretion of new bright galaxies. The transitional phase of the fossil status is also reported by other simulations like Kundert et al. [34]. These fluctuations in the magnitude gap could be related with the large-scale environment in which the systems are located. Indeed, Diaz-Gimenez et al. found that, in the Millenium Simulation, the environment was different for fossil and non-fossil systems with similar masses. They showed an increase in the local density profile of galaxies at $\sim$2.5 r_{vir} from the group centers. This increment was more noticeable in fossil than in non-fossil systems and was linked with the earlier formation time of fossil groups. We will discuss in Section 5 what is found in observations that can be linked to the transitional fossil phase.

The properties of the galaxy populations in fossil and non-fossil systems have also been analysed using cosmological simulations. Romeo et al. [38] reported from cosmological hydrodynamical simulations and semi-analytical models that fossil and non-fossil systems show different star forming rates at low z, being indistinguishable at $z > 0.5$. In contrast, Kundert et al. [34] found no differences in the stellar age, metallicity, and star formation rates of BCGs in fossil and non-fossil systems from the Illustris Simulation. Raouf et al. [36] analysed the properties of the black holes developed in the center of the BCGs for fossil and non-fossil systems. They found that the mass of the black holes hosted in BCGs is larger in dynamically evolved groups with a lower rate of mass accretion a result confirmed also in Khosroshahi et al. [39].

Kanagusuku et al. [40] found that fossil and non-fossil systems selected from the Millennium Simulation showed different galaxy populations. In particular, at early times, FGs comprised two large brightest galaxies surrounded by faint ones. At the faint end of the luminosity function, fossil systems turned to be denser at early times that non-fossil ones. This trend reverses at a later time and became similar before $z = 0$. This was caused by an increase at a constant rate of the number of faint objects in non-fossil systems. In contrast, the number of faint galaxies reached a plateau at z$\sim$0.6 in FGs, and then grows faster towards $z = 0$. The evolution of the galaxy luminosity function as a function of redshift for fossil and non-fossil systems was studied by Gozaliasl et al. [25]. They build up luminosity functions of galaxy aggregations based on the Millennium Simulation. They found that the bright end of the galaxy luminosity function strongly evolved for fossil systems from $z = 0.5$ to $z = 0$, with changes in $M^*\sim$1.2 mag. This suggests that the mergers of the M^* galaxies in fossil systems have a significant impact in the formation of the bright cluster galaxies. In contrast, the faint-end slope of the luminosity function shows no considerable redshift evolution in fossil systems, unlike in non-fossil ones where it grows by 25–42% towards low redshifts.

4. Observational Properties

In this section, we review the observational properties of FGs. In Section 4.1, we describe the properties of the intra-cluster medium; in particular, we present global scaling relations, mass and entropy profiles, cool cores, halo concentrations, and metallicities. In Section 4.2, we analyse the galaxy population and, in particular, FGs' luminosity functions, galaxy substructures, stellar populations, central galaxies, and large scale structures.

4.1. *Properties of the Halos*

The intra-cluster medium (ICM) is the largest baryonic component in galaxy clusters, responsible for $\sim$10% of the total mass of the cluster, e.g., in [41]. In fact, galaxy formation is inefficient and only $\sim$10% of the gas is converted in stars and galaxies [42], leaving the vast majority adrift in the intra-cluster space. This gas is trapped in the deep potential well of the cluster and heated to X-ray-emitting temperatures through shocks and adiabatic compressions [43].

Gravitational collapse predicts tight scaling relations between ICM and cluster mass, according to the so-called self-similar model [44,45]. Moreover, cosmological simulations predict that the scaled thermodynamical profiles of galaxy clusters are nearly universal, e.g., in [46]. For these reasons, the ICM is a powerful tool to study the formation and evolution of galaxy clusters: deviation from the gravitational collapse predictions can be used to investigate non-gravitational physics, such as cooling and feedback from supernovae and active galactic nuclei (AGN).

4.1.1. Global Scaling Relations

In the left panel of Figure 4, we show the correlation between X-ray temperatures and luminosities for various samples of FGs and non-FGs. It can be seen that both types of objects are found in the same, tight, correlation. We won't enter into a detailed discussion of purely X-ray relations and their meaning, we refer the reader to the companion review by Lovisari et al. for a description of X-ray scaling relations in galaxy groups. The result presented in the left panel of Figure 4 can be extended to all those relations that only involve X-ray data: FGs and non-FGs are usually found in the same correlations. This seems to indicate that FGs and non-FGs are formed in a similar way.

However, discussion arose in those studies focused on the comparison between X-ray and optical properties of FGs and non-FGs. In fact, Jones et al. [9] and Khosroshahi et al. [47] found FGs to be over luminous in X-rays using samples of five and seven FGs, respectively. They claimed FGs to be a factor 5–10 brighter than regular groups or clusters in the X-ray.

On the other hand, more recent studied found no differences in the statistical properties of FGs and non-FGs samples [14,48,49]. In particular, these authors claimed that it is crucial to use homogeneous data and procedures to analyse both the FG and the control samples. According to their statements, this could be the source of the excess of X-ray luminosity (or the lack of optical luminosity) found in the previous studies.

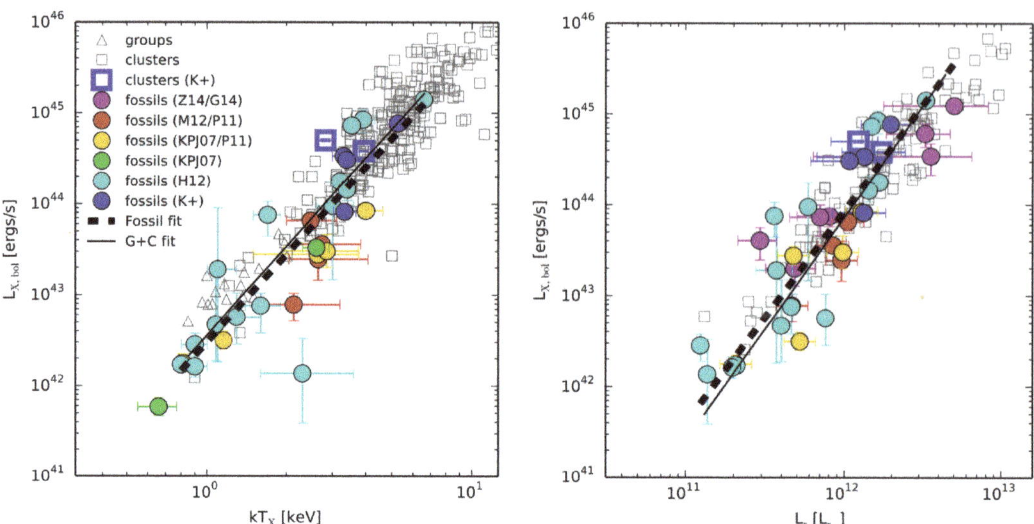

Figure 4. Left panel: The X-ray temperature versus the bolometric X-ray luminosity as presented in Kundert et al. [50]. Grey triangles and squares are groups and clusters, respectively. Systems labeled with K+ are taken from Kundert et al. [50], Z14 from Zarattini et al. [22], G14 from Girardi et al. [49], M12 from Miller et al. [51], P11 from Proctor et al. [52], KPJ from Khosroshahi et al. [47], and H12 from Harrison et al. [14]. The plotted lines are the orthogonal BCES fits to the fossil sample (dashed line) and to the sample of groups and clusters (solid line) computed in the same range of parameters. **Right panel**: the total $r-$band luminosity versus the bolometric X-ray luminosity as presented in Kundert et al. [50], with the same lines of colour code as in the left panel, see the Figure 5 of [50] for details.

Nevertheless, Khosroshahi et al. [53] discussed a sample of groups, one of which defined as fossil that lies above the $L_X - L_{opt}$ relation of non-fossil systems, reopening the debate on fossil system scaling relations.

The (possible) final point of this debate was put by Kundert et al. [50]: as well as demonstrating that no differences are found between fossils and non-fossils in their sample of 10 groups and clusters observed with specific Suzaku follow up, they recomputed the luminosities of the FGs from Khosroshahi et al. [47] and Proctor et al. [52] in a homogeneous way. In the right panel of Figure 4, we show the relation presented in Kundert et al. [50]. The authors concluded that the discrepancies in the literature can be reconciled if X-ray and optical luminosities were computed using the same bands and radii, pointing out that many differences could be due not only to the low statistics but also the lack of homogeneous datasets.

In parallel with the discussion on the $L_X - L_{opt}$ relation, a debate on the mass-to-light (M/L) ratio of FGs arose. In fact, if there is a chance that FGs are under luminous in the optical bands, they should have larger M/L ratios than non-FGs [47]. Again, the studies available in the literature are somewhat contradictory. Sun et al. [11], Khosroshahi et al. [12,16] found normal values for their samples of FGs, compatible with non-fossil systems, although if marginally darker for a fixed mass. On the other hand, Vikhlinin et al. [5] found a high M/L for their sample of overluminous elliptical galaxies (OLEGs, about three times larger). The same result was found by Proctor et al. [52] in analysing a sample of 10 FG candidates. The authors suggested that FGs are simply dark clusters: they are characterised by a mass and a central galaxy that are typical of galaxy clusters, but embedded in a poor environment, so that the richness and the total optical luminosity are below the non-fossil ones. Finally, Yoshioka et al. [15] also found high M/L ratios for their sample of "isolated X-ray overluminous elliptical galaxies." However, we suggest that inconsistencies in the measured M/L ratios could be due to differences in the methodology and quality of the data. For example, the optical luminosities of Vikhlinin et al. [5] and Khosroshahi et al. [12] are computed in the R band and within r_{200} for every object in the sample, those of Proctor et al. [52] in a variable radius of 500–1000 kpc, whereas Sun et al. [11] estimated a normal M/L ratio from the gas fraction profile out to 450 kpc, and Yoshioka et al. [15] uses the B-band luminosity of the BCG as the total luminosity of the system. Thus, it seems reasonable that the disagreement between different results could be healed only with a homogeneous study of a large sample of FGs. As we already mentioned for other topics, such a study is far from being performed.

4.1.2. Mass and Entropy Profiles

The study of mass profiles in clusters is an important tool to confirm the ΛCDM paradigm. In fact, this model predicts a universal mass profile that does not depend on the mass of the cluster, and it is usually assumed to have the shape of a Navarro–Frenk–White profile NFW [54].

Mass profiles were studied mainly for individual FGs, or small samples, making difficult to extrapolate general conclusions. Possibly the first mass profiles of FGs to be computed were those of NGC 6482 [16], RX J1416.4 + 2315 [12], and ESO 3060170 [11]. The first two systems shows a mass profile well described by an NFW, with a high central concentration that was interpreted as a sign of early formation. On the other hand, the mass profile of ESO 3060170 showed a flattening in the external regions not compatible with numerical simulations and also confirmed in Su et al. [55] out to the virial radius. Yoshioka et al. [15] studied the mass profiles of four FG candidates, finding no differences with normal groups/clusters. In addition, Gastaldello et al. [56] studied the mass profile of ESO 3,060,170, within a sample of 16 relaxed groups and clusters: they found a good agreement with a NFW profile, but it must be noticed that their data reached R∼200 kpc, whereas Sun et al. [11] and Su et al. [55] data reached R∼500 kpc and R∼1000 kpc, respectively. The three mass profiles are comparable within R∼200 kpc and differences with the standard NFW profiles rose in the most external regions. This clarifies again the complexity of the comparison when individual FGs are studied using data from different sources, within different radii, and treated with different techniques. Another two mass profiles for the FGs RXC J0216.7-4749 and RXC J2315.7-0222 were studied in Démoclès et al. [57]. Only the latter has a good profile, well fitted, out to R_{500}, by a NFW profile plus a central stellar component.

Entropy is also of great interest because it controls ICM global properties and records the thermal history of a cluster, since it is conserved in adiabatic processes. Entropy is therefore a useful quantity for studying the effects of feedback on the cluster environment and investigating any breakdown of cluster self-similarity. Most of the studies cited for the discussion of the mass profiles were also able to compute an entropy profile. Again, since only individual systems were studied, the results are somewhat controversial and it is not trivial to generalise the conclusions to the entire FG category. In particular, Démoclès et al. [57] found that the entropy profiles out to R_{500} of their two FGs show a considerable excess above the expectations from non-radiative simulations, especially for RXC J0216.7-4749. This is expected if significant non-gravitational processes affect the ICM. Su et al. [55] found and entropy profile that is in agreement with simulations out to $\sim 0.9\,R_{200}$ and then flattens in the outskirts, due to gas clumpiness and outward redistribution. Humphrey et al. [58] studied RXJ 1159+5531 combining Suzaku, Chandra, and XMM observations to find no evidence of the flattening in the entropy profile outside $\sim R_{500}$. A similar results was also found in Su et al. [59] for the same cluster: its entropy profile is consistent with predictions from gravity-only simulations.

The urge for a systematic study of a large sample of FGs in the X-rays appears as necessary to constrain the average mass and entropy profiles of these objects. However, it seems difficult to realise, due to the small number of nearby FGs that can be deeply observed with current X-ray facilities. An improvement on this side is expected with the next all-sky survey that will be taken by eROSITA and Athena missions. The former is expected to find $\sim 10^5$ X-ray clusters and groups, and the latter will take advantage of its high-resolution for better constraint radial profiles of, for example, temperature, density, and mass.

4.1.3. Cool Cores

Early observations of the gas in galaxy clusters found that it was so dense in the central regions that its cooling time was much shorter than the Hubble time, e.g., in [60]. The majority of the clusters studied in the literature show these cool cores, CC, e.g., in [61,62]. CCs are usually associated with relaxed clusters, since mergers easily erase them. For this reason, it appeared as natural to look for CCs in FGs, in order to confirm their *old and dynamically relaxed* status.

The first study on CCs for a sample of FG candidates was done by Vikhlinin et al. [5]. In their work, the authors studied four isolated elliptical galaxies selected from ROSAT X-ray data and confirmed as fossil systems with a dedicated optical follow up. It is worth noting that the authors suggested that these objects could be FGs, but, at the time, no operational definition was available, so no Δm_{12} is computed in their paper. However, at least three out of four were later confirmed as FGs in other publications, confirming the accuracy of their approach. Vikhlinin et al. [5] results indicated the presence of CCs in the central regions of these objects.

Later works were focused on deeper studies of individual FGs and found controversial results: Khosroshahi et al. [12,16] found no central drop in NGC 6482 and RX J1416.4 + 2315. On the other hand, Sun et al. [11] found a CC in ESO 3060170, as well as Démoclès et al. [57] for RXC J0216.7-4749 and RXC J2315.7-0222 and Su et al. [59] for RXJ1159 + 5531. It is interesting to note that Miraghaei et al. [63] studied three of the cited clusters (NGC 6482, RX J1416.4 + 2315, and ESO 3060170) using radio observations, finding signs of recent AGN activity only in the first two. However, the AGN power computed was not sufficient to remove CCs from these clusters.

The need for larger samples was partly satisfied only recently, when Bharadwaj et al. [64] studied a sample of 17 FGs for which Chandra archival data were available. They defined three different diagnostics to evaluate the presence of CCs, and they found that $\sim$80% of FGs showed clear hints of the presence of CCs (e.g., at least two diagnostics compatible with the CC).

It seems, thus, reasonable to claim that FGs are mostly cool-cored. However, the fraction of FGs with a CC is similar to that of non-FGs. For example, Hudson et al. [62] studied a large sample of 64 galaxy clusters for which high-quality X-ray data from Chandra were available, finding that $\sim$70% of their clusters host a CC. It thus seems that this is not a peculiar behaviour of FGs.

4.1.4. Halo Concentration

One of the main parameters of the NFW model is the concentration, usually computed as $c_\Delta = r_\Delta / r_s$, where r_Δ represents the radius of a sphere of mean interior density ρ_Δ and r_s is the scale radius of the NFW profile. Typical values of Δ are 200 (often assumed to be equivalent to the virial radius) or 500.

Navarro et al. [54] pointed out that the concentration parameter reflects the density of the Universe when the halo formed. In particular, older halos formed in higher-density environments and tend to have larger concentrations. Several theoretical studies found that FGs assembled half of their mass at earlier epochs than non-fossil ones, see, e.g., in [30,37]. In this framework, it is expected that FGs would be located in high concentrated halos. Different numerical models are used in the literature to compute halo concentration in clusters, e.g., in [65,66]. Thus, observational results can be easily compared with theoretical predictions to test the formation scenario of FGs.

From an observational point of view, various methods can be used to compute the c parameter. A first approach to derive the central concentration of DM halos is to measure the gas mass profile in the X-rays out to r_{200} or r_{500}, fit with an NFW profile and then use the derived r_s to compute the concentration. Using this approach, Khosroshahi et al. [16] Khosroshahi et al. [12] Khosroshahi et al. [47], and Buote [67] found concentration values higher than expected for their samples of individual FGs Other authors, like Démoclès et al. [57] and Pratt et al. [68], found normal concentrations in a total of six FGs. Again, comparing results with such small statistics and taking into account non-homogeneities in the analytic procedures makes it difficult to reach a final conclusion that can be applied to the mean FG population.

However, other approaches can be used to measure halo concentration. In particular, Vitorelli et al. [69] stacked $\sim$1000 systems from the CS82 survey in different magnitude-gap bins, the larger of which has mean $\Delta m_{12}\sim1.7$. They cross-correlate weak lensing measurements with NFW parametric mass profiles to measure masses and concentrations of their sample. They found that halos in the $\Delta m_{12}\sim1.7$ bin have a higher probability to be more concentrated and, thus, probably formed earlier.

Finally, the halo concentration can be estimated using the velocity of member galaxies as tracers of the underlying mass distribution. This was done in Zarattini et al. [70], where the authors analysed a sample of $\sim$100 clusters and groups, dividing them into different magnitude-gap bins. For each bin, they stacked all the available galaxies to increase the statistic. They found $c_{200} = 2.5 \pm 0.4$ for the bin with the largest magnitude gap (defined as $\Delta m_{12} > 1.5$). These values are in agreement, within the uncertainties, with their results in the other three magnitude-gap bins, as well as with other similar work in the literature of non-fossil clusters, e.g., in [71,72] and references therein. However, the large uncertainties typical of this observational technique prevented the authors from reaching a strong conclusion.

4.1.5. Metallicity

The hot intra-group gas contains elements that are typically synthesized in stars and SNe. For general details on this topic, we refer the reader to the companion review of Gastaldello et al. Here, we focus on the single study conducted on metal abundances in FGs, presented in Sato et al. [73]. The authors get Suzaku data out to 0.5 r_{180} for NCG 1550 and were able to confirm that the abundance ratios O/Fe, Mg/Fe, Si/Fe, and S/Fe are similar to those of other poor groups observed with the same satellite. Moreover, the number ratio of type-I and type-II SNe computed in Sato et al. [73] is also similar to that obtained for non-fossil groups. As a consequence, their work can be included in those that are not finding differences between FGs and non-FGs.

4.2. Galaxy Population

The study of the galaxy population in FGs is mainly done in the optical range. In this section, we will firstly discuss the observational properties of the central galaxies. We will then move to the luminosity functions, galaxy substructures, and the large-scale structure around FGs.

4.2.1. Central Galaxies: Formation Scenarios

Central galaxies in clusters are a unique class of objects. They are usually the largest and most luminous member galaxies, and they lie very close to the peak of the cluster X-ray emission, e.g., in [71,74]. Moreover, in the velocity space, they sit near the rest frame velocity of the cluster [75–77]. These characteristics imply that the BCGs are located at the minimum in the cluster potential well. Zarattini et al. [70] found that there is a dependence of the velocity segregation on the magnitude gap. This result means that BCGs in FGs are located closer to the minimum of the cluster's potential well, when compared to BCGs in non-fossil systems. This difference is not found for satellite galaxies, independent from their mass.

The formation of central galaxies in FGs is thought to be the end result of the group evolution [4,9]. The main actor, in this scenario, would be dynamical friction. However, Sommer-Larsen [29] suggested that the main difference between FGs and non-FGs has to be found in the initial velocity distribution. In particular, he found that satellite galaxies in FGs should be located on more radial orbits than in non-FGs. This would favour low angular momentum mergers, for which dynamical friction could be more effective [78].

Méndez-Abreu et al. [79] analysed the photometric properties of central galaxies in FGs. They studied the position of the central galaxies in FGs in the fundamental plane and its projections to explain the formation of these objects (see Figure 5). Central galaxies in FGs results in having large K-band luminosities. The K_s luminosity is a good proxy of the total stellar mass since the typical M/L $\sim$1 for an old stellar population [80]. Thus, it seems clear that central galaxies in FGs are amongst the most massive galaxies known. The left panel of Figure 5 shows the correlations between the K_s−band luminosities and the central velocity dispersion, Faber–Jakson relation in [81]. Similarly, the right panel of Figure 5 shows another projection of the fundamental plane, the K_s−band luminosity vs. effective radius (r_e). In both relations, BCGs in FGs are found slightly outside the correlations: this bend was found for the first time in Bernardi et al. [82] for early-type galaxies, and it can be interpreted in terms of the formation scenario of the BCGs. In particular, if major dissipationless mergers between galaxies are the main mechanisms to build up the mass of the BCGs, the final size is expected to increase, but not its central velocity dispersion; however, if minor dry mergers are the predominant mechanism, they are expected to change both the size and velocity dispersion of the BCGs. Thus, the results presented in Méndez-Abreu et al. [79] seem to favour the first scenario for these massive galaxies. This is also supported by new results obtained from numerical simulations. Kundert et al. [34] investigated the origin of FGs in the Illustris simulation. With respect to the stellar mass assembly of the BCGs, they found (see their Figure 6) that the accretion is similar at high-redshifts for fossil and non-fossil systems, whereas a clear difference starts to appear at $z\sim0.3$. From this point, BCGs systematically accreted more mass in FGs than in non-FGs.

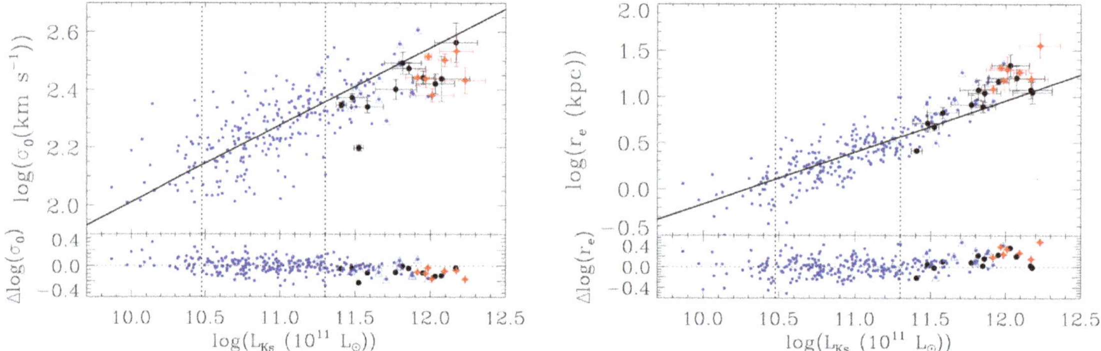

Figure 5. Left panel: distribution of the BCGs of Méndez-Abreu et al. [79] red stars and large black points and the early-type galaxies of Pahre et al. [83] small blue points in the log σ_0 vs log L_{ks} plane. The BCGs in the Pahre et al. [83] sample are marked by blue open triangles. **Right panel**: same as the left panel, but in the log r_e vs. log L_{Ks} plane. The solid line represents the best fit to the galaxies in the luminosity range $3 \times 10^{10} < L_{Ks}/L_{\odot} < 2 \times 10^{11}$. The bottom panels represent the residuals from the best fit. The original image can be found in Méndez-Abreu et al. [79] and corresponds to their Figures 9 and 10.

However, this formation scenario is in contradiction with the one proposed by Khosroshahi et al. [12]: in fact, these authors found disky isophotes in their central regions of BCGs, in contrast with most BCGs in non-fossil systems [84]. This result seems to favour a scenario in which mergers in FGs were rich in gas, thus including large M* spirals. A similar result was found also by Eigenthaler and Zeilinger [85], since they found many shells around central galaxies in FGs. These shells were likely formed recently via major mergers of spiral galaxies [86].

Signs of recent mergers were indeed found by Alamo-Martínez et al. [87] while analysing the surface brightness profiles of three FGs. However, this work was mainly focused on the study of globular clusters in FGs. These objects are powerful tools to study galaxy assembly, since they are old and dense enough to survive galactic interactions. Alamo-Martínez et al. [87] results seem to point out that globular clusters in BCGs formed in a similar way in fossil and non-fossil systems. In terms of the formation scenario of the BCGs, this can be take as a confirmation that similar processes are at work in the formation of BCGs in fossil and non-fossil systems, although more statistics is needed to generalise this conclusion. A similar result, in terms of the formation scenario, is found also in Madrid [88] and Madrid and Donzelli [89], in which the authors compared the properties of ultra-compact dwarf galaxies (UCDs) in FGs and in the Coma cluster. UCDs are considered the bright-and-massive tail of the globular cluster distribution. Their results showed that UCDs are likely to be a common occurrence in all environments.

The presence of AGNs can also be used to study the formation scenarios of BCGs. In fact, AGNs need major mergers to form, since they use the gas provided by the merger as their fueling mechanism. On the other hand, if no other merger occurred, they are destined to end this fuel and inactivate. Hess et al. [90] studied the sample of 34 FG candidates from Santos et al. [6] using radio observations in order to detect the presence of AGNs in FGs. They found that 67% of these FG candidates contain a radio-loud AGN. This result seems in contrast with the old formation expected for FGs: in fact, AGN should have run out of fuel since FGs's last major merger. For this reason, Hess et al. [90] suggested that other mechanisms, such as minor mergers, cooling flows, or late time accretion should be invoked to keep the AGNs alive in FGs. However, it is worth noting that the Santos et al. [6] sample was not pure and about half of the sample was formed by non-fossil systems [22].

4.2.2. Central Galaxies: Stellar Populations Properties

We already mentioned that the formation scenario proposed by Ponman et al. [4] supposed that FGs formed at high redshift, with few interactions with the large-scale structure along their lives. This would leave enough time for the M^* galaxies to merge with the BCG, thus forming the Δm_{12} gap. On the other hand, Mulchaey and Zabludoff [91] suggested the so-called *failed group* scenario, in which the BCG is formed as a local over density and no other bright galaxy formed within the group.

However, these two scenarios should also leave clear imprints in the stellar populations of the BCGs. In fact, in the *failed group* scenario, the central galaxy formed via monolithic collapse that is expected to create large radial metallicity gradients in the distribution of the stars. On the other hand, in the *merging scenario,* the BCG suffered various major mergers that have the power to erase those gradients, since they mix up the stars and gas during the merging process, e.g., in [92].

La Barbera et al. [7] were the first in using stellar populations to investigate difference between FG's BCGs and regular elliptical galaxies. They used spectra from the SDSS DR4 for their sample of 25 BCGs in FGs and 17 field elliptical galaxies that act as the control sample. They searched for the single-stellar-population model that best fits the spectra, thus computing mean ages, metallicities, and $\alpha-$enhancement for the two populations of BCGs. The authors showed that no significant difference is found in these parameters and concludes that BCGs in FGs did not form earlier than the other galaxies.

Harrison et al. [14] select a sample of 17 FG candidates by combining XMM observations with SDSS DR7 data. They analyse the stellar populations of the BCGs of these systems using SDSS spectra and the Starlight code [93]. Their results showed no significant differences in stellar star formation rates, age, and metallicities between FGs and their two control samples, one built by optically-selected BCGs, the other with X-ray-selected BCGs.

Eigenthaler and Zeilinger [94] studied the presence of such gradients in age and metallicity for a sample of six BGCs in FGs using longslit spectroscopy from the 4.2 m

William Herschel Telescope (WHT). They found that the metallicity gradient is flatter with respect to the predictions of the monolithic collapse (~ -0.2 instead of -0.5), thus indicating the presence of mergers during the life of the BCGs. On the other hand, the age gradient is, on average, negligible.

A similar study was performed by Proctor et al. [95] for a sample of two central galaxies in FGs, using longslit spectroscopy obtained with the 8 m Gemini North telescope. The authors found different results for the two BCGs: SDSS J073422.21 + 265133.9 showed a strong metallicity gradient and a slightly positive age gradient, suggesting a relatively recent episode of stellar formation in the centre. NGC 2484, on the other hand, showed an old stellar population ($\sim$10 Gyr) and a flat central metallicity that was interpreted as the evidence of an inside-out stellar formation, at least in the final episode of stellar formation.

Trevisan and Mamon [96] studied a large sample of 550 groups to characterise the dependence of the stellar populations of the BCGs and the second brightest galaxies with the magnitude gap. They did not find differences in the distribution of colours, star-formation rates, $\alpha-$enhancement, age, metallicities, and star-formation histories in systems with different Δm_{12}.

Corsini et al. [97] studied both the stellar populations and radial gradients for a sample of two BCGs in FGs using the 10.4 m Gran Telescopio Canarias (GTC) telescope. They confirmed the results of La Barbera et al. [7] and Eigenthaler and Zeilinger [94] out to the effective radius. They found an underlying and diffuse older stellar population, with a younger one located near the centre of the galaxies. This was interpreted as the sign of the last major merger with gas, which occurred $\sim$5 Gyr ago. Corsini et al. [97] also found a radial metallicity gradient in agreement with the Eigenthaler and Zeilinger [94] one.

Finally, Raouf et al. [98] divided a sample of groups from the Galaxy and the Mass Assembly (GAMA) survey into relaxed and unrelaxed, using Δm_{12} and the luminosity offset as the relaxation indicators. They found that BCGs in unrelaxed systems are bluer, more star forming, and with non-elliptical morphologies than those in relaxed systems. They conclude that the higher rate of recent mergers expected in unrelaxed groups could be responsible for these differences. A similar result was also found in Pierini et al. [99]. These authors claimed that there are few star-forming galaxies in FGs, making them more mature then coeval and similar mass groups.

Very recently, Raouf et al. [100] studied the kinematic of gas and stars in 154 central galaxies taken the Sydney-AAO Multi-object Integral field (SAMI) galaxy survey. In particular, they divided this sample into low and high luminosity gap system, with the latter that can be assimilated as FGs. They found that there is a weak statistical difference (at approx. $1-\sigma$ level) between the magnitude gap and the gas-star kinematics misalignment. In addition, a similar difference was observed between the magnitude gap and the regularity of the stellar rotation of the BCGs. In particular, systems with high magnitude gaps are found to be more regular rotators and with a smaller fraction of gas-star misaligned kinematics.

These studies did not find relevant differences in the stellar populations of central galaxies in fossil and non-fossil systems. The imprint of monolithic collapse is not found, all the observations point towards the creation of FGs via the *merging scenario*, in which the gap is created via major mergers of M* galaxies. The possibility of the existence of a fossil phase in the life of a cluster is supported also by the presence of younger stellar populations in the centre of galaxy, probably due to recent major mergers with gas. In the top panel of Figure 6, we show the relation between central velocity dispersion and central metallicity for a sample of ten FGs taken from Eigenthaler and Zeilinger [94], Proctor et al. [95], Corsini et al. [97] and compared with the sample of normal and dwarf early-type galaxies of Koleva et al. [101]. In the lower panel of the same figure, the comparison is done for the central ages of the same samples. It can be seen that in both cases FGs are found in the same correlations as normal early-type galaxies. This plot again confirms that central galaxies in FGs are amongst the most massive known in the Universe.

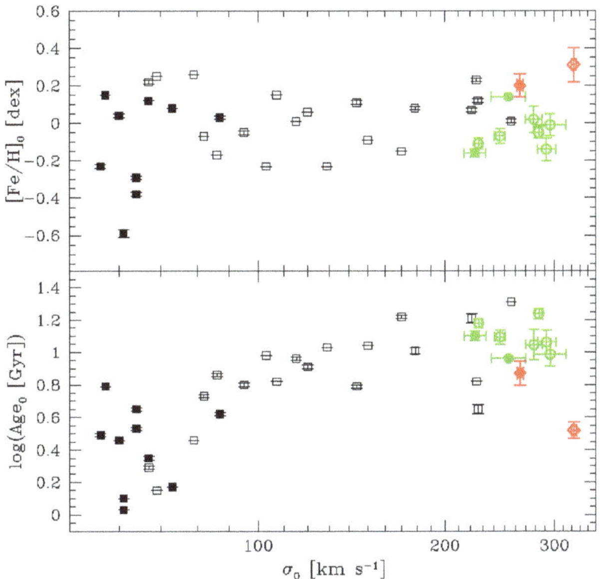

Figure 6. Central metallicity (**top panel**) and central age (**bottom panel**) as a function of the central velocity dispersion. Red diamonds are taken from Corsini et al. [97], green open circles from Eigenthaler and Zeilinger [94], green filled circles from Proctor et al. [95], open and filled squares are the early-type normal and dwarf galaxies with $\sigma > 50$ km s^{-1} from Koleva et al. [101]. The original image can be found in Corsini et al. [97] (see their Figure 8).

4.2.3. Luminosity Functions

The luminosity function (hereafter LF) is one of the most powerful tools to study the galaxy population of a group/cluster of galaxies. It is given by the number density of galaxies per luminosity interval and it is usually described parametrically with the Schechter function [102]. The main parameters are the characteristic magnitude (M*) and the faint-end slope (α). The former describes the bright part of the LF, whereas the latter is related to the dwarf galaxy population. A debate is ongoing on the universality of the LF: in fact, photometric studies found that LFs in clusters are steeper than in the field, $-2.0 < \alpha < -1.8$ in clusters and $-1.5 < \alpha < -1.3$ in the field, see [103,104]. On the other hand, spectroscopic studies found no differences in the α parameters of the field and clusters, finding a general value of $\alpha \sim -1.3$ [105] and references therein.

The study of LFs in FGs was mainly focused on individual FGs, due to their paucity. As a consequence, most of the first results were contradictory. Mendes de Oliveira et al. [18] found, for the FG called RX J1552.2+2013, $M^* = -21.18 \pm 0.57$ and $\alpha = -0.77 \pm 0.37$ using spectroscopically-confirmed members, or $M^* = -21.27 \pm 0.62$ and $\alpha = -0.64 \pm 0.30$ for photometrically-selected galaxies in the $r-$band. In the same year, Khosroshahi et al. [12] computed $M^* = -20.40 \pm 0.22$ and $\alpha = -1.23 \pm 0.28$ for RX J1416.4+2315, a fossil system with a mass similar to RX J1552.2 + 2013. In addition, Trentham et al. [106] presented an LF for a single FG (NGC 1407), finding $\alpha = -1.35$. The difference, especially in the faint-end slope, is important. In fact, $\alpha = -1$ indicates a flat LF, in which the number of dwarf galaxies is not changing with magnitude. On the other hand, in a steeper function like that of Khosroshahi et al. [12], the number of dwarf galaxies is rapidly growing and, in a flatter one like that of Mendes de Oliveira et al. [18], it is decreasing.

Zibetti et al. [107] also studied the photometric LF of a sample of five FGs. They found a faint-end slope in agreement with the one of regular clusters presented in Popesso et al. [103]. However, it is worth noting that Popesso et al. [103] found an upturn at fainter magnitudes, so that their LF can be fitted with a double Schechter function. This behaviour is not found on the already-cited LF of FGs because none of these are deep enough. However, Lieder et al. [108] computed a very deep LF for NGC 6482 using spectroscopic data from the Subaru/Suprime-Cam, finding $\alpha = -1.32 \pm 0.05$. They did not fit a double Schechter function; however, a change in the faint-end slope is present also in their Figure 12.

Aguerri et al. [13] presented the LF of RX J105453.3 + 552102, a massive FG at $z = 0.5$. They found $M^* = -20.86 \pm 0.26$ and $\alpha = -0.54 \pm 0.18$, thus confirming a flatter trend for the dwarf galaxy population. Adami et al. [109] also computed the LF for two FG, finding that their faint-end slope is relatively flat, but without giving numbers. Finally, Aguerri et al. [110] studied the spectroscopic LF of RXJ075243.6 + 455653, finding $\alpha = -1.08 \pm 0.33$.

The first systematic study of the dependence of the LF on the magnitude gap was presented in Zarattini et al. [111]. The authors selected ~100 clusters and groups spanning a wide Δm_{12} range and dividing their analysis in four bins of Δm_{12}. Their study was based on a hybrid method for computing the LF, in which the bright part was treated as a quasi-spectroscopic LF, whereas, in the faint end, photometric data were dominant. The authors computed a classical LF and one in which the magnitudes of each systems are referred to the magnitude of the central galaxy (e.g., $M_r - M_r, BCG$, called *relative* LFs). The latter permits to compare directly the differences due to the magnitude gap, and the authors found that this technique offers the best results for highlighting the differences between their four subsamples. These *relative* LFs are shown in the left panel of Figure 7. Zarattini et al. [111] found that both M* and α changes with the magnitude gap. In particular, systems with $\Delta m_{12} < 0.5$ have the brightest M* and the steepest α slope, whereas systems with $\Delta m_{12} > 1.5$ have the faintest M* and the flattest α. The differences are larger than 3σ between the two most-extreme cases and the trend with Δm_{12} is clearly visible in the right panel, where the values of the *relative* M* and α are shown for the four subsamples.

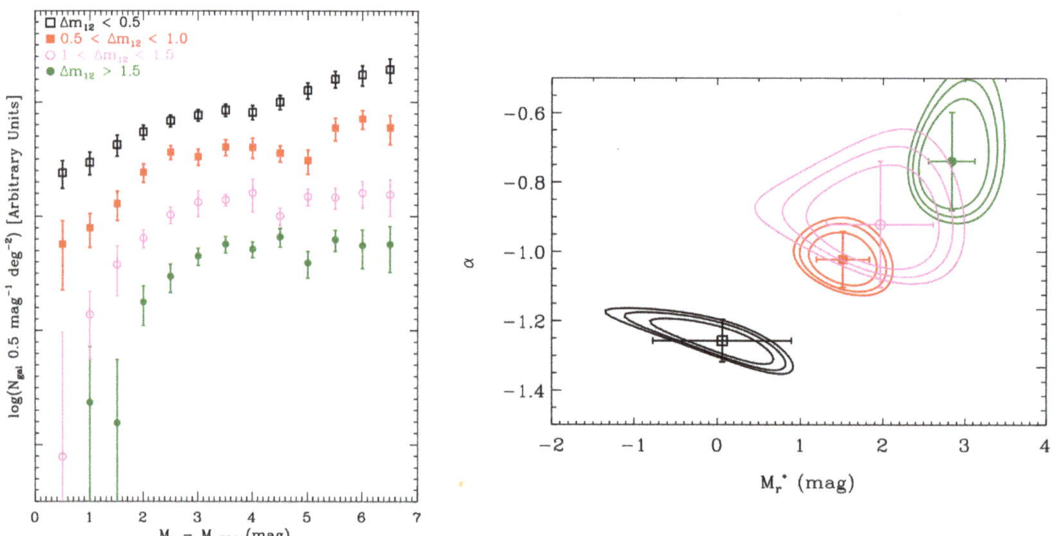

Figure 7. Left panel: *relative* LFs for the four subsamples presented in Zarattini et al. [111]. Empty black squares are systems with $\Delta m_{12} < 0.5$, filled red circles are systems with $0.5 < \Delta m_{12} < 1.0$, empty violet circles are systems with $1.0 < \Delta m_{12} < 1.5$, and filled green circles are systems with $\Delta m_{12} > 1.5$. **Right panel**: Uncertainty contours for the Schechter fits of LFs in the left panel. Contours represent 68%, 95%, and 99% c.l. and the colour and symbol codes are the same as in left panel. The original image can be found in Zarattini et al. [111]; in particular, the left panel corresponds to their Figure 7, the right panel to Figure 8.

The discussion is usually focused on the faint end of the LF, since differences in the bright part between fossils and non-fossils can be easily explained with the same mechanisms that are responsible for the creation of the magnitude gap. Moreover, the presence of the gap itself as a selection criteria implies differences in the bright part of the LFs.

On the other hand, the debate on the differences in the faint-end slope is more complex. In fact, dwarf galaxies can not be merged into the BCG in a reasonable time, since the merging time scale is inversely proportional to the mass of the satellite. A possible explanation is that dwarf galaxies are located in radial orbits passing close to the centre of the cluster. This will result in their disruption, accounting for a part of the missing population. This explanation was also invoked as a possible reason for the high merging rate of FGs. It can also justify the formation of the magnitude gap, since Lacey and Cole [78] showed that the merging timescale for satellite on radial orbits is shorter than for tangential ones. Another possibility is that FGs could lack dwarf galaxies for differences in their accretion time. In fact, Aguerri et al. [110] found that the large-scale environment of FGS03 is very rich, so the flat LF found could be explained if the dwarf populations are still trapped in nearby groups, awaiting to be merged with the FG. In this context, FGs would be systems in early stages of their mass assembly. We will discuss in more detail the large-scale structure of FGs in Section 4.2.5.

Differences in the accretion history of dwarf galaxies were indeed found also in Kanagusuku et al. [40] while studying the bright and dwarf galaxy populations in fossil and non-fossil clusters in the Millennium simulation. They found that FGs had a denser dwarf population at an early time ($z > 0.7$); then, the trend reverses ($0.5 < z < 0.3$), and, finally, it becomes similar at $z = 0$.

To finally solve the issue of the observed lack of dwarf galaxies in FGs, deep and extended spectroscopy would be needed. We will discuss in Section 6.2 the expected impact on this topic of the next-generation spectroscopic surveys.

4.2.4. Galaxy Substructures

If the *old and relaxed* model is correct, one should expect to find a smaller amount of galaxy substructures in FGs than in non-FGs. The only study on this topic was done in Zarattini et al. [112]. They analysed the sample of 34 FG candidates of the FOGO project to compute the fraction of FGs with signs of substructures using a variety of methods. In particular, the entire sample was studied with a two-dimensional approach, able to detect substructures in the projected space of the cluster. Moreover, for a subsample of candidates for which an extended spectroscopic follow up was available, they also applied a series of one- and three-dimensional tests (e.g., the Dressler–Schectman test).

Zarattini et al. [112] results depend critically on the adopted tests, but the comparison with a control sample shows that galaxy substructures are present in a similar fraction in fossil and non-fossil systems. This presence of substructures in FGs is hardly compatible with an old formation, followed by a passive evolution, with no major interactions with the surrounding large-scale structure. Indeed, the small number of genuine FGs in the sample prevent reaching a definitive conclusion.

4.2.5. Large Scale Environment

Differences in the large-scale environment and in the way in which it interacts with FGs were invoked as a possible cause of the different evolution of FGs and non-FGs [4,30,113]. Observational results are scarce on this topic, since only few individual FGs were studied so far.

Adami et al. [114] studied the large-scale structure around the FG RXJ1119.7 + 2126 using spectroscoic data. They conclude that this FG is located at the centre of a low galaxy density bubble.

Pierini et al. [99] found controversial results in their analysis of two FGs. One of the two is found in an isolated environment, whereas the second one is located in a dense environment, with 27 other groups or clusters in the surroundings.

In addition, Adami et al. [109] studied the environments of other three FGs using photometric and spectroscopic data. They found that one system (1RXS J235814.4 + 150524) is in a poor environment, though its galaxy density map shows a clear signature of the surrounding cosmic web. The second FG (RX J1119.7 + 2126) is very isolated, whereas the third one (NGC 6034) is embedded is a very rich environment.

Finally, Díaz-Giménez et al. [113] analysed the large-scale structure in FGs and non-FGs from both the theoretical and observational points of view. We already mentioned their theoretical results in Section 3, here we focus on their observational tests. In fact, they used four FGs selected from Voevodkin et al. [48] and coming from the 400d cluster survey [115] and a control sample of non-FGs from the same survey. Their observational results confirmed the peak found in numerical simulations in the local density profile of galaxies around groups as a function of the normalised group-centric distance. This peak, located at about $2.5r/r_{vir}$ at $z = 0$, is more prominent in FGs. However, the difference is clearer in numerical simulations than in observations, probably again due to the small sample of FGs available.

More extended studies on larger samples are thus generally needed to confirm if FGs are characterised by a special large-scale environment.

5. Past and Future of Fossil Systems Evolution

If FGs are the end product of groups/clusters evolution, a question should naturally arise: are their progenitors regular groups/clusters or do they belong to some particular class? A possible answer that we will discuss in the first part of this section is that the progenitors could be found in compact groups. In fact, in these systems, various bright galaxies are cooped up in a small area, making them ideal candidates for fast and efficient mergers.

However, von Benda-Beckmann et al. [37] suggested that the fossil status may be only a transitional phase in the life of a regular cluster. If this is the case, there is no need to find a special category of progenitors, since the acquisition of the fossil status could happen to any group/cluster in the period between the last major merger and the subsequent arrival of another bright galaxy from the cosmic web. We will discuss this topic from an observational point of view in the second part of this section.

5.1. Compact and Loose Groups as Progenitors of Fossil Systems

The discussion on the progenitors of FGs started even before the actual discovery of these systems. In fact, Ponman and Bertram [3] suggested that compact groups were the result of orbital decay in larger systems and that they should culminate in a final merger, in which all the bright galaxies would merge at the centre of the system. They also added that a new category of "fossil groups" were awaiting discovery in the ROSAT all-sky survey. A year after this claim, they announced the discovery of the first FG [4].

Since then, many studies looked for FG's progenitors, using a variety of techniques. Miles et al. [116] studied a sample of 25 clusters in the X-ray, finding that some of them are dimmed in luminosity. Their interpretation of the result was that, according to a specific toy simulation, groups with dimmed X-ray luminosity have lower velocity dispersion. This would lead to the formation of FGs, since low-velocity encounters between massive galaxies are the most efficient in terms of merging time scale. This result was also confirmed by the analysis of a compact group at $z = 0.22$ studied in Mendes de Oliveira and Carrasco [117]. In fact, these authors showed that this system has many characteristics in common with FGs and, in particular, the merging of the four brightest members would lead to the formation of a BCG of $M_r \sim -23$, a typical value of central galaxies in FGs. A similar result was also found in the study of galaxy pairs [118]. In this case, the author claims that E + S pairs could be the last step in the formation of FGs, thus a sort of transition between compact groups and FGs. In addition, Pierini et al. [99] found evidence that compact groups are favoured as progenitors over the early assembly. On the other hand, Yoshioka et al. [15] found a M/L ratio for FGs that is too high if compared with compact groups, thus claiming that these systems are not the ideal progenitors.

A novel technique to search for FG's progenitors is the use of strong gravitational lensing in galaxy groups. Johnson et al. [119] investigated the fraction of FGs in lensed and non-lensed galaxy groups, finding that the fraction of FGs is larger in lensed systems (13% versus 3%). They also identified 12 possible FG progenitors that were later investigated in detail in Johnson et al. [120] using Chandra and the Hubble Space Telescope. Their results showed that the X-ray temperatures of the candidate progenitors are higher than those of the control sample. They also find hints of differences in the LFs of FGs and non-FGs, but these differences are erased when BCGs are removed. Finally, Schirmer et al. [121] studied the strong-lensed FG J0454-0309 that is found behind a well-studied non-fossil poor cluster. Their analysis supports a scenario in which the fossil system is falling into the poor cluster and where the central galaxy of the FG will become the brightest galaxy of the new system.

Another approach was proposed by Tovmassian [122]: they compared the $K-$band absolute magnitudes of BCGs in regular clusters and FGs, finding that the latter are systematically fainter. The author concluded that FG progenitors are likely poor groups. Moreover, it is interesting to note that Tovmassian et al. [123] studied *"the properties of Hickson's compact groups and of the Loose Groups within which they are Embedded."* In this work, no link with FGs is suggested, but the result could be seen under a different light after the Tovmassian [122] study, confirming the idea that poor compact groups could be the ideal progenitors of FGs. However, this result apparently collides with what was presented in Section 4.2, where we cited various studies claiming that BCGs in FGs are amongst the most-massive galaxies in the Universe. On the other hand, Farhang et al. [124] analysed the mass assembly histories fo compact and fossil systems in the Millennium simulation and associated semi-analytical models. They found that only 30% of FGs could originate from compact groups. They conclude that most of the fossil and compact groups follow different evolutionary paths.

Finally, Buote and Barth [125] suggested that compact elliptical galaxies (CEGs) surrounded by an X-ray halo should be considered as FGs. In fact, detailed X-ray observations of two of these systems found that mass and entropy profiles and concentration are compatible with other FGs studied in the literature. In this case, the progenitors are expected to be the so-called "red-nuggets", compact galaxies found a $z{\sim}2$ [126]. This case is indeed peculiar: the authors did not study FGs in order to find their progenitors, but better suggested that a new type of galaxy should be considered in the fossil category.

5.2. Transitional Fossil Phase

We already mentioned in Section 3 that von Benda-Beckmann et al. [37] suggested that FGs are only a transitional phase in the life of a regular group/cluster. They claimed that this phase would happen just after a major merger and before other bright galaxies are accreted to the group. An example of such a process is found in Irwin et al. [127]: the Cheshire Cat galaxy group is formed by two smaller groups, dominated by one bright galaxy each. These groups are experiencing a line-of-sight merger that will end up in approximately one Gyr with the merging of the two structures. The authors suggested that the resulting structure will be a massive fossil group, dominated by a large $M_r = -24$ galaxy.

A similar case is the one presented in Aguerri et al. [110]: RX J075243.6 + 455653 was found to actually accomplish the fossil definition of $\Delta m_{12} > 2$ within half the virial radius; however, another galaxy almost as bright as the BCG is found just outside that radius. Depending on its orbit, RX J075243.6 + 455653 became fossils in the very last part of its life, or, in the opposite case, it will become non-fossils in the near future.

The existence of a fossil phase may thus explain some of the controversial results presented along this review. It is possible that the observational definition based on the magnitude gap alone is not sufficient to clearly separate the population of real FGs to that of non-FGs dominated by a massive central galaxy. This would confirm the results of Raouf et al. [32], since the authors suggested that other observational quantities (like the luminosity of the BCG and its separation from the luminosity centroid of the group) should be used to create a sample dominated by purely old FGs. We thus suggest to start using this new definition in the search for FGs as a way for creating a sample of old systems. However, using these additional observational constraints could dramatically reduce the number of identified systems. In Section 6.1, we will give a list of the most-secure FGs up to date: only one out of 18 FGs for which the absolute magnitude of the BCG is available will survive the application of the Raouf et al. [32] criterium ($\Delta m_{12} > 2.0$ and $M_{r,BCG} > -22.5$).

6. Discussion and Conclusions

In this section, we propose a sample of genuine FGs that can be the starting point for new follow ups of these objects. Then, we discuss what we presented along the review and draw our general conclusions.

6.1. Sample of Genuine Fossil Groups

In Table 1, we present a list of confirmed FGs in the literature. The goal is to offer to the reader a sample as pure as possible for future follow ups. The list is probably not complete, but we did our best to select FGs applying a rigorous criterium on the Δm_{12} parameter. In particular, we consider as fossils those systems with $\Delta m_{12} \geq 2.0$ within half the (projected) virial radius. Moreover, we exclude FGs for which membership was done using a fixed cut in Δz, except when no ambiguous galaxy was found within half the (projected) virial radius. The magnitude gap between the first and fourth brightest galaxies (Δm_{14}) is also given, when available, but it was not used for the selection, since it was computed only in the most recent studies. The absolute $r-$band magnitude of the BCG is included, when available, to simplify the application of the Raouf et al. [32] criterium. Moreover, we also list the mass of the system, when available. However, we note that the mass is computed in a very inhomogeneous way (different methods and radii), and our goal is to offer an at-a-glance reference to the reader. Finally, the redshift is given for all FGs listed in the table.

Table 1. A non-exhaustive list of confirmed FGs for which at least $\Delta m_{12} \geq 2$ (and eventually $\Delta m_{14} \geq 2.5$) is computed within half the virial radius in the literature.

Name	Δm_{12}	Δm_{14}	$M_{r,BCG}$	z	Mass [10^{14} M$_\odot$]	Reference
RX J1340.6 + 4018 *	2.3	/	/	0.171	0.28	Ponman et al. [4]
RXJ1119.7 + 2126	>2.5	/	−22.8	0.061	/	Jones et al. [9]
RXJ1331.5 + 1108	2.0	/	−23.6	0.081	/	Jones et al. [9]
RXJ1416.4 + 2315	2.4	/	−25.0	0.137	/	Jones et al. [9]
RXJ1552.2 + 2013 **	2.3	/	−24.7	0.135	/	Jones et al. [9]
NGC 6482	2.06	/	−22.7	0.0131	0.042	Khosroshahi et al. [16]
ESO 3060170	2.61	/	−24.4	0.0358	1–2	Sun et al. [11]
UGC 842	2.99	/	−23.0	0.045	0.4	Voevodkin et al. [128]
AWM 4	2.23	/	/	0.0317	1.4	Zibetti et al. [107]
J0454-0309	2.5	/	$-24.1^{\dagger}$	0.26	0.75–0.90	Schirmer et al. [121]
RXC J0216.7-4749	>2.21	/	/	0.064	0.8	Démoclès et al. [57]
CXGG 095951+0140.8	2.10	/	$-24.9^{\dagger}$	0.372	0.95	Pierini et al. [99]
CXGG 095951 + 0212.6	2.32	/	$-23.9^{\dagger}$	0.425	0.19	Pierini et al. [99]
SDSS J0906+0301	3.09	/	/	0.1359	1.3	Proctor et al. [52]
SDSS J1045+0420	2.00	/	/	0.1539	2.2	Proctor et al. [52]
1RXS J235814.4 + 150524	>2.0	/	/	0.178	/	Adami et al. [109]
DMM2008 IV	2.4	3.0	/	0.0796	/	Harrison et al. [14]
WHL J083454.9 + 553421	2.4	3.0	/	0.2412	/	Harrison et al. [14]
A0963	2.2	2.7	/	0.2056	/	Harrison et al. [14]
A1068	2.3	3.1	/	0.1381	/	Harrison et al. [14]
BLOX J1230.6 + 1113.3 ID	2.1	3.5	/	0.1169	/	Harrison et al. [14]

Table 1. *Cont.*

Name	Δm_{12}	Δm_{14}	$M_{r,BCG}$	z	Mass [10^{14} M$_\odot$]	Reference
XMMXCS J123338.5 + 374114.9	2.6	3.2	/	0.1023	/	Harrison et al. [14]
ZwCl 1305.4 + 2941	2.6	3.1	/	0.2406	/	Harrison et al. [14]
MaxBCG J197.94248 + 22.02702	2.1	2.7	/	0.1715	/	Harrison et al. [14]
XMMXCS J134825.6 + 580015.8	2.0	2.6	/	0.1274	/	Harrison et al. [14]
XMMXCS J141657.5 + 231239.2	2.8	3.1	/	0.1159	/	Harrison et al. [14]
XMMXCS J160129.8 + 083856.3	2.4	3.1	/	0.1875	/	Harrison et al. [14]
FG12	>2.0	/	/	0.089	0.6	La Barbera et al. [129]
FGS02	>2.21	>2.28	−25.0	0.23	18.7	Zarattini et al. [22]
FGS03	2.09	2.55	−22.6	0.052	0.42	Zarattini et al. [22]
FGS08	>2.12	>2.17	−24.2	0.409	/	Zarattini et al. [22]
FGS10	2.12	2.24	−25.3	0.468	8.32	Zarattini et al. [22]
FGS20	2.17	>2.46	−23.6	0.094	1.63	Zarattini et al. [22]
FGS28	>3.28	>3.68	−21.3	0.032	/	Zarattini et al. [22]
2PIGG 2515	3.4	/	−23.4	0.062	0.37	Khosroshahi et al. [53]
2PIGG 2868	2.5	/	−23.1	0.067	0.25	Khosroshahi et al. [53]

Notes. Column (1): System name, as presented in the cited publication. Column (2): Magnitude gap between the two brightest member galaxies. Column (3): Magnitude gap between the first and fourth brightest member galaxies. Column (4): $r-$band absolute magnitude of the BCG. (5): redshift. (6): Mass. Column (7) Reference paper. It is worth noting that we only cite minimal references for each FG and the same object can be also found in other publications. Moreover, we only cite systems for which Δm_{12} is strictly larger than 2, in order to propose a sample of genuine fossil systems. * This system is disqualified as a fossil in Mendes de Oliveira et al. [130] using $i-$ and $g-$bands. ** This system is disqualified as a fossil in Zibetti et al. [107]. † Computed in the $i-$band.

We note that, if contradictory information is available, we always choose to apply the Jones et al. [9] criteria in the most severe way. For example, in Proctor et al. [52], the sample of Miller et al. [51] was studied in more detail, computing r_{200} in two different ways: one obtained from weak lensing analysis and the other from X-ray data. The latter is found to be ∼50% larger than the former. As a consequence, the number of FGs found using the smallest radius is 10, a number that reduces to 3 if the largest radius is used. In order to provide the cleanest sample of genuine FGs, we include in Table 1 only the three obtained with the largest r_{200}, citing only Proctor et al. [52], even if most of the candidates were also present in Miller et al. [51]. This choice is done in order to direct the reader to the most up-to-date and/or relevant information.

It is worth noting that some famous FGs are excluded from the list. As an example, we discuss the prototype of this category, NGC 1132 [91], for which we were not able to find Δm_{12}. There are indeed information on the fainter galaxies, e.g., in [131], but not a clear computation of the magnitude gap. However, Kim et al. [131] claimed that the second brightest galaxy is NGC 1126, a spiral galaxy located at 8.4 arcminutes or 230 kpc in projection. The virial radius of this group is estimated to be r_{200}∼800 kpc; thus, this galaxy should be inside 0.5 r_{200}. The difference in the velocity space is $\Delta v = 438$ km s^{-1}, as obtained using the Nasa Extragalactic Database (NED), so the two galaxies can be part of the same group, although a precise dynamical study should be done to confirm the membership. In Kim et al. [131], NGC 1126 is described as "seven times fainter in B". We check in the SDSS DR16 the magnitudes of both NGC 1132 and NGC 1126: the former has $m_r = 12.20$, the latter $m_r = 14.01$, leading to a $\Delta m_{12} = 1.81$, rejecting it as a genuine FG. However, as we already mentioned, an accurate study of the membership of NGC 1126 to the group of NGC 1132 should be done and errors in SDSS magnitudes can not be excluded (NGC 1126 is flagged with "unreliable photometry", as many other bright galaxies, mainly due to an over estimation of the sky around bright objects). In conclusion, with this example, we aim at demonstrating that also the definition of the most commonly-accepted FGs may be not rigorous, or may need deeper studies to include them in a sample of genuine FGs.

6.2. Conclusions and Future Prospects

Along this review, we analysed the most-studied topics on FGs. The aim of these studies was to test the so-called *merging scenario*, which predicts that fossil systems formed earlier than non-fossils, having enough time to merge all the bright galaxies with the BCG and remaining somewhat isolated from the cosmic web (e.g., they did not receive other bright galaxies from the merging with other groups/clusters). However, the general framework that can be obtained from this review is that FGs probably formed and evolved in a similar way as non-FGs. In particular, we show that early differences reported in global properties such as the scaling relations and M/L ratios of the halos of fossil systems can be reconciled when homogeneous datasets are used. Probably, an analogous result would be obtained for the differences observed in the mass and the entropy profiles in some individual systems. A more homogeneus and large sample is required to be analysed in this case. Moreover, no differences are found in the fraction of galaxy substructures identified in FGs and non-FGs. This again indicates that the halos of FGs are not significantly older than those from non-FGs.

The central galaxies in fossil groups show similar stellar ages and metallicities than BCGs in the center of non-fossil systems. In addition, the location of these galaxies in the fundamental plane and its projections indicate a formation process driven by dissipationless mergers in a similar way to other bright early type galaxies.

The similarities found in the formation of fossil and non-fossil systems seem to indicate that the large magnitude gap could just be a transient phase in the evolution of groups and clusters, as reported by different numerical simulations. This magnitude gap would be more connected with recent major mergers rather than with an old formation.

If this is the case, one should find an explanation for those differences that can not be reconciled with inhomogeneities in the data. We already mentioned that Sommer-Larsen [29] proposed that more radial orbits for galaxies in fossils could be responsible for the formation of the gap. This idea is supported also by Lacey and Cole [78]: the merger timescale with the central halo is shorter for M* galaxies on radial orbits than for galaxies on tangential orbits (see their Equation (4.2)). From an observational point of view, hints of this difference are found in Zarattini et al. [132]. The authors studied the orbital structure of a sample of $\sim$100 groups and clusters, dividing them in four bins of Δm_{12}. Their larger magnitude gap bin ($\Delta m_{12} > 1.5$) shows the presence of radial orbits in the external regions (0.8–1 r_{200}) that is not found in the other three bins, all with $\Delta m_{12} < 1.5$. However, the results should be confirmed with a larger sample of genuine FGs, as we already mentioned along this review for a significant part of the discussed topics.

The other main topic that remains open is the difference found in the faint-end slope of FGs LFs. This is difficult to explain within the current models of formation and evolution of clusters. In fact, most of the studies point towards a sort of *global* value for the faint-end slope in clusters and groups. It is worth noting, however, that the majority of the studies of LFs in FGs used photometric data and that, in the literature, significant differences were found even in regular clusters when only photometric data were used. Thus, the next step in this discussion awaits the use of large spectroscopic datasets that will become available with the next generation of astronomical instruments (e.g., WEAVE, 4MOST, DESI). However, we can tentatively say that the presence of radial orbits could give an answer also to the problem of the faint-end slope of the LF: in fact, if massive galaxies on radial orbits have a shorter timescale to be merged within the central galaxy, dwarf galaxies could be more easily disrupted on such orbits, if they pass near the BCG [133]. Another possible explanation for the differences in the galaxy populations between FGs and non-FGs could be found in the surrounding environment, since hints of different large-scale structures are found in some individual studies. In particular, we can not exclude that FGs are still in the process of accreting dwarf galaxies, but the general picture remains to be clarified.

The creation of a large and strict sample of genuine FGs and a homogeneous follow up will be the key to the characterisation of FGs in the near future. For this reason, we gave in Section 6.1 a table with the most-secure FGs to date. The computation of the Δm_{12} is not a real issue with surveys like SDSS, DES, or Pan-STARRS1, already available for 3/4 of the sky. The arrival of new facilities will be useful for the confirmation of the FGs candidates found with these photometric surveys. New X-ray data will be available with the new all-sky surveys like eROSITA (we refer the reader to the companion review of Ekert et al. for a detailed description on the impact of eROSITA and other X-ray surveys on the study of galaxy groups). In addition, the spectroscopic follow up will be possible with extended spectroscopic surveys like WEAVE, 4MOST, and DESI or with precise photometric redshifts surveys like J-PAS. The firm identification of at least 50/100 FGs will be the main scientific goals in this field for the next decade. This will be easily achieved in the near future, since eROSITA is expected to find $\sim 10^5$ groups/clusters. For comparison, the REFLEX cluster catalogue has $\sim$1500 groups/clusters. Most of these new clusters will have dedicated spectroscopic follow ups with the next generation multi-object spectrographs.

Author Contributions: Both authors contributed equally to the paper. Both authors have read and agreed to the published version of the manuscript

Funding: S.Z. is funded by Padua University grant ARPE-DFA-2020. J.A.L.A. was founded by the project AYA2017-83204-P.

Data Availability Statement: The data used in this review were found in other published articles.

Acknowledgments: The authors thanks MNRAS, A&A, and the AAS, together with the authors of the corresponding publications, for granting permission for using images published in their journals.

Conflicts of Interest: The authors declare no conflict of interest.

References

1. White, S.D.M.; Rees, M.J. Core condensation in heavy halos: A two-stage theory for galaxy formation and clustering. *Mon. Not. R. Astron. Soc.* **1978**, *183*, 341–358. [CrossRef]
2. White, S.D.M.; Frenk, C.S. Galaxy Formation through Hierarchical Clustering. *Astrophys. J.* **1991**, *379*, 52. [CrossRef]
3. Ponman, T.J.; Bertram, D. Hot gas and dark matter in a compact galaxy group. *Nature* **1993**, *363*, 51–54. [CrossRef]
4. Ponman, T.J.; Allan, D.J.; Jones, L.R.; Merrifield, M.; McHardy, I.M.; Lehto, H.J.; Luppino, G.A. A possible fossil galaxy group. *Nature* **1994**, *369*, 462–464. [CrossRef]
5. Vikhlinin, A.; McNamara, B.R.; Hornstrup, A.; Quintana, H.; Forman, W.; Jones, C.; Way, M. X-Ray Overluminous Elliptical Galaxies: A New Class of Mass Concentrations in the Universe? *Astrophys. J. Lett.* **1999**, *520*, L1–L4. [CrossRef]
6. Santos, W.A.; Mendes de Oliveira, C.; Sodré, L., Jr. Fossil Groups in the Sloan Digital Sky Survey. *Astron. J.* **2007**, *134*, 1551. [CrossRef]
7. La Barbera, F.; de Carvalho, R.R.; de la Rosa, I.G.; Sorrentino, G.; Gal, R.R.; Kohl-Moreira, J.L. The Nature of Fossil Galaxy Groups: Are They Really Fossils? *Astron. J.* **2009**, *137*, 3942–3960. [CrossRef]
8. Dariush, A.; Khosroshahi, H.G.; Ponman, T.J.; Pearce, F.; Raychaudhury, S.; Hartley, W. The mass assembly of fossil groups of galaxies in the Millennium simulation. *Mon. Not. R. Astron. Soc.* **2007**, *382*, 433–442. [CrossRef]
9. Jones, L.R.; Ponman, T.J.; Horton, A.; Babul, A.; Ebeling, H.; Burke, D.J. The nature and space density of fossil groups of galaxies. *Mon. Not. R. Astron. Soc.* **2003**, *343*, 627–638. [CrossRef]
10. Dariush, A.A.; Raychaudhury, S.; Ponman, T.J.; Khosroshahi, H.G.; Benson, A.J.; Bower, R.G.; Pearce, F. The mass assembly of galaxy groups and the evolution of the magnitude gap. *Mon. Not. R. Astron. Soc.* **2010**, *405*, 1873–1887. [CrossRef]
11. Sun, M.; Forman, W.; Vikhlinin, A.; Hornstrup, A.; Jones, C.; Murray, S.S. ESO 3060170: A Massive Fossil Galaxy Group with a Heated Gas Core? *Astrophys. J.* **2004**, *612*, 805–816. [CrossRef]
12. Khosroshahi, H.G.; Ponman, T.J.; Jones, L.R. The central elliptical galaxy in fossil groups and formation of brightest cluster galaxies. *Mon. Not. R. Astron. Soc.* **2006**, *372*, L68–L72. [CrossRef]
13. Aguerri, J.A.L.; Girardi, M.; Boschin, W.; Barrena, R.; Méndez-Abreu, J.; Sánchez-Janssen, R.; Borgani, S.; Castro-Rodriguez, N.; Corsini, E.M.; Del Burgo, C.; et al. Fossil groups origins. I. RX J105453.3+552102 a very massive and relaxed system at z 0.5. *Astron. Astrophys.* **2011**, *527*, A143. [CrossRef]
14. Harrison, C.D.; Miller, C.J.; Richards, J.W.; Lloyd-Davies, E.J.; Hoyle, B.; Romer, A.K.; Mehrtens, N.; Hilton, M.; Stott, J.P.; Capozzi, D.; et al. The XMM Cluster Survey: The Stellar Mass Assembly of Fossil Galaxies. *Astrophys. J.* **2012**, *752*, 12. [CrossRef]
15. Yoshioka, T.; Furuzawa, A.; Takahashi, S.; Tawara, Y.; Sato, S.; Yamashita, K.; Kumai, Y. The dark matter halo structure of "fossil" groups candidates. *Adv. Space Res.* **2004**, *34*, 2525–2529. [CrossRef]
16. Khosroshahi, H.G.; Jones, L.R.; Ponman, T.J. An old galaxy group: Chandra X-ray observations of the nearby fossil group NGC 6482. *Mon. Not. R. Astron. Soc.* **2004**, *349*, 1240–1250. [CrossRef]

17. Ulmer, M.P.; Adami, C.; Covone, G.; Durret, F.; Lima Neto, G.B.; Sabirli, K.; Holden, B.; Kron, R.G.; Romer, A.K. Cl 1205+44: A Fossil Group at z = 0.59. *Astrophys. J.* **2005**, *624*, 124–134. [CrossRef]
18. Mendes de Oliveira, C.L.; Cypriano, E.S.; Sodré, L., J. The Luminosity Function of the Fossil Group RX J1552.2+2013. *Astron. J.* **2006**, *131*, 158–167. [CrossRef]
19. Adelman-McCarthy, J.K.; Agüeros, M.A.; Allam, S.S.; Anderson, K.S.J.; Anderson, S.F.; Annis, J.; Bahcall, N.A.; Bailer-Jones, C.A.L.; Baldry, I.K.; Barentine, J.C.; et al. Fifth Data Release Sloan Digit. Sky Surv. *Astrophys. J. Suppl.* **2007**, *172*, 634–644. [CrossRef]
20. Eisenstein, D.J.; Annis, J.; Gunn, J.E.; Szalay, A.S.; Connolly, A.J.; Nichol, R.C.; Bahcall, N.A.; Bernardi, M.; Burles, S.; Castander, F.J.; et al. Spectroscopic Target Selection for the Sloan Digital Sky Survey: The Luminous Red Galaxy Sample. *Astron. J.* **2001**, *122*, 2267–2280. [CrossRef]
21. Voges, W.; Aschenbach, B.; Boller, T.; Bräuninger, H.; Briel, U.; Burkert, W.; Dennerl, K.; Englhauser, J.; Gruber, R.; Haberl, F.; et al. The ROSAT all-sky survey bright source catalogue. *Astron. Astrophys.* **1999**, *349*, 389–405.
22. Zarattini, S.; Barrena, R.; Girardi, M.; Castro-Rodriguez, N.; Boschin, W.; Aguerri, J.A.L.; Méndez-Abreu, J.; Sánchez-Janssen, R.; Catalán-Torrecilla, C.; Corsini, E.M.; et al. Fossil group origins. IV. Characterization of the sample and observational properties of fossil systems. *Astron. Astrophys.* **2014**, *565*, A116. [CrossRef]
23. Tavasoli, S.; Khosroshahi, H.G.; Koohpaee, A.; Rahmani, H.; Ghanbari, J. A Statistical Study of the Luminosity Gap in Galaxy Groups. *Publ. Astron. Soc. Pac.* **2011**, *123*, 1. [CrossRef]
24. Makarov, D.; Karachentsev, I. Galaxy groups and clouds in the local (z$\sim$ 0.01) Universe. *Mon. Not. R. Astron. Soc.* **2011**, *412*, 2498–2520. [CrossRef]
25. Gozaliasl, G.; Finoguenov, A.; Khosroshahi, H.G.; Mirkazemi, M.; Salvato, M.; Jassur, D.M.Z.; Erfanianfar, G.; Popesso, P.; Tanaka, M.; Lerchster, M.; et al. Mining the gap: Evolution of the magnitude gap in X-ray galaxy groups from the 3-square-degree XMM coverage of CFHTLS. *Astron. Astrophys.* **2014**, *566*, A140. [CrossRef]
26. Moore, B.; Ghigna, S.; Governato, F.; Lake, G.; Quinn, T.; Stadel, J.; Tozzi, P. Dark Matter Substructure within Galactic Halos. *Astrophys. J. Lett.* **1999**, *524*, L19–L22. [CrossRef]
27. Mateo, M.L. Dwarf Galaxies of the Local Group. *Annu. Rev. Astron. Astrophys.* **1998**, *36*, 435–506. [CrossRef]
28. D'Onghia, E.; Lake, G. Cold Dark Matter's Small-Scale Crisis Grows Up. *Astrophys. J.* **2004**, *612*, 628–632. [CrossRef]
29. Sommer-Larsen, J. Properties of intra-group stars and galaxies in galaxy groups: 'normal' versus 'fossil' groups. *Mon. Not. R. Astron. Soc.* **2006**, *369*, 958–968. [CrossRef]
30. D'Onghia, E.; Sommer-Larsen, J.; Romeo, A.D.; Burkert, A.; Pedersen, K.; Portinari, L.; Rasmussen, J. The Formation of Fossil Galaxy Groups in the Hierarchical Universe. *Astrophys. J. Lett.* **2005**, *630*, L109–L112. [CrossRef]
31. Deason, A.J.; Conroy, C.; Wetzel, A.R.; Tinker, J.L. Stellar Mass-gap as a Probe of Halo Assembly History and Concentration: Youth Hidden among Old Fossils. *Astrophys. J.* **2013**, *777*, 154. [CrossRef]
32. Raouf, M.; Khosroshahi, H.G.; Ponman, T.J.; Dariush, A.A.; Molaeinezhad, A.; Tavasoli, S. Ultimate age-dating method for galaxy groups; clues from the Millennium Simulations. *Mon. Not. R. Astron. Soc.* **2014**, *442*, 1578–1585. [CrossRef]
33. Zhoolideh Haghighi, M.H.; Raouf, M.; Khosroshahi, H.G.; Farhang, A.; Gozaliasl, G. On the Reliability of Photometric and Spectroscopic Tracers of Halo Relaxation. *Astrophys. J.* **2020**, *904*, 36. [CrossRef]
34. Kundert, A.; D'Onghia, E.; Aguerri, J.A.L. Are Fossil Groups Early-forming Galaxy Systems? *Astrophys. J.* **2017**, *845*, 45. [CrossRef]
35. Raouf, M.; Khosroshahi, H.G.; Mamon, G.A.; Croton, D.J.; Hashemizadeh, A.; Dariush, A.A. Merger History of Central Galaxies in Semi-analytic Models of Galaxy Formation. *Astrophys. J.* **2018**, *863*, 40. [CrossRef]
36. Raouf, M.; Khosroshahi, H.G.; Dariush, A. Evolution of Galaxy Groups in the Illustris Simulation. *Astrophys. J.* **2016**, *824*, 140. [CrossRef]
37. von Benda-Beckmann, A.M.; D'Onghia, E.; Gottlöber, S.; Hoeft, M.; Khalatyan, A.; Klypin, A.; Müller, V. The fossil phase in the life of a galaxy group. *Mon. Not. R. Astron. Soc.* **2008**, *386*, 2345–2352. [CrossRef]
38. Romeo, A.D.; Kang, X.; Contini, E.; Sommer-Larsen, J.; Fassbender, R.; Napolitano, N.R.; Antonuccio-Delogu, V.; Gavignaud, I. A study on the multicolour evolution of red-sequence galaxy populations: Insights from hydrodynamical simulations and semi-analytical models. *Astron. Astrophys.* **2015**, *581*, A50. [CrossRef]
39. Khosroshahi, H.G.; Raouf, M.; Miraghaei, H.; Brough, S.; Croton, D.J.; Driver, S.; Graham, A.; Baldry, I.; Brown, M.; Prescott, M.; et al. Galaxy In addition, Mass Assembly (GAMA): A "No Smoking" Zone for Giant Elliptical Galaxies? *Astrophys. J.* **2017**, *842*, 81. [CrossRef]
40. Kanagusuku, M.J.; Díaz-Giménez, E.; Zandivarez, A. Fossil groups in the Millennium simulation. From the brightest to the faintest galaxies during the past 8 Gyr. *Astron. Astrophys.* **2016**, *586*, A40. [CrossRef]
41. Rosati, P.; Borgani, S.; Norman, C. The Evolution of X-ray Clusters of Galaxies. *Annu. Rev. Astron. Astrophys.* **2002**, *40*, 539–577. [CrossRef]
42. Voit, G.M. Tracing cosmic evolution with clusters of galaxies. *Rev. Mod. Phys.* **2005**, *77*, 207–258. [CrossRef]
43. Ghirardini, V.; Eckert, D.; Ettori, S.; Pointecouteau, E.; Molendi, S.; Gaspari, M.; Rossetti, M.; De Grandi, S.; Roncarelli, M.; Bourdin, H.; et al. Universal thermodynamic properties of the intracluster medium over two decades in radius in the X-COP sample. *Astron. Astrophys.* **2019**, *621*, A41. [CrossRef]
44. Kaiser, N. Evolution and clustering of rich clusters. *Mon. Not. R. Astron. Soc.* **1986**, *222*, 323–345. [CrossRef]

45. Bryan, G.L.; Norman, M.L. Statistical Properties of X-Ray Clusters: Analytic and Numerical Comparisons. *Astrophys. J.* **1998**, *495*, 80–99. [CrossRef]
46. Frenk, C.S.; White, S.D.M.; Bode, P.; Bond, J.R.; Bryan, G.L.; Cen, R.; Couchman, H.M.P.; Evrard, A.E.; Gnedin, N.; Jenkins, A.; et al. The Santa Barbara Cluster Comparison Project: A Comparison of Cosmological Hydrodynamics Solutions. *Astrophys. J.* **1999**, *525*, 554–582. [CrossRef]
47. Khosroshahi, H.G.; Ponman, T.J.; Jones, L.R. Scaling relations in fossil galaxy groups. *Mon. Not. R. Astron. Soc.* **2007**, *377*, 595–606. [CrossRef]
48. Voevodkin, A.; Borozdin, K.; Heitmann, K.; Habib, S.; Vikhlinin, A.; Mescheryakov, A.; Hornstrup, A.; Burenin, R. Fossil Systems in the 400d Cluster Catalog. *Astrophys. J.* **2010**, *708*, 1376–1387. [CrossRef]
49. Girardi, M.; Aguerri, J.A.L.; De Grandi, S.; D'Onghia, E.; Barrena, R.; Boschin, W.; Mé ndez-Abreu, J.; Sánchez-Janssen, R.; Zarattini, S.; Biviano, A.; et al. Fossil group origins. III. The relation between optical and X-ray luminosities. *Astron. Astrophys.* **2014**, *565*, A115. [CrossRef]
50. Kundert, A.; Gastaldello, F.; D'Onghia, E.; Girardi, M.; Aguerri, J.A.L.; Barrena, R.; Corsini, E.M.; De Grandi, S.; Jiménez-Bailón, E.; Lozada-Muñoz, M.; et al. Fossil group origins - VI. Global X-ray scaling relations of fossil galaxy clusters. *Mon. Not. R. Astron. Soc.* **2015**, *454*, 161–176. [CrossRef]
51. Miller, E.D.; Rykoff, E.S.; Dupke, R.A.; Mendes de Oliveira, C.; Lopes de Oliveira, R.; Proctor, R.N.; Garmire, G.P.; Koester, B.P.; McKay, T.A. Finding Fossil Groups: Optical Identification and X-Ray Confirmation. *Astrophys. J.* **2012**, *747*, 94. [CrossRef]
52. Proctor, R.N.; de Oliveira, C.M.; Dupke, R.; de Oliveira, R.L.; Cypriano, E.S.; Miller, E.D.; Rykoff, E. On the mass-to-light ratios of fossil groups. Are they simply dark clusters? *Mon. Not. R. Astron. Soc.* **2011**, *418*, 2054–2073. [CrossRef]
53. Khosroshahi, H.G.; Gozaliasl, G.; Rasmussen, J.; Molaeinezhad, A.; Ponman, T.; Dariush, A.A.; Sanderson, A.J.R. Optically selected fossil groups; X-ray observations and galaxy properties. *Mon. Not. R. Astron. Soc.* **2014**, *443*, 318–327. [CrossRef]
54. Navarro, J.F.; Frenk, C.S.; White, S.D.M. A Universal Density Profile from Hierarchical Clustering. *Astrophys. J.* **1997**, *490*, 493–508. [CrossRef]
55. Su, Y.; White, R.E., II; Miller, E.D. Suzaku Observations of the X-Ray Brightest Fossil Group ESO 3060170. *Astrophys. J.* **2013**, *775*, 89. [CrossRef]
56. Gastaldello, F.; Buote, D.A.; Humphrey, P.J.; Zappacosta, L.; Bullock, J.S.; Brighenti, F.; Mathews, W.G. Probing the Dark Matter and Gas Fraction in Relaxed Galaxy Groups with X-Ray Observations from Chandra and XMM-Newton. *Astrophys. J.* **2007**, *669*, 158–183. [CrossRef]
57. Démoclès, J.; Pratt, G.W.; Pierini, D.; Arnaud, M.; Zibetti, S.; D'Onghia, E. Testing adiabatic contraction of dark matter in fossil group candidates. *Astron. Astrophys.* **2010**, *517*, A52. [CrossRef]
58. Humphrey, P.J.; Buote, D.A.; Brighenti, F.; Flohic, H.M.L.G.; Gastaldello, F.; Mathews, W.G. Tracing the Gas to the Virial Radius (R_{100}) in a Fossil Group. *Astrophys. J.* **2012**, *748*, 11. [CrossRef]
59. Su, Y.; Buote, D.; Gastaldello, F.; Brighenti, F. The Entire Virial Radius of the Fossil Cluster RX J1159+5531: I. Gas Properties. *Astrophys. J.* **2015**, *805*, 104. [CrossRef]
60. Lea, S.M.; Silk, J.; Kellogg, E.; Murray, S. Thermal-Bremsstrahlung Interpretation of Cluster X-Ray Sources. *Astrophys. J. Lett.* **1973**, *184*, L105. [CrossRef]
61. Vikhlinin, A.; Markevitch, M.; Murray, S.S.; Jones, C.; Forman, W.; Van Speybroeck, L. Chandra Temperature Profiles for a Sample of Nearby Relaxed Galaxy Clusters. *Astrophys. J.* **2005**, *628*, 655–672. [CrossRef]
62. Hudson, D.S.; Mittal, R.; Reiprich, T.H.; Nulsen, P.E.J.; Andernach, H.; Sarazin, C.L. What is a cool-core cluster? a detailed analysis of the cores of the X-ray flux-limited HIFLUGCS cluster sample. *Astron. Astrophys.* **2010**, *513*, A37. [CrossRef]
63. Miraghaei, H.; Khosroshahi, H.G.; Klöckner, H.R.; Ponman, T.J.; Jetha, N.N.; Raychaudhury, S. IGM heating in fossil galaxy groups. *Mon. Not. R. Astron. Soc.* **2014**, *444*, 651–666. [CrossRef]
64. Bharadwaj, V.; Reiprich, T.H.; Sanders, J.S.; Schellenberger, G. Investigating the cores of fossil systems with Chandra. *Astron. Astrophys.* **2016**, *585*, A125. [CrossRef]
65. Dolag, K.; Bartelmann, M.; Perrotta, F.; Baccigalupi, C.; Moscardini, L.; Meneghetti, M.; Tormen, G. Numerical study of halo concentrations in dark-energy cosmologies. *Astron. Astrophys.* **2004**, *416*, 853–864. [CrossRef]
66. De Boni, C.; Ettori, S.; Dolag, K.; Moscardini, L. Hydrodynamical simulations of galaxy clusters in dark energy cosmologies - II. c-M relation. *Mon. Not. R. Astron. Soc.* **2013**, *428*, 2921–2938. [CrossRef]
67. Buote, D.A. The Unusually High Halo Concentration of the Fossil Group NGC 6482: Evidence for Weak Adiabatic Contraction. *Astrophys. J.* **2017**, *834*, 164. [CrossRef]
68. Pratt, G.W.; Pointecouteau, E.; Arnaud, M.; van der Burg, R.F.J. The hot gas content of fossil galaxy clusters. *Astron. Astrophys.* **2016**, *590*, L1. [CrossRef]
69. Vitorelli, A.Z.; Cypriano, E.S.; Makler, M.; Pereira, M.E.S.; Erben, T.; Moraes, B. On mass concentrations and magnitude gaps of galaxy systems in the CS82 survey. *Mon. Not. R. Astron. Soc.* **2018**, *474*, 866–875. [CrossRef]
70. Zarattini, S.; Aguerri, J.A.L.; Biviano, A.; Girardi, M.; Corsini, E.M.; D'Onghia, E. Fossil group origins. X. Velocity segregation in fossil systems. *Astron. Astrophys.* **2019**, *631*, A16. [CrossRef]
71. Lin, Y.T.; Mohr, J.J.; Stanford, S.A. K-Band Properties of Galaxy Clusters and Groups: Luminosity Function, Radial Distribution, and Halo Occupation Number. *Astrophys. J.* **2004**, *610*, 745–761. [CrossRef]

72. van der Burg, R.F.J.; Hoekstra, H.; Muzzin, A.; Sifón, C.; Balogh, M.L.; McGee, S.L. Evidence for the inside-out growth of the stellar mass distribution in galaxy clusters since z ~1. *Astron. Astrophys.* **2015**, *577*, A19. [CrossRef]
73. Sato, K.; Kawaharada, M.; Nakazawa, K.; Matsushita, K.; Ishisaki, Y.; Yamasaki, N.Y.; Ohashi, T. Metallicity of the Fossil Group NGC 1550 Observed with Suzaku. *Publ. Astron. Soc. Jpn.* **2010**, *62*, 1445. [CrossRef]
74. Jones, C.; Forman, W. The structure of clusters of galaxies observed with Einstein. *Astrophys. J.* **1984**, *276*, 38–55. [CrossRef]
75. Quintana, H.; Lawrie, D.G. On the determination of velocity dispersions for cD clusters of galaxies. *Astron. J.* **1982**, *87*, 1–6. [CrossRef]
76. Zabludoff, A.I.; Huchra, J.P.; Geller, M.J. The Kinematics of Abell Clusters. *Astrophys. J. Suppl.* **1990**, *74*, 1. [CrossRef]
77. Oegerle, W.R.; Hill, J.M. Dynamics of cD Clusters of Galaxies. IV. Conclusion of a Survey of 25 Abell Clusters. *Astron. J.* **2001**, *122*, 2858–2873. [CrossRef]
78. Lacey, C.; Cole, S. Merger rates in hierarchical models of galaxy formation. *Mon. Not. R. Astron. Soc.* **1993**, *262*, 627–649. [CrossRef]
79. Méndez-Abreu, J.; Aguerri, J.A.L.; Barrena, R.; Sánchez-Janssen, R.; Boschin, W.; Castro- Rodriguez, N.; Corsini, E.M.; Del Burgo, C.; D'Onghia, E.; Girardi, M.; et al. Fossil group origins. II. Unveiling the formation of the brightest group galaxies through their scaling relations. *Astron. Astrophys.* **2012**, *537*, A25. [CrossRef]
80. Bruzual, G.; Charlot, S. Stellar population synthesis at the resolution of 2003. *Mon. Not. R. Astron. Soc.* **2003**, *344*, 1000–1028. [CrossRef]
81. Faber, S.M.; Jackson, R.E. Velocity dispersions and mass-to-light ratios for elliptical galaxies. *Astrophys. J.* **1976**, *204*, 668–683. [CrossRef]
82. Bernardi, M.; Roche, N.; Shankar, F.; Sheth, R.K. Evidence of major dry mergers at $M_* > 2 \times 10^{11} M_\odot$ from curvature in early-type galaxy scaling relations? *Mon. Not. R. Astron. Soc.* **2011**, *412*, L6–L10. [CrossRef]
83. Pahre, M.A.; Djorgovski, S.G.; de Carvalho, R.R. Near-Infrared Imaging of Early-Type Galaxies. III. The Near-Infrared Fundamental Plane. *Astron. J.* **1998**, *116*, 1591–1605. [CrossRef]
84. Rest, A.; van den Bosch, F.C.; Jaffe, W.; Tran, H.; Tsvetanov, Z.; Ford, H.C.; Davies, J.; Schafer, J. WFPC2 Images of the Central Regions of Early-Type Galaxies. I. The Data. *Astron. J.* **2001**, *121*, 2431–2482. [CrossRef]
85. Eigenthaler, P.; Zeilinger, W.W. The properties of fossil groups of galaxies. *Astron. Nachrichten* **2009**, *330*, 978. [CrossRef]
86. Hernquist, L.; Spergel, D.N. Formation of Shells in Major Mergers. *Astrophys. J. Lett.* **1992**, *399*, L117. [CrossRef]
87. Alamo-Martínez, K.A.; West, M.J.; Blakeslee, J.P.; González-Lópezlira, R.A.; Jordán, A.; Gregg, M.; Côté, P.; Drinkwater, M.J.; van den Bergh, S. Globular cluster systems in fossil groups: NGC 6482, NGC 1132, and ESO 306-017. *Astron. Astrophys.* **2012**, *546*, A15. [CrossRef]
88. Madrid, J.P. Ultra-compact Dwarfs in the Fossil Group NGC 1132. *Astrophys. J. Lett.* **2011**, *737*, L13. [CrossRef]
89. Madrid, J.P.; Donzelli, C.J. Gemini Spectroscopy of Ultracompact Dwarfs in the Fossil Group NGC 1132. *Astrophys. J.* **2013**, *770*, 158. [CrossRef]
90. Hess, K.M.; Wilcots, E.M.; Hartwick, V.L. Fresh Activity in Old Systems: Radio AGNs in Fossil Groups of Galaxies. *Astron. J.* **2012**, *144*, 48. [CrossRef]
91. Mulchaey, J.S.; Zabludoff, A.I. The Isolated Elliptical NGC 1132: Evidence for a Merged Group of Galaxies? *Astrophys. J.* **1999**, *514*, 133–137. [CrossRef]
92. Pipino, A.; D'Ercole, A.; Chiappini, C.; Matteucci, F. Abundance gradient slopes versus mass in spheroids: Predictions by monolithic models. *Mon. Not. R. Astron. Soc.* **2010**, *407*, 1347–1359. [CrossRef]
93. Cid Fernandes, R.; Mateus, A.; Sodré, L.; Stasińska, G.; Gomes, J.M. Semi-empirical analysis of Sloan Digital Sky Survey galaxies - I. Spectral synthesis method. *Mon. Not. R. Astron. Soc.* **2005**, *358*, 363–378. [CrossRef]
94. Eigenthaler, P.; Zeilinger, W.W. Age and metallicity gradients in fossil ellipticals. *Astron. Astrophys.* **2013**, *553*, A99. [CrossRef]
95. Proctor, R.N.; Mendes de Oliveira, C.; Eigenthaler, P. Spatially resolved stellar population parameters in the BCGs of two fossil groups. *Mon. Not. R. Astron. Soc.* **2014**, *439*, 2281–2290. [CrossRef]
96. Trevisan, M.; Mamon, G.A. A finer view of the conditional galaxy luminosity function and magnitude-gap statistics. *Mon. Not. R. Astron. Soc.* **2017**, *471*, 2022–2038. [CrossRef]
97. Corsini, E.M.; Morelli, L.; Zarattini, S.; Aguerri, J.A.L.; Costantin, L.; D'Onghia, E.; Girardi, M.; Kundert, A.; Méndez-Abreu, J.; Thomas, J. Fossil group origins. IX. Probing the formation of fossil galaxy groups with stellar population gradients of their central galaxies. *Astron. Astrophys.* **2018**, *618*, A172. [CrossRef]
98. Raouf, M.; Smith, R.; Khosroshahi, H.G.; Dariush, A.A.; Driver, S.; Ko, J.; Hwang, H.S. The Impact of the Dynamical State of Galaxy Groups on the Stellar Populations of Central Galaxies. *Astrophys. J.* **2019**, *887*, 264. [CrossRef]
99. Pierini, D.; Giodini, S.; Finoguenov, A.; Böhringer, H.; D'Onghia, E.; Pratt, G.W.; Démoclès, J.; Pannella, M.; Zibetti, S.; Braglia, F.G.; et al. Two fossil groups of galaxies at $z \approx 0.4$ in the Cosmic Evolution Survey: Accelerated stellar-mass build-up, different progenitors. *Mon. Not. R. Astron. Soc.* **2011**, *417*, 2927–2937. [CrossRef]
100. Raouf, M.; Smith, R.; Khosroshahi, H.G.; Sande, J.v.d.; Bryant, J.J.; Cortese, L.; Brough, S.; Croom, S.M.; Hwang, H.S.; Driver, S.; et al. The SAMI Galaxy Survey: Kinematics of Stars and Gas in Brightest Group Galaxies—The Role of Group Dynamics. *Astrophys. J.* **2021**, *908*, 123. [CrossRef]
101. Koleva, M.; Prugniel, P.; De Rijcke, S.; Zeilinger, W.W. Age and metallicity gradients in early-type galaxies: A dwarf-to-giant sequence. *Mon. Not. R. Astron. Soc.* **2011**, *417*, 1643–1671. [CrossRef]

102. Schechter, P. An analytic expression for the luminosity function for galaxies. *Astrophys. J.* **1976**, *203*, 297–306. [CrossRef]
103. Popesso, P.; Biviano, A.; Böhringer, H.; Romaniello, M. RASS-SDSS Galaxy cluster survey. IV. A ubiquitous dwarf galaxy population in clusters. *Astron. Astrophys.* **2006**, *445*, 29–42. [CrossRef]
104. Blanton, M.R.; Lupton, R.H.; Schlegel, D.J.; Strauss, M.A.; Brinkmann, J.; Fukugita, M.; Loveday, J. The Properties and Luminosity Function of Extremely Low Luminosity Galaxies. *Astrophys. J.* **2005**, *631*, 208–230. [CrossRef]
105. Aguerri, J.A.L.; Girardi, M.; Agulli, I.; Negri, A.; Dalla Vecchia, C.; Domínguez Palmero, L. Deep spectroscopy in nearby galaxy clusters - V. The Perseus cluster. *Mon. Not. R. Astron. Soc.* **2020**, *494*, 1681–1692. [CrossRef]
106. Trentham, N.; Tully, R.B.; Mahdavi, A. Dwarf galaxies in the dynamically evolved NGC 1407 Group. *Mon. Not. R. Astron. Soc.* **2006**, *369*, 1375–1391. [CrossRef]
107. Zibetti, S.; Pierini, D.; Pratt, G.W. Are fossil groups a challenge of the cold dark matter paradigm? *Mon. Not. R. Astron. Soc.* **2009**, *392*, 525–536. [CrossRef]
108. Lieder, S.; Mieske, S.; Sánchez-Janssen, R.; Hilker, M.; Lisker, T.; Tanaka, M. A normal abundance of faint satellites in the fossil group NGC 6482. *Astron. Astrophys.* **2013**, *559*, A76. [CrossRef]
109. Adami, C.; Jouvel, S.; Guennou, L.; Le Brun, V.; Durret, F.; Clement, B.; Clerc, N.; Comerón, S.; Ilbert, O.; Lin, Y.; et al. Comparison of the properties of two fossil groups of galaxies with the normal group NGC 6034 based on multiband imaging and optical spectroscopy. *Astron. Astrophys.* **2012**, *540*, A105. [CrossRef]
110. Aguerri, J.A.L.; Longobardi, A.; Zarattini, S.; Kundert, A.; D'Onghia, E.; Domínguez-Palmero, L. Fossil group origins. VIII. RX J075243.6+455653 a transitionary fossil group. *Astron. Astrophys.* **2018**, *609*, A48. [CrossRef]
111. Zarattini, S.; Aguerri, J.A.L.; Sánchez-Janssen, R.; Barrena, R.; Boschin, W.; del Burgo, C.; Castro-Rodriguez, N.; Corsini, E.M.; D'Onghia, E.; Girardi, M.; et al. Fossil group origins. V. The dependence of the luminosity function on the magnitude gap. *Astron. Astrophys.* **2015**, *581*, A16. [CrossRef]
112. Zarattini, S.; Girardi, M.; Aguerri, J.A.L.; Boschin, W.; Barrena, R.; del Burgo, C.; Castro-Rodriguez, N.; Corsini, E.M.; D'Onghia, E.; Kundert, A.; et al. Fossil group origins. VII. Galaxy substructures in fossil systems. *Astron. Astrophys.* **2016**, *586*, A63. [CrossRef]
113. Díaz-Giménez, E.; Zandivarez, A.; Proctor, R.; Mendes de Oliveira, C.; Abramo, L.R. Fossil groups in the Millennium simulation. Their environment and its evolution. *Astron. Astrophys.* **2011**, *527*, A129. [CrossRef]
114. Adami, C.; Russeil, D.; Durret, F. The isolated fossil group RX J1119.7+2126. *Astron. Astrophys.* **2007**, *467*, 459–463. [CrossRef]
115. Burenin, R.A.; Vikhlinin, A.; Hornstrup, A.; Ebeling, H.; Quintana, H.; Mescheryakov, A. The 400 Square Degree ROSAT PSPC Galaxy Cluster Survey: Catalog and Statistical Calibration. *Astrophys. J. Suppl.* **2007**, *172*, 561–582. [CrossRef]
116. Miles, T.A.; Raychaudhury, S.; Forbes, D.A.; Goudfrooij, P.; Ponman, T.J.; Kozhurina-Platais, V. The Group Evolution Multiwavelength Study (GEMS): bimodal luminosity functions in galaxy groups. *Mon. Not. R. Astron. Soc.* **2004**, *355*, 785–793. [CrossRef]
117. Mendes de Oliveira, C.L.; Carrasco, E.R. The Compact Group-Fossil Group Connection: Observations of a Massive Compact Group at z=0.22. *Astrophys. J. Lett.* **2007**, *670*, L93–L96. [CrossRef]
118. Grützbauch, R.; Zeilinger, W.W.; Rampazzo, R.; Held, E.V.; Sulentic, J.W.; Trinchieri, G. Small-scale systems of galaxies. IV. Searching for the faint galaxy population associated with X-ray detected isolated E+S pairs. *Astron. Astrophys.* **2009**, *502*, 473–498. [CrossRef]
119. Johnson, L.E.; Irwin, J.A.; White, R.E., II; Wong, K.W.; Maksym, W.P.; Dupke, R.A.; Miller, E.D.; Carrasco, E.R. Using Strong Gravitational Lensing to Identify Fossil Group Progenitors. *Astrophys. J.* **2018**, *856*, 131. [CrossRef]
120. Johnson, L.E.; Irwin, J.A.; White, R.E., II; Wong, K.W.; Dupke, R.A. Chandra and HST Snapshots of Fossil System Progenitors. *Astrophys. J.* **2018**, *869*, 170. [CrossRef]
121. Schirmer, M.; Suyu, S.; Schrabback, T.; Hildebrandt, H.; Erben, T.; Halkola, A. J0454-0309: Evidence of a strong lensing fossil group falling into a poor galaxy cluster. *Astron. Astrophys.* **2010**, *514*, A60. [CrossRef]
122. Tovmassian, H. On the Precursors of Fossil Groups. *Rev. Mex. Astron. Astrofis.* **2010**, *46*, 61–66.
123. Tovmassian, H.; Plionis, M.; Torres-Papaqui, J.P. Physical properties of Hickson compact groups and of the loose groups within which they are embedded. *Astron. Astrophys.* **2006**, *456*, 839–846. [CrossRef]
124. Farhang, A.; Khosroshahi, H.G.; Mamon, G.A.; Dariush, A.A.; Raouf, M. Evolution of Compact and Fossil Groups of Galaxies from Semi-analytical Models of Galaxy Formation. *Astrophys. J.* **2017**, *840*, 58. [CrossRef]
125. Buote, D.A.; Barth, A.J. The Luminous X-Ray Halos of Two Compact Elliptical Galaxies. *Astrophys. J.* **2018**, *854*, 143. [CrossRef]
126. Ferré-Mateu, A.; Trujillo, I.; Martín-Navarro, I.; Vazdekis, A.; Mezcua, M.; Balcells, M.; Domínguez, L. Two new confirmed massive relic galaxies: Red nuggets in the present-day Universe. *Mon. Not. R. Astron. Soc.* **2017**, *467*, 1929–1939. [CrossRef]
127. Irwin, J.A.; Dupke, R.; Carrasco, E.R.; Maksym, W.P.; Johnson, L.; White, Raymond E., I. The Cheshire Cat Gravitational Lens: The Formation of a Massive Fossil Group. *Astrophys. J.* **2015**, *806*, 268. [CrossRef]
128. Voevodkin, A.; Miller, C.J.; Borozdin, K.; Heitmann, K.; Habib, S.; Ricker, P.; Nichol, R.C. X-Ray Observations of Optically Selected Giant Elliptical-Dominated Galaxy Groups. *Astrophys. J.* **2008**, *684*, 204–211. [CrossRef]
129. La Barbera, F.; Paolillo, M.; De Filippis, E.; de Carvalho, R.R. Characterizing the nature of fossil groups with XMM. *Mon. Not. R. Astron. Soc.* **2012**, *422*, 3010–3018. [CrossRef]
130. Mendes de Oliveira, C.L.; Cypriano, E.S.; Dupke, R.A.; Sodré, L., Jr. An Optical and X-Ray Study of the Fossil Group RX J1340.6+4018. *Astron. J.* **2009**, *138*, 502–509. [CrossRef]

131. Kim, D.W.; Anderson, C.; Burke, D.; Fabbiano, G.; Fruscione, A.; Lauer, J.; McCollough, M.; Morgan, D.; Mossman, A.; O'Sullivan, E.; et al. Disturbed Fossil Group Galaxy NGC 1132. *Astrophys. J.* **2018**, *853*, 129. [CrossRef]
132. Zarattini, S.; Biviano, A.; Aguerri, J.A.L.; Girardi, M.; D'Onghia, E. Fossil group origins XI. Galaxy orbits in fossil systems. *Astron. Astrophys.* **2021**, Submitted.
133. Smith, R.; Sánchez-Janssen, R.; Beasley, M.A.; Candlish, G.N.; Gibson, B.K.; Puzia, T.H.; Janz, J.; Knebe, A.; Aguerri, J.A.L.; Lisker, T.; et al. The sensitivity of harassment to orbit: Mass loss from early-type dwarfs in galaxy clusters. *Mon. Not. R. Astron. Soc.* **2015**, *454*, 2502–2516. [CrossRef]

MDPI
St. Alban-Anlage 66
4052 Basel
Switzerland
Tel. +41 61 683 77 34
Fax +41 61 302 89 18
www.mdpi.com

Universe Editorial Office
E-mail: universe@mdpi.com
www.mdpi.com/journal/universe

www.ingramcontent.com/pod-product-compliance
Lightning Source LLC
LaVergne TN
LVHW070114170726
843515LV00007B/1690

* 9 7 8 3 0 3 6 5 1 7 7 3 5 *